Cosmological Inflation and Large-Scale Structure

This textbook provides graduate students with a thorough and up-to-date introduction to the inflationary cosmology. Enormous progress has been made in this area in the past few years and this book is the first to provide a modern and unified overview. It covers all aspects of inflationary cosmology – from the origin of density perturbations during the inflationary epoch of the very early Universe, through the evolution of the perturbations, up to the present for a range of possible cosmologies – and carefully compares predictions with the latest observations, including those of the cosmic microwave background, the clustering and velocities of galaxies, and the epoch of structure formation. To help the student to develop a thorough understanding, problems are provided at the end of each chapter, and numerical answers and hints are included at the end of the book.

With the host of international experiments being performed and planned for the near future (including NASA's Microwave Anisotropy Probe satellite and ESA's Planck mission), inflationary cosmology promises to be one of the most exciting and fruitful topics of research in science in the next decade. This book provides graduate students with the ideal introduction.

Andrew Liddle is professor of astrophysics at the University of Sussex, and has also worked as a lecturer at Imperial College, London. He received his Ph.D. from the University of Glasgow in 1989 and has published more than seventy-five papers in refereed journals, mostly on the topics covered in this book. He travels widely in support of his research, with collaborators in Europe, the United States, Canada, Japan, and Australia.

David Lyth is a senior lecturer at the University of Lancaster. His research career began in 1962 on what was then the very young subject of particle theory. He is author of around 100 published papers and reviews. He has played a leading role in several areas of particle cosmology, especially inflation. Dr. Lyth has been a visiting fellow at the Isaac Newton Institute at Cambridge University and a regular visitor to the University of California at Berkeley.

Cosmological Inflation and Large-Scale Structure

ANDREW R. LIDDLE
University of Sussex

DAVID H. LYTH
University of Lancaster

CAMBRIDGE
UNIVERSITY PRESS

CAMBRIDGE UNIVERSITY PRESS
Cambridge, New York, Melbourne, Madrid, Cape Town, Singapore, São Paulo, Delhi

Cambridge University Press
32 Avenue of the Americas, New York, NY 10013-2473, USA

www.cambridge.org
Information on this title: www.cambridge.org/9780521660228

First published 2000
Reprinted 2002, 2004, 2006

A catalog record for this publication is available from the British Library

Library of Congress Cataloging in Publication data
Liddle, Andrew R.
Cosmological inflation and large-scale structure / Andrew R.
Liddle and David H. Lyth.
p.   cm.
Includes bibliographical references and index.
ISBN 0-521-66022-X — ISBN 0-521-57598-2 (pbk.)
1. Inflationary universe.   2. Large scale structure (Astronomy).
I. Lyth, D. H. (David Hilary)   II. Title.
QB991.I54L53   2000
523.1 — dc21   99-14940
CIP

ISBN 978-0-521-66022-8 hardback
ISBN 978-0-521-57598-0 paperback

Transferred to digital printing 2008

To Amanda
and
to Margaret, John, and Duncan

Contents

Frequently used symbols

Frequently used symbols and their place of definition

Symbol	Definition	Page
M_{Pl}	Reduced Planck mass $(= 2.436 \times 10^{18} \text{ GeV})$	13
G	Gravitational constant $(= 1/8\pi M_{\text{Pl}}^2)$	13
a	Scale factor of the Universe	13
t	Time	13, 23
τ	Conformal time	14
H	Hubble parameter	14
H_0, h	Present Hubble parameter	15
z	Redshift	15
ρ, P	Density and pressure	15
Λ	Cosmological constant	15
K	Spatial curvature	16
ρ_{c}	Critical density	17
Ω	Density parameter	17
T	Temperature	18
n_g	Number density of any species g	18, 351
g_*	Effective number of particle species	18
w	Pressure-to-density ratio ($w = P/\rho$)	21
r_{hor}	Particle horizon distance	24
ϕ	Scalar field	41
$V(\phi)$	Scalar field potential	41
ϵ, η	Slow-roll parameters	42
N	Number of e-foldings	43
$\epsilon_{\text{H}}, \eta_{\text{H}}$	Slow-roll parameters	51
k	Comoving wavenumber	59
δ	Density contrast	61
$T_g(k)$	Transfer function of any quantity g	62
$\mathcal{P}_g(k)$	Spectrum of any quantity g	65
$\xi(R)$	Correlation function	66
ℓ, m	Spherical expansion variables	68
$Y_{\ell m}$	Spherical harmonics	68
$W(kR)$	Window function	71

Continued ...

Symbol	Definition	Page
$\sigma_g(R)$	Dispersion of any quantity g	72
n	Spectral index (density perturbations)	75, 187
v	Peculiar velocity	76
Φ, Ψ	Gravitational potentials	79, 101, 342
V	Peculiar velocity (irrotational part)	83, 338
c_s	Sound speed	84, 349
t_{pr}	Proper time	93, 318
$\mathcal{R}$	Curvature perturbation	98, 341
Σ_{ij}, Π_{ij}	Anisotropic stress	101, 322, 337
$\delta_H(k)$	Density perturbation amplitude	106
$T(k)$	Cold dark matter transfer function	107
Γ	Shape parameter	108
C_ℓ	Spectrum of the cosmic microwave background anisotropy	116
Θ	Photon brightness function	123, 359
Q, U	Stokes' parameters	126
E, B	Polarization components	128
κ	Optical depth	131
$g(\Omega)$	Growth suppression factor	144, 148
n_{grav}	Spectral index (gravitational waves)	154, 193
n_{iso}	Spectral index (isocurvature)	159
S	Action	164
$\mathcal{L}$	Lagrangian density	165
r	Tensor-to-scalar ratio	193
b, b_I	Optical and infrared bias parameters	264, 265
$T_{\mu\nu}$	Stress–energy tensor	320, 331

Preface

The 1990s have seen substantial consolidation of theoretical cosmology, coupled with dramatic observational advances, including the emergence of an entirely new field of observational astronomy – the study of irregularities in the cosmic microwave background radiation. A key idea of modern cosmology is *cosmological inflation*, which is a possible theory for the origin of all structures in the Universe, including ourselves! The time is ripe for a new book describing this field of research.

This book is based loosely on our 1993 *Physics Reports* article. We have widened the range of discussion and have made much of the material more pedagogical. We believe that this book will prove useful to starting graduate students in cosmology, to active researchers specializing in the field, and to all levels in between.

Our view of the inflationary cosmology and its consequences has been influenced by many people over the years. ARL especially thanks Alfredo Henriques and Gordon Moorhouse for showing the way into this research area. DHL would like particularly to acknowledge a long-term collaboration with Ewan Stewart. Much thanks is due to all our collaborators on the topics within this book, namely Mark Abney, Domingos Barbosa, Tiago Barreiro, John Barrow, Marco Bruni, Ted Bunn, Ed Copeland, Laura Covi, George Ellis, Mary Gaillard, Juan Garcìa-Bellido, Anne Green, Louise Griffiths, Ian Grivell, Rocky Kolb, Andrew Laycock, Jim Lidsey, Andrei Linde, Anupam Mazumdar, Milan Mijič, Manash Mukherjee, Hitoshi Murayama, Paul Parsons, Antonio Riotto, Dave Roberts, Leszek Roszkowski, Bob Schaefer, Franz Schunck, Douglas Scott, Qaisar Shafi, Ewan Stewart, Will Sutherland, Michael Turner, Pedro Viana, David Wands, Martin White, and Andrzej Woszczyna. Apart from our collaborators, we have had useful conversations with many others, far too many to mention. We hope they know who they are!

We are extremely grateful to Andrei Linde, Martin White, and especially Gordon Moorhouse for their careful reading of the manuscript. The figures for Chapter 12 were made by Pedro Viana, and the compilation of cosmic microwave background anisotropy data shown in Figures 5.9 and 9.2 was kindly provided by Martin White. Many figures were made using the superb publically available CMBFAST code (Seljak and Zaldarriaga 1996), which we strongly recommend everyone to get.

Although we wrote most of the book at our home institutes, occasionally we were somewhere more glamorous. ARL would like to thank the Università di Padova, the University of New South Wales, and the Aspen Center for Physics, and DHL the University of California at Berkeley. ARL acknowledges the generous support of the Royal Society throughout this endeavour.

Of course, we have done our best to ensure that the contents of this book are accurate; however, some errors may have slipped through. We would be very grateful if readers would inform us of any they spot. We plan to keep an up-to-date record of any errors, accessible at the book's World Wide Web Home Page at

`http://star-www.cpes.sussex.ac.uk/~andrewl/infbook.html`

which can be used to check for errors we already know about.

Andrew R. Liddle and David H. Lyth
October 1998

1 Introduction

1.1 This book

The study of the early Universe came into its own as a research field during the 1980s. Though there had been occasional forays during the seventies and even before that, it was during the 1980s that a wide range of topics, united by the adoption of modern particle physics ideas in a cosmological context, were investigated in detail. This era of study culminated with the publication in 1990 of the classic book *The Early Universe* by Kolb and Turner, in which the authors described ideas across the whole range of what had become known as **particle cosmology** or **particle astrophysics**, including such topics as topological defects, inflationary cosmology, dark matter, axions, and even quantum cosmology.

Although all these topics matured during the 1980s, if we look back at the papers of that era, we are struck by the rarity with which any detailed comparison with observations could be made. In that regard, particle cosmology in the nineties and onward has become a very different subject from what it was during the eighties because, for the first time, there are observations of a quality that seriously constrains some of the possible physics of the early Universe. Those observations are of structure in the Universe, and a starring role among them is played by the first detection of microwave background anisotropies by the Cosmic Background Explorer (COBE) satellite, announced in 1992. These were the first observations that could be more or less directly interpreted as constraints on early Universe physics. As we will see, by the middle of the first decade of the twenty-first century, we should have a wealth of data constraining our conceptions of what may have occurred during the Universe's earliest stages, and, most likely, several of the ideas described in Kolb and Turner's book will have been banished from serious discussion.

This book is not about particle cosmology as a whole, but rather is about a single topic, **inflationary cosmology**, introduced in a seminal paper by Guth (1981). This has been a research field of lasting popularity; more papers have been written about inflation than any other area of early Universe cosmology, and one of Guth's favourite transparencies in review talks charts the rise of the publication count. Although introduced to resolve problems associated with the initial conditions needed for the Big Bang cosmology, its lasting prominence is owed to a property discovered soon after its introduction: It provides a possible explanation for the initial inhomogeneities in the Universe that are believed to have led to all the structures we see, from the earliest objects formed to the clustering of galaxies to the observed irregularities in the microwave background.

Our aim is twofold. First, we wish to give a unified view of the entire process of modelling the inflationary epoch, predicting the small irregularities that it generates, and evolving these irregularities using linear equations that are valid as long as the irregularities remain small. The resulting theoretical structure, starting with the quantum fluctuations of a free field, continuing with general-relativistic gas dynamics, and ending with the free fall of photons and matter, is perhaps one of the most beautiful and complete in the entire field of physics. Certainly, it lies at the opposite extreme from ad hoc models, not of course confined to physics, whose only merit is sometimes to make the author feel better than if the desired result had been written down immediately. Let us hope that the theory is true as well as beautiful!

Second, we wish to describe the state of the art, with respect to both inflation model-building and the confrontation of theory with observation. In the former area, Kolb and Turner's above mentioned book and Linde's *Particle Physics and Inflationary Cosmology* were both written in 1990, and since then, there have been many developments in the theoretical modelling of inflation, including major shifts in the perception of which ideas are the most relevant. Techniques for generating predictions for generic inflationary models also have come some way during that period. The latter area is in the process of being revolutionized by observations of the cosmic microwave background (cmb) anisotropy; the approval of two separate satellite experiments, Microwave Anisotropy Probe (**MAP**) by the National Aeronautics and Space Administration and **Planck** by the European Space Agency, to explore anisotropies down to angular scales of a few arc-minutes promises data of a quality that will be hard to surpass when it comes to constraining or excluding the inflationary cosmology. The rapid progress in both areas means that we are providing something resembling a snapshot of the current situation, though we believe that it will provide a useful orientation for at least some years to come.

As mentioned already, our discussion focuses on the evolution of *small* irregularities. Because inflation ultimately is supposed to provide the origin of all structure, potentially any measure of that structure can, in principle, be used to constrain inflation. This provides a connection to a research area known variously as large-scale structure or physical cosmology, which on its own is a much vaster research area than all of particle cosmology put together. Much has been written on this topic; for example, four books produced after the crucial COBE observations are those of Padmanabhan (1993), Peebles (1993), Coles and Lucchin (1995), and Peacock (1999). By restricting ourselves almost entirely to the linear regime, our focus is both narrower and deeper.

Incidentally, we use the phase "large-scale structure" to refer only to irregularities in matter density, such as the galaxy distribution and motions. The term usually does not include microwave background anisotropies, the exception being the title of this book!

1.2   The Universe we see

Extensive discussion of the nature of the observed Universe has been given in the recent textbooks just mentioned, and so, we will be brief in this introduction.

A description of our observed Universe can be broken into two parts: the global description of the Universe, which is given in terms of a set of parameters that we call the **cosmological parameters**, and the irregularities observed in the Universe.

The cosmological parameters tell us about the geometry of the Universe, and about the material contained within it. These parameters are defined in Chapter 2. The dynamics of an expanding Universe are characterized by two quantities: the expansion rate, given by the Hubble parameter, and the spatial curvature. The latter, in fact, is determined by the amounts of different types of material in the Universe. Direct observation shows that the Universe contains quite a significant amount of baryonic matter, of which we are made, and also contains quite a bit of radiation in the form of the cmb, which can be characterized by a thermal distribution at a temperature $T_0 = 2.728$ K. These are the only two forms of matter that are observed directly. However, on the basis of standard particle physics, it is assumed that there is also a cosmic neutrino background, contributing about the same energy density as the radiation. Beyond that, there is substantial circumstantial evidence (though, as we write, no direct detection of it) that the Universe contains a large (and probably dominant) amount of nonbaryonic dark matter, of some as yet unknown form. The details of how the Universe, and particularly any irregularities within it, will evolve depends on the nature of this dark matter. To get structure formation models to work, it normally is assumed that there must be at least some so-called **cold dark matter**, comprising particles with negligible velocity. However, there also may be a component of **hot dark matter** (particles whose velocities are relativistic for at least some of their evolution) or something more exotic yet. Another possibility, for which there is increasing observational support, is that the Universe might possess a nonzero cosmological constant.

Determination of the various cosmological parameters is a key goal in cosmology, but one in which much progress remains to be made. Of all those just listed, only the present microwave background temperature is known to a satisfying level of accuracy. Other parameters, such as the Hubble constant or the density parameter, remain the subjects of much controversy. In Chapter 2, we briefly review the current observational status. We hope that, in the near future (and for you the reader maybe even the recent past), the situation will become much more definite; in particular, satellite measurements of cmb anisotropies promise to pin down many of the cosmological parameters to a high degree of accuracy.

The second aspect of the observed Universe is the long-established realization that material within it is distributed irregularly. Such irregularities are known as **density perturbations**. An understanding of the origin and evolution of structure in the Universe is the outstanding problem in cosmology at the moment, and this book is primarily about this topic in the context of the inflationary cosmology. Measures of structure in the Universe now come from a variety of sources. Historically, the distribution of galaxies was the most studied, popularized though large galaxy redshift surveys such as the CfA survey in the mid-1980s. Nowadays, we have access to a much more diverse range of measures. The many observations of anisotropies in the cmb, across a range of angular scales, tell us about structure in the Universe long ago when the microwave background was created. The velocities of galaxies can be determined quite accurately, telling us about the gravitational attraction they experience. The abundance of different types of object probes the size of the irregularities in the density of the Universe – at the present epoch, clusters of galaxies are a useful probe, and the study of very distant objects such as quasars can tell us about structure when the Universe was younger, as can observations of distant galaxies with technology such as the Hubble Space Telescope and the Keck telescope on Hawaii. All of these are discussed in Chapters 9 through 12.

Within the context of inflation, all of these structures can be quantified by a small number of parameters describing the initial perturbations, whose subsequent evolution is determined by the cosmological parameters. Such parameters could be called the **inflationary parameters**. They include as a minimum the overall amplitude and scale dependence of the density perturbations; this might be, for example, a power law requiring two parameters that we aim to fix through observation. In fact, the amplitude is already rather well determined by the COBE satellite. In the simplest inflation models, this is all we need, but in more complicated versions, some additional parameters might be necessary. If so, they too in principle can be determined from observations.

1.3   Overview: From cosmological inflation to large-scale structure

This book can be divided loosely into four parts. In Chapters 2 and 3, we introduce the homogeneous Universe and the role that inflation plays in setting its initial conditions. The second part, from Chapters 4 through 8, concerns the development of inhomogeneities in the Universe, from their inception during inflation up to the present. The third part, from Chapters 9 through 13, concerns observations and the way in which they constrain the theoretical development in the first eight chapters. This part ends with an overview. Finally, the last two chapters, separated from the main flow of the book, give a more advanced treatment of inhomogeneities in the Universe, including complete derivations of some results that were assumed for the simpler treatment in the main body of the book.

1.3.1   Hot Big Bang cosmology

We begin our discussion proper in Chapter 2 with a rather rapid summary of the Hot Big Bang theory. This sets down some of our notation and allows us quickly to summarize the results that we use later. Anyone desiring a more leisurely account will find one in any of the books mentioned in Section 1.1. We collect quite a range of different results; in particular, we analyze low-density Universes, both with and without a cosmological constant, as well as the case of a spatially flat Universe with a critical matter density. The last case is the simplest but is disfavoured by observation; we show that there are both theoretical and observational reasons for also considering the low-density cases. By contrast, there is little motivation from either theory or observation to consider closed Universes, where there is greater than a critical density of matter, and we do not concern ourselves with that situation.

1.3.2   Inflation

In Chapter 3, we move on to a discussion of inflationary cosmology (Guth 1981), looking at the general properties rather than at specific models. The definition of inflation is extraordinarily simple: it is any period of the Universe's evolution during which the scale factor, describing the size of the Universe, is accelerating. This leads to a very rapid expansion of the Universe, though perhaps a better way of thinking of this is that the characteristic scale of the Universe, given by the Hubble length, is *shrinking* relative to any fixed scale caught up in the rapid expansion. In that sense, inflation is actually akin to zooming in on a small part of the initial Universe.

Inflation does not in any way replace the Hot Big Bang theory, but rather is an accessory attached during its earliest stages. Inflation certainly cannot proceed forever; the great successes of the Big Bang theory, such as nucleosynthesis (the formation of light elements) and the origin of the thermal microwave background radiation, require the standard evolutionary progression from radiation domination to matter domination, and it is assumed that inflation must end some considerable time before that to allow generation of observed properties such as the baryon–antibaryon asymmetry of the Universe.

As we see later, a sufficiently long period of inflation can resolve certain concerns about the initial conditions necessary for the Big Bang cosmology to lead to a Universe such as our own. In particular, it can explain why the Universe should be close to spatial flatness and why it should appear homogeneous, at least on large scales. It was these problems that motivated the original introduction of inflation by Guth (1981); although accelerated expansion, in fact, already had been considered, most notably by Starobinsky (1980) but also much earlier, it was the strong connection Guth made between rapid expansion and these problems that was the true beginning of inflationary cosmology.

Nevertheless, these problems can no longer be regarded as the strongest motivation for inflationary cosmology because it is not at all clear that they could ever be used to falsify inflation. In fact, they have even been eroded to some extent; for example, it formerly was thought that inflation necessarily gave a spatially flat Universe if it gave homogeneity, but there now exist inflationary models that can give a homogeneous open Universe as well (see Chapter 8). Linde in particular has been vocal (e.g., Linde 1997) in suggesting that the idea of inflation as a theory of initial conditions may be very hard to exclude, and indeed only a few possible observational signals, such as a global rotation of the observable Universe, would be in conflict with inflation in this context (Albrecht 1997; Barrow and Liddle 1997).

By contrast to inflation as a theory of initial conditions, the model of inflation as a possible origin of structure in the Universe is a powerfully predictive one. Different inflation models typically lead to different predictions for the observed structures, and observations can discriminate strongly between them. Future observations certainly will exclude most of the models currently under discussion, and they are also capable of ruling out all of them. Inflation as the origin of structure is therefore very much a proper science of prediction and observation, meriting detailed examination. It is true that even if inflation fails as a model for structure formation, one may be left with the possibility of inflation to fix the initial conditions and some other mechanism for the origin of structure [topological defects being the only known candidate, and a rather unpromising one at that; see Allen et al. (1997) and Pen et al. (1997)], but if we learn that much, we have already learned a lot.

All the standard models of inflation are based on a type of matter known as a scalar field; scalar fields are, among other things, thought to be responsible for the physics of symmetry breaking. Particle physics has yet to offer a definitive view on the detailed properties of such fields and, in particular, has not specified the potential energy, which, it turns out, is responsible for driving the inflationary expansion. The freedom exists to build a wide range of different inflationary models, based on different choices of the potential energy and perhaps different motivations for its particle physics origin. We reserve discussion of specific models until Chapter 8, to be able to discuss them in relation to their predictions for structure formation. In Chapter 3, we develop the machinery needed to deal with scalar fields in an expanding Universe, including extensive discussion of an analytical scheme known as the slow-roll approximation, which is used widely throughout the book. We also briefly discuss the end of inflation, an epoch known as reheating, though its details are not important when considering structure formation from inflation.

1.3.3   Simplest model of structure formation

The key idea in studying structure in the Universe is that of **gravitational instability**. Stated simply, this notes that if the material in the Universe is distributed irregularly, then the overdense regions provide extra gravitational attraction and draw material toward them, thus becoming more overdense. That is, under the action of gravity, irregularities become more pronounced as time passes. At the present epoch, we find that on moderate scales (e.g., less than 10 Mpc), the material in our Universe is very unevenly distributed, in the form of galaxies and clusters of galaxies. On larger scales, it begins to appear homogeneous. On the other hand, at very early times, as sampled by the cmb anisotropies, the Universe is distributed much more evenly. Gravitational instability provides a mechanism to get from a fairly smooth distribution at that time to the more irregular present Universe. It is a dramatic success that this simple picture goes a long way to explaining what is observed, and current attention is focused entirely on the details of the gravitational instability process. This depends on the nature of the Universe as a whole, for example, on how rapidly it is expanding and on how much material is in it to provide the gravitational attraction, and it also depends on the form of the initial irregularities. As we see, inflation provides the most promising theory for the origin of these initial irregularities.

The detailed study of cosmological perturbations is a highly technical topic, and we have chosen to give a simplified treatment within the main body of the book, in order to keep it at roughly the same technical level as the rest of the book. Ideas from general relativity are avoided as far as possible. From time to time, our simplified approach requires us to quote and use results without proper mathematical justification. Because an understanding of cosmological perturbations is so central to current developments in cosmology, we also provide two advanced chapters, 14 and 15, at the end of the book. For readers who are interested, these give a fully self-contained and mathematically rigorous general-relativistic treatment of cosmological perturbations, in which all the results quoted within the book are derived. These can be studied either in their own right or used to fill in the gaps of the earlier discussion.

We begin our discussion of structure formation in Chapters 4 and 5 by setting up some of the machinery for the description of perturbations in an expanding Universe. Our strategy is to keep the discussion as simple as possible, and so we focus on the simplest model, the

cold dark matter (CDM) model. In this model the Universe contains a critical density of material (making it spatially flat), all of the dark matter is cold, and the density perturbations are of a type known as adiabatic. In these chapters, we carry out an analysis with minimal reference to general relativity, really only needing the idea of a locally inertial frame in which the laws of special relativity apply. Further, in certain circumstances perturbations on small scales are amenable to a treatment using only Newtonian gravity. Here, small means relative to the characteristic length scale of an expanding Universe, the **Hubble length**.

In these chapters, we do not concern ourselves with the origin of the perturbations, deferring that until Chapter 7. We simply assume that there is an initial spectrum of perturbations that can be taken to have power-law form. We discuss the statistical nature of the perturbations, their description via their spectrum, and their evolution. This leads ultimately to predictions for the present form of the perturbations, and for the anisotropies in the microwave background. In addition to the temperature anisotropies, we discuss the polarization of the microwave background, which carries additional valuable information, as well as the effect on the microwave background if the atoms in the Universe are reionized at some epoch well before the present, enabling scattering of the microwave photons from the liberated electrons.

1.3.4 Extensions to the simplest model

Although theoretically the simplest scenario, a model in which the density is critical and all dark matter is cold is not the only possibility, and we study extensions in Chapter 6. There is considerable observational evidence that the density is less than critical, and the dark matter need not all be cold. In particular, moving to a low-density Universe, either with or without a cosmological constant, brings better concordance with large-scale structure observations.

Concerning the initial perturbations, an alternative to an adiabatic perturbation is an isocurvature perturbation, where the relative amounts of different materials are perturbed while leaving the total density constant. However, this gives much larger microwave anisotropies for a given size of density perturbation, and most likely cannot be the sole source of perturbations, though they may accompany the usual adiabatic perturbation.

In Chapter 6, we also discuss gravitational wave perturbations. These are inevitable at some level in all inflationary models. The amplitude of gravitational waves reduces rapidly once they come within the Hubble length, and they can be important only on large angular scales in the microwave background (those scales being larger than the Hubble radius at the time the background was formed). As we write there is no way of telling whether a significant fraction of the anisotropies that COBE sees are due to gravitational waves rather than density perturbations, a possibility that has been given substantial attention in the literature [see Lidsey et al. (1997) for a review]. However, in most models of inflation, and in particular within the context of a class of models known as hybrid inflation, the gravitational waves have a negligible effect.

We do not consider magnetic fields, either their possible generation during inflation or their possible effects on observations of structure such as early structure formation and the polarization of the microwave background radiation.

1.3.5   Scalar fields and the vacuum fluctuation

In Chapter 7 we carry out a detailed calculation of the density perturbations produced by inflation. Inflation gives rise to irregularities in the Universe because we live in a quantum world, not a classical one. Inflation is assumed to be driven by a scalar field and, classically, the result of the accelerated expansion is to drive the observable Universe toward a state of perfect homogeneity. However, in a quantum world, perfect homogeneity cannot be attained; we are always left with some residual fluctuations in the scalar field. The typical size of these fluctuations is a property of quantum mechanics, which means that they can be predicted using standard techniques, as we demonstrate in detail in Chapter 7. In particular, they do not depend on the initial conditions before inflation, and so, the theory is highly predictive.

It is an extravagant claim that quantum fluctuations, normally associated with microscopic phenomena, can lead to structures such as clusters of galaxies. The quantum fluctuations occurring in the space between you and this page certainly do not do anything of that sort. And, although it is true that in the early Universe the quantum fluctuations are much bigger than those we normally consider, because the timescales are so much shorter and the energy scales higher, they still will turn out to be small in the sense of being only a very minor perturbation on the classical behaviour. The crucial difference is rather that the Universe is accelerating, which means that the quantum fluctuations can be caught up in the rapid expansion and stretched to huge sizes, orders of magnitude larger than the Hubble scale, which sets the scale of causal physics. Once the fluctuation is taken to such a large scale, it is unable to evolve and becomes frozen-in; crucially, scales are pulled outside the horizon with such swiftness that the amplitude has no chance to tend to zero, but instead is frozen-in at a fixed nonzero value.

In early work on inflationary perturbations, they normally were described as giving a type of density perturbations known as a scale-invariant or Harrison–Zel'dovich spectrum (basically, because physical conditions do not change much as perturbations on all relevant scales are produced). Since then, observations have developed to such an extent that this approximation can no longer be used, and must be replaced by something more accurate. For the observations described in this book, it normally proves adequate to approximate the perturbations by a power-law spectrum, with different inflation models leading to different spectral indices. Future observations can measure the spectral index very precisely, and hence discriminate strongly between different models of inflation.

By the same mechanism, inflationary models inevitably produce gravitational waves at some level. In some models, these may be significant enough to be detectable.

Although the inflationary perturbations typically are calculated in terms of the scalar field perturbation, this leads to a perturbation in the total energy density of the Universe, and hence in the spatial curvature of the Universe. This last is the most useful when one comes to consider how the perturbations will evolve; the scalar field is not useful because it decays long before the present. In the standard scenario, where the perturbations are adiabatic, the perturbation in the spatial curvature remains constant as long as the scale is larger than the Hubble length. This allows us to evolve the perturbations forward in time until the Hubble length grows to encompass them in the postinflationary Universe; for the scales of interest for structure formation, this happens in the recent past, where the standard Big Bang model applies.

1.3.6 Models of inflation

The literature contains a large number of different models of inflation. A model of inflation amounts to a choice for the potential of the scalar field (or fields) driving inflation, plus a means of ending inflation. By this point in the book, we are able to predict the perturbations arising in different models, using the results of Chapter 7, which enables a proper discussion to be made in Chapter 8.

The main discussion focuses on two rather different paradigms. Throughout the early 1990s, discussion was dominated by what we call **single-field models**. In these models, the scalar-field potential often is chosen to be some convenient simple function, such as a monomial or exponential, and the initial conditions are chosen such that the scalar field is well displaced from any minimum (only obeying the condition that its energy density be less than the Planck energy, where quantum gravity is thought to become important). In several such models, gravitational waves may be produced at quite a significant level.

In the mid-1990s, however, this paradigm was challenged by a new wave of inflationary model building, based on particle physics motivation such as the theories of supersymmetry, supergravity, and superstrings. In the context of the last two of these, we do not expect inflation to be possible for field values exceeding a Planck mass, regardless of whether the potential energy there is larger than the Planck energy, because supergravity corrections tend to generate a steep potential that is unable to sustain inflation. This consideration, among others, has led to the popularity of a new class of models known generically as **hybrid inflation**, which rely on interactions between two scalar fields and utilize the flat potentials expected in supersymmetry theories. These models can give substantial inflation for a much more modest evolution of the scalar field, remaining always much less than the Planck mass.

Although, as we write, the hybrid inflation models are a recent innovation under intense investigation, we feel that they are the most important development in inflationary model building for a considerable time, and we devote a large amount of space to them. If indeed they secure their place as the modern inflationary paradigm, supplanting the single-field models described earlier, they will lead to an important rethinking of the observational implications of inflation. For example, in these models the gravitational wave production is expected to be negligible.

We end our discussion of inflationary models with two less standard possibilities. First, we briefly discuss models based on extended gravitational theories, the aim being to obtain the scalar field driving inflation from the gravitational sector of the theory rather than from the matter sector as we have assumed thus far. In most regards, however, these models are simply a rewriting of the single-field inflation models, and share their drawbacks in comparison to the hybrid scenarios.

Second, we discuss models of inflation leading to an open Universe. This was long thought impossible, the belief being that, if a nonflat Universe was obtained it would show strong inhomogeneities. However, it turns out that quantum tunnelling can lead to a homogeneous open Universe. This is interesting because there is quite a lot of observational evidence supporting low-density Universes, and it previously had been thought necessary to introduce a cosmological constant to restore spatial flatness in order to be consistent with inflation. This is no longer the case; there is a whole class of inflationary models consistent with an open Universe. It is even possible to realize the open inflation scenario within a hybrid inflation context.

1.3.7   Observations of structure

Having explored the theoretical side, we move on to comparison with observational data. This retrospectively justifies the favoured models on which the earlier discussion focuses; they were chosen, of course, to give a good representation of the observational data.

For the most part, we have sought to keep the discussion of specific observations short because the observational situation doubtless will improve. The exception to this is the microwave background on large angular scales. The crucial COBE observations are the easiest to interpret in the context of inflation, and they are also more or less definitive because, on the largest angular scales, their accuracy is limited not by instrument noise, but by the statistical uncertainty known as **cosmic variance**, arising from our having only a single microwave sky to look at. Because the COBE experiment has reached its conclusion with the release of the full four-year data set, the inflationary implications merit very detailed discussion and this is done in Chapter 9.

In the future, it is expected that new microwave background anisotropy satellites will provide the best way to constrain cosmological parameters, including those describing the initial perturbations. This is partly due to observational advances permitting a huge amount of useful data to be taken, and partly because the theoretical interpretation is simple because it is all linear perturbation theory. In anticipation of the future launch of the MAP and Planck satellites, we make a detailed discussion of the prospects, also in Chapter 9.

Chapter 10 discusses galaxy motions and clustering. This material can be found in greater detail in, for example, the books of Padmanabhan (1993) and Coles and Lucchin (1995). We write at the commencement of two large galaxy redshift surveys – the 2dF survey and the Sloan Digital Sky Survey – which will dramatically increase our knowledge of the geography of the nearby Universe.

Chapter 11 is the only part of the book where we leave the linear regime, in order to discuss the formation of gravitationally bound objects. Our main aim is to describe some of the methods available to tackle this situation, the most prominent being the Press–Schechter theory, which enables an estimate of the number density of objects as a function of their mass. The most important constraint arising from this comes from the abundance of galaxy clusters, which tightly constrains the spectrum on a scale of around 10 Mpc. In combination with COBE, it leads to powerful conclusions, including the exclusion of the once-popular Standard CDM model because it predicts far too many rich galaxy clusters. Press–Schechter theory also can be used to estimate the abundance of high-redshift objects, such as quasars or the damped Lyman-alpha systems seen in quasar absorption spectra. Finally, it can be used to say something about when the Universe might have been reionized by radiation from the early stages of structure formation; reionization that occurred sufficiently early, by a redshift of 20, for example, could have quite a significant effect on microwave background anisotropies.

Simply as an illustration, in Chapter 12 we combine observations from a range of sources. This data compilation completes a circle in that it motivates a lot of the earlier discussion, in particular, highlighting the failure of the Standard CDM model and indicating the most desirable directions for modification. These detailed results doubtless will be superseded quickly, but the general techniques for comparing theory and observation should continue to be useful. In Chapter 13, we look toward the future.

1.3.8 Advanced topic: Perturbations in detail

We end the book with an advanced treatment of cosmological perturbations. The technical level in these chapters is some way beyond that of the rest of the book, which is why we have located this material away from the main flow, but the topic is of such central importance to modern cosmology that we felt its inclusion imperative. These chapters use the full formalism of general relativity and fully justify some of the results used, but left unproven, in Chapters 4 through 6.

The formalism set up is necessary for the computation of the present matter and radiation spectra to sufficient accuracy to compare with presently available observations. Fortunately, to work in these research areas, it is not necessary to understand fully their complexity because there exist fitting functions and publically available computer programs [e.g., CMBFAST by Seljak and Zaldarriaga (1996)] that can carry out these tasks. These can be treated as a black box – you feed in the input and out comes the desired answer. Such tools are vital for the day-to-day task of carrying out research in cosmology.

1.4   Notes on examples

Most chapters end with a few examples to allow the reader to practice applying the information given within the chapters. Several of these examples require some simple numerical calculations for their solution; in cosmology these days, it is practically impossible to avoid carrying out some numerical work at some stage. A typical task is the numerical computation of an integral that cannot be done analytically, or the evaluation of some special functions. These can be done via specially written programs, using library packages [e.g., *Numerical Recipes* by Press et al. (1992), which is also an invaluable source of general information on scientific computation], or a computer algebra package such as Mathematica or Maple.

At the end of the book, we list numerical answers to the examples as well as provide some hints as to how we think they should be solved.

2 The Hot Big Bang cosmology

The central premise of modern cosmology is that, at least on large scales, the Universe is homogeneous and isotropic.[1] This is borne out by a variety of observations, most spectacularly the nearly identical temperature of microwave background radiation coming from different parts of the sky. Despite the belief in homogeneity on large scales, it is all too apparent that in nearby regions the Universe is highly inhomogeneous, with material clumped into stars, galaxies, and galaxy clusters. It is believed that these irregularities have grown over time, through gravitational attraction, from a distribution that was more homogeneous in the past.

It is convenient then to break up the dynamics of the Universe into two parts. The large-scale behaviour of the Universe can be described by assuming a homogeneous and isotropic background. On this background, we can superimpose the short-scale irregularities. For much of the evolution of the Universe, these irregularities can be considered to be small perturbations on the evolution of the background Universe, and can be tackled using linear perturbation theory; we discuss this extensively, starting in Chapter 4. It is also possible to continue beyond the realm of linear perturbation theory, via a range of analytic and numerical techniques, which we discuss only briefly, in Chapter 11. In this chapter and the next, we concern ourselves solely with the evolution of the background, isotropic Universe. This usually is called the **Robertson–Walker Universe**, often with Friedmann and occasionally with Lemaitre also named.

A cornerstone of modern cosmology is the theory of **nucleosynthesis** (Pagel 1997; Schramm and Turner 1998), explaining the primordial abundance of the very light elements. It relies on the **Hot Big Bang** model of the Universe, and its success assures us that the model gives an essentially correct description of the Universe starting at some epoch before (though not necessarily long before) nucleosynthesis takes place. The Hot Big Bang model is described in many books (e.g., Kolb and Turner 1990; Padmanabhan 1993; Peebles 1993; Coles and Lucchin 1995). Starting at some epoch well before nucleosynthesis, the Universe consists of a hot gas, which cools with the expansion of the Universe. It has several components, corresponding to the various particle species present, among which are the photons now observed as the cosmic microwave background radiation. At first the energy density is dominated by the relativistic particle species (including photons), collectively termed **radiation**, and later it is dominated by the nonrelativistic species, termed **matter**.

---

[1] By Universe, strictly speaking, we mean the region around us that can be explored by observation, limited by the distance that light has been able to travel. We have no certain knowledge of more distant regions because light from them is still on its way toward us. It is certainly reasonable to expect homogeneity and isotropy to persist for quite some distance, but there is no reason why it should persist indefinitely and, as we see in Chapter 7, there exist eternal inflation models saying that it definitely does not.

In sharp contrast to this well-defined picture, we have no certain knowledge of the Universe well before nucleosynthesis. According to current thinking, based on the Standard Model[2] of particle physics and its most commonly considered extensions, the Universe is gaseous back to some very early epoch, except possibly at brief phase transitions. In a loose sense, this entire gaseous era often is termed the Big Bang, the adjective Hot now being omitted because it need not be in thermal equilibrium or dominated by radiation. Before this, there is thought to have been an era of **inflation**, during which the energy density of the Universe was dominated by the potential of the scalar fields. Inflation is supposed to determine the initial conditions for the Big Bang, including the perturbations.

In this chapter, we give some basic results for the Robertson–Walker Universe, focusing particularly on the Hot Big Bang and the subsequent matter-dominated era. The aim is to provide a brief summary, before moving on to inflation in the following chapter.

In keeping with conventional notation in cosmology, we set the speed of light c equal to one, so that all velocities are measured as fractions of c. Where relevant, we also set the Planck constant $\hbar$ to one, so that there is only one independent mechanical unit. In particular, the phrases "mass density" and "energy density" become interchangeable. Often it is convenient to take this unit as energy, and we usually set the Boltzmann constant k_B equal to 1 so that temperature too is measured in energy units.

Newton's constant G can be used to define the **reduced Planck mass** $M_{Pl} = (8\pi G)^{-1/2}$. Thought of as a mass, $M_{Pl} = 4.342 \times 10^{-6}$ g, which converts into an energy of 2.436×10^{18} GeV. We use the reduced Planck mass throughout, normally omitting the word "reduced." It is a factor $\sqrt{8\pi}$ less than the alternative definition of the Planck mass (e.g., as in Kolb and Turner 1990), never used in this book, which gives $m_{Pl} = 1.22 \times 10^{19}$ GeV. We use M_{Pl} and G interchangeably, depending on the context. Inserting appropriate combinations of $\hbar$ and c, we also can obtain the reduced Planck time $T_{Pl} \equiv \hbar/c^2 M_{Pl} = 2.70 \times 10^{-43}$ s and reduced Planck length $L_{Pl} \equiv \hbar/c M_{Pl} = 8.10 \times 10^{-33}$ cm.

2.1 The expanding Universe

2.1.1 Scale factor and Hubble parameter

If the Universe is homogeneous and isotropic, the distance between any two comoving points is proportional to a universal scale factor $a(t)$, where t is cosmic time. A comoving point is one moving with the expansion of the Universe, defined formally as the location of an observer who measures zero momentum density. In most of the book, we deal with the case of a flat (Euclidean) spatial geometry, and in that case it is convenient to normalize the scale factor to be unity at the present epoch. Throughout, a subscript 0 indicates the present epoch.

The distance of a given comoving point, measured from our location, can be written $r(t) = x a(t)$. The constant x is called the **comoving distance** and is equal to the physical distance at the present epoch.

---

[2] Particle physics and cosmology are plagued by "standard models" of various sorts. When capitalized as here, the phrase always refers to the Standard Model of particle interactions. This is distinct from, for example, the Standard Cold Dark Matter model of structure formation.

A slightly different time variable, known as **conformal time**, often is useful. This is defined by

$$d\tau = \frac{dt}{a(t)}. \tag{2.1}$$

For a freely moving particle with velocity $c = 1$, the coordinate distance travelled during a conformal time interval $\Delta\tau$ is simply $\Delta\tau$. We apply this result to photons and massless neutrinos.

At any epoch, the rate of expansion of the Universe is given by the **Hubble parameter** $H \equiv \dot{a}/a$. The Hubble time H^{-1} and the Hubble distance or length cH^{-1} (equal to H^{-1} with our chosen unit $c = 1$) are of crucial importance. The latter often is called the **horizon** because it provides an estimate of the distance that light can travel while the Universe expands appreciably. (Here, "light" indicates an idealized carrier of information, travelling at speed $c = 1$ without collisions.) Later, we consider two more idealized quantities: the **particle horizon**, which is the distance that light could have travelled since the beginning of the Universe at $a = 0$; and the **event horizon**, which is the distance that light will be able to travel in the future. Of these three, the Hubble distance is the most important, which is why it has come to be called simply "the horizon."

For most purposes, we can ignore the expansion of the Universe in a region much smaller than the Hubble distance, during a time interval much less than the Hubble time (in other words, in a region of space-time that is small on the Hubble scale). In particular, causal processes such as the propagation of waves and the establishment of thermal equilibrium occur as if there were no expansion. Such processes cannot occur on much bigger scales.

Later in this book, when we study the inhomogeneous Universe, a crucial question is how to quantify distance scales and, in particular, how a given scale compares with characteristic scales such as the Hubble length. Perturbations usually are analyzed, as we see in Chapter 4, by carrying out a Fourier expansion in comoving wavenumber k. The inverse wavenumber defines a length scale corresponding to a particular mode of the inhomogeneities; at least while perturbations have a small amplitude, they are stretched by the expansion, and comoving units are the most appropriate ones to use. We can compare the modes with the comoving Hubble length by forming the ratio k/aH; if this is greater than 1, the mode is said to be inside the horizon, whereas if it is less than 1, the mode is outside the horizon, which means that the scale is too large for causal processes to affect it. The key properties of inflation are due to the way in which scales evolve in comparison to the Hubble scale; a scale can begin inside the horizon and be stretched outside the horizon during inflation.

The relative velocity of a pair of nearby comoving observers, separated by distance $dr \ll H^{-1}$, is $v = Hdr \ll 1$. With our chosen unit $c = 1$, this is equal to the redshift $d\lambda/\lambda$ of a photon passing between the observers. It is also equal to the fractional increase da/a in the scale factor, and the wavelength λ of a photon as seen by a sequence of comoving observers stretches with the scale factor.

At the present epoch, the redshift z of light from a cosmological source is defined by

$$1 + z \equiv \frac{\lambda_{\text{obs}}}{\lambda_{\text{emit}}}, \tag{2.2}$$

where λ_{obs} is the observed wavelength and λ_{emit} is the wavelength at the point of emission. For $z \ll 1$, the redshift is given by Hubble's law as $z = Hdr$, which would allow an accurate determination of the present value H_0 if the distances of galaxies were known accurately. Despite decades of observations, this still is not the case. The uncertainty in H_0 usually is parameterized by a quantity h defined by

$$H_0 = 100h \text{ km} \cdot \text{s}^{-1} \text{Mpc}^{-1} \simeq \frac{h}{3000} \text{ Mpc}^{-1}, \tag{2.3}$$

where the last equality uses $c = 1$. As we will see in Section 2.5.1, observations suggest that it lies between 0.5 and 0.8. The present Hubble time and Hubble distance are

$$H_0^{-1} = 9.78h^{-1} \text{ Gyr}, \tag{2.4}$$

$$cH_0^{-1} = 2998h^{-1} \text{ Mpc}. \tag{2.5}$$

Whether or not z is small, the redshift z of light emitted at time t_1 is given by

$$1 + z = \frac{a(t_0)}{a(t_1)}. \tag{2.6}$$

Rather than years or megaparsecs, it is often desirable to specify both times and distances in terms of redshift. When redshift is used to refer to a time, it simply means the time at which the scale factor was a fraction $1/(1+z)$ of its present value. When used to refer to a distance, it means the distance that light can have traveled since that time. Because the Universe expands as the light propagates, the redshift distance is not equal to the redshift time multiplied by the speed of light.

In cosmology, distances tend to be measured via redshifts, and because the Hubble parameter is uncertain by a factor h, the true physical distances are uncertain by a factor h^{-1} even if, as is usually the case, the recession velocities are very well measured. To indicate this, distances normally are given in units such as h^{-1} Mpc.

2.1.2 Gravitational acceleration and the continuity equation

If gravity is negligible, $\ddot{a} = 0$ and the expansion neither accelerates nor decelerates. According to Einstein's theory of general relativity, the acceleration (positive or negative) once gravity is taken into account is

$$\frac{\ddot{a}}{a} = -\frac{\rho + 3P}{6M_{Pl}^2} + \frac{\Lambda}{3}, \tag{2.7}$$

where, instead of Newton's constant G, we are using $M_{Pl} = (8\pi G)^{-1/2}$. Here, ρ is the energy density of the Universe and P is its pressure. In addition, we include Λ, the cosmological constant. Many textbooks ignore the possibility of including such a term, but, increasingly, observational evidence has come to point in favour of a cosmological constant in our Universe. We therefore include it, when present always taking it to have the value necessary to make the Universe spatially flat.

The time dependence of ρ is given by the continuity (or fluid) equation,

$$\dot{\rho} = -3\frac{\dot{a}}{a}(\rho + P).\tag{2.8}$$

This also can be written

$$a\frac{d\rho}{da} = -3(\rho + P).\tag{2.9}$$

In turn, this is equivalent to the energy conservation law for adiabatic expansion, $dE = -P d\mathcal{V}$, where $E = \mathcal{V}\rho$ is the energy in a comoving volume $\mathcal{V} \propto a^3$. The expansion of an isotropic Universe is indeed adiabatic because heat cannot flow.

After inflation, the Universe is assumed to be a gas, except at possible phase transitions, which are supposed to be brief. If the constituents of the gas have mean-square velocity v^2, the pressure is $P = \rho v^2/3$ (remember that we are setting $c = 1$). This applies separately to each component of the gas, and a given component usually has either $v \ll 1$ (called matter or, sometimes, dust) or $v \simeq 1$ (called radiation). For the contribution of matter, the continuity equation gives $\rho_M \propto a^{-3}$, which expresses mass conservation. For the contribution of radiation, it gives $\rho_R \propto a^{-4}$, the extra factor a^{-1} coming from the redshift of particle energy between collisions. In the absence of particle creation or annihilation, $\rho_M/\rho_R \propto a$, and so the early Universe is dominated by radiation.

It is useful to regard the cosmological constant as a possible time-independent contribution to the energy density and pressure of the vacuum, so that

$$\rho_{\text{total}} = \rho + \rho_{\text{vac}},\tag{2.10}$$
$$P_{\text{total}} = P + P_{\text{vac}},\tag{2.11}$$

with $\rho_{\text{vac}} = -P_{\text{vac}} = M_{\text{Pl}}^2 \Lambda$.

According to present beliefs, $\rho + P$ is never negative, which means that, as we go back in time, ρ is always increasing. As a result, the cosmological constant is negligible in the early Universe, even if it is significant now. After inflation, gravity decelerates the expansion rate. In contrast, as we see in the next chapter, P typically is close to $-\rho$ during inflation so that gravity accelerates the expansion rate. Indeed, the formal definition of inflation is as an era of accelerating expansion, corresponding to $P < -\rho/3$.

2.1.3   The Friedmann equation

Using the continuity equation, the equation for $\ddot{a}$ can be replaced by the **Friedmann equation**

$$H^2 = \frac{\rho}{3M_{\text{Pl}}^2} + \frac{\Lambda}{3} - \frac{K}{a^2},\tag{2.12}$$

where $H = \dot{a}/a$ is the Hubble parameter. This equation can be confirmed by multiplying it by a^2 and differentiating it using the continuity equation to obtain Eq. (2.7). The constant K is related to the spatial geometry of the Universe. The Universe is flat (Euclidean) if $K = 0$, finite or closed if $K > 0$, and infinite or open if $K < 0$. (It is also infinite if $K = 0$, but, by custom, one reserves the term open for the case of negative K.)

From the Friedmann equation, we see that, for a given value of the Hubble parameter, there is a particular density, known as the **critical density** ρ_c, for which the Universe is spatially flat in the absence of a cosmological constant. This is given by

$$\rho_c = 3 M_{Pl}^2 H^2 \tag{2.13}$$

and is a function of time. Its present value is $\rho_{c,0} = 1.88 h^2 \times 10^{-29}$ g $\cdot$ cm^{-3}, which in astrophysical units corresponds to

$$\rho_{c,0} = 2.775 h^{-1} \times 10^{11} M_\odot / (h^{-1} \text{ Mpc})^3, \tag{2.14}$$

where $M_\odot = 1.99 \times 10^{33}$ g is the solar mass. This also can be written in particle physics units as

$$\rho_{c,0} = (3.000 \times 10^{-3} \text{ eV})^4 h^2. \tag{2.15}$$

It is usually simplest to measure the energy density as a fraction of the critical density, defining the **density parameter** $\Omega = \rho / \rho_c$. This can be applied separately to different components of matter in the Universe, such as nonrelativistic matter, radiation, and baryons. One also can include a contribution $\Omega_\Lambda \equiv \Lambda / 3H^2$ corresponding to the cosmological constant, so that $\Omega_{total} = \Omega + \Omega_\Lambda$. Then the Friedmann equation can be rewritten as

$$\Omega_{total} - 1 = \frac{K}{a^2 H^2}. \tag{2.16}$$

Normally, Ω_{total} is time dependent, but if it equals 1, corresponding to the spatially flat case, then it retains this value forever.

The present value of Ω is denoted Ω_0. As we will see in Section 2.5, observations suggest a value of Ω_0 between 0.2 and 0.5, and may well support the existence of a cosmological constant, too. As we go back in time, Ω_Λ converges rapidly to 0 and Ω converges rapidly to 1 (unless and until we reach the era of inflation),[3] making K and Λ negligible in the Friedmann equation.

For $K = \Lambda = 0$, the Friedmann equation is solved easily during radiation or matter domination. We have, respectively,

$$\rho_R \propto a^{-4}, \qquad a \propto t^{1/2} \propto \tau, \tag{2.17}$$

$$\rho_M \propto a^{-3}, \qquad a \propto t^{2/3} \propto \tau^2. \tag{2.18}$$

More generally, if we know the dependence of ρ on a, the relation $\dot{a} = aH$ can be integrated using

$$t = C + \int_0^a \frac{da}{aH}, \tag{2.19}$$

where $H(a, K, \Lambda)$ is given by the Friedmann equation. The constant C is taken to be 0, so that $t = 0$ when $a = 0$.

---

[3] This statement could be untrue in the recent past if there are both a cosmological constant and nonzero spatial curvature, but this case is rarely considered.

2.2   **Epochs**

2.2.1   Radiation domination

We start with the era of radiation domination that contains nucleosynthesis, which is what we generally mean by the term "Hot Big Bang."

Thermodynamics during the Hot Big Bang has been much discussed (e.g., by Kolb and Turner 1990), and we only cover the topic briefly in this introduction. A more detailed discussion appears in Chapter 15. For definiteness we assume that there is a single epoch of radiation domination.

Provided that their momentum distribution has the blackbody form, the energy density and temperature of a collection of photons are related by

$$\rho_\gamma = \frac{\pi^2 g_*^\gamma}{30} T^4. \tag{2.20}$$

Here, $g_*^\gamma = 2$ is the number of spin states of the photon, and the Boltzmann constant has been set equal to 1 (in normal units, it is $8.618 \times 10^{-5} \, \text{eV K}^{-1}$). The corresponding photon number density is

$$n_\gamma = \frac{\zeta(3) g_*^\gamma}{\pi^2} T^3, \tag{2.21}$$

where the zeta function evaluates to $\zeta(3) = 1.202$.

During the Hot Big Bang, photons and other relativistic species are in thermal equilibrium at the same temperature, with zero chemical potential. This leads to blackbody distributions for the photons and any other bosons, and the fermionic analogue for fermions. The energy and number densities of each boson species are given by expressions (2.20) and (2.21), whereas for fermions they are multiplied by 7/8 and 3/4, respectively. In both cases, the mean energy per particle is of order T and these expressions apply to a given species until it becomes nonrelativistic at the epoch $T \sim m$, where m is the mass. Then the energy and number density fall exponentially until thermal equilibrium fails, and interactions effectively cease. During radiation domination, the energy density is dominated by that of the relativistic species, given by

$$\rho_R = \frac{\pi^2}{30} g_*(T) \, T^4, \tag{2.22}$$

where

$$g_*(T) \equiv \sum_{\text{bosons}} g_*^i + \frac{7}{8} \sum_{\text{fermions}} g_*^i \tag{2.23}$$

and the summation runs over all relativistic species.

The number of particle species, $g_*(T)$, depends on the underlying particle physics model being considered. In the Standard Model of particle physics, at high temperature, $g_* = 106.75$; extensions such as supersymmetry or Grand Unified Theories may increase this to several hundred. At low temperatures, the number of degrees of freedom drops as particles become nonrelativistic. This is shown in Figure 2.1.

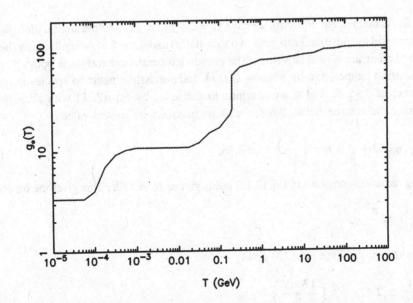

Fig. 2.1. The effective number of relativistic species $g_*(T)$ for the standard model, taking T as the photon temperature (following Kolb and Turner 1990). Supersymmetry will roughly double g_* at $T \gtrsim 100$ GeV. After the epoch $T \sim 1$ MeV, the electrons fall out of thermal equilibrium and the neutrinos acquire a different temperature, and g_* is defined by Eq. (2.22) with T the photon temperature.

During the Hot Big Bang, the entropy density s can be derived via the second law of thermodynamics, $dE = T dS - P dV$, where $V \propto a^3$ is a comoving volume with energy $E = \rho V$ and entropy $S = sV$. This can be rewritten as

$$d\rho = (sT - \rho - P)\frac{dV}{V} + T\,ds, \tag{2.24}$$

and, because ρ depends only on T, this implies that

$$s = \frac{(\rho + P)}{T}. \tag{2.25}$$

It is dominated by the relativistic species and therefore is given by

$$s = \frac{2\pi^2}{45} g_* T^3 \simeq 1.8\, g_* n_\gamma. \tag{2.26}$$

That is, the entropy is measured by the number density of photons. Because there is no heat transfer in an isotropic Universe, the entropy $S = a^3 s$ in a comoving volume is constant, and therefore $T \propto g_*^{-1/3}/a$. Ignoring the small variation of the prefactor, T falls like $1/a$.

All of this holds until the neutrinos fall out of thermal equilibrium, at $T \sim 1$ MeV. Their momentum distribution keeps the same form (as long as they are massless) with effective temperature $T_\nu \propto 1/a$. Shortly afterward, the electrons and positrons become nonrelativistic and annihilate, except for the one electron per proton that maintains electrical neutrality. Ignoring the latter and applying entropy conservation to the electrons, positrons, and photons (the species in thermal equilibrium), we have $g_* = 2 + 7/2 = 11/2$ before annihilation (because the electron and the positron each have two spin states) and $g_* = 2$ after annihilation (because only the photons are now relativistic). The photon temperature therefore is boosted relative to the neutrino temperature by a factor $\sqrt[3]{11/4}$. Each massless neutrino species now has a temperature of 1.95 K, and if we continue to define g_* by Eq. (2.22) with T the photon temperature and assume that all three species are massless, its present value is

$$g_{*,0} = 2 + \frac{7}{8} \times 6 \times \left(\frac{4}{11}\right)^{4/3} = 3.36. \tag{2.27}$$

During radiation domination, Eq. (2.17) holds giving $H = 1/2t$. This gives the timescale

$$\frac{1}{4t^2} = \frac{\pi^2}{90} g_* \frac{T^4}{M_{\rm Pl}^2}. \tag{2.28}$$

Substituting in values (see page 13) gives

$$\frac{t}{1\,{\rm s}} \simeq 2.42 g_*^{-1/2} \left(\frac{1\,{\rm MeV}}{T}\right)^2. \tag{2.29}$$

2.2.2   From radiation to matter domination

The only particle species whose present density is known to high accuracy is the photon. The energy density in photons is dominated by the energy in the microwave background, a blackbody with $T = 2.728$ K (Fixsen et al. 1996), which gives

$$\Omega_{\gamma,0} h^2 = 2.48 \times 10^{-5} \tag{2.30}$$

with negligible uncertainty. Assuming in addition three massless species of neutrino, the total energy density in relativistic particles (radiation) is

$$\Omega_{R,0} = 4.17 \times 10^{-5} h^{-2}. \tag{2.31}$$

Note that the number of massless species is being inferred from standard particle physics models and is not being measured directly. In particular, one or more neutrino species may have significant mass, changing $\Omega_{R,0}$ somewhat. Accepting Eq (2.31) as a working hypothesis, the redshift of **matter–radiation equality** is given by

$$1 + z_{\rm eq} = \frac{\Omega_0}{\Omega_{R,0}} = 24\,000\,\Omega_0 h^2. \tag{2.32}$$

The corresponding temperature is

$$T_{\rm eq} = T_0(1 + z_{\rm eq}) = 65\,500\,\Omega_0 h^2 \,{\rm K}. \tag{2.33}$$

The definition of the density parameter indicates that at any epoch the matter component obeys

$$\Omega_{\rm M}(1 + z)^{-3} H^2 = {\rm const} = \Omega_0 H_0^2, \tag{2.34}$$

where the second equality assumes matter domination today. By definition, at equality the radiation density matches the matter density, and the Friedmann equation gives the Hubble parameter at equality as

$$\frac{H_{eq}}{H_0} = \sqrt{2}\,\Omega_0^{1/2}(1 + z_{eq})^{3/2} = 5.25 \times 10^6 h^3 \Omega_0^2, \tag{2.35}$$

yielding

$$\frac{a_{eq}H_{eq}}{a_0 H_0} = 219\,\Omega_0 h. \tag{2.36}$$

The comoving Hubble length at matter–radiation equality was therefore

$$(a_{eq}H_{eq})^{-1} = 14\Omega_0^{-1}h^{-2}\ \text{Mpc}. \tag{2.37}$$

Now let us consider the evolution of the scale factor between radiation domination and matter domination. If we take $P = w\rho$ with constant w, then the continuity equation (2.8) can be rewritten

$$\frac{d}{dt}\left(\rho a^{3(w+1)}\right) = 0 \tag{2.38}$$

and solved to give

$$\rho = \frac{\rho_0}{a^{3(w+1)}}, \qquad a = \left(\frac{t}{t_0}\right)^{2/3(w+1)} = \left(\frac{\tau}{\tau_0}\right)^{2/(3w+1)}, \tag{2.39}$$

where the solutions are shown for both cosmic time t and conformal time τ. In particular, this includes the classic solutions

$$a \propto t^{2/3} \propto \tau^2 \quad \text{(matter)}, \tag{2.40}$$

$$a \propto t^{1/2} \propto \tau \quad \text{(radiation)}. \tag{2.41}$$

Including both matter and radiation, the Friedmann equation becomes

$$H^2 = \frac{\rho_{eq}}{3M_{Pl}^2}\left[\left(\frac{a_{eq}}{a}\right)^3 + \left(\frac{a_{eq}}{a}\right)^4\right], \tag{2.42}$$

where a_{eq} is the scale factor when the densities of matter and radiation are the same, and ρ_{eq} is the density in each component at that time. This can be solved exactly using conformal time:

$$\frac{a(\tau)}{a_{eq}} = (2\sqrt{2} - 2)\left(\frac{\tau}{\tau_{eq}}\right) + (1 - 2\sqrt{2} + 2)\left(\frac{\tau}{\tau_{eq}}\right)^2, \tag{2.43}$$

with

$$\tau_{eq} = \frac{2\sqrt{2} - 2}{a_{eq}}\sqrt{\frac{3M_{Pl}^2}{\rho_{eq}}}. \tag{2.44}$$

We see a smooth rollover between the two characteristic behaviours of radiation domination and matter domination. There is no good way to rewrite this in terms of cosmic time.

2.2.3   Matter domination

Now we consider the matter-dominated era. We have already seen the famous solution $a \propto t^{2/3}$ for the critical-density case. We now look at the low-density case, both without and with a cosmological constant.

Matter domination in an open Universe

Because the curvature term in the Friedmann equation falls off only as a^2, whereas the matter energy density falls as a^3, domination by the curvature term is a stable situation with late-time solution $a \propto t$. However, we know that the present Universe is not a long way from the critical density, and we should consider both terms. Again, use of conformal time facilitates a solution; fixing $K = -1$ gives

$$a(\tau) = \frac{\rho_{M,0}}{6 M_{Pl}^2} \left(\cosh \tau - 1 \right) = \frac{\Omega_0 H_0^2}{2} \left(\cosh \tau - 1 \right). \tag{2.45}$$

Here, τ_0 must be chosen to give the scale factor its present value; from Eq. (2.16), this is $a(\tau_0) = H_0^{-1}/\sqrt{1 - \Omega_0}$.

Unfortunately, in terms of cosmic time an analytic solution for the evolution of the scale factor can be given only parametrically, as

$$\frac{a(t)}{a(t_0)} = (\cosh \psi - 1) \frac{\Omega_0}{2(1 - \Omega_0)}, \tag{2.46}$$

$$H_0 t = (\sinh \psi - \psi) \frac{\Omega_0}{2(1 - \Omega_0)^{3/2}}, \tag{2.47}$$

where ψ is a parameter whose present value again must give the correct $a(t_0)$. The range of ψ is $[0, \infty)$. A useful equation giving the evolution of Ω with redshift, assuming that the Universe only contains nonrelativistic matter, is

$$\Omega(z) = \Omega_0 \frac{1+z}{1 + \Omega_0 z}. \tag{2.48}$$

Example 2.3 indicates how this is derived. Provided that $1 + z \gg 1/\Omega_0 - 1$, spatial curvature can be ignored.

The cosmological constant

Many observations suggest that the density of matter in the Universe is less than the critical density. If this is so, then an alternative to believing that the Universe has negative spatial curvature is to introduce a cosmological constant Λ to restore spatial flatness. Then the Friedmann equation reads

$$H^2 = \frac{\rho}{3 M_{Pl}^2} + \frac{\Lambda}{3}. \tag{2.49}$$

In much of the literature, one has to read rather closely to discern whether the phrase "low-density Universe" does or does not include a cosmological constant to restore spatial flatness.

In a Universe dominated by the cosmological constant, the solution is an exponential expansion rate

$$a(t) \propto \exp\left(\sqrt{\frac{\Lambda}{3}}\, t\right), \tag{2.50}$$

the effective density remaining constant by definition.

Even if our Universe does possess a cosmological constant, it cannot be dominated completely by it and we should consider both matter and Λ contributions. Again, invertible analytic solutions are not available. Similarly to the open case, we can calculate the redshift dependence of the matter density, which now is given by

$$\Omega(z) = \Omega_0 \frac{(1+z)^3}{1 - \Omega_0 + (1+z)^3 \Omega_0}. \tag{2.51}$$

The Universe is close to critical density at $(1+z)^3 \gg 1/\Omega_0 - 1$.

2.3   Scales

2.3.1   Characteristic scales

In general, a homogeneous and isotropic Universe is characterized by two length scales: the Hubble scale and the curvature scale.

The Hubble scale is just the reciprocal of the Hubble parameter, H^{-1}. It can indicate either time or distance (length scales being converted to timescales through division by c, which we have set to 1). The Hubble time represents the characteristic timescale of evolution of the scale factor; during a Hubble time, $a(t)$ grows by a factor e (in the context of inflation, this often is referred to as an e-folding).[4] Because during normal evolution the scale factor is decelerating, the Hubble time is usually a good estimate of the age of the Universe. Consequently, the Hubble radius is normally a reasonable estimate of the size of the observable Universe, given by the distance that light could have travelled since the Big Bang.

The curvature scale arises because the constant K in the Friedmann equation is related to the spatial geometry. If K vanishes, the Universe is flat (Euclidean geometry); otherwise there is spatial curvature, which becomes significant if we survey the Universe out to a distance of order of the **curvature scale**

$$r_{\text{curv}} = a\,|K|^{-1/2}. \tag{2.52}$$

This precise expression for the curvature scale comes from the form of the metric for a homogeneous Universe, the **Robertson–Walker metric**

$$ds^2 = -dt^2 + a^2(t)\left[\frac{d\tilde{x}^2}{1 - K\tilde{x}^2} + \tilde{x}^2(d\theta^2 + \sin^2\theta\, d\phi^2)\right], \tag{2.53}$$

---

[4] The continuous definition of time in Hubble units, including the variation of H, is given by $t_{\text{hub}} = \int_i^f dt/H^{-1} = \ln(a_f/a_i)$.

which also can be written in a more convenient form by redefinition of the radial coordinate as

$$ds^2 = -dt^2 + a^2(t)\left[dx^2 + |K|^{-1}\sinh^2(|K|^{1/2}x)(d\theta^2 + \sin^2\theta\,d\phi)\right] \tag{2.54}$$

for an open Universe. For a closed Universe, sinh is replaced by sin. For $r \ll r_{\text{curv}}$, we see that the Euclidean spatial geometry is recovered. If K vanishes, as is suggested by inflation, the curvature scale is infinite, leaving only the Hubble scale.

The ratio of the two characteristic scales is related to Ω via the Friedmann equation:

$$\sqrt{|\Omega - 1|} = \frac{H^{-1}}{r_{\text{curv}}}. \tag{2.55}$$

In open Universes the Hubble length is always less than the curvature scale, approaching it in the limit $\Omega \to 0$, which is the typical late-time behaviour. In a closed Universe, the Hubble length can exceed the curvature scale (though in that case it has no particular interpretation as a length). Equation (2.55) needs slight modification for the contracting phase of a closed Universe.

2.3.2   Particle horizon

The coordinate distance travelled by a freely moving photon, emitted at time t and observed by us now, is

$$x(t) = \int_t^{t_0} \frac{dt}{a}. \tag{2.56}$$

This integral converges rapidly as we go back through the matter-dominated era, and continues to do so as we go further back into the Big Bang era, though of course photons really are not moving freely during that era. Ignoring this fact and making the idealization that the Big Bang started at the epoch $a = 0$, the present distance of an object emitting light at that epoch is

$$r_{\text{hor}}(t_0) \equiv a_0 \int_0^{t_0} \frac{dt}{a}. \tag{2.57}$$

This is called the **particle horizon**, and the same term is applied to the corresponding quantity, with t_0 replaced by an arbitrary epoch.

Evaluated at the present epoch, the particle horizon is crucially important because it defines the size of the observable Universe. Assuming critical density and approximating the Universe as always matter dominated, it is given by $r_{\text{hor}}(t_0) = 2H_0^{-1}$.

The particle horizon changes if we move to a low-density Universe. Keeping the excellent approximation of matter domination back to the Big Bang, then in an open Universe,

$$r_{\text{hor}}(t_0) = \frac{H_0^{-1}}{\sqrt{1 - \Omega_0}} \cosh^{-1}\left[\frac{2 - \Omega_0}{\Omega_0}\right]. \tag{2.58}$$

The prefactor to the cosh is just the curvature distance; the particle horizon distance exceeds the curvature distance if $\Omega_0 < 0.79$.

In a cosmological-constant model, even one giving spatial flatness, the answer is not available analytically, but is given in integral form by (Carroll et al. 1992)

$$r_{hor}(t_0) = H_0^{-1} \int_0^\infty \frac{dz}{\sqrt{1 + 3\Omega_0 z + 3\Omega_0 z^2 + \Omega_0 z^3}}. \tag{2.59}$$

Hu and White (1997a) quote a fitting function that proves accurate to better than 1 percent in cases of interest, which (ignoring a correction for the time of the matter–radiation transition and correcting a typo in the sign of the prefactor of the logarithm) is

$$r_{hor}(t_0) \simeq 2H_0^{-1} \frac{1 + 0.084 \ln \Omega_0}{\sqrt{\Omega_0}}. \tag{2.60}$$

In summary, and taking the critical-density case as an example, we have three important scales:

(1) t_0 Age of the Universe,
(2) $H_0^{-1} = 3t_0/2$ Hubble distance/horizon,
(3) $r_{hor}(t_0) = 3t_0 = 2H_0^{-1}$ Particle horizon.

Note that the distance that light travels is greater than we would get by naïvely multiplying the age of the Universe by the speed of light, because the Universe was physically smaller when the light first set out.

2.3.3 Predicted ages

As we will see in Section 2.5.1, observations seem to require $t_0 > 10\,\mathrm{Gyr}$. Taking $\Omega_0 = 1$, the age of the Universe is

$$t_0 = \frac{2}{3} H_0^{-1} = 2.06h^{-1} \times 10^{17}\,\mathrm{s} = 6.52h^{-1}\,\mathrm{Gyr}. \tag{2.61}$$

The consensus is that this can be in agreement with observations only if the Hubble parameter h is near the low end of the measurements made.

For a given h, the Universe can be older if we move to a low-density Universe, either with or without a cosmological constant. In an open Universe, the age becomes, from Eqs. (2.46) and (2.47),

$$t_0 = H_0^{-1} \left[\frac{1}{1 - \Omega_0} - \frac{\Omega_0}{2(1 - \Omega_0)^{3/2}} \cosh^{-1}\left(\frac{2 - \Omega_0}{\Omega_0}\right) \right]. \tag{2.62}$$

As Ω_0 tends to 1, the square bracket tends to 2/3, as it must. As Ω_0 is decreased, the square bracket increases monotonically, reaching unity in the limit $\Omega_0 \to 0$. So, an open Universe can be, at most, 1.5 times as old as a flat one for the same h, giving an age equal to the Hubble time. For a spatially flat Universe with a cosmological constant, the boost to the age is more dramatic, given by (Carroll et al. 1992)

$$t_0 = \frac{2H_0^{-1}}{3} \frac{1}{\sqrt{1 - \Omega_0}} \ln\left(\frac{1 + \sqrt{1 - \Omega_0}}{\sqrt{\Omega_0}}\right), \tag{2.63}$$

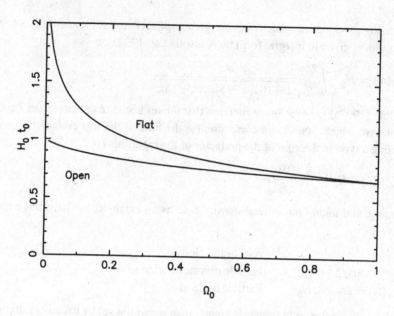

Fig. 2.2. Predicted ages for open Universes and for flat Universes with a cosmological constant.

which diverges in the low Ω_0 limit. In place of the logarithm, the last factor also can be written as $\sinh^{-1}(\sqrt{(1-\Omega_0)/\Omega_0})$. For $\Omega_0 = 0.2$, about the smallest conceivable value, the age is 1.6 times the $\Omega_0 = 1$ value, for a given h. For $\Omega_0 = 0.26$, the age equals the Hubble time. Ages are shown in Figure 2.2.

2.4 The cosmic microwave background

Many of the most dramatic observational results in cosmology in recent years have related to the cmb and, in particular, its anisotropies. The anisotropies are discussed extensively later in the book. Here, we review some of the properties of the microwave background itself (Hogan et al. 1982; White et al. 1994).

The details of last scattering have been long understood. We can model the time of last scattering by a visibility function, which measures the probability that a particular photon last scattered in a redshift interval dz. Conveniently, this proves to be well approximated by a Gaussian at mean redshift $z_{rec} = 1100$ with width $\Delta z \simeq 80$, pretty much independent of all cosmological parameters (Jones and Wyse 1985).[5]

Even if the density parameter of the Universe is less than unity today, it would have been very close to 1 at the time of last scattering. Because matter–radiation equality is at $1 + z_{eq} =$

---

[5] This assumes that there is no very early reionization of the intergalactic medium by ultraviolet photons emitted by the earliest generations of stars and quasars, which could create free electrons that can scatter the microwave photons. In inflation-based models, it is thought very unlikely that reionization could be early enough to rescatter all of the microwave background photons to create a new last-scattering surface, but it may well be that some modest fraction are rescattered, altering the anisotropy pattern. We discuss this further in Chapters 5 and 11.

$24\,000\,\Omega_0 h^2$, then for normal parameters, matter domination occurs a reasonable amount of time before recombination. The time of recombination therefore can be estimated from the matter domination result of Eq. (2.34) as

$$t_{rec} = \frac{2}{3} H_{rec}^{-1} = \frac{2}{3} \frac{\Omega_0^{-1/2} H_0^{-1}}{(1 + z_{rec})^{3/2}} = 1.8 \times 10^5 \Omega_0^{-1/2} h^{-1} \text{ yr.} \tag{2.64}$$

The comoving particle horizon size at this time was $180\Omega_0^{-1/2} h^{-1}$ Mpc, the comoving Hubble length was half that size at $90\Omega_0^{-1/2} h^{-1}$ Mpc, and the thickness of the last-scattering surface was $7\Omega_0^{-1/2} h^{-1}$ Mpc.

The corresponding angular sizes on the last-scattering surface are also of interest. In an open Universe, the distance to the last-scattering surface is given by Eq. (2.58), and the "circumference" of the sphere at that distance is obtained most directly from the metric (2.54). The trigonometric functions conveniently destroy each other to give the circumference as

$$2\pi r_{eff} = \frac{4\pi H_0^{-1}}{\Omega_0}. \tag{2.65}$$

This is to be reduced by a small amount if we use the precise redshift of last scattering rather than taking it to be at infinity; for example, if $\Omega_0 = 1$, then the distance to the last-scattering surface is $r_{lss} = 5820 h^{-1}$ Mpc, rather than $2H_0^{-1} = 6000 h^{-1}$ Mpc. The angle subtended by the Hubble scale at last scattering is $0.86\Omega_0^{1/2}$ deg, and that by the scale corresponding to the thickness of the last-scattering surface is $4\Omega_0^{1/2}$ arc-minutes. Finally, the curvature scale itself subtends an angle of $\Omega_0/2\sqrt{1-\Omega_0}$ rad. Experiments that probe scales larger than 1 deg, such as COBE, are studying scales that were still larger than the horizon scale at the time the microwave background was formed.

A cosmological-constant model has the standard Euclidean geometry, but the distance to the horizon now is given by Eq. (2.60). Ignoring the logarithm, we see that the angle subtended by the Hubble scale at last scattering is 0.86 deg and the thickness of the last-scattering surface corresponds to 4 arc-minutes, both independent of Ω_0 to a reasonable approximation.

2.5 Ingredients for a model of the Universe

We now give a brief summary of the observational status of the various parameters required to specify a cosmological model. This breaks up naturally into two parts: the parameters that describe the global state of the Universe and the parameters needed to describe the matter in the Universe, in particular, the dark matter assumed to pervade the Universe. These two in combination are what we are calling the **cosmological parameters**. When we come to study models of structure formation, a third set of information is required, which is a description of the initial perturbations in the Universe from which structure grows.

As we write, there is still considerable uncertainty concerning an accurate description of the Universe. In fact, there is only a single cosmological parameter that is uncontroversially well determined to an accuracy that permits its variation to be neglected when we study models of structure in the Universe, that being the present temperature of the microwave background, $T_0 = 2.728 \pm 0.004$ K. All other cosmological parameters may be varied, consistently with observations, sufficiently to affect large-scale structure predictions. We do not aim to provide a detailed discussion here, but rather a rough guide to what is possible and a list of references that the reader can pursue.

2.5.1   Hubble parameter and age

In recent years, progress appears finally to have been made toward a consensus on the value of the Hubble constant, but the story remains some way from being closed. Many measurements now are coming in at around $h = 0.70$, in particular, the Hubble Space Telescope observations of Virgo Cepheids (Freedman 1997) and the type Ia supernovae method (Riess et al. 1996), with uncertainties around 15 percent. On the other hand, values below $h = 0.50$ still appear from time to time (e.g., Schechter et al. 1997). For reviews, see Jacoby et al. (1992) and Hendry and Tayler (1996).

The age of the Universe was long thought problematic in the context of a critical-density Universe, with many estimates of globular-cluster ages, based on stellar evolution, being upward of 15 Gyr. This changed dramatically when the Hipparcos satellite measured a large number of stellar parallaxes. The stars turned out to be farther, hence brighter and younger, than previously thought, bringing down the age limits (Reid 1997; Feast and Catchpole 1997; Chaboyer et al. 1998). (This also lowered the Hubble parameter estimated by distance ladder techniques.) A safe lower limit from globular-cluster ages seems to be about 10 Gyr, with perhaps an extra gigayear to be added to allow them to form in the first place.

A critical-density Universe with $h = 0.70$ has an age below 10 Gyr, incompatible with even the lowest estimates of the age of the Universe. When theorists consider the critical-density case, the usual choice is therefore $h = 0.50$, which comfortably passes the age constraint. The advantages for structure formation of an even lower value have been spelled out by Bartlett et al. (1995), but such values sit uneasily with direct observations.

An interesting result is a demonstration that the local and global Hubble constants cannot differ greatly, using type Ia supernovae (Kim et al. 1997). This counters arguments that if we live in a significantly underdense region, the local expansion rate in the surrounding tens of megaparsecs, which most measurement techniques sample, might be much higher than the true expansion rate. Although such an effect may persist at the 10-percent level or so, it cannot be large enough to reconcile a very high locally measured value with the age of a critical-density Universe.

In conclusion, the view that we take is that the "reasonable" range for h is between 0.5 and 0.8, and although values outside that range are not inconceivable, they are disfavoured.

2.5.2   Density parameter

It has long been accepted that most of the mass in the Universe neither emits nor absorbs radiation; it is dark as opposed to luminous. But observation of the luminous matter probes, through gravity, the total mass density, and this allows us to estimate the density parameter Ω_0 in a number of ways. In a rich galaxy cluster, observation of the luminous matter enables us to infer the cluster mass density and hence the total cluster mass. As will be discussed in connection with Eq. (2.68), this leads to one way of estimating the density parameter. The growth of perturbations in the Universe is affected by the mean mass density, and we will see that this too provides useful information about the density parameter. Finally, observations of supernovae at high redshift give information about the expansion rate of the Universe, and hence about both the density parameter and the cosmological constant. Collectively, this body of evidence suggests a value of Ω_0 in the range 0.2 to 0.5.

2.5.3 Cosmological constant

Direct measurements of the density parameter say nothing about the cosmological constant, but compatibility with the simplest models of inflation requires a flat spatial geometry. The simplest way to make a low-density Universe compatible with these models is to add a cosmological-constant term. We only consider the cosmological constant with the value chosen to restore spatial flatness.

A large cosmological constant has a very significant influence on the expansion rate at late times, which can be probed observationally in several ways. The most important at present is the magnitude–redshift relation for distant type Ia supernovae, which are believed to be good standard candles once a correction for rate of decay of light emission is taken into account. To many people's surprise, the evidence is strongly in favour of a cosmological constant (Perlmutter et al. 1998; Schmidt et al. 1998). In particular, for spatially flat cosmologies, the best-fit Ω_0 is about 0.3, with Λ comprising the remainder. One possibility is that this cosmological constant may be an effective one rather than a true one, and indeed may be varying slowly with time. Such behaviour can be modelled via the potential energy of a scalar field, in exactly the same way as inflation is.

Another way of constraining Λ is via gravitational lensing (Kochanek 1996; Falco et al. 1998), which gives $\Omega_\Lambda < 0.74$ at 2-σ confidence. We find that large-scale structure models tend to support the lensing limit. If these measurements are all correct, then Ω_0 must be not far from 0.3 to 0.4 or so.

2.5.4 Baryonic matter

In cosmology, ordinary matter is referred to as baryonic matter or simply baryons, because the protons and neutrons (baryons) rather than the electrons account for practically all of its density, denoted Ω_b.

Nucleosynthesis [Alpher et al. 1948; Hoyle and Tayler 1964; Walker et al. 1991; see Pagel (1997) and Schramm and Turner (1998) for reviews] provides the best estimate of the baryon density. Although the theoretical calculations are well developed and accurate, the observational data require considerable interpretation. Consequently, although the theory of nucleosynthesis is undoubtedly one of the great successes of cosmology, the limits on the baryon density are not as tight as we would like. A typical example (Copi et al. 1995a,b) gives

$$0.010 \leq \Omega_b h^2 \leq 0.022, \tag{2.66}$$

with even higher values obtained by some authors (Hata et al. 1995; Kernan and Sarkar 1996). This range seems intended to correspond to the 95 percent confidence range (though without any implication that the error bars can be treated as Gaussian within that range). When we refer to the nucleosynthesis value of Ω_b, we mean the central value of the Copi et al. range, namely

$$\Omega_b^{\text{nuc}} h^2 = 0.016. \tag{2.67}$$

Observations (Tytler et al. 1996) of the deuterium abundance in quasar absorption systems suggest that the true value may well be toward or even beyond the top of the quoted range. (Results often are quoted in terms of the ratio of baryon and photon numbers η; the conversion is $\eta = 2.68 \times 10^{-8} \Omega_b h^2$.)

The ratio Ω_b / Ω_0 can be estimated by looking to clusters of galaxies, an approach popularized by White et al. (1993b). Originally labelled the "baryon catastrophe," this is now normally given the less emotive title of the "cluster baryon fraction." The majority of the baryons in clusters tend to be in the form of hot X-ray–emitting gas, the distribution of which also can be used to measure the total mass of the cluster from the condition for hydrostatic equilibrium in the gravitational potential well, admittedly with some modelling uncertainties. Because clusters are such large objects, we expect (with support from simulations) the balance of material within them to be representative of the Universe as a whole. White and Fabian (1995) compiled data for thirteen clusters, and obtained a baryon fraction

$$\frac{\Omega_b}{\Omega_0} = 0.14^{+0.08}_{-0.04} \left(\frac{h}{0.5} \right)^{-3/2} \tag{2.68}$$

at the 95 percent level.

Given the nucleosynthesis value of Ω_b, this implies that the Universe has a low density, perhaps around 0.4. The cluster baryon fraction is inconsistent with nucleosynthesis in a critical-density Universe, unless the theory of nucleosynthesis is somehow flawed, either through the interpretation of observations (which seems rather unlikely) or through the intervention of new physics (e.g., Gyuk and Turner 1994; Mathews et al. 1996). Alternatively, the cluster mass determinations might be wrong, and there is evidence that X-ray mass determinations may be too low (e.g., Gunn and Thomas 1996; Wu and Fang 1997), presumably because the true gas distribution is more complicated than the modelling currently allows. Finally, we might worry whether the earlier statement, that the cluster baryon fraction should accurately represent the universal value, stands up to examination because determinations in different clusters do not appear completely consistent (Loewenstein and Mushotzky 1996); it is not clear at present whether the differences can be explained away through systematic errors (implying that the uncertainties typically have been underestimated), or whether they genuinely undermine the claim that clusters fairly sample the material content of the Universe.

An ambitious goal is to derive the baryon density from first principles, a topic known as **baryogenesis**. It cannot be derived from decoupling from thermal equilibrium, as can the massive neutrino case discussed shortly, because the density is determined not thermally but because there is a conserved quantity, the baryon number, that is nonzero in our present Universe. A common assumption is that, at high energies, the Universe has zero net baryon number, and that the baryon number is generated by some baryon-number violating processes in the early Universe. Within the Standard Model, only nonperturbative weak interactions violate baryon number. Many extensions to the Standard Model violate baryon number. See Dolgov (1992) for a review. At present, there is not even an order-of-magnitude theoretical understanding of the baryon number of the Universe.

The baryonic dark matter could reside in a number of forms, including cold gas and compact objects, the latter having been detected via microlensing observations and therefore being present in some reasonable abundance. A detailed account of the possibilities can be found in Carr (1994).

2.5.5 Cold and hot dark matter

Particle physics has offered no shortage of nonbaryonic dark matter candidates (e.g., Jungman et al. 1996a), from massive neutrinos to axions or the lightest supersymmetric particle (Section 6.1.2). As far as large-scale structure is concerned, it is more fruitful to classify dark matter by its random motion than by its particle physics origin.

CDM refers to particles with no significant random motion, at least as far as structure formation is concerned. To be precise, their random motion is nonrelativistic when even the smallest cosmologically interesting scales come inside the horizon. It is the simplest option because, by definition, the individual particle properties are unimportant, so that their cosmological influence is specified fully by the density, Ω_{CDM}. CDM candidates can be elementary particles (the lightest supersymmetric particle being the favourite candidate) or much more massive objects such as primordial black holes. As we see later, structure formation models based on inflationary perturbations nearly always feature at least some CDM, and they often are referred to generically as CDM models.

At the other extreme is hot dark matter, which is still relativistic when galaxy scales enter the horizon, and between the two extremes lies warm dark matter. These are more complex possibilities because the cosmological effects of such matter depend not only on the density but also on the nature of the random motion. In particular, we need to know the freestreaming length, measuring the typical distance that a particle is able to move.

Commonly, the need to introduce extra parameters describing the particle motions is avoided by making a specific assumption, for example, that the hot dark matter takes the form of a massive neutrino. With this assumption, the relation between the particle mass (which determines its random motion) and its density is fixed, namely

$$\Omega_\nu h^2 = \frac{\sum_i m_{\nu_i}}{94\,\text{eV}}, \tag{2.69}$$

where the sum is over any neutrino types with mass less than about 1 MeV, implying that they are relativistic at decoupling. That the density scales linearly with mass is easy to understand; because the particles are relativistic at decoupling, the final number density is independent of the particle mass. The detailed answer arises as follows (see, e.g., Kolb and Turner 1990). We have $n_\nu = (3/4) \times (4/11) n_\gamma$, where the 4/11 comes from the difference in temperature (Section 2.2.1) and the 3/4 because neutrinos are fermions and the photons are bosons. From Eqs. (2.20) and (2.21) the mean photon energy is $2.7 T_\gamma$, and the present energy density in radiation is given by Eq. (2.30). Putting all of this together gives Eq. (2.69).

For neutrinos of mass above 1 MeV, the assumption of decoupling while relativistic breaks down and a much more sophisticated analysis becomes necessary; see, for example, Kolb and Turner (1990). The total energy density begins to fall below the linear extrapolation, because Boltzmann suppression reduces the number density of neutrinos before they decouple. There is, in fact, a second solution in which the neutrinos would give a critical density at a mass of a few giga-electron volts; however, this mass exceeds the laboratory bounds on all three of the known neutrino species.

For a hot component of sufficient density to be interesting, we expect a mass in the range (depending somewhat on h) from a few electron volts to a few tens of electron volts. This is around the current experimental bounds on the electron neutrino mass, but well below the limits on the other two known species. (Neutrino oscillation experiments are providing increasing evidence of nonzero neutrino masses, but only give the difference in the mass squared, rather than the actual masses.) With such masses, the neutrinos will be relativistic early on but will have become nonrelativistic by the present. Neutrinos are the only particles in the Standard Model that, through its extension to allow them masses, also can be the dark matter. If they are used in this way, we must be careful to make the appropriate modifications, in particular to the effective number of massless species in the Universe.

There are hot (or warm) dark matter candidates other than neutrinos and, in that case, the necessity of introducing extra parameters to describe them cannot be evaded. And that is only the start of an increasing sequence of complexity; for example, the dark-matter particles could decay (presumably into another form of invisible matter) or even annihilate. So far, however, these options have not received a huge amount of attention; the standard assumption is that the dark matter is either entirely cold or is a mixture of cold and hot. We return to some of the more exotic examples in Chapter 6.

2.6   History of our Universe

We end this chapter with an overview of the history of the Universe, including the speculative era before nucleosynthesis.

Nucleosynthesis can be considered as beginning at a time of order 1 s, corresponding to a temperature of order 1 MeV and an energy density of order $(1 \text{ MeV})^4$. The regime of much bigger energy density is the realm of the particle cosmologist. There is little prospect of ever going more than a couple of orders of magnitude beyond 100 GeV through direct terrestrial investigation (though probes of extremely high energy cosmic rays may have something to say). Within this realm, it is not completely unreasonable to imagine that present ideas in particle physics can be extrapolated, and particle physicists have not been short of ideas as to how the Standard Model might be extended. Many speculative ideas have been suggested for what might happen in this regime, one of which is cosmological inflation. Although a test of such an idea is outside the reach of particle accelerators, the observational consequences in cosmology may be profound, as we see throughout this book. The early Universe thus offers the possibility of acting as a distinctive probe of very high energy physics.

The observable Universe may have been created with energy density below the Planck scale. More usually, it is supposed that the Universe does have a history extending all the way back to the Planck scale. Whenever its history began, there may at first have been additional space dimensions, which compactify only later.

There is virtually no understanding of what might happen beyond the Planck scale. A very speculative possibility, which utilizes the duality properties of superstring theory, is the **pre-Big-Bang** scenario (Gasperini and Veneziano 1993a,b). Here, a contracting Universe undergoes some kind of tunnelling event to become an expanding Universe as the contracting phase approaches the Planck scale. Such a process may even be able to produce density perturbations, though it remains to be seen if they can have the right form – in models so far, both the gravitational waves and the adiabatic perturbations are too far from scale invariance, but isocurvature perturbations may be created with a more reasonable shape (Copeland et al. 1997). We do not discuss the pre-Big-Bang scenario further here; for a review, see Gasperini (1997).

Taking the history to begin at the Planck scale, we can divide it loosely into a series of epochs:

- $M_{\rm Pl} > \rho^{1/4} > 100\,{\rm GeV}$
In this very wide regime of densities, there is little indication as to the appropriate underlying physics (100 GeV being the electroweak energy scale, specified by the masses of the W and Z particles). In this regime, the standard tool is speculative extrapolations from the physics that we do understand, with the assumption that the general techniques and principles still apply. It is in this regime that cosmological inflation would occur, and we are assuming that it does.

After inflation ends, the Universe is dominated by nonrelativistic particles (matter dominated) for some time; we might refer to this era as a **Cold Big Bang**. Eventually, the nonrelativistic particles decay into relativistic particles and produce a radiation-dominated Universe. This process is called **reheating**. In the simplest scenario, radiation then dominates until the onset of the present matter-dominated era.

In more complicated scenarios, radiation domination temporarily may give way to an early era of matter domination, which ends when the relevant particles decay to form radiation. Just before such a matter-dominated era, there might be an era of so-called thermal inflation (Section 3.9), but this type of inflation would not generate the inhomogeneities that are our main concern in this book. Whether thermal inflation occurs or not, an early era of matter domination reduces the abundance of particles that were produced earlier, which may be welcome. We may regard this reduction as coming from the entropy that is produced by the particle decays signalling the end of the matter-dominated era.

- $100\,\text{GeV} > \rho^{1/4} > 10\,\text{eV}$

Barring very late decaying particles or other exceptional circumstances, the Hot Big Bang begins at a temperature higher than $100\,\text{GeV}$, and we proceed on this assumption. The relevant physics is described by the Standard Model, or some supersymmetric extension of it. The first phenomenon is the electroweak phase transition, at which electromagnetism and the weak interactions first gain their separate identities. If the Hot Big Bang indeed begins before $100\,\text{GeV}$, this is the last possible time at which the baryon asymmetry could be generated. Later, in the quark–hadron phase transition at the QCD scale ($\sim 10^{12}\,\text{K}$), it becomes energetically favourable for the free quarks that existed in equilibrium with the radiation sea to condense in hadrons such as protons, neutrons, and pions.

In principle, these two events could be avoided if the start of the Hot Big Bang were delayed, for example, by long-lived massive particles from an earlier epoch. Such particles, or the related scalar fields, would have to play some role in creating the baryon asymmetry as well as delaying the onset of the Hot Big Bang.

At the very latest, the Hot Big Bang must begin by $T \sim 10\,\text{MeV}$ because otherwise the standard nucleosynthesis calculation is invalidated. Nucleosynthesis occurs at $T \sim 0.1\,\text{MeV}$, when the protons and neutrons bind together into atomic nuclei, primarily hydrogen and helium-4.

At around $10\,\text{eV}$, corresponding to $10^5\,\text{K}$, the nonrelativistic matter content in the Universe reaches the same density as the relativistic, and the radiation era ends.

- $10\,\text{eV} > \rho^{1/4}$

With the onset of matter domination, the expansion rate increases to $a(t) \propto t^{2/3}$. From that point on, the temperature decreases more quickly than the (fourth root of the) energy density. The first significant event of this epoch occurs at around $3000\,\text{K}$, when decoupling of the radiation from matter occurs. The first event in this sequence is recombination (somewhat of a misnomer because the objects in question were never previously combined), in which the majority of the electrons bind with nuclei to form atoms, the photon energies being insufficient to dislodge them. At this point the photons remain coupled to the ionized electrons that remain. Shortly afterward, as the ionized fraction falls further, radiation decouples completely from the matter, and the photons propagate freely. We see these today as the microwave background.

Between decoupling and the present, structure must form in the Universe, including the first large gravitationally bound systems. Before this time, the Universe is close to homogeneity on all scales, with linear perturbation theory easily adequate to deal with any irregularities that exist. According to the inflationary cosmology, structure formation is brought on by perturbations laid down in the early history of the Universe.

In Universes that are not flat, another transition occurs when the horizon scale is of the order of the curvature scale, and the Universe becomes curvature dominated. In a closed Universe, this is followed by collapse.

Table 2.1 provides a short summary of the most important epochs in the standard model for the history of the Universe.

Table 2.1. *A history of the Universe*

t	$\rho^{1/4}$	T	Event[a]
10^{-42} s	10^{18} GeV	~ 0	Inflation begins?
$10^{-32\pm6}$ s	$10^{13\pm3}$ GeV	~ 0	Inflation ends, Cold Big Bang begins?
$10^{-18\pm6}$ s	$10^{6\pm3}$ GeV	$10^{6\pm3}$ GeV	Hot Big Bang begins?
10^{-10} s	100 GeV	100 GeV	Electroweak phase transition?
10^{-4} s	100 MeV	100 MeV	Quark–hadron phase transition?
10^{-2} s	10 MeV	10 MeV	γ, v, e, $\bar{e}$, n, and p all in thermal equilibrium.
1 s	1 MeV	1 MeV	v decoupling, $e\bar{e}$ annihilation.
100 s	0.1 MeV	0.1 MeV	Nucleosynthesis.
10^4 yr	1 eV	1 eV	Matter–radiation equality.
10^5 yr	0.1 eV	0.1 eV	Atom formation, photon decoupling.
$\sim 10^9$ yr	10^{-3} eV	10^{-4} eV	First bound structures form.
Now	$3 \times 10^{-3} h^{1/2} \Omega_0^{1/4}$ eV	2.728 K	The present.

[a] The Hot Big Bang might be delayed or interrupted. If so, electroweak symmetry might not be restored (no electroweak phase transition) and free quarks might never be present (no quark–hadron phase transition). However, the success of the standard nucleosynthesis calculation shows that an uninterrupted Hot Big Bang must be under way before $T \sim 1$ MeV.

Examples

2.1 How many neutrino families would there have to be to make matter–radiation equality and decoupling coincide, assuming $\Omega_0 = 1$ and $h = 0.5$?

2.2 Can an open Universe evolve into a closed one?

2.3 For a Universe containing only nonrelativistic matter (and no cosmological constant), show that the Friedmann equation can be rewritten as

$$H(z) = H_0(1 + z)(1 + \Omega_0 z)^{1/2}.$$

Use this to derive Eq. (2.48).

2.4 Compute the Hubble radius at nucleosynthesis, assuming that it occurs at a temperature of 0.1 MeV. To what scale does this correspond today? What was the energy within a Hubble volume at nucleosynthesis (expressed in solar mass units), and what mass is contained within that comoving volume at the present epoch?

3 Inflation

3.1 Motivation for inflation

Primarily, this is a book about cosmological inflation, and one in which we intend to give as up-to-date a viewpoint as possible. In the modern view, by far the most important property of inflation is that it can generate irregularities in the Universe, which may lead to the formation of structure. This allows the possibility of testing various aspects of the inflationary scenario, and the bulk of our book is devoted to these topics.

However, the historical motivation for inflation was rather different, and arose largely on more philosophical grounds concerning the question of whether the initial conditions required for the Hot Big Bang seem likely or not. In this chapter, we begin by briefly discussing those aspects of the historical motivation for inflation, before moving on to a description of the classical dynamics of inflation.

3.1.1 Flatness problem

As we saw in Chapter 2, the Friedmann equation can be written as an equation for the density parameter Ω, Eq. (2.16). Ignoring the cosmological constant (or, if you like, including it in Ω), it is

$$\Omega - 1 = \frac{K}{a^2 H^2}.$$
(3.1)

If the Universe is flat ($\Omega = 1$), then it remains so for all time. Otherwise, the density parameter evolves. The flatness problem is simply that during radiation or matter domination, the combination aH is a decreasing function of time.[1] For example, in a nearly flat, matter-dominated Universe, we have $|1 - \Omega| \propto t^{2/3}$, and in a nearly flat, radiation-dominated Universe, we have $|1 - \Omega| \propto t$. We know observationally that, at the present, Ω_0 is not hugely different (certainly not more than an order of magnitude) from unity, which implies that at much earlier times it must have been extremely close to 1. To obtain our present Universe, then at nucleosynthesis, for example, when the Universe was around 1 s old, we require that

$$|\Omega(t_{\text{nuc}}) - 1| \lesssim 10^{-16}.$$
(3.2)

At early times, Ω must be yet closer to 1.

---

[1] The combination aH also decreases in more general circumstances, such as during a thermal phase transition.

The flatness problem states that such finely tuned initial conditions seem extremely unlikely. Almost all initial conditions lead either to a closed Universe that recollapses almost immediately, or to an open Universe that very quickly enters the curvature-dominated regime and cools to below 3 K within the first second of its existence. For this reason the flatness problem also is phrased sometimes as an age problem – how did our Universe get to be so old?

3.1.2 Horizon problem

In Section 2.4, we saw that, in the Hot Big Bang model, the comoving distance over which causal interactions can occur before the microwave background is released ($180\Omega_0^{-1/2}h^{-1}$ Mpc) is considerably less than the comoving distance that the radiation travels after decoupling [$5820h^{-1}$ Mpc in a flat Universe, and somewhat more in a low-density Universe in accord with Eq. (2.58) or Eq. (2.60)].

This means that microwaves coming from regions separated by more than the horizon scale at last scattering, which typically subtends about a degree, cannot have interacted before decoupling. The Hot Big Bang model therefore offers no prospect of explaining why the temperature seen in different regions of the sky is so accurately the same; the homogeneity must form part of the initial conditions.

A similar situation exists at nucleosynthesis: To preserve the success of the standard theory, the Universe must be homogeneous on scales much larger than the horizon size at that time; if there were fluctuations in the density from point to point then, because of the nonlinearity of the nucleosynthesis process, the presently observed values would not be reproduced when all these separate regions later were added together.

3.1.3 Unwanted relics

If the Hot Big Bang begins at a very high temperature, relics that are forbidden by observation may survive to the present.

Perhaps the most problematic relic, from the modern viewpoint, is the gravitino. This particle occurs in supergravity as the spin-$\frac{3}{2}$ partner of the graviton, and has only gravitational-strength interactions. In most versions of supergravity, the mass of the gravitino is of order 100 GeV, in which case nucleosynthesis is upset if the Hot Big Bang begins before $T \gtrsim 10^9$ GeV (Ellis et al. 1986).

Another very troublesome class of relics comprises the moduli occurring in superstring theory (de Carlos et al. 1993; Banks et al. 1994). These are spin-0 particles, corresponding to the fields that parameterize the vacuum in the absence of supersymmetry breaking. Their masses and lifetimes are typically of the same order as those of the gravitino but, being associated with scalar fields, they are even more likely to be overproduced in the early Universe.

Depending on the theory, there also may be unwanted topological defects; for reviews, see Vilenkin and Shellard (1994) and Hindmarsh and Kibble (1995). If the symmetry of a Grand

Unified Theory is restored in the early Universe, magnetic monopoles are produced when it is broken spontaneously. Their abundance typically is higher than observation allows, unless they are connected by strings. Historically, getting rid of unwanted monopoles was one of the main motivations for inflation. Cosmic strings, textures, and domain walls also may be problematic, but that is more model dependent. Textures present no problem if their energy scale is well below 10^{16} GeV, and neither do cosmic strings if the same is true of their energy per unit length. Global strings need to be a couple of orders of magnitude lighter to avoid problems. Stable domain walls are fatal unless their symmetry-breaking scale is less than a million electron volts or so (Zel'dovich et al. 1975; Vilenkin and Shellard 1994), but they are destabilized easily by giving one of the two vacua a slightly bigger energy density than the other.

3.1.4   Homogeneity and isotropy

The horizon problem tells us that the large-scale homogeneity and isotropy of the Universe must be part of the initial conditions. However, in practice, we know that the Universe is not perfectly homogeneous, though it comes very close to it on large scales. A vital question then is whether we can develop a theory of the origin of the inhomogeneity, or whether it too must be consigned to the realm of initial conditions.

Within the Hot Big Bang model, the situation is not yet completely clear-cut. The simplest interpretation of the anisotropies in the microwave background seen by the Cosmic Background Explorer (COBE) satellite is that they correspond to irregularities at the surface of last scattering. Then their corresponding scale is much larger than the horizon size at that time. With the Hot Big Bang model, such perturbations could not be generated causally, and again would have to have been part of the initial conditions (Hu et al. 1994b; Liddle 1995). However, it remains possible that these large-angle anisotropies are generated by gravitational effects much closer to us than last scattering; this is what happens, for example, in topological defect theories of structure formation.

So, the Hot Big Bang theory is unable to explain the large-scale homogeneity of the Universe. It yet may be able to explain the generation of inhomogeneities, but as we see, the most attractive way to generate the inhomogeneities is to go beyond the standard Hot Big Bang model.

3.2   Inflation in the abstract

The inflationary cosmology (Guth 1981; Albrecht and Steinhardt 1982; Linde 1982, 1983) is not a replacement for the Hot Big Bang model, but rather an add-on that occurs at very early times without disturbing any of its successes.

The precise definition of inflation is simply any epoch during which the scale factor of the Universe is accelerating:

$$\boxed{\text{INFLATION} \quad \Longleftrightarrow \quad \ddot{a} > 0.}$$

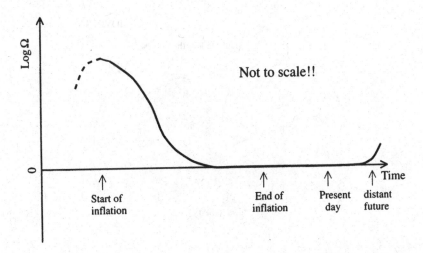

Fig. 3.1. Schematic of the inflationary solution to the flatness problem. Whether there is a definite start time to inflation or not is irrelevant. By definition, during inflation Ω is driven toward 1, and with sufficient inflation it will finish so close by the time inflation ends that in all the subsequent evolution up to the present it remains indistinguishably close. Only in the distant future will it move away again.

Inflation sometimes is described just as a rapid expansion, though it is not very clear with respect to what the expansion is supposed to be rapid.

There is an equivalent alternative expression of the condition for inflation that gives it a more physical interpretation:

$$\text{INFLATION} \quad \Longleftrightarrow \quad \frac{d}{dt}\frac{H^{-1}}{a} < 0.$$

Because H^{-1}/a is the comoving Hubble length, the condition for inflation is that the comoving Hubble length, which is the most important characteristic scale of the expanding Universe, is decreasing with time. Viewed in coordinates fixed with the expansion, the observable Universe actually becomes *smaller* during inflation because the characteristic scale occupies a smaller and smaller coordinate size as inflation proceeds.

If inflation occurs, then it is possible for all the aforementioned problems of the Big Bang model to be solved. The easiest one to see is the flatness problem; the condition for inflation is precisely the condition that Ω is driven toward 1 rather than away from it [Eq. (3.1)]. Figure 3.1 shows a schematic of the desired behaviour. Relic abundances can be reduced to a satisfactory level by the expansion during inflation, provided that they are produced before the inflationary epoch. The horizon problem can be solved because of the dramatic reduction in the comoving Hubble length during inflation, which allows our present observable Universe to originate from a tiny region that was well inside the Hubble radius early on during inflation, as shown in Figure 3.2. A much more detailed account of these resolutions can be found in Kolb and Turner (1990) and Linde (1990a).

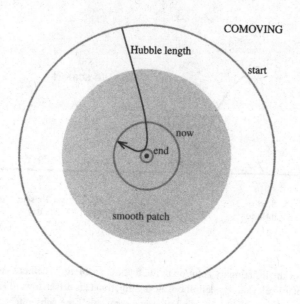

Fig. 3.2. Schematic of the inflationary solution to the horizon problem. The plot is in comoving units. During inflation, the comoving Hubble radius decreases dramatically, allowing our entire observable region (the ring marked "now" around the central dot indicating our position) to lie within a region (the shaded smooth patch) that was well inside the Hubble radius at the start of inflation. Any initial inhomogeneities (lying outside the smooth patch) finish up on scales vastly larger than our observable Universe.

Assuming that we work within general relativity, the condition for inflation can be rewritten as a requirement on the material driving the expansion. Directly from the acceleration equation (2.7), with $\Lambda = 0$ or absorbed into ρ and P, we find

$$\boxed{\text{INFLATION} \quad \Longleftrightarrow \quad \rho + 3P < 0.}$$

Because we always assume ρ to be positive, it is necessary for P to be negative to satisfy this condition, which is independent of the curvature of the Universe.

3.3 Scalar fields in cosmology

To obtain inflation, we need material with the unusual property of a negative pressure. Such a material is a scalar field, describing scalar (spin-0) particles. Although, as yet, there has been no direct observation of a fundamental scalar particle (such as the Higgs particle), such particles proliferate in modern particle theories. They play a crucial role in bringing about symmetry breaking between the fundamental forces. Scalar fields were introduced by particle physicists long before particle cosmology came into being as a subject, but were pounced upon by the cosmology community because of the range of interesting phenomena in which they may partake. Inflationary cosmology is one arena where they play a vital role; they possess the unusual feature of a potential energy that may redshift extremely slowly as the Universe expands. This corresponds to an effective equation of state with a negative pressure, which we have just seen is exactly what we need for inflation. The scalar field responsible for inflation is often called the **inflaton**.

The standard way to specify a particle theory is via its Lagrangian, from which the equations of motion can be derived. We discuss this in detail for a scalar field in Chapters 7 and 14. For now, as our starting point, we adopt expressions for the energy density and pressure of a homogeneous scalar field $\phi \equiv \phi(t)$, which are

$$\rho_\phi = \frac{1}{2}\dot{\phi}^2 + V(\phi), \tag{3.3}$$

$$P_\phi = \frac{1}{2}\dot{\phi}^2 - V(\phi). \tag{3.4}$$

These eventually are derived as Eqs. (14.87) and (14.88). The term $V(\phi)$ is the **potential** of the scalar field, which we might hope to derive from some particle physics motivation. For now, we treat it more or less as a free function. Different inflationary models, as described in Chapter 8, correspond to different choices for the potential.

Note that, although the scalar field acts as a perfect fluid, it does not possess an equation of state relating P_ϕ and ρ_ϕ because the same energy density can correspond to different values of the pressure if the energy density is distributed differently between the potential and kinetic terms.

The equations of motion now can be derived directly by substituting these relations into the Friedmann and continuity equations (2.12) and (2.8), respectively. Assuming a spatially flat Universe, we obtain

$$H^2 = \frac{1}{3M_{\rm Pl}^2}\left[V(\phi) + \frac{1}{2}\dot{\phi}^2\right] \tag{3.5}$$

and

$$\ddot{\phi} + 3H\dot{\phi} = -\frac{dV}{d\phi}, \tag{3.6}$$

which also is called the scalar wave equation. We remind the reader that throughout this book we are using the reduced Planck mass $M_{\rm Pl}$, defined on page 13.

From the forms of the effective energy density and pressure, the condition for inflation is satisfied, provided that $\dot{\phi}^2 < V(\phi)$. With a suitably flat potential, then even if this condition is not obeyed initially, it very quickly comes to be satisfied, provided that the scalar field is displaced away from the minimum of its potential.

Once inflation gets under way, then by definition the curvature term in the Friedmann equation becomes less and less important. Normally, it is assumed negligible from the start; if it is not, then the beginning stages of inflation will render it so.

3.4   Slow-roll inflation

The standard approximation technique for analyzing inflation is the **slow-roll approximation**. This approximation throws away the last term of Eq. (3.5) and the first term of Eq. (3.6), leaving

$$H^2 \simeq \frac{V(\phi)}{3M_{\mathrm{Pl}}^2}, \tag{3.7}$$

$$3H\dot{\phi} \simeq -V'(\phi), \tag{3.8}$$

where we have introduced the notation that primes are derivatives with respect to the scalar field ϕ, and that $\simeq$ indicates that quantities are equal within the slow-roll approximation.

For this approximation to be valid, it is necessary for two conditions to hold. These are

$$\epsilon(\phi) \ll 1, \qquad |\eta(\phi)| \ll 1, \tag{3.9}$$

where the **slow-roll parameters** ϵ and η are defined by (Liddle and Lyth 1992, 1993a)

$$\epsilon(\phi) = \frac{M_{\mathrm{Pl}}^2}{2} \left(\frac{V'}{V} \right)^2, \tag{3.10}$$

$$\eta(\phi) = M_{\mathrm{Pl}}^2 \frac{V''}{V}. \tag{3.11}$$

Notice that ϵ is positive by definition. The slow-roll parameters prove to be a very useful way of quantifying the predictions of inflation and we use this notation throughout the book.

That these conditions are necessary for the slow-roll approximation to be valid can be found easily by substitution. However, note that they are not *sufficient* conditions because they only restrict the form of the potential. Because the full scalar wave equation is second order, the value of $\dot{\phi}$ can be chosen freely and, in particular, can be chosen so as to violate the slow-roll approximation. It is therefore an additional "assumption" that the solution for a given potential satisfies Eq. (3.8). However, we show in Section 3.7 that this assumption can be proven, by considering the attractor behaviour of solutions to the equations of motion. In fact, the attractor behaviour is vital for the success of the slow-roll approximation; otherwise, the fact that the approximation reduces the order of the system by 1 would make it unable to represent generic solutions.

Practically everything we need follows from Eqs. (3.8) and (3.9), which we refer to simply as the **slow-roll conditions**. In particular, Eq. (3.7) is an immediate consequence of Eq. (3.8) and $\epsilon \ll 1$.

The slow-roll parameters make it is easy to see where inflation might occur on a given potential. For example, for $V(\phi) = m^2\phi^2/2$, they are satisfied provided that $\phi^2 > 2M_{\mathrm{Pl}}^2$. For such a potential, inflation proceeds until the scalar field gets too close to the minimum for the slow-roll conditions to be maintained, and inflation comes to an end.

Only for a very few simple choices of potential can the full equations for the scalar field evolution be solved exactly. On the other hand, from the slow-roll equations, it is easy to find solutions, provided only that the relevant integrations can be performed analytically. However, in most circumstances it is not even necessary to find solutions to the equations of motion, as we see later.

3.4.1 Relation between inflation and slow roll

The slow-roll approximation is a sufficient condition for inflation. To see this, rewrite the condition for inflation as

$$\frac{\ddot{a}}{a} = \dot{H} + H^2 > 0. \tag{3.12}$$

This is obviously satisfied if $\dot{H}$ is positive.[2] Otherwise, we require

$$-\frac{\dot{H}}{H^2} < 1. \tag{3.13}$$

Meanwhile, substitution of the slow-roll equations yields

$$-\frac{\dot{H}}{H^2} \simeq \frac{M_{\text{Pl}}^2}{2} \left(\frac{V'}{V}\right)^2 = \epsilon. \tag{3.14}$$

Consequently, if the slow-roll approximation is valid ($\epsilon \ll 1$), then inflation is guaranteed.

As before, this condition is sufficient but not necessary, because the validity of the slow-roll approximation is required in its derivation. It is therefore possible in principle for inflation to continue even if the slow-roll conditions are violated, though in practice the amount of inflation that occurs under this circumstance is very small.

An inflation model consists of a potential and a way of ending inflation. One way for inflation to end is by violation of the slow-roll conditions as the field approaches a minimum with zero or negligible potential energy. In such cases, it can be assumed that inflation comes to an end when $\epsilon(\phi)$ reaches unity. This directly tells us the value of ϕ where inflation comes to an end and, because ϵ diverges as the field approaches the minimum of the potential, this endpoint will be displaced somewhat from the minimum. In models such as hybrid inflation, where extra physics intervenes to end inflation, inflation can end while the slow-roll conditions are still well respected.

3.4.2 Amount of inflation

The amount of inflation that occurs normally is quantified by the ratio of the scale factor at the final time to its value at some initial time. Because this typically is a vast quantity, the logarithm is taken to give the number of e-foldings N:

$$N(t) \equiv \ln \frac{a(t_{\text{end}})}{a(t)}, \tag{3.15}$$

---

[2] This would require $P < -\rho$ in a general relativity theory and hence cannot be caused by a scalar field, though it can occur in extended theories of gravity, which we explore in Chapter 8.

where t_{end} is the time at the end of inflation. This measures the amount of inflation that still has to occur after time t, with N decreasing to 0 at the end of inflation. To solve the horizon and flatness problems, around seventy e-foldings of inflation are required.

As we saw earlier, the best characterization of inflation is that the comoving Hubble length $1/aH$ is decreasing. Therefore, there is a case to be made (Liddle et al. 1994) for quantifying the amount of inflation by a slightly different quantity, that being the ratio of the initial comoving Hubble length to the final one

$$\tilde{N}(t) = \ln \frac{a(t_{end})H(t_{end})}{a(t)H(t)}. \tag{3.16}$$

Although technically more accurate, during inflation $a(t)$ typically varies much faster than $H(t)$, and so, the difference is not very important. We stick to the usual convention and use $N(t)$ to quantify inflation.

For most purposes, the only knowledge we need is how much more inflation will occur from a given scalar field value ϕ, rather than from a given time. This can be calculated immediately via the slow-roll approximation, without any need to solve the equations of motion for the expansion:

$$N \equiv \ln \frac{a(t_{end})}{a(t)} = \int_t^{t_{end}} H dt \simeq \frac{1}{M_{Pl}^2} \int_{\phi_{end}}^{\phi} \frac{V}{V'} d\phi, \tag{3.17}$$

where ϕ_{end} is defined by $\epsilon(\phi_{end}) = 1$ if inflation ends through violation of the slow-roll conditions.

3.4.3   Evolution of scales

When we later discuss the production and evolution of density perturbations, we will be interested in the history of each comoving wavenumber. An important question concerning a given scale is whether it is larger or smaller than the horizon; strictly speaking, we always should refer to the Hubble length, but we follow common practice in using "horizon" and "Hubble length" interchangeably, with the understanding that the latter is intended. Density perturbations normally are identified by their comoving wavenumber k, arising from a Fourier decomposition of the density perturbation. We define a scale to be equal to the horizon when $k = aH$.

Recall that, by definition, during inflation the comoving Hubble length is decreasing whereas at all other times it is increasing. A fixed comoving length scale k^{-1} therefore may begin its evolution considerably smaller than H^{-1}/a, and by the end of inflation be considerably larger. For any scale that does cross the horizon during inflationary evolution, an important epoch to quantify is the time that it equals the Hubble radius, $k = aH$. With some simple assumptions, this can be related to the number of e-foldings of inflation that occur after that time. The evolution of a scale k, which we might imagine, for example, to be the scale presently equalling the Hubble radius, $k = a_0 H_0$, is shown in Figure 3.3.

To make the identification of scales, we require a model for the complete evolution of the Universe up to the present. For example, in the simplest viable cosmology, we can break the

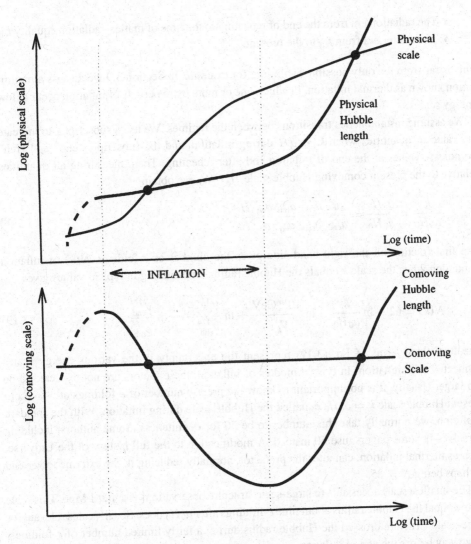

Fig. 3.3. Two views of the behaviour of a comoving scale k relative to the Hubble length (horizon scale). By definition, the comoving Hubble length $1/aH$ is decreasing during inflation and otherwise increasing. The upper panel shows physical coordinates, the lower one comoving. The vertical axis covers many powers of 10 in scale. A scale starts well inside the horizon, then crosses outside some time before the end of inflation, reentering long after inflation is over.

evolution up into chunks as follows:

- From the time the scale k^{-1} equals the Hubble radius to the end of inflation.
- From the end of inflation until the Hot Big Bang is restored. This epoch is considered in Section 3.8. Here, we assume that the Universe behaves as if matter dominated during it.

- The radiation era from the end of reheating to the time of matter–radiation equality t_{eq}.
- The matter era from t_{eq} to the present.

This is far from the only possible evolution; for example, in Section 3.9 we discuss a modification known as thermal inflation, in which one or more extra periods of inflation occur at low energy scales.

We assume instantaneous transitions between the regimes. We use a subscript k to indicate the value of quantities when $k = aH$ during inflation, and the subscripts "end" and "reh" to indicate values at the end of inflation and after reheating. Then, measuring all quantities relative to the present comoving Hubble scale H_0^{-1}/a_0, we obtain

$$\frac{k}{a_0 H_0} = \frac{a_k H_k}{a_0 H_0} = \frac{a_k}{a_{end}} \frac{a_{end}}{a_{reh}} \frac{a_{reh}}{a_{eq}} \frac{a_{eq}}{a_0} \frac{H_k}{H_0}. \tag{3.18}$$

The first fraction on the right-hand side gives the number of e-foldings $N(k)$ of inflation remaining when the scale k equals the Hubble radius. Substituting in typical values gives

$$N(k) = 62 - \ln \frac{k}{a_0 H_0} - \ln \frac{10^{16}\,\text{GeV}}{V_k^{1/4}} + \ln \frac{V_k^{1/4}}{V_{end}^{1/4}} - \frac{1}{3} \ln \frac{V_{end}^{1/4}}{\rho_{reh}^{1/4}}. \tag{3.19}$$

The final three terms of Eq. (3.19) represent the uncertainty in the various energy scales connected with inflation. In typical models of inflation, these factors are not expected to be too large. Usually, it is not important to know the precise number of e-foldings at which our present Hubble scale $k = a_0 H_0$ equalled the Hubble scale during inflation. With the standard evolution, we normally take this number to be 50 for definiteness. Some authors, including ourselves in some papers, use 60 instead. A modification to the full history of the Universe, such as thermal inflation, can alter the prefactor, normally reducing it, the extreme lower end perhaps being $N = 25$.

The smallest scale accessible to large-scale structure observations is about 1 Mpc. This scale would equal the Hubble radius about nine e-foldings after $a_0 H_0$ does. So, all scales relevant for large-scale structure crossed the Hubble radius during a fairly limited number of e-foldings, some way before the end of inflation.

3.4.4 Initial conditions for inflation

Though not compulsory, it normally is imagined that an era of inflation begins at the Planck scale, corresponding to $V^{1/4} \sim M_{Pl}$. This is indeed desirable for two reasons. One is to prevent the Universe from collapsing within a Planck time or so, if Ω is initially bigger than 1 (without being fine-tuned to a value extremely close to 1). Note though that this need not be a problem in a very chaotic situation where parts of the Universe are open and parts are effectively closed; the latter may form black holes but the former remain viable initial conditions.

The other, which applies also to the case $\Omega < 1$, is that inflation protects an initially homogeneous region from invasion by its inhomogeneous surroundings. This invasion will propagate with speed of order $c = 1$, and a patch that is homogeneous at time t will survive until time t_2 only if its initial size is bigger than

$$r(t) = a(t) \int_t^{t_2} \frac{dt}{a} = a(t) \int_a^{a_2} \frac{da}{a^2 H}. \tag{3.20}$$

If inflation begins promptly, and we take t_2 as the end of inflation, the integral is dominated by the lower limit giving $r(t) \sim H^{-1}(t)$. In words, the invasion travels about a Hubble distance in the first Hubble time, but then stops so that the initially homogeneous patch need not be much bigger than the Hubble distance. In contrast, if inflation is delayed, the integral is dominated by the upper limit (whereas t_2 is before the beginning of inflation), leading to

$$r(t) \sim \frac{a(t)H(t)}{a(t_2)H(t_2)} H^{-1}(t) \gg H^{-1}(t). \tag{3.21}$$

In words, the invasion continues indefinitely, and the initial patch needs to be huge to survive until inflation eventually protects it.[3]

Slow-roll inflation is very effective at erasing memory of what went before it and, as a result, our Universe retains no memory of the era before it left the horizon. However, this does not mean that we should ignore this era completely. On the contrary, a complete model of inflation should specify not only a potential, but a reasonable way in which the inflaton field can find itself slow-rolling down the appropriate part of this potential, when our Universe leaves the horizon.

Let us start by asking what happens at the Planck scale. The almost universally accepted proposal, due to Linde (1983, 1990a), is that conditions at this era are what he termed **chaotic**. This does not refer directly to chaos theory; rather, it is intended to indicate that, initially, the field takes on a wide range of different values in different regions of the Universe. One imagines the Universe emerging from the Planck era with the scalar field well displaced from any minimum, such that the typical energy is of order of the Planck energy. In regions where the field has suitable values, inflation is able to begin. Some numerical investigation of the details has been carried out, for example, testing to what extent spatial gradients in the scalar field might inhibit the onset of inflation, with the conclusion that they cannot prevent inflation from happening in at least some regions (Albrecht et al. 1985; Kung and Brandenberger 1990; Yi and Vishniac 1993).

Because this early era of inflation does not lead to any directly observable consequences, there is no necessity for the perturbations to be small, and indeed, it is perfectly possible for the predicted density perturbations to be of order of unity even if the energy density of the field is well below the Planck scale. Such a situation leads to a phenomenon known as **eternal inflation** (Linde 1986; Linde et al. 1994), in which the quantum fluctuations in the scalar field can dominate over the classical behaviour, allowing the field to diffuse up the potential as well as to roll and diffuse down. Because the Universe expands more rapidly if the energy density is

---

[3] In the mathematical limit $t_2 \to \infty$, $r(t)$ is called the **event horizon**, as opposed to the particle horizon introduced earlier. During inflation, it makes sense to pretend that the event horizon exists (taking t_2 to be any late time before the end of inflation, regarded as infinity). After inflation, it makes sense to pretend that the particle horizon exists (taking the initial time to be any early time after the end of inflation, regarded as 0).

higher, the physical volume may be dominated by regions moving up the potential. In this situation, parts of the Universe continue to inflate forever, emitting a constant stream of regions where the field moves away from the eternal regime and enters a conventional inflationary period. Any one of these regions could house our own Universe. Eternal inflation can occur either at large field values or near a maximum of the potential, and we discuss it further (page 185) after we have derived formulae for the density perturbations.

Inflation need not be a common occurrence near the Planck scale. All we need is for our Universe to be an inflating region and, particularly with eternal inflation, the huge expansion of the inflating region makes that in some sense likely, though admittedly this idea is hard to quantify.

Various things might happen between the Planck scale and the epoch when our Universe leaves the horizon. The simplest possibility, and the one usually envisaged nowadays, is that scalar fields continue to dominate the energy density of the Universe throughout this era, with some form of inflation occurring most of the time. (It might perhaps be interrupted by one or more periods when some field is oscillating around a minimum of the potential.) An alternative, which used to be very popular, is that the Universe became dominated by radiation in thermal equilibrium; in other words, a hot big bang begins, though not *the* Hot Big Bang as we have defined it (page 12).

Whatever happens, our observable Universe is assumed to eventually undergo an era of inflation, which starts while it is within the horizon and ends some tens of *e*-folds after it leaves the horizon. This inflation must be of the slow-roll variety, if it is to explain the observed large-scale structure. The field either will be rolling away from a maximum of the potential or coming in from large field values; either way, it is quite likely that inflation is initially of the eternal variety. If the field is rolling away from the maximum, an attractive possibility is that the maximum corresponds to a fixed point under the symmetries of the model. This is the usual case, normally signalled by the convention that the maximum is at $\phi = 0$. In that case, the field can be driven to the maximum by its interaction with other scalar fields, or by thermal effects if there is thermal equilibrium. Alternatively, one might imagine that the initial field values are chaotic, just as was postulated near the Planck scale, with our Universe happening to be near the maximum.

We provide examples of these different mechanisms for starting inflation when we review the range of inflation models in Chapter 8.

3.5   Exact solutions

For almost all known inflationary models, the slow-roll approximation works so well that nothing more is needed. Even though it must by definition fail toward the end of inflation, this typically only gives a small misestimation of the number of *e*-foldings that occur, and this uncertainty is subdominant to the uncertainties in this quantity from the various energy scales in Eq. (3.19).[4] Nevertheless, it is useful to have some exact solutions to the full equations of motion in order to study their properties, and several are known.

---

[4] In a specific inflationary model, V_k and V_{end} can be computed fairly accurately, but the physics of reheating is less well understood at present.

3.5.1 Power-law inflation

The most prominent exact solution is **power-law inflation** (Lucchin and Matarrese 1985), which, as we see later, has the extra advantage that the equations for the generation of density perturbations also can be solved exactly.

Power-law inflation arises when the potential is chosen to take the exponential form

$$V(\phi) = V_0 \exp\left(-\sqrt{\frac{2}{p}}\frac{\phi}{M_{\mathrm{Pl}}}\right), \tag{3.22}$$

where V_0 and p are constants. The spatially flat equations of motion then have the particular solution

$$a = a_0 t^p, \tag{3.23}$$

$$\frac{\phi}{M_{\mathrm{Pl}}} = \sqrt{2p} \ln\left(\sqrt{\frac{V_0}{p(3p-1)}}\frac{t}{M_{\mathrm{Pl}}}\right). \tag{3.24}$$

The general homogeneous solution, with one extra initial condition, also can be found in a parametric form (Salopek and Bond 1990), but solutions for any initial conditions rapidly approach this particular solution, in accordance with the inflationary attractor behaviour.

Provided that $p > 1$, this solution satisfies the condition for inflation. The slow-roll parameters are simply $\epsilon = \eta/2 = 1/p$, and are independent of ϕ. With this potential, inflation never comes to an end (though we discuss ways of circumventing this problem in Chapter 8).

Provided that the scalar field is the only matter in the Universe, it acts as a perfect fluid with $w = 2/3p$, using the language of Eq. (2.7). This correspondence will break down if another type of matter is also present, because the scale-factor evolution feeds into the effective equation of state.

3.5.2 Other exact solutions

The technology exists to compute a variety of exact solutions, via the basic strategy of beginning with the solution and deriving the potential necessary to support it. In this type of approach, there is no control over the type of potential that emerges from such a procedure, and normally the result does not take on a simple form.

The most investigated such solution is the **intermediate inflation** model (Barrow 1990; Muslimov 1990). This gives rise to an expansion

$$a(t) \propto \exp(At^f), \quad 0 < f < 1, \quad A > 0. \tag{3.25}$$

It arises from the potential

$$V(\phi) \propto \left(\frac{\phi}{M_{\mathrm{Pl}}}\right)^{-\beta}\left(1 - \frac{\beta^2}{6}\frac{M_{\mathrm{Pl}}^2}{\phi^2}\right), \tag{3.26}$$

where $\beta = 4(f^{-1} - 1)$. The expansion is faster than any power law and slower than exponential. As with power-law inflation, there is no natural end to inflation in this model. Note that the asymptotic potential is just the power-law form $V_{\text{asymp}}(\phi) \propto \phi^{-\beta}$.

The literature contains a variety of other exact inflationary solutions. There is only a single case at present (Easther 1995) in which a model has been constructed so that the density perturbation equations also can be solved, as they can for power-law inflation. In all other cases, it is only the classical scalar field dynamics that have been solved exactly.

3.6   Hamilton–Jacobi formulation of inflation

The Hamilton–Jacobi formulation (Salopek and Bond 1990) is a powerful way of rewriting the equations of motion, which allows an easier derivation of many inflation results. The formalism has applications to the general inhomogeneous situation, though we concentrate here on the homogeneous version as applied to spatially flat cosmologies.

The formulation can be derived by considering the scalar field itself to be the time variable. This can be carried out during any slow-rolling epoch in which the scalar field varies monotonically with time. In its simplest form, it will break down during the oscillatory epoch that ends inflation (though it is usually possible to patch together solutions from separate monotonic epochs).

For definiteness, throughout we take $\dot{\phi} > 0$. If this is not satisfied, it can be brought about by redefining $\phi \rightarrow -\phi$.

Differentiating Eq. (3.5) with respect to t and substituting in Eq. (3.6) gives

$$2\dot{H} = -\frac{\dot{\phi}^2}{M_{\text{Pl}}^2}. \tag{3.27}$$

Dividing both sides by $\dot{\phi}$, permitted by the monotonicity assumption, gives

$$\dot{\phi} = -2 M_{\text{Pl}}^2 H'(\phi), \tag{3.28}$$

which gives the relation between ϕ and t. This allows us to write the Friedmann equation in the first-order form

$$[H'(\phi)]^2 - \frac{3}{2M_{\text{Pl}}^2} H^2(\phi) = -\frac{1}{2M_{\text{Pl}}^4} V(\phi). \tag{3.29}$$

Equation (3.29) is the Hamilton–Jacobi equation. It allows us to consider $H(\phi)$, rather than $V(\phi)$, as the fundamental quantity to be specified. Because H, unlike V, is a geometric quantity, inflation is described more naturally in that language (Muslimov 1990; Salopek and Bond 1990; Lidsey 1991). Once $H(\phi)$ has been specified, we immediately can obtain the corresponding potential from Eq. (3.29). Also, Eq. (3.28) gives the relation between ϕ and t, which enables us to obtain $H(t)$, which, if it is desired, can be integrated to $a(t)$. Therefore, this is the most direct route to obtaining a large set of exact inflationary solutions; for example,

$H(\phi) \propto \phi^{-\beta/2}$ gives intermediate inflation, and

$$H(\phi) \propto \exp\left(-\sqrt{\frac{1}{2p}}\frac{\phi}{M_{\text{Pl}}}\right) \tag{3.30}$$

gives power-law inflation.

We can use the Hamilton–Jacobi formalism to write down a slightly different version of the slow-roll approximation than we did earlier, defining slow-roll parameters ϵ_{H} and η_{H} as

$$\epsilon_{\text{H}} = 2M_{\text{Pl}}^2 \left(\frac{H'(\phi)}{H(\phi)}\right)^2, \tag{3.31}$$

$$\eta_{\text{H}} = 2M_{\text{Pl}}^2 \frac{H''(\phi)}{H(\phi)}. \tag{3.32}$$

In the slow-roll limit, $\epsilon_{\text{H}} \to \epsilon$ and $\eta_{\text{H}} \to \eta - \epsilon$.

Some manipulation allows these to be written in various ways, such as

$$\epsilon_{\text{H}} = 3\frac{\dot{\phi}^2/2}{V + \dot{\phi}^2/2} = -\frac{d\ln H}{d\ln a}, \tag{3.33}$$

$$\eta_{\text{H}} = -3\frac{\ddot{\phi}}{3H\dot{\phi}} = -\frac{d\ln\dot{\phi}}{d\ln a} = -\frac{d\ln H'}{d\ln a}. \tag{3.34}$$

Consequently, the smallness of ϵ_{H} and η_{H} is precisely the condition for neglecting the unwanted terms in Eqs. (3.5) and (3.6); however, the derivation of these conditions is exact, whereas that using ϵ and η required the slow-roll approximation to be valid. We also can consider $\epsilon_{\text{H}} \ll 1$ to be the condition to neglect the first term in Eq. (3.29), and $\eta_{\text{H}} \ll 1$ to be the condition to neglect the first term of its ϕ derivative.

With these new parameters, many results that were approximate in terms of $V(\phi)$ become exact. First, the definition of inflation now is given *precisely* by

$$\ddot{a} > 0 \iff \epsilon_{\text{H}} < 1, \tag{3.35}$$

and the equation for the number of *e*-foldings becomes

$$N \equiv \ln\frac{a(t_{\text{end}})}{a(t)} = \int_t^{t_{\text{end}}} H\,dt = -\frac{1}{2M_{\text{Pl}}^2}\int_\phi^{\phi_{\text{end}}} \frac{H}{H'}\,d\phi. \tag{3.36}$$

Notice that, in terms of the new slow-roll parameters, the validity of the slow-roll approximation does not depend on any additional assumptions regarding attractor behaviour. This is because when we discuss $H(\phi)$, we are dealing with the solution directly.

3.7 Inflationary attractor

If inflation is to be truly predictive, the evolution when the scalar field is at some given point on the potential has to be independent of the initial conditions. Otherwise, any result, such as the amplitude of density perturbations, would depend on the unknowable initial conditions. However, the scalar wave equation is a second-order equation, implying that $\dot{\phi}$, in principle, can take on any value anywhere on the potential, and so, there certainly is not a unique solution at each point on the potential. Inflation therefore can be predictive only if the solutions exhibit an attractor behaviour, where the differences between solutions of different initial conditions rapidly vanish. As we now see, the inflationary equations do indeed possess this vital property, though it has not been discussed often in the literature (Salopek and Bond 1990; Liddle et al. 1994).

Related to this, we have noted already that the slow-roll approximation reduces by one the order of the equations describing inflation. Written in the original form, this arises from the dropping of the $\ddot{\phi}$ term in Eq. (3.6), and in the Hamilton–Jacobi form the dropping of the H' term in Eq. (3.29). This means that some initial value of $\dot{\phi}(t)$ [or equivalently, $H(\phi)$], instead of being a free parameter, is determined by the slow-roll equations. The attractor behaviour is necessary if the slow-roll solution is to have any chance of representing the entire one-parameter family of solutions it replaces.

To demonstrate the attractor behaviour, we use the Hamilton–Jacobi formalism, which greatly simplifies the analysis. This was carried out first by Salopek and Bond (1990). We restrict ourselves to linear homogeneous perturbations, which is all that is needed because, classically at least, inflation does indeed generate large smooth patches. For simplicity, we also assume that the perturbations do not reverse the sign of $\dot{\phi}$, though the result holds under more general circumstances.

Because we have chosen ϕ to be increasing with time, our aim is to show that all solutions rapidly approach one another as ϕ increases. Suppose $H_0(\phi)$ is any solution to Eq. (3.29), which can be either inflationary or noninflationary. Add to this a linear homogeneous perturbation $\delta H(\phi)$; the attractor condition will be satisfied if it becomes small as ϕ increases. Substituting $H(\phi) = H_0(\phi) + \delta H(\phi)$ into Eq. (3.29) and linearizing, we find that the perturbation obeys

$$H_0' \, \delta H' \simeq \frac{3}{2M_{\mathrm{Pl}}^2} H_0 \, \delta H, \qquad (3.37)$$

which has the general solution

$$\delta H(\phi) = \delta H(\phi_i) \exp\left(\frac{3}{2M_{\mathrm{Pl}}^2} \int_{\phi_i}^{\phi} \frac{H_0(\phi)}{H_0'(\phi)} \, d\phi \right), \qquad (3.38)$$

where $\delta H(\phi_i)$ is the value at some initial point ϕ_i. Because H_0' and $d\phi$ have opposing signs, the integrand within the exponential term is negative definite, and hence all linear perturbations do indeed die away.

When $H_0(\phi)$ is an inflationary solution, the behaviour is particularly dramatic because the condition for inflation, $\epsilon_{\mathrm{H}} < 1$, bounds the integrand away from 0. We obtain

$$\delta H(\phi) < \delta H(\phi_i) \exp\left(-\frac{3}{\sqrt{2}} \frac{\phi - \phi_i}{M_{\mathrm{Pl}}} \right). \qquad (3.39)$$

That is, if there is an inflationary solution, all linear perturbations approach it *at least* exponentially fast as the scalar field rolls (Liddle et al. 1994).

Whether inflationary or not, the solution can be expressed in terms of the amount of expansion because the term inside the integral is related to the number of e-foldings of expansion as given by Eq. (3.36). This gives the following precise result (Salopek and Bond 1990):

$$\delta H(\phi) = \delta H(\phi_0) \exp[-3(N_i - N)], \qquad (3.40)$$

where the e-foldings are evaluated for the background solution $H_0(\phi)$.

Neither the assumption of linearity nor the assumption that $\dot{\phi}$ does not change sign is very restrictive. The latter case can matter only if the perturbation takes the field over the top of a maximum in the potential because otherwise it will simply roll up, reverse its direction, and pass back down through the same point, where it can be regarded as a perturbation on the original solution with the same sign of $\dot{\phi}$. If the perturbation is nonlinear, then the solution is made more complicated but, because the full equation is only first order, it is easy to see that solutions are compelled to approach one another regardless of whether the perturbation is linear or not.

Notice that the slow-roll solution is not precisely the attractor solution that all solutions to the full equations approach. Generically, though, it is a good approximation to it whenever the slow-roll conditions are satisfied.

Returning to the original equations of motion (3.5) and (3.6), the attractor behaviour that we have demonstrated indicates that, regardless of initial conditions, the late-time solutions are the same up to a time shift, which cannot be measured.

3.8 Reheating: Recovering the Hot Big Bang

Reheating is the process whereby the period of inflationary expansion gives way to the standard Hot Big Bang evolution. For the main focus of this book – density perturbations – the epoch of reheating is not of particular importance. Although it does contribute an uncertainty in relating present scales to the inflationary epoch through Eq. (3.19), this typically has little impact on the predictions from the inflationary scenario. On the other hand, an understanding of reheating is crucial to our understanding of various other questions, such as whether topological defects can be produced after inflation, whether gravitinos, for example, might be overproduced, and whether baryogenesis can be brought about successfully.

The topic of reheating has seen some important developments during the 1990s, which have led to a significant change of view since books such as Kolb and Turner's (1990) were written. For full accounts, see Kofman et al. (1994, 1997), Shtanov et al. (1995), and Boyanovsky et al. (1996).

There are typically three parts to the reheating process:

(1) noninflationary scalar field dynamics,
(2) decay of inflaton particles, and
(3) thermalization of the decay products.

It is the theory of the second of these stages that has changed recently.

3.8.1   Scalar field oscillations

Once inflation is over, the scalar field begins to move rapidly on the Hubble timescale, and begins to oscillate about the minimum of the potential. This is a coherent oscillation, the phase being the same at all points in the large homogeneous region created by inflation. If there are no rapid particle decays (a situation that we will see exists if the only decay channels are into fermions), then this oscillating phase can last for some considerable time because the particle decay time still may be much longer than the Hubble time. Such a situation can be described by looking at the time-averaged behaviour of the scalar field. For a potential that can be approximated as ϕ^2 near its minimum, the equation is just that of a harmonic oscillator, and the average energy $\bar{\rho}_\phi = \langle \dot{\phi}^2 \rangle_t$ obeys the equation

$$\dot{\bar{\rho}}_\phi + 3H\bar{\rho}_\phi = 0. \tag{3.41}$$

This is exactly the equation for the density of nonrelativistic matter, and so, during the coherent oscillation phase, the energy density falls as $1/a^3$, represented by a decay of the amplitude of the oscillations. This was the result used to obtain the e-folding relation (3.19).

3.8.2   Coherent inflaton decays

The next step is to include the decay of inflaton particles, which will happen once the Hubble time (i.e., the age of the Universe) reaches the decay time. One way of treating this is to insert a phenomenological decay term $\Gamma_\phi \dot{\phi}$ directly into the left-hand side of Eq. (3.6). However, this turns out not to be a valid way of introducing particle decays, even in the slow-decay case in which only fermionic decays are available. First, as noted in a cryptic footnote by Kolb and Turner (1990), it cannot be applied away from the oscillating phase. Second, it is not correct to insert such a term directly into the equation for ϕ anyway, as noted by Kofman et al. (1994). However, in the slow-decay case, such an equation is correct, provided that it refers only to the time-averaged scalar field

$$\dot{\bar{\rho}}_\phi + (3H + \Gamma_\phi)\bar{\rho}_\phi = 0. \tag{3.42}$$

So, such an equation can be used to describe the "envelope" of the oscillations, if only fermionic decay routes are available.

Much more interesting is the situation in which the inflaton may decay into bosonic particles. Such a situation allows a decay by parametric resonance (Traschen and Brandenberger 1990), which in many models can be broad (Kofman et al. 1994, 1997). This permits an extremely rapid decay of the inflaton particles, conceivably so rapid that the oscillating phase ends nearly as soon as it has begun. This dramatically rapid decay has been termed **preheating** to distinguish it from the later stage of particle decay and thermalization. The decays can be into a second bosonic field, or into quanta of the inflaton field itself.

The occupation numbers generated by the parametric resonance typically are huge, so that the bosons created are far from thermal equilibrium. The large occupation number explains why preheating does not occur if the only decay routes are to fermions; the Pauli exclusion principle will prevent further decays once the energy states are filled.

3.8.3 Decay and thermalization

Once parametric resonance has created the high occupation number states, or if parametric resonance is ineffective, the remainder of reheating can proceed according to the standard slow-decay picture (Abbott et al. 1982; Dolgov and Linde 1982). The bosonic particles should decay, interact, and finally reach thermal equilibrium. The details will be strongly dependent on the field theory adopted, which ultimately will determine the temperature at which the Universe can be said to have reached thermal equilibrium, reentering the standard Hot Big Bang behaviour. In this final regard, the new theory of reheating does not seem to give much change to the final answer because the decay products of the resonance quickly become subdominant to the energy density remaining in the oscillations after the resonance turns off (Kofman et al. 1997).

3.9 **Thermal inflation**

The idea of **thermal inflation** (Lyth and Stewart 1995, 1996a) is somewhat tangential to the rest of the discussion in this book, but has one important consequence concerning the e-folding relation (3.19), and so, we mention it here.

Thermal inflation is a short period of inflation that may occur *in addition to* the period of inflation that we have been discussing. It takes place while a light scalar field ϕ (with, for example, $m \sim 100 \, \text{GeV}$), with nonzero vacuum expectation value, is trapped by thermal effects in the false vacuum at $\phi = 0$. The requirement for inflation is $T^4 \lesssim V_0$ and the requirement for trapping is $T \gtrsim m$, and so, thermal inflation takes place if $V_0 \gg m^4$ and it occurs in the regime $m \lesssim T \lesssim V_0^{1/4}$. Because $a \propto 1/T$ during inflation, there are $\ln(V_0^{1/4}/m)$ e-folds of thermal inflation, of order 10 for a typical value $V_0^{1/4} \sim 10^6 \, \text{GeV}$.

Thermal inflation is desirable because it may solve relic abundance problems not solved by the original inflationary epoch. In particular, it can solve the moduli problem of Section 3.1.3; normal inflation is unable to do this because its energy scale is required to be too high to generate the right density perturbations. A fairly short period of thermal inflation can produce an adequate dilution of these unwanted relics.

Because the effective mass during thermal inflation is $T \gg H$, there is no significant vacuum fluctuation. Thermal inflation does have a modest impact on large-scale structure though, because it affects the correspondence between a comoving scale k and the number of e-foldings before the end of inflation at which it crossed outside the horizon, given by Eq. (3.19). The derivation of that relation depends on a model of the entire evolution of the Universe from the end of inflation to the present, and thermal inflation is a serious revision of that evolution. It stretches scales outside the horizon by a factor $\exp(N_{\text{thermal}})$, while keeping the energy density more or less fixed. The net effect is to reduce the prefactor of Eq. (3.19) by N_{thermal}, which is expected to be about 10. So, with thermal inflation, the scales we observe correspond to a later stage of inflation than they would have had thermal inflation not occurred.

In general, there may be even more than one period of thermal inflation, driven by a different scalar field. If there are several periods, then the perturbations that we see could correspond to quite close to the end of inflation. (Indeed, the fact that we observe perturbations limits the amount of thermal inflation that could have occurred.) Except in models of the hybrid inflationary type discussed extensively in Chapter 8, the closer to the end of inflation that we are, the greater should be the deviations from a scale-invariant density perturbation spectrum because the slow-roll parameters will become large.

Examples

3.1 Using the critical density, we can write the Friedmann equation as

$$H^2 = H^2 \Omega - \frac{k}{a^2}.$$

Assume that the Universe contains only relativistic matter, and that we are allowed to freely choose an initial density parameter at a very early epoch $t = 10^{-40}$ s.

Choose Ω at that time to be 0.99. Assuming the Universe expands as radiation dominated whenever the first term in the Friedmann equation dominates, and as curvature dominated whenever the second term dominates, estimate the age of the Universe when Ω becomes smaller than 0.01.

3.2 Immediately after the big bang, a ray of light is emitted from point **A**. It is received at point **B** at the time of decoupling $[(180\,000\Omega_0^{-1/2}h^{-1}$ years, from Eq. (2.64)]. Assuming a critical-density Universe that remained radiation dominated all the way from the big bang to decoupling, calculate the physical separation of **A** and **B** at the time of decoupling. If decoupling occurs at a redshift of 1100, what is the physical separation of **A** and **B** today?

Imagine **A** and **B** to be at two locations on the microwave background, located about $6000h^{-1}$ Mpc away from us. What would be their apparent angular separation?

The area around point **A** of this angular separation is the largest region that can have influenced **A** before the microwave background radiation from **A** was emitted. How many separate such regions are there on the microwave sky?

3.3 A Universe that possesses a cosmological constant Λ but no other matter evolves according to the Friedmann equation

$$H^2 + \frac{k}{a^2} - \frac{\Lambda}{3} = 0.$$

Demonstrate that this gives the same solutions for the scale factor as a $p = -\rho$ perfect fluid, and relate ρ to Λ.

As the Universe expands, the curvature term rapidly becomes unimportant compared to the density term. Indicate the evolution of the Hubble length in comoving coordinates. Compare this with the evolution of the Hubble length in a matter-dominated Universe in comoving coordinates. Name the principal qualitative difference.

3.4 A massless free scalar field is one for which the potential is identically zero. Find the general solutions for homogeneous, spatially flat cosmologies containing such a scalar field (and no other matter).

Does such a Universe become curvature dominated more or less easily than a matter-dominated Universe?

3.5 Suppose monopoles form at a temperature of 10^{28} K (equivalently, $3 \times 10^{-4} M_{Pl}$) with a mass of $10^{-3} M_{Pl}$, and that the Universe behaves as in the standard Hot Big Bang. Assume that annihilations between monopoles and antimonopoles can be neglected, and take the present-day limit on monopoles to be $\Omega_{mon} < 10^{-6}$ (this is known as the Parker bound; see, e.g., Kolb and Turner 1990), where Ω_{mon} indicates the fraction of the total energy density residing in monopoles. Calculate an upper bound on the number of monopoles per horizon volume at formation, assuming $g_* \sim 100$ at that time and that monopole annihilation is negligible.

Suppose that monopoles form with a density of order 1 per horizon volume. If exponential inflation occurs after the monopoles have formed, how many e-foldings of inflation are required to satisfy the Parker bound as stated. Compare this with the number of e-foldings required to solve the horizon problem and briefly explain the origin of the difference.

3.6 Demonstrate that the two conditions in Eq. (3.9) are necessary conditions for the slow-roll approximation to be valid.

3.7 Consider $V = \lambda \phi^4$, where λ is the self-coupling. Assume that the field rolls toward $\phi = 0$ from the positive side. Calculate the value of ϕ where each of the slow-roll conditions in Eq. (3.9) first break down. Do they break down at the same place?

Assuming that inflation ends when $\epsilon = 1$, calculate the number of e-foldings of inflation that occur for an initial value ϕ_i, using Eq. (3.17).

Demonstrate that the slow-roll solutions with $\phi = \phi_i$ and $a = a_i$ at $t = t_i$ are

$$\phi = \phi_i \exp\left[-\sqrt{\frac{32\lambda M_{Pl}^2}{6}}\, (t - t_i) \right],$$

$$a = a_i \exp\left(\frac{\phi_i^2}{8 M_{Pl}^2} \left\{ 1 - \exp\left[-\sqrt{\frac{64\lambda M_{Pl}^2}{3}}\, (t - t_i) \right] \right\} \right).$$

Use the solution for ϕ to calculate the time that inflation ends. Demonstrate that the number of e-foldings calculated using the solution for a is the same as that which you calculated above.

Expand the solution for a at small $t - t_i$ to demonstrate that the inflation is approximately exponential at the initial stage. Calculate the time constant κ [from $a \sim \exp(\kappa t)$] and demonstrate that it equals the (slow-roll) Hubble parameter during inflation.

3.8 For the potential $V(\phi) = m^2 \phi^2 / 2$, calculate the ratio of the effective pressure P_ϕ to the energy density ρ_ϕ sixty e-foldings before the end of inflation.

4 Simplest model for the origin of structure I

4.1 Introduction

Thus far, we have seen how inflation can generate a flat and homogeneous Universe, from a wide range of initial conditions, through the classical evolution of a Universe dominated by the inflaton field. The true merit of inflation, however, is that it provides a theory of *inhomogeneities* in the Universe, which may explain the observed structures. These inhomogeneities arise from the quantum fluctuations in the inflaton field about its vacuum state, in other words, by the vacuum fluctuation.

The vacuum fluctuation generates a primeval density perturbation, of the type that cosmologists call *Gaussian* and *adiabatic*, and whose spectral index is close to 1. Such a primeval perturbation was regarded, even before the advent of inflation, as a viable candidate for the origin of large-scale structure and the then-unobserved cosmic microwave background (cmb) anisotropy (Peebles 1980). To understand the evolution of the primeval perturbation to the present, we need to know the nature and amount of the nonbaryonic dark matter, as well as the value of the cosmological constant. The simplest possibility is to have zero cosmological constant, and cold nonbaryonic dark matter giving critical density. The result fairly can be said to be the simplest plausible model for the origin of large-scale structure and the cmb anisotropy. It is called the **cold dark matter (CDM) model** (Peebles 1982; Blumenthal et al. 1984; Davis et al. 1985, 1992a).

In this chapter and the next, we study the simplest model, before describing some possible extensions of it in Chapter 6. At this stage, we pay little attention to the inflationary origin of the primeval density perturbation, largely reserving that topic for Chapters 7 and 8. Instead, we explain in some detail what is meant by a Gaussian, adiabatic primeval density perturbation with a spectral index close to 1. Then, we go on to give a compact account of the theory of the subsequent evolution of the perturbations and, toward the end of Chapter 5, a very brief comparison with observation.

Regarding the inflationary origin, we content ourselves at this stage with listing the features of inflation that give rise to the simplest model:

(1) *While cosmologically interesting scales are leaving the horizon, there is slow-roll inflation. The inflaton field on those scales starts out in the vacuum state (no inflaton particles with the corresponding momenta), and its vacuum fluctuation has negligible interaction with itself and other fields.*

This leads to a Gaussian adiabatic density perturbation, with a spectral index that is close to 1.

(2) *The inflaton field has only one component.*

This ensures that the perturbation to the spatial curvature is constant while outside the horizon, leading to standard predictions for the spectrum and the spectral index.

(3) *The vacuum fluctuation of fields other than the inflaton has no significant effect after inflation.*

This implies that the adiabatic density perturbation is not accompanied by an isocurvature density perturbation.

(4) *The gravitational waves generated as a vacuum fluctuation have a negligible effect on the cmb anisotropy.*

(5) *The nonbaryonic dark matter is cold, the Universe has critical density, and there is no cosmological constant.*

This restates our assumptions about the Universe after inflation, which fix the evolution of the perturbations once H_0 and Ω_b are specified.

Nature may or may not have chosen the simplest model, and indeed, as we see later in this book, there is quite a lot of evidence from observation that at least the last assumption cannot be quite right. It appears that either the dark matter is not purely cold or the Universe has a subcritical matter density (most likely with, but possibly without, a cosmological constant). These possibilities are considered in detail in Chapter 6, as is the possible relaxing of the other assumptions.

4.2 Sequence of events

Because some of the calculations are quite long, we begin with an overview of the sequence of events. At this stage, the main results are stated, without any justification except in the simplest cases. The next four chapters put the flesh on this skeletal description.

4.2.1 Vacuum fluctuation

During inflation, classical physics predicts that the inflaton field ϕ becomes homogeneous and isotropic on scales well inside the horizon. However, we live in a quantum Universe, and at the quantum level, there remains the vacuum fluctuation $\delta\phi$. It is useful to make a Fourier expansion in a comoving box with sides of comoving length L (physical length aL):

$$\delta\phi(\mathbf{x}, t) = \sum_{\mathbf{k}} \delta\phi_{\mathbf{k}}(t) e^{i\mathbf{k}\cdot\mathbf{x}}. \tag{4.1}$$

As usual, $\mathbf{x}$ is related to the physical position $\mathbf{r}$ by $\mathbf{r} = a(t)\mathbf{x}$, and so, the physical wavenumber is k/a. The possible values of $\mathbf{k}$ form a cubic lattice, with spacing

$$\Delta k = \frac{2\pi}{L}. \tag{4.2}$$

Carrying out the expansion in a box imposes an artificial periodicity, but this will not matter as long as the box is much larger than any scale in which we are interested.

The inverse wavenumber a/k defines a distance scale carried along with the expansion, which is specified conveniently by its present value $1/k$. A scale is said to be inside the horizon if aH/k is less than 1, and outside the horizon if it is bigger. Scales of interest leave the horizon at some epoch during inflation, and reenter it long after inflation ends, as we saw in Figure 3.3 on page 45.

There is a vacuum fluctuation for each Fourier component $\delta\phi_k(t)$, which evolves independently of the others. A few *e*-folds (Hubble times) after horizon exit, it can be regarded as a classical quantity, with an almost constant value, which we denote as $\delta\phi_k(t_*)$. Many *e*-folds after horizon exit, $\delta\phi_k$ may have had time to change significantly, but it will become clear that we are interested only in its value at the epoch t_*, taken to be only a *few e*-folds after horizon exit. The reason is that the curvature perturbation, which is the important thing, will have become frozen-in to a constant value by then.

4.2.2   Linear evolution of cosmological perturbations

The perturbation $\delta\phi(\mathbf{x}, t)$ is not the only departure from homogeneity and isotropy. It leads to a perturbation in the energy density $\delta\rho(\mathbf{x}, t)$ and hence the metric of space-time, and after inflation, when the inflaton field decays into conventional matter, there will be inherited perturbations $\delta\rho_i(\mathbf{x}, t)$ in the densities of each individual particle species. There will be more complicated perturbations too, such as the perturbation $\Theta(t, \mathbf{x}, \mathbf{n})$ in the function specifying the number of photons at position $\mathbf{x}$, with momentum in the direction $\mathbf{n}$. At our position and at the present epoch, this is the cmb anisotropy.

All perturbations are determined by $\delta\phi_k(t_*)$. Let us consider a generic perturbation $g(\mathbf{x}, t)$ (in the case of Θ, we focus on a fixed $\mathbf{n}$), and make a Fourier expansion

$$g(\mathbf{x}, t) = \sum_k g_k(t)e^{i\mathbf{k}\cdot\mathbf{x}}. \tag{4.3}$$

As long as they are small, the time dependence of the perturbations for a given $\mathbf{k}$ is given by a set of linear differential equations, with no coupling between different $\mathbf{k}$. For a given $\mathbf{k}$-mode of any quantity, the solution of these equations is determined by $\delta\phi_k(t_*)$ and, because the equations are linear, the solution is of the form

$$g_k(t) = T_g(t, k)\,\delta\phi_k(t_*). \tag{4.4}$$

The **transfer function** $T_g(t, k)$ is fixed by the cosmological model under consideration; in the case we are considering, we just need to specify H_0 and Ω_b. Note that T_g is independent of the direction of $\mathbf{k}$ because the evolution equations are invariant under rotations.

Depending on the context, various prefactors are pulled out before defining the transfer function. In particular, it often is defined to approach 1 on large scales.

4.2.3 Approach of horizon entry and curvature perturbation $\mathcal{R}$

As we present in great detail, the magnitude and statistical nature of the vacuum fluctuation at the epoch t_* can be calculated in a given model of inflation. To confront the calculation with observation, we need the appropriate transfer functions evaluated at the present epoch. How do we bridge the gap between inflation and the present?

In its usual form, cosmological perturbation theory starts at an initial epoch, taken to be somewhat before the epoch when scales of cosmological interest start to enter the horizon. This initial epoch is well after nucleosynthesis, which means that we know the material content of the Universe, except for the nonbaryonic dark matter. Assuming that the latter is cold, we have photons, massless neutrinos, ordinary matter, and CDM. We discuss in Chapters 14 and 15 how cosmological perturbation theory determines the evolution of all perturbations after the initial epoch, if we know the initial energy density perturbation of every species.

We want to know how to calculate the transfer functions, which determine these initial density perturbations from the vacuum fluctuation $\delta\phi_k(t_*)$. We can only speculate about *most* of what goes on between t_* and the initial epoch, but it turns out that we *can* calculate the required transfer functions. The reason is that, on scales well outside the horizon, each region of the Universe evolves essentially as an independent unperturbed Universe, with no causal contact between different regions. Moreover, by virtue of assumption 3 on page 59, the separate Universes are identical, up to the convention for synchronizing their clocks.

We denote the energy densities of the four species by ρ_γ, ρ_ν, ρ_b, and ρ_c. Instead of the density itself, it is better to work with the **density contrast** $\delta \equiv \delta\rho/\rho$. At the initial epoch, we find that the separate density contrasts and the total density contrast are related in the following way:

$$\frac{1}{3}\delta_{kb} = \frac{1}{3}\delta_{kc} = \frac{1}{4}\delta_{k\gamma} = \frac{1}{4}\delta_{k\nu} \left(= \frac{1}{4}\delta_k \right). \tag{4.5}$$

This relation among the separate density contrasts is called the **adiabatic condition**. It is derived in Section 4.8.1. The relation to the total density contrast δ_k comes from the fact that the initial epoch is in the radiation-dominated era.

A set of density contrasts satisfying the adiabatic condition is called an **adiabatic density perturbation**. We will show that the vacuum fluctuation of the inflaton field generates an adiabatic density perturbation. The most general possible set of density perturbations is a linear combination of an adiabatic density perturbation and some **isocurvature density perturbation**, which is defined as a set of individual density perturbations that add up to give zero $\delta\rho$. We see in Chapter 6 that an isocurvature density perturbation may be generated by the vacuum fluctuation of a field other than the inflaton but, in the simplest model that we are considering at the moment, it is assumed to be absent.

All that remains is to specify the transfer function relating the initial value of δ to the vacuum $\delta\phi_k(t_*)$. At this point, we introduce a crucial perturbation, which we denote $\mathcal{R}(\mathbf{x}, t)$ and call the **curvature perturbation**. It plays a key role in this book. Its formal definition does not appear until Eq. (14.132), where it is given in terms of the spatial curvature perturbation seen by comoving observers. Equivalently, $-(2/3)k^2\mathcal{R}_k(t)$ can be defined as the Fourier component of the parameter K appearing in a locally defined version of the Friedmann equation [Eqs. (4.165) and (4.167)].

The great virtue of $\mathcal{R}_k(t)$ is that it is constant, provided that the pressure perturbation is negligible, which we see from Eq. (4.166). By virtue of assumptions 2 and 3 on page 59, this turns out to be the case on scales well outside the horizon, regardless of the (perhaps unknown) matter content of the Universe. As a result, $\mathcal{R}_k(t)$ has a constant value between the epoch t_* and the initial epoch, which we call its **primordial** value. Most of the time, we take $\mathcal{R}_k$ to denote the primordial value. It is related to the inflaton perturbation by

$$\mathcal{R}_\mathbf{k} = -\left[\frac{H}{\dot\phi}\, \delta\phi_\mathbf{k}\right]_{t=t_*}. \tag{4.6}$$

We show in Section 4.6.4 how this equation can be rephrased as a statement about space-time geometry, which is proved in Section 14.6. This equation encodes all the information we need concerning the perturbations from inflation (in the absence of isocurvature modes), and in Section 7.4 we compute the inflaton field perturbation $\delta\phi_\mathbf{k}(t_*)$ that stands on its right-hand side.

The curvature perturbation $\mathcal{R}_\mathbf{k}(t)$ is a more useful quantity than the inflaton field perturbation $\delta\phi_\mathbf{k}(t)$ because it is constant outside the horizon, whereas the latter is not. Also, it remains well defined after the scalar field decays and $\delta\phi_\mathbf{k}$ ceases to exist. It is useful to think of the transfer functions as being defined in terms of $\mathcal{R}_\mathbf{k}$, so that, instead of Eq. (4.4), a generic perturbation $g_\mathbf{k}(t)$ is given in terms of the primordial curvature perturbation by

$$g_\mathbf{k}(t) = T_g(t, k)\, \mathcal{R}_\mathbf{k}. \tag{4.7}$$

At the moment, we are focusing on the total density contrast δ at the initial epoch. Ignoring neutrino freestreaming, we find in Section 4.8.2 the simple result

$$\delta_\mathbf{k} = \frac{4}{9}\left(\frac{k}{aH}\right)^2 \mathcal{R}_\mathbf{k}. \tag{4.8}$$

In Section 15.6, we show that neutrino freestreaming multiplies this by a numerical factor of order 1, which must be included in an accurate calculation.

4.2.4   Large-angle cmb anisotropy

From the initial condition, we can calculate presently observable quantities. The simplest prediction is for the cmb anisotropy observed by the Cosmic Background Explorer (COBE) satellite. The angular resolution of COBE is a few degrees, which corresponds to scales that are well outside the horizon at last scattering (see Section 2.4). On such scales, causal effects have not had time to operate, and to a good approximation the anisotropy is given by the remarkable formula

$$\frac{\delta T}{T} = -\frac{1}{5}\mathcal{R}(\mathbf{x}_{ls}), \tag{4.9}$$

where $\mathcal{R}_k$ is the primordial curvature perturbation. The subscript ls stands for last scattering, so that $x_{ls} = 2H_0^{-1}e$, where e is the unit vector in the direction of observation and $2H_0^{-1}$ is a good estimate of the distance to the last-scattering surface.

This is called the **Sachs–Wolfe effect** (Sachs and Wolfe 1967). It usually is written in terms of the perturbation in the gravitational potential, which is given (during matter domination) by $\Phi_k = -3\mathcal{R}_k/5$, leading to

$$\frac{\delta T(e)}{T} = \frac{1}{3}\Phi(x_{ls}). \tag{4.10}$$

We derive it in two different ways, in Sections 5.2.6 and 15.3.1.

4.2.5   The Smaller-scale cmb anisotropy and structure formation

On smaller scales, we have to consider the causal effects that operate after horizon entry. There is a competition between gravity, which tries to increase the density contrast by attracting more matter to the overdense regions, and random particle motion.

For **massless neutrinos** the motion wins and their density contrast falls exponentially. For CDM, random motion is negligible, and its density contrast grows. The growth is only logarithmic during radiation domination, but proportional to $a(t)$ after matter domination.

For **baryons** and **photons** the story is more complicated. By baryons, in this context we mean nuclei *and* electrons, because the Coulomb interaction ensures that the number densities of electrons and protons are practically equal at each point in space. We give an oversimplified account at this point, for orientation. Until photon decoupling, there are free electrons and nuclei, with frequent Thomson scattering of the photons and electrons. Because of the pressure, the density contrast of this tightly coupled baryon–photon fluid oscillates as a standing **acoustic wave** after horizon entry. At decoupling, the electrons bind into atoms and Thomson scattering practically ceases, removing the pressure support. The acoustic oscillation then ends, and the photons travel freely to become the cmb. As a function of angular scale the cmb anisotropy exhibits a series of peaks and troughs, usually called acoustic or Doppler peaks, which provide a snapshot of the acoustic oscillation just before it ends.

After decoupling, the baryons fall into the potential wells already created by the CDM, acquiring the same density contrast. Ultimately, the density perturbation on a given scale may evolve beyond the linear regime, and structures of the corresponding size become gravitationally bound. On scales larger than about $8h^{-1}$ Mpc, this has yet to happen, and we can compare the linearly evolved theory with observation. Most of this book concerns the linear evolution. On smaller scales, we have to rely on numerical simulation or on semianalytic approximation to the nonlinear evolution, which we discuss briefly in Chapter 11.

4.3   **Gaussian perturbations**

Now we begin to discuss the generation and evolution of the perturbations. First, we must set up some formalism and, in particular, we must consider the statistical nature of the vacuum fluctuation. It is crucial and forms the whole basis for the comparison of theory with observation.

The vacuum fluctuation is studied in detail in Chapter 7. We show there that the perturbations generated by the vacuum fluctuation are Gaussian, which, roughly speaking, means that their Fourier components are uncorrelated. The Gaussian property greatly simplifies the discussion and is taken for granted throughout except for a brief discussion in Chapter 6. On the other hand, most of the results continue to apply in the non-Gaussian case; in particular, we still can define a spectrum and a correlation function, though they no longer provide a complete description of the statistical properties. There are a variety of observational tests that can probe possible non-Gaussianity, though the only one we mention is the topology of the galaxy distribution in Section 10.1.4. As yet, all tests carried out are consistent with initial perturbations, which are Gaussian, though the strength of this statement is unclear given the lack so far of well-motivated and calculable non-Gaussian models.

4.3.1 Vacuum fluctuation

In the quantum vacuum state, at any instant, the Fourier coefficients $\delta\phi_{\mathbf{k}}$ do not have well-defined values. The vacuum state, however, can be expanded in terms of states in which they do have well-defined values, and the probability of finding a given set of values is the modulus squared of the relevant coefficient in the expansion.

In ordinary applications of quantum physics the vague phrase "probability of finding" can be replaced by "probability that a measurement yields," but this appears unreasonable for the Universe. The density perturbation, for example, surely exists independently of whether we observe it. So, one instead asserts that our Universe corresponds to a typical member of the ensemble of possible Universes, obtained when we expand the vacuum state into states with a definite inflaton-field perturbation. Perhaps surprisingly, this allows us to make definite predictions, which can be compared with observation.

Taking the usual attitude to quantum mechanics, we do not address the issue of how a particular member of the ensemble comes to have been chosen. This is the Schrödinger cat paradox, with the Universe replacing the cat. There is a substantial body of research on possible solutions to the paradox, both for the cat and for the Universe, but it has not yet entered the mainstream of physics.

According to quantum field theory, the real and imaginary parts of each component $\delta\phi_{\mathbf{k}}$ have the dynamics of a harmonic oscillator. In the vacuum state, each real and imaginary part has a Gaussian probability distribution, with no correlation between them except for the reality condition.[1] The variances (mean squares) of the probability distributions can be calculated from the model of inflation. They are independent of the direction of $\mathbf{k}$, and for a given $\mathbf{k}$ they are the same for the real and imaginary parts. Equivalently, for a given $\mathbf{k}$, the phase of $\delta\phi_{\mathbf{k}}$ is drawn randomly from a uniform distribution.

---

[1] In fact, nothing depends on the shapes of the probability distributions of the Fourier coefficients; what matters is their independence. The reason is that, in the limit of large box size, we will be summing over an infinite number of values of $\mathbf{k}$, within an infinitesimal cell of k-space. According to the central limit theorem, the probability distribution of the sum will be Gaussian for any (reasonable) probability distributions of the individual terms. Thus we lose no generality in writing the probability distributions as Gaussian. That the Gaussianity *for each individual* $\mathbf{k}$ is predicted by quantum field theory seems to be an accident with no deep meaning.

These properties are inherited by the Fourier components $g_\mathbf{k} = T_g(t, k)\delta\phi_\mathbf{k}(t_*)$ of a generic perturbation, and they define what is called a **Gaussian perturbation**. To be more precise, it is the fact that the Fourier coefficients have independent probability distributions that makes the perturbation what mathematicians call Gaussian. The additional fact that the variances are independent of the direction of $\mathbf{k}$ and are equal for the real and imaginary parts gives the perturbation the additional property that its stochastic properties are invariant under translations and rotations. Cosmologists take this additional property for granted, taking the term Gaussian to include it. We follow that practice.

The stochastic properties of a Gaussian perturbation are simple, as we now explain. Our discussion is at the informal level that is usual in physics, and the relationship of this discussion to quantum field theory is given in Chapter 7. A rigorous discussion can be found in mathematics texts, such as Karlin and Taylor (1975) for the elementary aspects or Adler (1981) for an advanced discussion.

4.3.2 Basic properties of Gaussian perturbations

Recall that we are carrying out the expansion in a large box, so that $\mathbf{k}$ takes on discrete values $\mathbf{k}_1, \mathbf{k}_2, \ldots$. Let us denote the real part of $g_{\mathbf{k}_n}$ by R_n and its imaginary part by I_n. The probability of finding R_n in a given interval is $\mathbf{P}(R_n)d(R_n)$, where

$$\mathbf{P}(R_n) = \frac{1}{\sqrt{2\pi}\,\sigma_n} \exp\left(-\frac{1}{2}\frac{R_n^2}{\sigma_n^2}\right). \tag{4.11}$$

Here $\sigma_n^2 = \langle R_n^2 \rangle$ is the ensemble average of R_n^2, and the prefactor ensures that the total probability is unity. The quantity σ_n is known as the **dispersion**, and σ_n^2 is known as the **variance** or **mean square**. Identical expressions hold for the imaginary part I_n, and it is predicted to have the same variance:

$$\sigma_n^2 = \langle R_n^2 \rangle = \langle I_n^2 \rangle = \frac{1}{2}\langle |g_{\mathbf{k}_n}|^2 \rangle. \tag{4.12}$$

The phase of $g_{\mathbf{k}_n}$ is random, with a uniform probability distribution.

We also can write

$$\frac{1}{2}\langle g_{\mathbf{k}_n}^* g_{\mathbf{k}_{n'}} \rangle = \delta_{nn'}\sigma_n^2. \tag{4.13}$$

This is the same as the preceding expression, except that it encodes both the independence of the distributions and the random phase; the random phase sets to zero the term $\langle g_{\mathbf{k}_{-n}}^* g_{\mathbf{k}_n} \rangle = \langle g_{\mathbf{k}_n}^2 \rangle$, which otherwise would be present. Using it, we can calculate the ensemble average of $g^2(\mathbf{x})$:

$$\langle g^2(\mathbf{x}) \rangle = \sum_n \langle |g_{\mathbf{k}_n}|^2 \rangle = 2\sum_n \sigma_n^2. \tag{4.14}$$

Note that it is independent of $\mathbf{x}$, and therefore invariant under translations and rotations.

In the limit of large box size, σ_n^2 is predicted to be independent of the direction of $\mathbf{k}_n$. It is more convenient to consider a quantity $\mathcal{P}_g$, known as the **spectrum** of g:

$$\mathcal{P}_g(k) \equiv \left(\frac{L}{2\pi}\right)^3 4\pi k^3 \langle |g_\mathbf{k}|^2 \rangle. \tag{4.15}$$

Using Eq. (4.27) below to go to the limit of large box size in Eq. (4.14), and using the volume element $d^3k = 4\pi k^2\, dk$, where $k \equiv |\mathbf{k}|$, we find

$$\sigma_g^2(\mathbf{x}) \equiv \langle g^2(\mathbf{x}) \rangle = \int_0^\infty \mathcal{P}_g(k)\frac{dk}{k}. \tag{4.16}$$

The definition of $\mathcal{P}_g$ was chosen to make this expression simple.

An alternative definition, used only for the matter density perturbation, is

$$P_g \equiv L^3 \langle |g_\mathbf{k}|^2 \rangle = \frac{2\pi^2}{k^3}\mathcal{P}_g(k), \tag{4.17}$$

leading to

$$\sigma_g^2(\mathbf{x}) = \frac{1}{2\pi^2}\int P_g(k)k^2\, dk. \tag{4.18}$$

We also can work out the two-point correlation function, defined by

$$\xi(\mathbf{x}_1, \mathbf{x}_2) \equiv \langle g(\mathbf{x}_1)g(\mathbf{x}_2)\rangle. \tag{4.19}$$

It is given in the infinite box limit, using Eqs. (4.3), (4.13), and (4.15), by

$$\begin{aligned}
\xi(\mathbf{x}_1, \mathbf{x}_2) &= \int \frac{d^3k}{4\pi k^3}\mathcal{P}_g(k)e^{i\mathbf{k}\cdot(\mathbf{x}_1 - \mathbf{x}_2)} \\
&= \frac{1}{2}\int_0^\infty \frac{dk}{k}\mathcal{P}_g(k)\int_{-1}^1 d(\cos\theta)e^{ikx\cos\theta} \\
&= \int_0^\infty \mathcal{P}_g(k)\frac{\sin(kx)}{kx}\frac{dk}{k},
\end{aligned} \tag{4.20}$$

where $x = |\mathbf{x}_1 - \mathbf{x}_2|$. Again, we see that it is invariant under translations and rotations.

The invariance under translations and rotations holds for all stochastic properties of the ensemble. In general terms, it is a consequence of the invariance of the vacuum state. To be specific, it follows from the randomness of the phases and the independence of the variances $\langle |g_\mathbf{k}|^2 \rangle$ on the direction of $\mathbf{k}$. To see this schematically, return to the finite box and consider the probability of finding the R_n and I_n in given intervals. It is $\mathbf{P}\prod_n dR_n\, dI_n$, where, from Eq. (4.11),

$$\begin{aligned}
\mathbf{P} &= \prod_n N_n \exp\left(-\frac{1}{2}\frac{R_n^2}{\sigma_n^2}\right)\exp\left(-\frac{1}{2}\frac{I_n^2}{\sigma_n^2}\right) \\
&= \exp\left[-\frac{1}{2}\sum_n (\ln N_n)\frac{R_n^2 + I_n^2}{\sigma_n^2}\right],
\end{aligned} \tag{4.21}$$

where $N_n = (\sqrt{2\pi}\sigma_n)^{-2}$. A translation of the origin of coordinates multiplies each Fourier coefficient by a phase, and a rotation mixes the Fourier coefficients with a given k value, but not those with different k. The mixing is an orthogonal (i.e., real and unitary) transformation, which leaves the sum of squares of the relevant coefficients invariant. However, because σ_n depends only on k, the probability distribution (4.21) involves just this sum of squares, and so, it is invariant too. (To make this argument fully rigorous, we would need to go to the limit of an infinite box.)

The probability distribution of $g(\mathbf{x})$ at a given position is Gaussian, by virtue of the central limit theorem. Thus the probability $\mathbf{P}dg$ of finding g in a given interval at the position $\mathbf{x}$ is

$$\mathbf{P}(g) = \frac{1}{\sqrt{2\pi}\,\sigma_g} \exp\left(-\frac{g^2}{2\sigma_g^2}\right). \tag{4.22}$$

We also can write down simple expressions for the N-point correlation functions, and for the probability distribution of a product $g(\mathbf{x}_1)\cdots g(\mathbf{x}_N)$, but we do not need them.

Fourier integral

It is often more convenient to take the limit of large box size, so that the Fourier series becomes a Fourier integral:

$$g(\mathbf{x}) = \frac{1}{(2\pi)^{3/2}} \int g(\mathbf{k}) e^{i\mathbf{k}\cdot\mathbf{x}} d^3k. \tag{4.23}$$

The same symbol can serve to denote both the quantity and its Fourier transform, with the argument $\mathbf{x}$ or $\mathbf{k}$ showing which is which. The numerical prefactor is convenient because it makes the basis functions orthonormal instead of just orthogonal, since

$$\frac{1}{(2\pi)^3} \int e^{i(\mathbf{k}-\mathbf{k}')\cdot\mathbf{x}} d^3x = \delta^3(\mathbf{k}-\mathbf{k}'), \tag{4.24}$$

where δ^3 here is the Dirac delta function. The inverse is

$$g(\mathbf{k}) = \frac{1}{(2\pi)^{3/2}} \int g(\mathbf{x}) e^{-i\mathbf{k}\cdot\mathbf{x}} d^3x. \tag{4.25}$$

The spectrum is given by

$$\langle g^*(\mathbf{k})g(\mathbf{k}')\rangle = \delta^3(\mathbf{k}-\mathbf{k}')\frac{2\pi^2}{k^3}\mathcal{P}_g(k). \tag{4.26}$$

We can check that this definition is equivalent to the earlier one, Eq. (4.15), by working out $\langle g^2(\mathbf{x})\rangle$ and checking that it agrees with Eq. (4.16).

In going between the Fourier sum and Fourier integral, the following correspondences are useful. First, from Eq. (4.2),

$$\left(\frac{2\pi}{L}\right)^3 \sum_{\mathbf{k}} \rightarrow \int d^3k. \tag{4.27}$$

From this we learn that

$$\left(\frac{L}{2\pi}\right)^3 g_{\mathbf{k}} \rightarrow \frac{1}{(2\pi)^{3/2}} g(\mathbf{k}) \tag{4.28}$$

and

$$\left(\frac{L}{2\pi}\right)^3 \delta_{nn'} \to \delta^3(\mathbf{k} - \mathbf{k'}). \tag{4.29}$$

The last relation can be verified by summing over n on the left-hand side and integrating over d^3k on the right-hand side.

Spherical expansion

Rather than using a Cartesian expansion, we can use a spherical expansion. The spherical expansion provides the best way of understanding the cmb anisotropy because the last-scattering surface is a sphere. Also, in contrast with the Fourier series, it can be generalized readily to the case of an open or closed Universe; all that happens is that the radial functions become different – see Section 6.3. Here we give the flat-Universe version.

The expansion is of the form

$$g(\mathbf{x}) = \int_0^\infty dk \sum_{\ell m} g_{\ell m}(k) Z_{k\ell m}(x, \theta, \phi), \tag{4.30}$$

where $g_{\ell m}(k)$ are the expansion coefficients and

$$Z_{k\ell m}(x, \theta, \phi) \equiv \sqrt{\frac{2}{\pi}} k \, j_\ell(kx) Y_{\ell m}(\theta, \phi). \tag{4.31}$$

Here, j_ℓ is the spherical Bessel function, (θ, ϕ) is the direction of $\mathbf{x}$, and $Y_{\ell m}$ is the spherical harmonic defined in terms of the associated Legendre function P_ℓ^m by (see, e.g., Press et al. 1992)

$$Y_{\ell m} = \left[\frac{2\ell + 1}{4\pi} \frac{(\ell - m)!}{(\ell + m)!}\right]^{1/2} P_\ell^m(\cos\theta) \, e^{im\phi}. \tag{4.32}$$

The spherical harmonics satisfy $Y_{\ell m}^* = (-1)^m Y_{\ell, -m}$, giving a reality condition $g_{\ell m} = (-1)^m \times g_{\ell, -m}^*$. $Y_{\ell 0}$ is independent of ϕ and given by

$$Y_{\ell 0}(\theta) = \sqrt{\frac{2\ell + 1}{4\pi}} P_\ell(\cos\theta), \tag{4.33}$$

with $P_\ell(1) = 1$.

The basis functions $Z_{k\ell m}$ are orthonormal,

$$\int Z_{k\ell m}^* Z_{k'\ell'm'} d^3x = \delta(k - k')\delta_{\ell\ell'}\delta_{mm'}, \tag{4.34}$$

where $d^3x = x^2 \sin\theta \, d\theta \, d\phi \, dx$. This is equivalent to the pair of relations

$$\int Y_{\ell m} Y_{\ell'm'}^* \, d\Omega = \delta_{\ell\ell'}\delta_{mm'}, \tag{4.35}$$

$$\int_0^\infty \left[\sqrt{\frac{2}{\pi}} k j_\ell(kx)\right]\left[\sqrt{\frac{2}{\pi}} k' j_\ell(k'x)\right] x^2 \, dx = \delta(k - k'). \tag{4.36}$$

The spherical expansion is equivalent to the Fourier integral of Eq. (4.23) because of the identity[2]

$$\exp(i\mathbf{k} \cdot \mathbf{x}) = 4\pi \sum_{\ell,m} i^\ell j_\ell(kx) Y_{\ell m}(\hat{\mathbf{x}}) Y_{\ell m}(\hat{\mathbf{k}}), \tag{4.37}$$

where the hats denote unit vectors. Substituting this into Eq. (4.23) and doing the angular integration, we find

$$g_{\ell m}(k) = k i^\ell \int g(k, \hat{\mathbf{k}}) Y_{\ell m}(\hat{\mathbf{k}}) \, d\Omega_\mathbf{k}. \tag{4.38}$$

This shows that the transformation from $g_{\ell m}(k)$ to $kg(k, \hat{\mathbf{k}})$ is unitary, as we also can see by comparing the normalizations (4.24) and (4.34) and remembering that $d^3k = k^2 \, d\Omega_k \, dk$. The transformation therefore preserves the sum of squares in Eq. (4.21), which means that the real and imaginary parts of each coefficient $g_{\ell m}(k)$ have a Gaussian probability distribution, with no correlation between different coefficients except for the reality condition. Because the transformation is unitary, it preserves the form of Eq. (4.26), leading to

$$\langle g_{\ell m}^*(k) g_{\ell' m'}(k') \rangle = \frac{2\pi^2}{k^3} \mathcal{P}_g(k) \delta(k - k') \delta_{\ell \ell'} \delta_{mm'}. \tag{4.39}$$

A more pedestrian way of verifying this formula is to work out $\langle g^2(\mathbf{x}) \rangle$ at $\mathbf{x} = 0$ in the spherical expansion and equate it with the one obtained from the Fourier integral Eq. (4.16). The calculation in the spherical expansion is trivial because all j_ℓ vanish at the origin except $j_0 = 1$, and $Y_{00} = 1/\sqrt{4\pi}$.

In this account we relied on the Fourier expansion, but it is possible to do everything without reference to it, which becomes essential if we go to an open or closed Universe. The important point is that the functions $Z_{k\ell m}$ are a complete orthonormal set of eigenfunctions of the Laplacian operator ∇^2, with eigenvalues $-(k/a)^2$. Using the spherical expansion to calculate the vacuum fluctuation, we find that the probability distribution of the real and imaginary parts of each $g_{\ell m}(k)$ are independent Gaussians, whose variance depends only on k. Because the $Z_{k\ell m}$ are an orthonormal basis, the change of basis corresponding to a shift in the origin or orientation of the coordinates will be unitary:

$$Z_{k\ell m} \to \sum_{\ell' m'} U_{\ell m \ell' m'}(k) Z_{k\ell' m'}. \tag{4.40}$$

(The coordinate change does not affect k because $-(k/a)^2$ is an eigenvalue of ∇^2, which is invariant under the change.) This coordinate change has no effect on the scalar g, and so, its coefficients must transform inversely[3]:

$$g_{k\ell m} \to \sum_{\ell' m'} U_{\ell m \ell' m'}^{-1}(k) g_{k\ell' m'}. \tag{4.41}$$

---

[2] A more familiar form for this identity is obtained by summing over m, using the addition theorem (4.45).
[3] In this and the preceding equation, U is regarded as a matrix whose rows and columns are labelled by the pair of indices (ℓ, m).

By the argument after Eq. (4.21), the coefficients still have the same probability distribution after the change of coordinates. As a result, there is no change in the stochastic properties of the perturbation $g(\mathbf{x})$.

A special case is that of a rotation, which leads to a unitary transformation, mixing only the m components:

$$Y_{\ell m} \rightarrow \sum_{m'} U_{mm'}(\ell)\, Y_{\ell m'}, \tag{4.42}$$

$$g_{k\ell m} \rightarrow \sum_{m'} U_{mm'}^{-1}(\ell)\, g_{k\ell m'}. \tag{4.43}$$

Here $U_{mm'}(\ell)$ is the rotation matrix $\mathcal{D}_{m'm}^{\ell}(\alpha, \beta, \gamma)$, specified by the Euler angles α, β, and γ. For the stochastic properties of $g(\mathbf{x})$ to be invariant under rotations, it is enough to have the spectrum independent of m. Its ℓ-independence gives the additional property of invariance under translations.

For future reference, we note that, by virtue of the unitarity,

$$\sum_m Y_{\ell m}^*(\theta_1, \phi_1) Y_{\ell m}(\theta_2, \phi_2) = \sum_m Y_{\ell m}^*(\theta_1', \phi_1') Y_{\ell m}(\theta_2', \phi_2'), \tag{4.44}$$

where the prime denotes a different choice of the polar coordinate system. This sum therefore depends only on the angle θ_{12} between the two points. It is evaluated conveniently by taking one of them to be at the pole because only $Y_{\ell 0}$ is nonvanishing there. Using Eq. (4.33), we find

$$\sum_m Y_{\ell m}^*(\theta_1, \phi_1) Y_{\ell m}(\theta_2, \phi_2) = Y_{\ell 0}^*(0) Y_{\ell 0}(\theta_{12}) = \frac{2\ell + 1}{4\pi} P_{\ell}(\cos\theta_{12}). \tag{4.45}$$

A similar trick works for the functions $Z_{k\ell m}$. The sum

$$\sum_{\ell m} Z_{k\ell m}^*(x_1, \theta_1, \phi_1) Z_{k\ell m}(x_2, \theta_2, \phi_2)$$

is invariant, and taking one of the points at the origin kills everything except Z_{k00}. This leads directly to Eq. (4.20) for the correlation function.

4.3.3 Smoothing and the ergodic property

From Eq. (4.16), we see that the mean square diverges if the spectrum fails to vanish in the limit of either large or small k.

Smoothing

Failure to vanish in the limit of large k is no problem. It simply indicates that there is a lot of structure on small scales, and we can get rid of such structure by **smoothing**. Instead of the perturbation $g(\mathbf{x})$, we consider a smoothed quantity

$$g(R, \mathbf{x}) = V^{-1} \int W(|\mathbf{x}' - \mathbf{x}|/R)\, g(\mathbf{x}')\, d^3x'. \tag{4.46}$$

That is, we replace the perturbation g with a weighted average over nearby points. In this expression, the **window function** $W(y)$ that governs the weighting is to fall off rapidly for $y > 1$, and V is its volume:

$$V \equiv \int d^3x \, W(x/R) = 4\pi R^3 \int y^2 W(y) \, dy. \tag{4.47}$$

The exact choice of window function is a matter of convenience. The simplest is the **top hat**, defined by $W = 1$ for $y \le 1$ and $W = 0$ for $y > 1$. Another simple possibility is the **Gaussian**, $W(y) = \exp(-y^2/2)$.

The Fourier component of the smoothed quantity is

$$g(R, \mathbf{k}) = \frac{1}{(2\pi)^{3/2}} \frac{1}{V} \int W(|\mathbf{x}' - \mathbf{x}|/R) \, g(\mathbf{x}') \, e^{i\mathbf{k}\cdot(\mathbf{x}'-\mathbf{x})} e^{-i\mathbf{k}\cdot\mathbf{x}'} d^3x \, d^3x', \tag{4.48}$$

where we have inserted a cancelling pair of factors $e^{\pm i\mathbf{k}\cdot\mathbf{x}'}$. Doing the x integration first, we find the convolution theorem

$$g(R, \mathbf{k}) = W(kR) \, g(\mathbf{k}). \tag{4.49}$$

Here, $W(kR)$ is the Fourier transform of $W(x/R)/V$:

$$W(kR) \equiv \frac{\int d^3x \, W(x/R) \, e^{i\mathbf{k}\cdot\mathbf{x}}}{\int d^3x \, W(x/R)}, \tag{4.50}$$

and $g(\mathbf{k})$ is the Fourier transform of $g(\mathbf{x})$ with the normalization of Eq. (4.25). If we change the variables to $\mathbf{y} \equiv \mathbf{x}/R$ and $\mathbf{q} \equiv \mathbf{k}R$, this expression becomes $W(kR) = \widetilde{W}(q)/\widetilde{W}(0)$, where $\widetilde{W}(q)$ is the three-dimensional Fourier transform of $W(y)$,

$$\widetilde{W}(q) = \int W(y) e^{i\mathbf{y}\cdot\mathbf{q}} \, d^3y. \tag{4.51}$$

We see that the window function $W(kR)$ is equal to 1 at $k = 0$ and falls off rapidly at $kR > 1$. For the top hat, we can perform the angular integration in Eq. (4.51) first [cf. Eq. (4.20)] to find

$$W(kR) = 3 \left[\frac{\sin(kR)}{(kR)^3} - \frac{\cos(kR)}{(kR)^2} \right]. \tag{4.52}$$

For the Gaussian, we can use Cartesian coordinates to find

$$W(kR) = \exp\left(-\frac{k^2 R^2}{2}\right). \tag{4.53}$$

These two filters are shown in Figure 4.1

For each window function, it is useful to define a mass $M = V\rho_{0c}$, which is the mass of matter enclosed in comoving volume V. For the top hat,

$$M = 1.16 \times 10^{12} h^{-1} \left(\frac{R}{h^{-1}\,\mathrm{Mpc}}\right)^3 M_\odot; \tag{4.54}$$

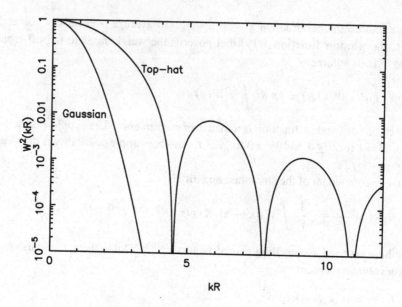

Fig. 4.1. Top-hat and Gaussian filters in Fourier space.

and for the Gaussian,

$$M = 4.37 \times 10^{12} h^{-1} \left(\frac{R}{h^{-1}\,\mathrm{Mpc}} \right)^3 M_\odot. \tag{4.55}$$

The smoothing removes structure on scales $\lesssim R$ without affecting structure on much bigger scales. Correspondingly, it filters out the Fourier components with $kR \gtrsim 1$ without significantly affecting those with $kR \ll 1$. Typical observational procedures automatically include such smoothing, for example, through finite resolution.

From Eq. (4.49), the smoothed quantity has mean square

$$\langle g^2(R, \mathbf{x}) \rangle \equiv \sigma_g^2(R) = \int_0^\infty W^2(kR) \mathcal{P}_g(k) \frac{dk}{k}. \tag{4.56}$$

If $\mathcal{P}_g(k)$ is increasing with k, then $W^2 \mathcal{P}_g$ typically will have a maximum at $k_R \sim 1/R$, giving roughly

$$\sigma_g^2(R) \sim \mathcal{P}_g(k_R). \tag{4.57}$$

So far we have focussed on a fixed time. Usually, we take R to be time independent, corresponding to a smoothing scale that is expanding with the Universe. Then, if the evolution of the original quantity is given on all scales by the linear equations of cosmological perturbation theory, that of the smoothed quantity certainly will be. On the other hand, by choosing R big enough, we can make the mean-square $\sigma^2(R)$ as small as desired and, after making it small enough, we can expect the smoothed quantity to evolve linearly even if the evolution is nonlinear below the smoothing scale. An application of this folk theorem to the bottom-up picture of structure formation is described in Section 5.1.3, in connection with the presently observed density perturbation.

Ergodic property

After any necessary smoothing, the mean square will be finite provided that the spectrum vanishes sufficiently quickly on large scales, $k \to 0$. Because we took the box size to infinity, this requirement is essential from a purely mathematical viewpoint. However, if the spectrum is roughly constant, we can make some sense of the physics by taking a large finite box. Physically, there ultimately may be a very-large-scale cutoff set by the horizon at the start of inflation, though that scale typically is vast.

As we see in the next section, the spectrum of the density perturbation is expected to vanish on large scales. On the other hand, the spectra of both the curvature perturbation and the inflaton field are roughly constant, the latter definitely increasing with scale.

We consider first the case in which the spectrum does vanish on large scales. Consider the correlation function $\xi(x)$ defined by Eq. (4.19). If the spectrum becomes negligible for $1/k$ bigger than some scale $1/k_*$, the correlation function will fall off rapidly above the distance $x_* = 1/k_*$. This distance is called the **correlation length**, and the limit of large box size will be attained if L is much bigger than x_*. Then, because the Fourier coefficients g_k have independent probability distributions, the average of $|g_k|^2$ over a cell in k space will be the same as the ensemble average. As a result, the spectrum of the ensemble, defined by Eq. (4.15), can be deduced by examining just one of its realizations. By checking that the g_k have random phases, we also check the Gaussianity.

Because there is no significant correlation between widely separated points in the box, the brackets in the expression $\langle g^2 \rangle$ can be interpreted as an average over all space. Then, using orthonormality

$$\frac{1}{L^3} \int d^3x \, e^{i\mathbf{k}_n \cdot \mathbf{x}} e^{-i\mathbf{k}_m \cdot \mathbf{x}} = \delta_{nm}, \tag{4.58}$$

Eq. (4.14) becomes Parseval's theorem

$$\int d^3x \, g^2(\mathbf{x}) = L^3 \sum_n \left| g_{\mathbf{k}_n} \right|^2. \tag{4.59}$$

Similarly, the correlation function $\xi(|\mathbf{x}_1 - \mathbf{x}_2|) = \langle g(\mathbf{x}_1)g(\mathbf{x}_2)\rangle$ can be interpreted as an average over $\mathbf{x}_1$ with, for example, $\mathbf{x}_2 - \mathbf{x}_1$ fixed. (In addition, we could average over the orientation of $\mathbf{x}_2 - \mathbf{x}_1$.)

We can take this further. With the size L of the box fixed (and, for simplicity, its orientation too), we can throw it down in random locations within a region whose size is much bigger than L. Under weak assumptions about the spectrum, one can show that this is equivalent to taking random samples of the ensemble for a fixed location of the box. This property of Gaussian random fields is called the **ergodic** property.[4]

---

[4] The term ergodic is a mathematical one, and its most familiar physics application comes in statistical mechanics, where, given enough time, a closed system randomly samples all of phase space.

The density contrast $\delta \equiv \delta\rho/\rho$, at least of the galaxy distribution, is being measured in the region around us with a radius of order $100h^{-1}$ Mpc, and these ideas are used routinely when interpreting the observations. Taking the correlation function of δ to be defined as a spatial average, we find a correlation length of only about $5h^{-1}$ Mpc.[5] We therefore can deduce the spectrum by computing the Fourier coefficients and identifying the brackets in Eq. (4.15) as the average over a cell in k space. Alternatively, we can deduce the spectrum from other expressions. One is the inverse of Eq. (4.20), giving the spectrum as the Fourier transform of the correlation function. Another is the approximate inverse of Eq. (4.56). These methods are complementary in practice, though equivalent in principle.

Now, suppose instead that the spectrum is roughly constant on large scales. Cases of interest are the curvature perturbation and the inflaton-field perturbation, whose spectra are roughly constant on scales outside the horizon. After smoothing on the horizon scale, the mean square within a region of size aL much larger than the horizon will be of order

$$\langle g^2 \rangle \sim \int_{1/L}^{aH} \mathcal{P}_g(k) \frac{dk}{k}. \tag{4.60}$$

For aL not too many orders of magnitude bigger than H^{-1}, this will, very roughly, give

$$\langle g^2 \rangle \sim \mathcal{P}_g, \tag{4.61}$$

where $\mathcal{P}_g$ is evaluated at, for example, the horizon scale $k = aH$. For instance, if $\mathcal{P}_g$ is constant to good accuracy, we will have $\langle g^2 \rangle \sim \mathcal{P}_g \ln(aL/H^{-1})$. Using these considerations, we can discuss the stochastic properties of the perturbation in a useful way, even though the things may go wrong if we take the box size too large (Lyth 1992).

4.3.4 Spectrum of the primordial curvature perturbation

The spectrum $\mathcal{P}_g$ of a generic perturbation g is given in terms of the spectrum $\mathcal{P}_\mathcal{R}$ of the primordial curvature perturbation[6] by $\mathcal{P}_g(t) = T_g^2(t)\mathcal{P}_\mathcal{R}$, where T_g is the transfer function appearing in Eq. (4.7). The calculation of $\mathcal{P}_\mathcal{R}(k)$ in a given model of inflation provides the foundation for all subsequent comparison of the model with observation, and is done from first principles in Chapter 7.

In the slow-roll approximation, the result can be stated simply. From Eq. (4.6),

$$\mathcal{P}_\mathcal{R}(k) = \left[\left(\frac{H}{\dot{\phi}} \right)^2 \mathcal{P}_\phi(k) \right]_{t=t_*}, \tag{4.62}$$

where t_* is a few Hubble times after the epoch of horizon given by exit $k = aH$. We are assuming that $\delta\phi_k$ is a practically free field (assumption 1 on page 58). In slow-roll inflation,

---

[5] The relevant observations, which we discuss in detail later, relate to the distribution and motion of galaxies and galaxy clusters. The particular number quoted is actually the correlation length of the number density of optically-selected galaxies, a rather well-determined quantity.

[6] Recall that $\mathcal{R}_k(t)$ has a constant value well outside the horizon, which we denote as $\mathcal{R}_k$ and call the primordial curvature perturbation.

$\delta\phi_{\mathbf{k}}$ also turns out to have negligible mass, at least until the epoch t_*. As we see in Chapter 7, *any* massless free field acquires, during slow-roll inflation, a Gaussian vacuum fluctuation. A few Hubble times after horizon exit, the fluctuation settles down to a constant value, with a spectrum given by

$$\mathcal{P}_\phi(k) = \left(\frac{H}{2\pi}\right)^2\bigg|_{k=aH}. \tag{4.63}$$

As indicated, H is to be evaluated at horizon exit, though it would make no significant difference if it were evaluated at the epoch t_* because, in slow-roll inflation, H has negligible variation in a few Hubble times. Inserting this into Eq. (4.62) gives the spectrum of the primordial curvature perturbation,

$$\mathcal{P}_\mathcal{R}(k) = \left[\left(\frac{H}{\dot\phi}\right)\left(\frac{H}{2\pi}\right)\right]^2_{k=aH}. \tag{4.64}$$

The factor $(H/\dot\phi)$ comes from Eq. (4.62), but it can be evaluated at the epoch of horizon exit $k = aH$ because, in the slow-roll approximation, $\dot\phi$ also has negligible variation in a few Hubble times.

In almost all models of inflation, the spectrum can be taken to be a power law with scale

$$\mathcal{P}_\mathcal{R}(k) \propto k^{n-1}. \tag{4.65}$$

The constant n is the spectral index,[7] and the value $n = 1$ corresponds to what is known as the scale-invariant or Harrison–Zel'dovich spectrum (Harrison 1970; Zel'dovich 1970). The reason for the simple power-law behaviour is that scales typically cross outside the horizon very quickly during inflation, and there is not much chance for physical conditions, which govern the size of the perturbations, to change very much. In Chapter 7, we show how to calculate n in a given model of inflation, and in Chapter 8, we survey the models that have been proposed so far. In the foreseeable future, cmb anisotropy satellites are expected to measure n with an accuracy $\Delta n \sim 0.01$, which will discriminate strongly between different models of inflation.

The normalization of $\mathcal{P}_\mathcal{R}$ on large scales already is known accurately from the COBE observations, as described in detail in Chapter 9. This too is a powerful constraint on models of inflation, though one that tends to be taken for granted because the rough order of magnitude was already known, from large-scale structure, when inflation was first proposed. An understanding of the order of magnitude is the best that particle theory can aspire to in its present state, and even this is proving quite elusive.

4.4 The density perturbation: Newtonian treatment

We now turn to the time dependence of the perturbations, starting with a Newtonian treatment of the matter density perturbation and related quantities. This treatment is valid on scales well

---

[7] The form $n - 1$ is an historical accident, originating from the older definition (4.17) of the spectrum $P_\delta(k)$ of the density contrast. With this definition, $P_\delta \propto k^n$ on scales well outside the horizon.

within the horizon, and after matter domination. It therefore deals with the very end of the story we outlined earlier. For larger scales and earlier times, one has to use general relativity instead of Newtonian physics.

4.4.1   Cosmic fluid

We begin with some basic ideas that apply in both the Newtonian and relativistic settings.

The Universe is modelled as a fluid, so that all relevant quantities are smoothly varying functions of position. This means that cosmic strings and other topological defects are assumed to play no role in structure formation, because they are not amenable to such a description. According to present ideas, the fluid after inflation is likely to have been a gas except during phase transitions. Certainly, this must be true from nucleosynthesis onward. After bound structures form (e.g., galaxies and galaxy clusters), they should be taken to be the "particles" of the cosmic "gas," along with those genuine particles that are not bound. The fluid concept applies only after smoothing on a comoving scale containing many "particles."

A key concept of cosmology, which applies whether the cosmic fluid is gaseous or not, is that of a **comoving observer**. Loosely speaking, a comoving observer is one moving with the expansion of the Universe, including the effect of its inhomogeneities. To be precise, comoving observers measure zero momentum density at their own location. Seen by any other observer, they are moving with the flow of energy. If we are dealing with a gas, comoving observers move with the average flow of the "particles."

4.4.2   Peculiar velocity

Now we adopt a Newtonian reference frame, specified by Cartesian space coordinates $\mathbf{r}$ and a universal time coordinate t. At each point in space, the **fluid velocity** is the velocity of the comoving observer, given by $\mathbf{u} = d\mathbf{r}/dt$, where $\mathbf{r}(t)$ is the physical position of the comoving observer. This defines the fluid velocity field $\mathbf{u}(t, \mathbf{r})$. In the unperturbed Universe, we can take $\mathbf{r}(t) = a(t)\mathbf{x}$, with $\mathbf{x}$ time-independent, and $\mathbf{u} = H\mathbf{r}$. Taking into account the perturbations, the comoving observer has position

$$\mathbf{r}(t) = a(t)\mathbf{x}(t),$$

(4.66)

with *time-dependent* comoving position $\mathbf{x}(t)$. The Hubble parameter $H(t)$, and hence the scale factor $a(t)$, can be defined by looking at very large $\mathbf{r}$, where the effect of the perturbation becomes negligible. However, this procedure cannot be exact, if only because the Newtonian viewpoint breaks down at $r \gtrsim H^{-1}$.

In the presence of the perturbations, it is useful to imagine that, in addition to the comoving observers, the Universe is populated by **uniform-expansion** observers with position $\mathbf{r}(t) = a(t)\mathbf{x}$ (fixed $\mathbf{x}$). This is illustrated in Figure 4.2. The **peculiar velocity** $\mathbf{v}(\mathbf{x})$ of a comoving observer is its velocity as measured by the uniform-expansion observer at the same point. The total velocity of the comoving observer is obtained by differentiating Eq. (4.66):

$$\mathbf{u}(\mathbf{x}, t) = H(t)\mathbf{r} + \mathbf{v}(\mathbf{x}, t).$$

(4.67)

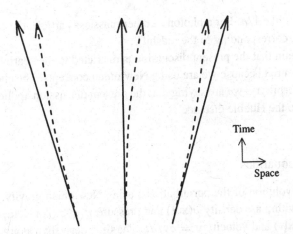

Fig. 4.2. The solid lines represent the trajectories of uniform-expansion observers. The time axis has been stretched to eliminate their gravitational deceleration, making them straight lines. The dashed lines represent the trajectories of comoving observers, which at some initial time coincide with the uniform-expansion observers. The comoving observers are being drawn toward an overdensity located between the central and rightmost observers.

The Newtonian reference frame is chosen so that the spatial average of $\mathbf{v}$ vanishes, which must be possible (at some useful level of accuracy) if we are to regard the Universe as homogeneous and isotropic on large scales. At the present epoch, this **cosmic rest frame** is defined by the matter as opposed to the radiation, because the matter dominates the momentum and therefore determines the motion of comoving observers. However, theory predicts that the radiation (both the neutrinos and the photons) will have practically the same rest frame. This agrees with observation in the case of the cmb.

The acceleration of the comoving observer is

$$\frac{d\mathbf{u}}{dt} = \ddot{a}\mathbf{x} + \mathbf{g}, \tag{4.68}$$

where the first term is the acceleration of the local uniform-expansion observer. The second term is the acceleration of the comoving observer relative to this observer, called the **peculiar acceleration**. From Eq. (4.67), it is given by

$$\mathbf{g} = \frac{d\mathbf{v}}{dt} + H\mathbf{v} = \frac{1}{a}\frac{d(a\mathbf{v})}{dt}. \tag{4.69}$$

If there is no peculiar acceleration, the peculiar velocity $\mathbf{v}$ decays like $1/a$. The decay reflects the fact that $\mathbf{v}(t)$ is the velocity measured by a succession of observers, which are moving in the same direction as the particle with ever-increasing velocity.

Although we focused on comoving observers, the peculiar velocity of any object can be defined in the same way. In particular, Earth has a peculiar velocity, the diurnal average of which is the peculiar velocity of the Sun (equal to $v = 371\ \text{km}\cdot\text{s}^{-1}$ as measured by the dipole of the cmb anisotropy). For a relativistic object, a similar analysis shows that the relativistic momentum **p** decays like $1/a$. For a photon or other massless particle, the energy $E = p$ has the same behaviour, corresponding to the redshift.

We emphasize again that the present discussion is restricted to the nearby Universe, where $Hr \ll 1$. This is essential because we are using Newtonian concepts. The choice of the origin $r = 0$ is, of course, arbitrary; we are saying that the present discussion applies in a region that is small compared to the Hubble distance.

4.4.3   Fluid-flow equations

Now we study the evolution of the perturbations, using Newtonian gravity. The Universe is modelled as a gas, with mass density $\rho(\mathbf{x}, t)$ and pressure $P \ll \rho$. An element of the gas has position $\mathbf{r}(t) = a(t)\mathbf{x}(t)$ and velocity $\mathbf{u} = d\mathbf{r}/dt$. The time derivative along the trajectory of the element is

$$\frac{d}{dt} = \frac{\partial}{\partial t} + \frac{dx_i}{dt}\frac{\partial}{\partial x_i}, \tag{4.70}$$

with summation over x_i understood.

The acceleration of the element is given by the **Euler equation**

$$\frac{d\mathbf{u}}{dt} = -\frac{1}{\rho}\nabla P - \nabla \Phi_{\text{gr}}. \tag{4.71}$$

Here $\Phi_{\text{gr}}(\mathbf{x}, t)$ is the gravitational potential, which satisfies the **Poisson equation**

$$\nabla^2 \Phi_{\text{gr}} = 4\pi G\rho. \tag{4.72}$$

(We use G rather than M_{Pl} here for familiarity; recall that, with our chosen units of $\hbar = c = 1$, $8\pi G \equiv 1/M_{\text{Pl}}^2$.) The gradient operator has components

$$\nabla_i = \frac{\partial}{\partial r^i} = a^{-1}\frac{\partial}{\partial x^i}. \tag{4.73}$$

We also need the mass conservation or **continuity** equation

$$\frac{d\rho(\mathbf{x}, t)}{dt} = -3H(\mathbf{x}, t)\rho(\mathbf{x}, t). \tag{4.74}$$

In this equation, $H(\mathbf{x}, t)$ is a locally defined Hubble parameter, which measures the rate of expansion of the Universe even though that expansion is inhomogeneous. It is defined by

$$H(\mathbf{x}, t) = \frac{1}{3}\nabla \cdot \mathbf{u}. \tag{4.75}$$

If $\mathcal{V}$ is the volume of an element of the gas at position $\mathbf{x}$ and time t, we can integrate this using the divergence theorem to find

$$3H\mathcal{V} = \int_{\mathcal{V}} \mathbf{u} \cdot d\mathbf{S}. \tag{4.76}$$

The right-hand side is just dV/dt, and so, we see that $H(\mathbf{x}, t)$ indeed measures the rate of expansion

$$\frac{1}{V}\frac{dV}{dt} = 3H(\mathbf{x}, t). \tag{4.77}$$

This shows that the continuity equation is equivalent to mass conservation.

In the unperturbed Universe, we choose the origin $\mathbf{x} = 0$ to move with the fluid. Then, Φ_{gr} is spherically symmetric, and from Eq. (4.72) it is equal to $2\pi r^2 \rho/3$. Also, $d\mathbf{u}/dt = \ddot{a}\mathbf{x} = (\ddot{a}/a)\mathbf{r}$. Using this in Eq. (4.71) and taking the divergence gives the Newtonian deceleration equation

$$\frac{\ddot{a}}{a} \equiv \dot{H} + H^2 = -\frac{4\pi G}{3}\rho. \tag{4.78}$$

As we note in Section 2.1, this plus the continuity equation Eq. (4.74) give the Friedmann equation

$$H^2 = \frac{8\pi G}{3}\rho - \frac{K}{a^2}. \tag{4.79}$$

In this chapter and the next, we focus on the case of critical density, $K = 0$, but the evolution equations that we develop hold for any K.

We introduce perturbations by writing

$$\rho(\mathbf{x}, t) = \rho(t) + \delta\rho(\mathbf{x}, t), \tag{4.80}$$

$$P(\mathbf{x}, t) = P(t) + \delta P(\mathbf{x}, t), \tag{4.81}$$

$$H(\mathbf{x}, t) = H(t) + \delta H(\mathbf{x}, t), \tag{4.82}$$

$$\Phi_{gr}(\mathbf{x}, t) = \frac{2\pi}{3}(ax)^2\rho(t) + \Phi(\mathbf{x}, t), \tag{4.83}$$

$$\mathbf{u} = H(t)\mathbf{r} + \mathbf{v}. \tag{4.84}$$

We have denoted the perturbation in Φ_{gr} as Φ because this is the usual notation for the corresponding relativistic quantity that we encounter later. We call Φ the **peculiar gravitational potential**.

The unperturbed quantities satisfy the Friedmann equation, and to first order the perturbations satisfy the linear equations.

$$\delta H = \frac{1}{3}\nabla \cdot \mathbf{v}, \tag{4.85}$$

$$\dot{\mathbf{v}} + H\mathbf{v} \equiv \mathbf{g} = -\nabla\Phi - \rho^{-1}\nabla\delta P, \tag{4.86}$$

$$\nabla^2\Phi = 4\pi G\delta\rho, \tag{4.87}$$

$$(\delta\rho)\dot{} = -3\rho\delta H - 3H\delta\rho, \tag{4.88}$$

where an overdot means $\partial/\partial t$ at fixed $\mathbf{x}$. The second equation is just Eq. (4.69), and the others require no comment. Because spatial gradients appear, it is often convenient to write each quantity as a Fourier series as in Eq. (4.1). Then, for a given $\mathbf{k}$, we can make the replacements

$$\nabla \to i\frac{\mathbf{k}}{a}; \qquad \nabla^2 \to -\left(\frac{k}{a}\right)^2. \tag{4.89}$$

Instead of $\delta\rho$, it is more useful to work with the **density contrast** $\delta \equiv \delta\rho/\rho$. Remembering that $\rho \propto 1/a^3$, we find that Eq. (4.88) becomes

$$\dot\delta = -3\,\delta H. \tag{4.90}$$

This, together with Eqs. (4.85), (4.86), and (4.87), provides the starting point for our calculations. For future reference, we note that, for critical density, Eq. (4.87) can be written using the Fourier expansion as

$$\delta_\mathbf{k} = -\frac{2}{3}\left(\frac{k}{aH}\right)^2 \Phi_\mathbf{k}. \tag{4.91}$$

More generally,

$$\delta_\mathbf{k} = -\frac{2}{3}\,\Omega\left(\frac{k}{aH}\right)^2 \Phi_\mathbf{k}. \tag{4.92}$$

4.4.4 Density perturbation

Let $(x_1, x_2, x_3) \equiv \mathbf{x}$ be the space coordinates, and write Eq. (4.85) as

$$\delta H = \frac{1}{3}a^{-1}\frac{\partial v_i}{\partial x_i}, \tag{4.93}$$

with a summation over i understood. Differentiating with respect to t gives

$$(\delta H)\dot{} = -H\delta H + \frac{1}{3}a^{-1}\frac{\partial}{\partial t}\frac{\partial v_i}{\partial x_i}. \tag{4.94}$$

On the other hand, taking the divergence of Eq. (4.86) gives, to first order,

$$a^{-1}\frac{\partial}{\partial x_i}\frac{\partial}{\partial t}v_i + Ha^{-1}\frac{\partial v_i}{\partial x_i} = -\nabla^2\Phi - \rho^{-1}\nabla^2\delta P. \tag{4.95}$$

Combining Eqs. (4.94) and (4.95), remembering Eq. (4.87), we find that

$$(\delta H)\dot{} = -2H\delta H - \frac{4\pi G}{3}\delta\rho - \frac{1}{3}\rho^{-1}\nabla^2\delta P. \tag{4.96}$$

Substituting Eq. (4.90) and its derivative into Eq. (4.96) gives

$$\ddot\delta_\mathbf{k} + 2H\dot\delta_\mathbf{k} - 4\pi G\rho\,\delta_\mathbf{k} + \left(\frac{k}{a}\right)^2\frac{\delta P_\mathbf{k}}{\rho} = 0. \tag{4.97}$$

Except on small scales before matter–radiation equality, the pressure is negligible, so that

$$\ddot\delta_\mathbf{k} + 2H\dot\delta_\mathbf{k} - 4\pi G\rho\,\delta_\mathbf{k} = 0. \tag{4.98}$$

This equation does not involve k. It has a growing solution, usually denoted by $D_1(t)$, and a decaying solution denoted by $D_2(t)$. The most general solution is $\delta_\mathbf{k} = f_{1\mathbf{k}}D_1 + f_{2\mathbf{k}}D_2$, with the f arbitrary. The decaying part can be ignored, at least when $\delta_\mathbf{k}$ is generated by the vacuum fluctuation.

At the moment, we are interested in the case of critical density. Accordingly, $H = 2/3t$ and $3H^2 = 8\pi G\rho$. Then, we can check that the growing solution of Eq. (4.98), giving the growth rate of the density contrast, is

$$D_1 \propto t^{2/3}. \tag{4.99}$$

Using Eq. (4.91) and $a \propto t^{2/3}$, we find that this corresponds to a *time-independent* peculiar gravitational potential $\Phi_\mathbf{k}$. The decaying solution is $D_2 \propto t^{-1}$.

We have derived Eqs. (4.97) and (4.98) assuming Newtonian gravity, which requires matter domination and no cosmological constant. However, it turns out that, as long as we restrict ourselves to scales well within the horizon, they also describe a perturbation in the matter density in the presence of a cosmological constant or a smooth background of, for example, radiation. This happens because, in all cases, general relativity practically reduces to Newtonian gravity near a free-falling observer whose motion is nonrelativistic with respect to the cosmic fluid. The only difference (Peebles 1980) is that the right-hand side of the Poisson equation becomes $4\pi G(\rho + 3P) - \Lambda$. This changes the unperturbed gravitational potential [so that it gives Eq. (4.78)], but not its perturbation, provided that the latter is due to the matter alone. As a result, Eqs. (4.97) and (4.98) remain valid if ρ denotes only the matter density, and δ the matter density contrast.

4.4.5 Constant curvature perturbation

There is a clever trick for finding the solution of Eq. (4.98) after matter domination (Peebles 1980). To arrive at it, consider the equation

$$\dot{H}(\mathbf{x}, t) + H^2(\mathbf{x}, t) = -\frac{4\pi G}{3}\rho(\mathbf{x}, t) + \frac{1}{3}\nabla \cdot \mathbf{a}, \tag{4.100}$$

where $\mathbf{a} = -\rho^{-1}\nabla\delta P$ is the nongravitational contribution to the peculiar gravitational acceleration. This equation, valid to first order in the perturbations, is equivalent to Eqs. (4.78) and (4.96). It is the Newtonian version of what in general relativity is called the (linearized) **Raychaudhuri equation** (Raychaudhuri 1955, 1979).

In the absence of a pressure perturbation, the Raychaudhuri equation has the same form as the unperturbed deceleration equation (4.78). The continuity equation (4.74) also has the same form as in the unperturbed case. This means that, in the absence of a pressure perturbation, each comoving region evolves like an unperturbed Universe. In particular, the deceleration and continuity equations imply a locally defined Friedmann equation

$$H^2(\mathbf{x}, t) = \frac{8\pi G}{3}\rho(\mathbf{x}, t) - \frac{K}{a^2(\mathbf{x}, t)} - \frac{\delta K}{a^2(t)}. \tag{4.101}$$

Here, δK is some constant that satisfies $\delta K \ll (aH)^2$, so that it represents a perturbation. It causes a perturbation $\delta\Omega \ll 1$ in the locally measured density parameter.

In contrast with the other equations, this one contains the locally defined scale factor (such that $\dot{a}/a$ is the locally defined Hubble parameter). As indicated, the unperturbed quantity $a(t)$ still can be used when it multiplies the perturbation δK.

The continuity equation implies $\rho \propto a^{-3}$ for the locally defined quantities, leading to $\delta = 3\delta a/a$. On the other hand, at each point in space, a is a solution of the Friedmann equation, and we know that the most general solution of this equation during matter domination is given by Eq. (2.19). It depends on K, and on the constant C specifying the origin of time.

The conclusion is that Eq. (4.98) has two independent solutions, given by

$$D_1 = a\frac{\partial a(C, K, t)}{\partial K} = a\frac{da}{dt}\frac{\partial t(C, K, a)}{\partial K}, \tag{4.102}$$

$$D_2 = a\frac{\partial a(C, K, t)}{\partial C} = a\frac{da}{dt}\frac{\partial t(C, K, a)}{\partial C}. \tag{4.103}$$

Using $da/dt = aH$ and Eq. (2.19), these solutions are

$$D_1 = H^{-1}\int_0^a \frac{da}{(aH)^3}, \tag{4.104}$$

$$D_2 = H^{-1}. \tag{4.105}$$

The first is the growing solution of Eq. (4.98), and the second is the decaying one. They give the evolution of the density perturbation in the presence of a cosmological constant and/or noncritical density.

This trick does not work for a matter density perturbation living in a smooth background of, for example, radiation, because the matter and radiation are not flowing together. This means that the locally defined Hubble parameter appearing in the Raychaudhuri equation, identical with the one appearing in the continuity equation for the *total* energy density, is different from the one appearing in the continuity equation for the matter density. The trick works only for the special case that the smooth background is a cosmological constant, equivalent to constants $\rho_{vac} = -P_{vac}$, which do not contribute to either side of the continuity equation.

4.4.6   Velocity gradient and vorticity

The **velocity gradient** of the cosmic fluid is defined by

$$u_{ij}(\mathbf{r}, t) \equiv \frac{\partial u_i}{\partial r_j}, \tag{4.106}$$

where, as always, $\mathbf{r} = a(t)\mathbf{x}$ is the physical coordinate system. As for any fluid, it is of the form

$$u_{ij} = H\delta_{ij} + \omega_{ij} + \sigma_{ij}, \tag{4.107}$$

where the **vorticity** ω_{ij} is antisymmetric, the **shear** σ_{ij} is traceless and symmetric, and H is the locally defined Hubble parameter that we already discussed. The inverse relations are

$$2\omega_{ij} = \frac{\partial u_i}{\partial r_j} - \frac{\partial u_j}{\partial r_i}, \tag{4.108}$$

$$2\sigma_{ij} = \frac{\partial u_i}{\partial r_j} + \frac{\partial u_j}{\partial r_i} - 2\delta_{ij}\delta H, \tag{4.109}$$

$$3H = \nabla \cdot \mathbf{u}. \tag{4.110}$$

The vorticity corresponds to the quantity $\nabla \times \mathbf{u}$. It measures the speed of rotation of a fluid element, whereas the shear measures the anisotropy in its expansion rate. (The rate averaged over all directions is measured by H.)

In the unperturbed Universe, $\mathbf{u} = H\mathbf{r}$, leading to $u_{ij} = H(t)\delta_{ij}$ with zero vorticity and shear. Including the perturbation, $\mathbf{u}$ is given by Eq. (4.67). Then, reverting to the comoving coordinate system $\mathbf{x}$, the velocity gradient is

$$u_{ij} = H(t)\delta_{ij} + \frac{1}{a}\frac{\partial v_i}{\partial x_j}, \tag{4.111}$$

where $\mathbf{v}$ is the peculiar velocity. The perturbation $\delta u_{ij} = a^{-1}(\partial v_i/\partial x_j)$ is given by

$$\delta u_{ij} = \delta H(\mathbf{x}, t)\,\delta_{ij} + \omega_{ij}(\mathbf{x}, t) + \sigma_{ij}(\mathbf{x}, t), \tag{4.112}$$

and in terms of the peculiar velocity

$$\omega_{ij} = \frac{1}{2a}\left[\frac{\partial v_i}{\partial x_j} - \frac{\partial v_j}{\partial x_i}\right], \tag{4.113}$$

$$\sigma_{ij} = \frac{1}{2a}\left[\frac{\partial v_i}{\partial x_j} + \frac{\partial v_j}{\partial x_i}\right] - \delta_{ij}\delta H, \tag{4.114}$$

$$3\delta H = \frac{1}{a}\sum_i \frac{\partial v_i}{\partial x_i}. \tag{4.115}$$

Like any vector field that has a Fourier expansion, $\mathbf{v_k}$ can be uniquely decomposed into a scalar part and a vector part:

$$\mathbf{v_k} = \mathbf{v_k^{sc}} + \mathbf{v_k^{vec}}, \tag{4.116}$$

where the scalar part $\mathbf{v_k^{sc}}$ is parallel to $\mathbf{k}$ whereas the vector part $\mathbf{v_k^{vec}}$ is perpendicular to it. The scalar part is so called because it can be written as the gradient of a scalar (the velocity potential) whereas the vector part cannot. It is convenient, instead of the velocity potential itself, to consider a quantity V such that

$$\mathbf{v_k^{sc}} = -\frac{i\mathbf{k}}{k}V_\mathbf{k}. \tag{4.117}$$

Then the magnitude of $\mathbf{v_k}$ is just $|V_\mathbf{k}|$. Note that

$$3\delta H_\mathbf{k} = \left(\frac{k}{a}\right)V_\mathbf{k}. \tag{4.118}$$

The terms "scalar" and "vector" are the usual ones in relativistic cosmological perturbation theory, and are the ones we adopt here. More generally, the scalar and vector parts often are called "longitudinal" and "transverse" (to the $\mathbf{k}$ direction).

The vorticity comes entirely from the vector part of $\mathbf{v}$ and, conversely, is determined by it (assuming that everything can be expanded in Fourier series). This leads to yet another terminology where the scalar and vector parts are called "irrotational" and "rotational." The vacuum fluctuation does not generate vorticity, and we assume that it is absent, so that the peculiar velocity is $\mathbf{v} = \mathbf{v^{sc}}$. We occasionally call V the peculiar velocity as well.

Equations (4.90) and (4.85) give $\nabla \cdot \mathbf{v} = -\dot{\delta}$, or

$$\frac{k}{a} V_{\mathbf{k}} = -\dot{\delta}_{\mathbf{k}}. \tag{4.119}$$

With critical density $a \propto t^{2/3}$, and with negligible pressure gradient, Φ is time independent. Then, from Eq. (4.86),

$$\mathbf{v} = -t \nabla \Phi \tag{4.120}$$

or

$$V_{\mathbf{k}} = \frac{2}{3} \frac{k}{aH} \Phi_{\mathbf{k}} = -\frac{aH}{k} \delta_{\mathbf{k}}. \tag{4.121}$$

4.5   The baryon density contrast: Newtonian treatment

In the preceding section, we studied the total density perturbation, making no distinction between CDM and baryons. We also ignored the pressure in the end, which is not correct for baryons on small scales. In this section, we treat the baryons and the CDM separately. We estimate the **Jeans scale**, above which the pressure is negligible but below which it dominates the effect of gravity. Above the Jeans scale, the baryon density contrast grows to match that of the CDM, whereas below this scale it oscillates as a standing wave; by analogy with the case of sound waves, this is called an **acoustic oscillation**.

4.5.1   Multifluid evolution equations

Quite generally, if the Universe consists of several fluids, interacting only through gravity, the Newtonian evolution equations for the ith fluid are

$$\ddot{\delta}_{\mathbf{k}i} + 2H\dot{\delta}_{\mathbf{k}i} + \left(\frac{k}{a}\right)^2 \frac{\delta P_{\mathbf{k}i}}{\rho_i} = 4\pi G\rho\, \delta_{\mathbf{k}}. \tag{4.122}$$

Here, $\delta_i \equiv \delta\rho_i/\rho_i$, and the derivation of these equations mimics that of the almost identical Eq. (4.97). Each species feels its *own* pressure gradient, but the *total* gravitational acceleration. It is assumed that the stress of each fluid is isotropic, so that it is defined by the pressure.

For the baryons and CDM, we have

$$\ddot{\delta}_{\mathbf{k}}^{\mathrm{b}} + 2H\dot{\delta}_{\mathbf{k}}^{\mathrm{b}} + c_{\mathrm{s}}^2(t) \left(\frac{k}{a}\right)^2 \delta_{\mathbf{k}}^{\mathrm{b}} = 4\pi G\rho\, \delta_{\mathbf{k}}, \tag{4.123}$$

$$\ddot{\delta}_{\mathbf{k}}^{\mathrm{c}} + 2H\dot{\delta}_{\mathbf{k}}^{\mathrm{c}} = 4\pi G\rho\, \delta_{\mathbf{k}}. \tag{4.124}$$

In the first equation, to invoke the adiabatic relation $\delta P_{\mathrm{b}}/\delta\rho_{\mathrm{b}} = dP_{\mathrm{b}}/d\rho_{\mathrm{b}} = c_{\mathrm{s}}^2$, we use the fact that heat flow is negligible in the baryon fluid.

Let us estimate the speed of sound c_{s}. We are dealing with a monatomic gas because there are only helium and hydrogen, with the latter monatomic at the relevant epochs; so,

$$c_{\mathrm{s}}^2 = \frac{5T_{\mathrm{b}}}{3m}, \tag{4.125}$$

with atomic mass $m \simeq 1$ GeV because hydrogen atoms dominate. Until $z \sim 100$, residual free electrons keep the baryon temperature T_b close to the photon temperature T_γ. Afterward, we have

$$\frac{T_b}{T_\gamma} \propto \frac{1}{1+z}. \tag{4.126}$$

4.5.2   Growth of the baryon density contrast

At decoupling, $\delta_c \gg \delta_b$, because δ_c has been growing since horizon entry, whereas δ_b has been either oscillating or decaying. We now show that, after decoupling, δ_b grows to match δ_c, provided that the baryon pressure is negligible.

Subtracting the baryon equation from the CDM equation, we find

$$\ddot{S}_{cb} + 2H\dot{S}_{cb} = 0, \tag{4.127}$$

where $S_{cb} \equiv \delta_c - \delta_b$. (This is the difference between the CDM and baryon entropy perturbations, as defined in Section 6.6.) The solution of this equation is $S_{cb} = A + Bt^{-1/3}$, and the solution of the equation for δ_c is $Ct^{2/3} + Dt^{-1}$. To determine the precise initial conditions, we would need to solve the equations before the baryon pressure becomes negligible. However, we initially will have $\delta_c \gg \delta_b$ because the growth of δ_c is not suppressed by the pressure. Also, there will be no strong cancellation between the growing and decaying modes because both $\dot{\delta}_b/\delta_b$ and $\dot{\delta}_c/\delta_c$ initially will be of order H. (This is the only scale in the problem, because the pressure term starts to become insignificant precisely when it is of the same order as the gravitational term.) As a result, we will have $A \sim Ct_{\text{initial}}^{2/3}$, with the decaying terms becoming negligible after, at most, a few Hubble times.

It follows that, after a few Hubble times, $|S_{cb}| \ll |\delta|$. Then $\delta = \delta_b = \delta_c$ to high accuracy because $\delta = f_b\delta_b + f_c\delta_c$, where $f_b = \rho_b/\rho$ is the baryon mass fraction, and similarly for f_c.

4.5.3   Jeans mass

The above calculation assumes that the last term on the left-hand side of Eq. (4.123) is negligible compared with the right-hand side, right up to the point where $\delta_b \simeq \delta_c$. The Jeans wavenumber k_J is defined as the value of k for which these terms are equal, and the corresponding wavelength λ_J is called the **Jeans length**,

$$\lambda_J = 2\pi c_s (4\pi G\rho)^{-1/2}. \tag{4.128}$$

The **Jeans mass** M_J is defined as the mass of matter in a sphere of radius $\lambda_J/2$,

$$M_J = \frac{\pi^{5/2}}{6} \frac{c_s^3}{G^{3/2}\rho^{1/2}}. \tag{4.129}$$

Using Eq. (4.125), we obtain

$$M_J \sim 10^6 h^{-1} \left(\frac{T_b}{T_\gamma}\right)^{3/2} M_\odot. \tag{4.130}$$

Until $z \sim 100$, $T_b = T_\gamma$ and $M_J \sim 10^6 M_\odot$. We see that, on the relevant scales, the density contrast becomes of order 1 at $z \sim 10$. By that time, Eq. (4.126) gives $M_J \sim 10^5 M_\odot$, which provides an estimate of the mass of the lightest baryonic objects that can form through gravitational collapse.

There is a cruder way of estimating the Jeans mass, which has wider applicability. Consider a spherically symmetric density enhancement with radius R (defined, for example, as the distance at which the density contrast has fallen by one-half). At a distance r from the centre, the pressure gives an outward acceleration $-[\partial(\delta P)/\partial r]/\rho$, and the inward peculiar gravitational acceleration is $G \, \delta M / r^2$, where δM is the excess mass within radius r. For an estimate, we can set $r \sim R$, $\delta M \sim \delta \rho R^3$, and $\partial(\delta P)/\partial r \sim \delta P / R$. An estimate of the Jeans length is provided by the value of R for which pressure and gravity balance:

$$\lambda_J \sim \sqrt{\frac{1}{G\rho} \frac{\delta P}{\delta \rho}}. \tag{4.131}$$

The pressure perturbation is given by $\delta P = \delta \rho \overline{v^2}/3$, where v is the random motion of the particles. This gives an alternative expression

$$\lambda_J \sim G^{-1/2} \rho^{-1/2} (\overline{v^2})^{1/2}. \tag{4.132}$$

This formula applies even if diffusion or freestreaming causes the stress to be anisotropic, because the components of the stress are still of order ρv^2, and the criterion is still that the acceleration due to the stress should balance that due to the gravity. (In the case of freestreaming, one also can regard the formula as giving the velocity required to escape from the overdense region.) We use it later when considering massive neutrinos.

4.5.4   Acoustic oscillations of the baryon fluid

On scales well below the Jeans scale, the pressure gradient dominates the effect of gravity. As a result, Eq. (4.97) applies to the baryon component of the gas with no gravity term

$$\ddot{\delta}_{\mathbf{k}}^{b} + 2H\dot{\delta}_{\mathbf{k}}^{b} + c_s^2 \left(\frac{k}{a}\right)^2 \delta_{\mathbf{k}}^{b} = 0. \tag{4.133}$$

Remember that, ultimately, everything is derived via linear equations from the initial curvature perturbation, so that $\delta_{\mathbf{k}}^{b}$ will be of the form

$$\delta_{\mathbf{k}}^{b} = T(k, t)\mathcal{R}_{\mathbf{k}}, \tag{4.134}$$

where T is a *real* linear transfer function. (The transfer function is always real because the differential equations leading to it do not involve i.) The time dependence of T is given by Eq. (4.133) and, typically, it will oscillate, with slowly varying frequency and amplitude. Pairs of modes with opposite $\mathbf{k}$ combine to give a standing wave,

$$\delta_{\mathbf{k}}^{b}(t)e^{i\mathbf{k}\cdot\mathbf{x}} + \delta_{-\mathbf{k}}^{b}(t)e^{-i\mathbf{k}\cdot\mathbf{x}} = A_{\mathbf{k}}\cos(\omega_k t + \alpha_k)\cos(\mathbf{k}\cdot\mathbf{x} + \beta_{\mathbf{k}}). \tag{4.135}$$

The spatial phase $\beta_\mathbf{k}$ is random, but the temporal phase α_k of the transfer function can be calculated.

This **acoustic oscillation** actually starts as soon as the scale enters the horizon, which for such short scales occurs during radiation domination. The previously discussed Newtonian treatment, on the other hand, is valid only well after matter domination, and well after the scale enters the horizon. To describe the oscillation from the beginning, so as to calculate the phase α_k, one needs the general-relativistic equations of Chapter 15.

4.6 Cosmological perturbation theory

In the preceding two sections, we used Newtonian physics to follow the evolution of the matter density perturbation. The result is valid on scales well within the horizon, after matter domination. Now we extend the treatment to bigger scales and earlier times, where general relativity is the appropriate framework.

We adopt what usually is called the covariant treatment, which is a direct extension of the Newtonian treatment. The basic philosophy is to start with evolution equations for locally defined quantities along each comoving worldline, which closely resemble the Newtonian ones. Only afterward do we compare different worldlines.

The evolution equations along a given worldline are derived from a starting point consisting of four equations. These are a relation between proper time and coordinate time, together with general-relativistic versions of the Euler, continuity, and Raychaudhuri equations. At this stage, we write down the four equations without proof, so dispensing with most of the complicated formalism of general relativity. A full treatment of cosmological perturbation theory, employing the more usual metric perturbation approach, is given in Chapter 14, where we also derive the starting point for the present approach.

4.6.1 Geodesics, coordinates, and locally inertial frames

Although we have no need of the general relativity formalism, we do need some of its basic ideas.

General relativity is required when we consider an extended region of space-time, such that gravity cannot be ignored. Special relativity is the local description, valid in a region of space-time small enough that gravity can be ignored.

As illustrated in Figure 4.3, there is a lightcone through each point in space-time, with timelike lines going from the past cone to the future cone and spacelike lines lying outside the cones. A timelike line represents the worldline of a possible observer. It is said to be orthogonal to a spacelike line intersecting it, if events on the spacelike line are simultaneous as seen by the observer. For an intersecting pair of spacelike lines, orthogonality has its usual geometric meaning. (One way of defining simultaneity is to bounce a light signal between nearby observers, as in Figure 4.4. The analogous construction for spacelike lines is to bisect a triangle with two equal angles.) A spacelike line is said to be a geodesic if it corresponds to the shortest distance between nearby points. A timelike line is said to be a geodesic if it corresponds to the longest proper time, and it corresponds to an object in free fall.

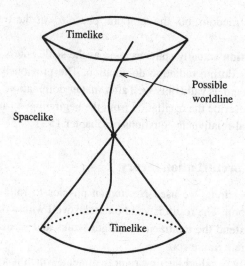

Fig. 4.3. The worldline of a possible observer lies within the lightcone, which divides timelike separations from spacelike separations.

In special relativity, there are globally preferred coordinate systems, the inertial frames, in which the laws of physics become particularly simple. In general relativity the laws of physics, which include the description of gravity, do not pick out any globally preferred coordinate systems.[8] One therefore should consider a generic coordinate system, defined simply as a smooth labeling of space-time points. There is a time coordinate t and three space coordinates (r_1, r_2, r_3). The time-coordinate lines (lines along which only t varies) are timelike, and the space-coordinate lines (lines along which only one of the r_i varies) are spacelike.

Near a given point in space-time, and more generally along the path of a given observer, things become simpler if one uses a **locally inertial** coordinate system, defined as one whose coordinate lines near the observer are geodesics. Such a system usually is called a locally inertial frame.

In special relativity, we makes the idealization that the coordinate lines can be chosen to be geodesics everywhere, thus defining a globally inertial coordinate system. According to general relativity, such a choice is impossible. This is analogous to the statement that Cartesian coordinates generally fail to exist on a curved surface, and so we say that, according to general relativity, space-time is curved. It can be regarded as flat in the vicinity of any given space-time point, and special relativity makes the idealization that it is flat everywhere.

Gravitational effects vanish in a locally inertial frame, as we approach the observer. This is the **equivalence principle**, which is a consequence of general relativity.

---

[8] The description of a given system might become simpler in particular coordinate systems. In particular, this happens for a system with symmetry, if the coordinates share the symmetry. The usual coordinate system $(\mathbf{x}, t)$ has this property for the unperturbed Universe, and we see later that there are only a few simple ways of generalizing it to include perturbations. We focus for the moment on the fundamental laws of physics.

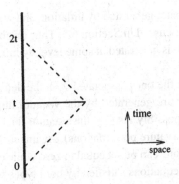

Fig. 4.4. The vertical line represents an observer equipped with a clock, and a device for bouncing a photon off a nearby object. This observer defines the half-time event as simultaneous with the bounce.

4.6.2 Basic features of cosmological perturbation theory

The aim is to derive linear equations, which determine the evolution of small perturbations away from homogeneity and isotropy. As in the case of quantum mechanical perturbation theory, each perturbation is multiplied by a common parameter. Then, in the exact evolution equations, we consider the power-series expansion in this parameter and drop all terms beyond the linear term. For clarity, we do not exhibit the parameter, but simply remember that it is supposed to multiply all perturbations. (In our notation, a perturbation usually, though not always, has a δ in front of it.)

The equations are coupled linear partial differential equations in the perturbations, the derivatives being with respect to space and time. Because the Universe is expanding, the coefficients contain explicit functions of time t, but because the Universe is homogeneous (to zero order in the perturbations), they do *not* contain explicit functions of $\mathbf{x}$. That is why the Fourier expansion is so useful. For each $\mathbf{k}$, the Fourier coefficients of the perturbations satisfy a set of coupled ordinary differential equations, containing time derivatives but no space derivatives. There is no coupling between modes with different $\mathbf{k}$.

The Fourier expansion is performed within a comoving box, which should be a few orders of magnitude bigger than any scale of interest so that its boundary has no effect on the physics. From a physical viewpoint, it should not be exponentially bigger, still less infinite, as we saw on page 74. On the other hand, when doing the mathematics, we usually pretend that the infinite limit exists.

The possible perturbations that might exist in the Universe do not all couple to each other. Rather, they are of three types, usually called **scalar**, **vector**, and **tensor**, with the perturbations of each type satisfying a separate set of linear evolution equations. This means that instead of the single parameter of the first paragraph, we actually can introduce a separate parameter for each of the three types. When considering a given type, we can assume that the other two types are absent, whether or not this is true in Nature.

The vector perturbations are not generated by inflation, and we assume that they are absent. The tensor perturbations are discussed in Section 6.5. They originate from a primordial gravitational wave amplitude, which is generated at some level by inflation, though in most models it is too weak to observe.

The scalar perturbations are the ones associated with the density perturbation, responsible for large-scale structure. They are generated by the vacuum fluctuation of the inflaton field (adiabatic perturbations) and possibly also by the vacuum fluctuation of a noninflaton field such as that of the axion (isocurvature perturbations). Again, each type satisfies a separate set of evolution equations, so that one can be set equal to zero while considering the other. For the moment though, we deal with equations satisfied by both types so that no separation need be made.

4.6.3   Gauges

The Newtonian view of space-time is adequate only for scales well inside the horizon. On larger scales, one has to worry about the choice of coordinates, and in the presence of perturbations there is no uniquely preferred choice.

Let us continue to denote the coordinates by the same symbols that we used for the Newtonian case, namely t and $\mathbf{x} \equiv (x_1, x_2, x_3)$. A choice of coordinates defines a **threading** of space-time into lines (corresponding to fixed $\mathbf{x}$) and a **slicing** into hypersurfaces (corresponding to fixed t). The lines of fixed $\mathbf{x}$ are chosen to be timelike, so that they are the worldlines of possible observers. Also, the slices are chosen to be spacelike.

We can imagine that the Universe is populated by observers corresponding to the threading. On each of the slices one can ask whether geometry is Euclidean. If it is, the coordinates $\mathbf{x}$ can be chosen to be Cartesian, and the space defined by the slices is said to be flat. Otherwise it is said to have (intrinsic) curvature.

Consider first the limiting case of the unperturbed Universe. In that case there are indeed preferred coordinates, distinguished on several grounds. The threading corresponds to the motion of comoving observers, defined as those who see zero momentum density at their own position. The comoving observers are free-falling, and the expansion defined by them is isotropic. The slicing is orthogonal to the threading, and on each slice, the Universe is homogeneous. The time coordinate is chosen to correspond to proper time along each worldline. In the critical-density case the slices are flat, and the space coordinates can be chosen to be Cartesian. This formidable array of properties makes the preferred coordinates so valuable that no others are ever considered, except that one sometimes uses spherical polar coordinates instead of Cartesian coordinates.

In the presence of perturbations, it is impossible to find coordinates satisfying all of these properties, and there is no uniquely preferred choice. The only universally accepted constraint is that the coordinates must reduce to the standard ones in the limit where the perturbations vanish. A choice of coordinates satisfying this constraint is called a **gauge**, and there is no *unique* preferred gauge, though there are a few simple choices. In particular, we can find gauges that satisfy some but not all of the conditions listed in the preceding paragraph. Well

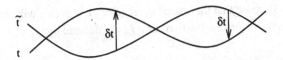

Fig. 4.5. The lines indicate slices defined by two different coordinate systems, whose time coordinates t and $\tilde{t}$ have the same numerical value. The time displacement between such slices is supposed to be of first order in the perturbations, and then its value $\delta t(\mathbf{x})$ is the same in every coordinate system. Without essential loss of generality, we can assume that its spatial average vanishes, so that the slices coincide on the average.

after horizon entry, all gauges ever suggested become the same, corresponding to the Newtonian description of space-time.

As will become clear, the only function of a gauge choice in cosmological perturbation theory is to define a slicing and threading of space-time. The slicing and threading in turn define the perturbations. For a perturbation $g(\mathbf{x}, t)$, the space-time coordinates can be regarded as those of unperturbed space-time because to include their perturbation would be a second-order effect. Once defined, the perturbations live in unperturbed space-time!

We can consider the gauge choices for scalar, vector, and tensor perturbations separately. It turns out that the evolution equations for the tensor perturbations are gauge independent, and we are taking the vector perturbations to vanish, which means that we actually are concerned only with the gauge choice for scalar perturbations.

It is perfectly valid to write down an equation that relates quantities defined in different gauges. Indeed, several of our equations are of that type. For the reason explained in Section 14.6.5, a formalism using such equations usually is said to be gauge invariant.

4.6.4 Changing the slicing

A slicing is needed to specify perturbations in quantities such as the energy density ρ, which have nonzero (and time-varying) values in the unperturbed Universe.

Take for definiteness the energy density. Given a slicing, we write

$$\rho(\mathbf{x}, t) = \rho(t) + \delta\rho(\mathbf{x}, t). \tag{4.136}$$

The first term is the unperturbed part, and it causes no confusion to use the same symbol for it and the full quantity. As we have said, once $\delta\rho$ has been defined the coordinates $\mathbf{x}, t$ can be identified with the standard ones in unperturbed space-time because the inclusion of *their* perturbations would be a second-order effect. In other words, *the perturbations live in unperturbed space-time.*

Now consider a different slicing, as illustrated in Figure 4.5. There will be a new time coordinate

$$\tilde{t}(\mathbf{x}, t) = t + \delta t(\mathbf{x}, t). \tag{4.137}$$

The quantity δt is taken to be of first order in the perturbations and, as is the case for any perturbation, we can suppose that the spatial average of δt vanishes without any essential loss of generality. The perturbations on the two slicings are defined by

$$\rho(\mathbf{x}, t) = \rho(t) + \delta\rho(\mathbf{x}, t), \tag{4.138}$$
$$\tilde{\rho}(\mathbf{x}, \tilde{t}) = \tilde{\rho}(\tilde{t}) + \widetilde{\delta\rho}(\mathbf{x}, \tilde{t}). \tag{4.139}$$

If t and $\tilde{t}$ have the same numerical values, the first terms on the right-hand side are the same because they correspond to the unperturbed quantity.

To first order,

$$\tilde{t}[\mathbf{x}, t - \delta t(\mathbf{x}, t)] = \tilde{t}(\mathbf{x}, t) - \frac{d\tilde{t}}{dt}\delta t(\mathbf{x}, t)$$
$$= \tilde{t}(\mathbf{x}, t) - \delta t(\mathbf{x}, t). \tag{4.140}$$

This shows that, to first order, $\delta t(\mathbf{x}, t)$ is the time displacement in going from a slice of fixed $\tilde{t}$ to a slice of fixed t with the same numerical value (Figure 4.5). To the same order, we can regard δt as referring to the time coordinate of the unperturbed Universe. Subtracting Eq. (4.138) from Eq. (4.139) gives, to first order,

$$\widetilde{\delta\rho}(\mathbf{x}, t) = \delta\rho(\mathbf{x}, t) - \dot{\rho}(t)\delta t(\mathbf{x}, t). \tag{4.141}$$

Again, t can be taken to be the time coordinate of the unperturbed Universe.

In this analysis we have not specified the motion of the observer who, at a given space-time point, is supposed to measure $\rho(\mathbf{x}, t)$. In the unperturbed Universe, ρ is to be measured by comoving observers. Including the perturbations, ρ can be measured by the observers corresponding to the threading of any gauge choice. We do not need to specify the threading because observers with relative velocity v measure ρ that differ only by an amount of order v^2 (Section 14.2).

We can summarize this situation by saying that, in the context of cosmological perturbation theory, ρ is a Lorentz scalar (a quantity invariant under both rotations and changes in the observer's velocity). The analysis leading to Eq. (4.141) holds for any quantity that is a scalar in this sense, including the pressure (defined, for example, as the trace of the stress tensor), as well as scalar fields that are genuine Lorentz scalars.

For a scalar, the only effect of a change in threading is to change the values of the space coordinates of each point. However, the spatial gradient of the scalar is already of first order in the perturbation, and the shift is a second-order effect that is to be ignored. The *perturbation in a scalar quantity is independent of the threading*.

Because the effect of a change in slicing depends on the time derivative of the unperturbed quantity, it too vanishes if the unperturbed quantity is time independent. In particular, it vanishes if the unperturbed quantity vanishes. This happens in some cases but it is atypical.

An important slicing is the **comoving slicing**, orthogonal to the worldlines of comoving observers. In this book, the density and pressure perturbations, $\delta\rho$ and δP, always are defined on the comoving slicing, unless otherwise stated, and so is the perturbation δH in the locally defined Hubble parameter. The curvature perturbation $\mathcal{R}$ is specified, in terms of $\delta\rho$ and δP on comoving slices, by Eqs. (4.165) and (4.167). Equivalently, it is specified in terms of the spatial curvature perturbation on comoving slices by Eq. (14.132).

The perturbation in the inflaton field vanishes on comoving slices, as we see on page 179. To define a nonzero perturbation, the best choice is the **spatially flat slicing** and, in this book, $\delta\phi$ always denotes the perturbation on that slicing.

Equation (4.6) gives $\mathcal{R}_\mathbf{k}$ in terms of $\delta\phi_\mathbf{k}(t_*)$, which means that it relates quantities defined on different slicings. Let us use Eq. (4.141), with ρ replaced by ϕ. Remembering that the inflaton-field perturbation vanishes on comoving slices, we learn that on flat slices

$$\delta\phi_\mathbf{k} = -\dot{\phi}\,\delta t, \tag{4.142}$$

where δt is the time displacement going from flat to comoving slices. As a result, Eq. (4.6) is equivalent to

$$H\delta t = \mathcal{R}, \tag{4.143}$$

where, as always, H defines the rate of expansion of the comoving worldlines. We need not include the perturbation in H because it would be a second-order effect. This remarkable relation is a statement about space-time geometry, which is valid at all epochs. It is derived on page 341.

4.7 Evolution equations

4.7.1 No anisotropic stress

We now derive equations for the evolution of the energy density perturbation. For the moment we continue to assume that stress is isotropic, so that it is described solely by the pressure. This is not actually valid before matter domination, owing to particle diffusion and freestreaming, but we see in Section 4.8.4 that anisotropic stress is unimportant for the present purpose.

Our starting point consists of the general-relativistic versions of the continuity, Euler, and Raychaudhuri equations. They are

$$\frac{d\rho}{dt_{\mathrm{pr}}} = -3H(\rho + P) \qquad \text{(continuity)}, \tag{4.144}$$

$$a = -\frac{\nabla P}{\rho + P} \qquad \text{(Euler)}, \tag{4.145}$$

$$\dot{H}(\mathbf{x}, t) + H^2(\mathbf{x}, t) = -\frac{4\pi G}{3}\rho(\mathbf{x}, t) + \frac{1}{3}\nabla \cdot \mathbf{a}. \tag{4.146}$$

The equations are valid at each point in space-time, and refer to quantities measured by a free-falling observer, equipped with a locally inertial frame, who is instantaneously at rest with respect to the fluid. The observer's time increment is dt_{pr}, which is also the proper time increment along a comoving worldline. (Proper time is defined formally on page 318.) At the observer's position the fluid velocity $\mathbf{u}$ vanishes, and its acceleration is $\mathbf{a}$. Nearby, $\mathbf{u}$ is nonzero and $3H = \nabla \cdot \mathbf{u}$ is the locally defined Hubble parameter.

The continuity equation is the same as in the unperturbed case, corresponding to the energy conservation condition $dE = -P\,dV$. The Raychaudhuri equation is the same as in the Newtonian case, and so is the Euler equation, except that the inertial mass (force divided by acceleration) is $\rho + P$ instead of ρ.

In the above equations, ρ, P, and H are locally defined quantities. Given a slicing, they can be split into an unperturbed part and a perturbation:

$$\rho(\mathbf{x}, t) = \rho(t) + \delta\rho(\mathbf{x}, t), \tag{4.147}$$

$$P(\mathbf{x}, t) = P(t) + \delta P(\mathbf{x}, t), \tag{4.148}$$

$$H(\mathbf{x}, t) = H(t) + \delta H(\mathbf{x}, t). \tag{4.149}$$

We use the comoving slicing so that, in the Euler equation, ∇ acts only on the pressure perturbation δP. Then, to first order,

$$a_i(\mathbf{x}, t) = -\frac{1}{a}\frac{\partial_i \delta P(\mathbf{x}, t)}{\rho(t) + P(t)} \quad \text{(Euler)}, \tag{4.150}$$

where $\partial_i \equiv \partial/\partial x_i$ is the derivative with respect to the comoving coordinate x_i. (Note that the scale factor a converting x_i into physical distance has nothing to do with the components a_i of the acceleration!) Here, as in any equation involving the perturbations, the coordinates t and x_i can be identified with the ones used to describe the unperturbed Universe.

Inserting Eq. (4.150) into the Raychaudhuri equation (4.146) gives

$$\frac{dH}{dt_{\mathrm{pr}}} + H^2 = -\frac{4\pi G}{3}(\rho + 3P) - \frac{1}{3}\frac{\nabla^2\delta P}{\rho + P}, \tag{4.151}$$

where $\nabla^2 = a^{-2}\sum \partial/\partial x_i^2$. We call this the Raychaudhuri equation as well.

We obtain the desired evolution equations by extracting the perturbations from this equation, and from the continuity equation. However, here, we encounter an important difference from the Newtonian case, which is that the coordinate time t labelling the slices generally cannot be identified with the proper time t_{pr} along a comoving worldline. Rather, as we show in Chapter 14, the two are related by

$$\nabla\left(\frac{dt}{dt_{\mathrm{pr}}}\right) = -\mathbf{a}\left(\frac{dt}{dt_{\mathrm{pr}}}\right). \tag{4.152}$$

This relation is illustrated in Figure 4.6. Using the Euler equation for $\mathbf{a}$ and taking the divergence gives, to first order,

$$\nabla^2\left(\frac{dt}{dt_{\mathrm{pr}}}\right) = \nabla^2\left(\frac{\delta P}{\rho + P}\right). \tag{4.153}$$

However, to zero order, $dt/dt_{\mathrm{pr}} = 1$, and ∇^2 acts only on the perturbation $[(dt/dt_{\mathrm{pr}}) - 1]$. Also, we are expanding everything in Fourier series, which means that $\nabla^2 f = 0$ implies $\delta f = 0$ for any f. It follows that, to first order,

$$\frac{dt}{dt_{\mathrm{pr}}} = 1 + \frac{\delta P}{\rho + P}, \tag{4.154}$$

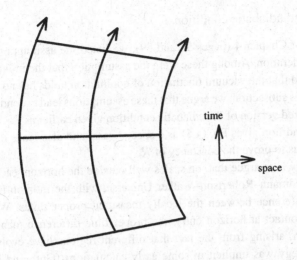

Fig. 4.6. This space-time diagram shows the threads and slices of an orthogonal coordinate system. Coordinate time t labels the slices and, because the threads are accelerating, its rate of variation with respect to proper time t_{pr} along a given worldline becomes slower in the direction of the acceleration [Eq. (4.152)]. In considering the sign of this effect, remember that in a space-time diagram a *longer* timelike interval is represented by a *shorter* line on the paper.

leading to

$$\delta\left(\frac{df}{dt_{pr}}\right) = (\delta f)\dot{} + \frac{\delta P}{\rho + P}\dot{f} \tag{4.155}$$

for a generic quantity f. (As always in this chapter, an overdot means $\partial/\partial t$ at fixed $\mathbf{x}$. We only use this notation when t can be regarded as an unperturbed coordinate to first order.)

Using Eq. (4.155) to extract the perturbations from the continuity and Raychaudhuri equations, we find

$$(\delta\rho_k)\dot{} = -3(\rho + P)\delta H_k - 3H\delta\rho_k, \tag{4.156}$$

$$(\delta H_k)\dot{} = -2H\delta H_k - \frac{4\pi G}{3}\delta\rho_k + \frac{1}{3}\left(\frac{k}{a}\right)^2\frac{\delta P_k}{\rho + P}. \tag{4.157}$$

In what follows, we use only Eq. (4.156).

4.8 Outside the horizon

We now define the curvature perturbation and show that it is constant well outside the horizon. We also show that the adiabatic condition (4.5) holds in this regime.

4.8.1 Generalized adiabatic condition

At the beginning of Chapter 4 (pages 58 and 59), we listed the assumptions that define the model under consideration. Among these were the assumption that the inflaton field has only one component, and that the vacuum fluctuation of noninflaton fields has no significant effect after inflation. In this subsection, we argue that these assumptions lead to condition Eq. (4.160), which is a generalized version of the adiabatic condition given earlier as Eq. (4.5). The generalized adiabatic condition gives Eq. (4.5) as a special case, and also leads to an equation for δP that will allow us to prove the constancy of $\mathcal{R}$.

Our basic strategy is to argue that, on scales well outside the horizon, each region evolves like a separate Friedmann–Robertson–Walker Universe, with the inflaton-field perturbation determining the difference between the locally measured proper times. When the regions come into causal contact at horizon entry, the proper time difference manifests itself as a density perturbation, arising from the fact that different regions have evolved by different amounts. This strategy was implicit in some early calculations (Guth and Pi 1982; Hawking 1982; Starobinsky 1982a). The precise version presented here is related to the work of Starobinsky (1985b) and Salopek (1995), but follows most closely that of Sasaki and Stewart (1996). On scales far outside the horizon, this approach is much more powerful than if we try to rely exclusively on cosmological perturbation theory, as for instance in the work of Mukhanov and Chibisov (1981, 1982), Bardeen et al. (1983), Lyth (1985), and Mukhanov et al. (1992).

The first step is to argue that the following conditions will hold outside the horizon:

(1) After the Universe has been smoothed on a comoving scale much larger than the horizon size, each comoving region somewhat smaller than the smoothing scale can be regarded as an unperturbed Universe.

(2) These unperturbed Universes are actually identical, if their clocks are synchronized on slices of constant ϕ a few Hubble times after horizon exit during inflation.

The first condition is to be expected in any reasonable model. There needs to be *some* smoothing scale that makes the perturbations negligible or it would not make sense to talk about a Robertson–Walker Universe. The horizon scale will be big enough, unless there is dramatic new physics on a much bigger scale.[9] The second condition, stating the irrelevance of anything except the single inflaton field, is true by virtue of assumption 3 on page 59.

Given a detailed model of the Universe between horizon exit and nucleosynthesis, we could check these two conditions. To do this, we would develop a closed system of equations describing the cosmological perturbations. If the Universe were a gas at all times, this would be essentially the system that we develop in Chapters 14 and 15. A closed system also can be developed during slow-roll inflation (it is trivial if the inflaton has only one component, as we are presently assuming, but becomes less so in the case of a multicomponent inflaton). We are doing without such a system, which means that we do not need to specify how

---

[9] Our later discussion of the absence of an observed Grishchuk–Zel'dovich effect (Section 5.2.5) or tilted-Universe effect (Section 6.6) more or less assures us that there is no such scale.

inflation ends, nor what type of matter is present between then and the onset of nucleosynthesis. In particular, there is no need to make the usual assumption that we are dealing with a gas.

Taken together, the two conditions listed above imply that a given quantity g has practically the same evolution along each comoving worldline, as long as the smoothing scale is well outside the horizon. It is a function $g(t_{pr})$ of the proper time t_{pr} along the worldline, the same for all worldlines if t_{pr} for different worldlines is synchronized on slices of constant ϕ a few Hubble times after horizon exit.

At any later epoch, we can choose a slicing that corresponds to constant t_{pr}, and then every quantity will be practically homogeneous. This slicing will not, however, be a convenient one for calculations. With a different slicing, separated by a proper time interval $\delta t(t, \mathbf{x})$, there will be a perturbation given by Eq. (4.141):

$$\delta g = -\dot{g}(t)\delta t(t, \mathbf{x}). \tag{4.158}$$

It follows that the ratio $\delta g/\dot{g}$ is the same for every quantity g. As a result, the perturbation in every quantity is determined by, for example, the energy density perturbation,

$$\frac{\delta g}{\dot{g}} = \frac{\delta \rho}{\dot{\rho}}. \tag{4.159}$$

So far, we have been discussing smoothed quantities, but smoothing on a given scale has no effect on Fourier modes well above that scale. Thus we have, for each Fourier mode well outside the horizon,

$$\frac{\delta g_{\mathbf{k}}}{\dot{g}} = \frac{\delta \rho_{\mathbf{k}}}{\dot{\rho}}. \tag{4.160}$$

We call this the **generalized adiabatic condition**. The term "adiabatic" is appropriate because the condition relates to the evolution of the unperturbed Universe, which is indeed adiabatic because there can be no heat flow in an isotropic Universe.

Taking g to be the density ρ_i of a given particle species during radiation domination leads to Eq. (4.5), which we call simply the "adiabatic condition". Let us see how. It comes about because each of the species separately satisfies the continuity equation $\dot{\rho}_i = -3H(\rho_i + P_i)$, expressing energy conservation [Eq. (2.8)].[10] As a result, Eq. (4.160) becomes

$$\frac{1}{1 + P_i/\rho_i}\delta_i = \frac{1}{1 + P/\rho}\delta. \tag{4.161}$$

Assuming radiation domination, $P/\rho = 1/3$. For a relativistic species (radiation), this gives

$$\delta_i = \delta \quad \text{(radiation).} \tag{4.162}$$

---

[10] The neutrinos have negligible interaction with other species, and so does the CDM. The photons and baryons are in thermal equilibrium, which in this context ensures that there is no energy exchange between them.

For matter it gives

$$\delta_i = \frac{3}{4}\delta \qquad \text{(matter)}. \tag{4.163}$$

Equation (4.5) states these results for the special cases of CDM, baryons, photons, and massless neutrinos.

Taking g to be the total pressure, Eq. (4.160) becomes

$$\frac{\delta P_{\mathbf{k}}}{\dot{P}} = \frac{\delta \rho_{\mathbf{k}}}{\dot{\rho}}. \tag{4.164}$$

We show in the next subsection that this leads to the constancy of the curvature perturbation $\mathcal{R}_{\mathbf{k}}$.

Later, Eq. (4.164) is encountered in an entirely different context. This is when it applies to a single component of the fluid, which has a well-defined adiabatic equation of state $P(\rho)$ even in the presence of the perturbations. Because the evolution in an unperturbed Universe is also adiabatic, this immediately leads to Eq. (4.164), now valid even on scales inside the horizon.

4.8.2   The curvature perturbation

At last we are ready to define the curvature perturbation $\mathcal{R}$, and to show that on each scale it is time independent well outside the horizon. We consider only the case of critical energy density, corresponding to $K = 0$ in the Friedmann equation.

The curvature perturbation $\mathcal{R}$ is defined by

$$H^2(\mathbf{x}, t) \equiv \frac{8\pi G}{3}\rho(\mathbf{x}, t) + \frac{2}{3}\nabla^2 \mathcal{R}(\mathbf{x}, t), \tag{4.165}$$

where, as usual, $\nabla^2 \equiv a^{-2}\sum_i \partial^2/\partial x_i^2$. (An equivalent definition, relating $\mathcal{R}$ to the spatial curvature seen by comoving observers, is given on page 341.)

This is equivalent to introducing a perturbation δK in the constant K appearing in the Friedmann equation, and then writing $\frac{2}{3}a^2\nabla^2\mathcal{R} = \delta K$. For the Fourier components, $-\frac{2}{3}k^2\mathcal{R}_{\mathbf{k}} = \delta K_{\mathbf{k}}$, and $\mathcal{R}_{\mathbf{k}}$ is constant if $\delta K_{\mathbf{k}}$ is constant. As we noted in Section 4.4.5, δK is indeed constant if the pressure perturbation vanishes. The reason is that, with $\delta P = 0$, the Raychaudhuri equation (4.151) is the same as in the unperturbed case, so that together with the continuity equation it implies the Friedmann equation with constant K.

We now argue that δP is negligible on scales far outside the horizon, so that $\mathcal{R}$ is practically constant on such scales.

Let us multiply the local Friedmann equation (4.165) by a^2, and take its time derivative using the continuity equation and the Raychaudhuri equation. If δP vanishes, we know that this will give $\dot{\mathcal{R}} = 0$. It follows that

$$\dot{\mathcal{R}}_{\mathbf{k}} = -H\frac{\delta P_{\mathbf{k}}}{(\rho + P)}. \tag{4.166}$$

We want to argue that $\dot{\mathcal{R}}_k$ is negligible in magnitude, compared with $H\mathcal{R}_k$, so that $\mathcal{R}_k$ does not change significantly even in a large number of Hubble times. To do this, we need to show that δP_k is sufficiently small compared with $\mathcal{R}_k$.

Perturbing the local Friedmann equation, we find

$$2H\delta H_k \equiv \frac{8\pi G}{3}\delta\rho_k - \frac{2}{3}\left(\frac{k}{a}\right)^2 \mathcal{R}_k. \tag{4.167}$$

In the relativistic setting, it is still useful to define the peculiar gravitational potential Φ by Eq. (4.91). Using it, we can write Eq. (4.166) as

$$\frac{1}{H}\frac{\partial \ln \mathcal{R}_k}{\partial t} = \frac{2}{3}\frac{\delta P_k}{\delta\rho_k}\left[\left(\frac{k}{aH}\right)^2 \frac{\Phi_k}{(1+w)\mathcal{R}_k}\right], \tag{4.168}$$

where again $w \equiv P/\rho$. This gives the fractional change in $\mathcal{R}_k$ per Hubble time. By virtue of the adiabatic condition (4.164), $|\delta P_k/\delta\rho_k|$ is of order 1 or, at any rate, scale independent, on scales far outside the horizon. As a result, $\mathcal{R}_k$ will have negligible variation far outside the horizon, provided that $|\Phi_k| \lesssim |\mathcal{R}_k|$, and we now argue that the latter condition indeed will be satisfied.

Using Eq. (4.167), we can eliminate δH_k in Eq. (4.156) in favour of $\mathcal{R}_k$. Using Eq. (4.90), this gives

$$\frac{2}{3}H^{-1}\dot{\Phi}_k + \frac{5+3w}{3}\Phi_k = -(1+w)\mathcal{R}_k. \tag{4.169}$$

During any era when w is constant, this equation has the growing solution

$$\Phi_k = -\frac{3+3w}{5+3w}\mathcal{R}_k. \tag{4.170}$$

During inflation, $1+w = \dot{\phi}^2/V$, which is both slowly varying and small for slow-roll inflation. It follows that[11]

$$\Phi_k = -\frac{3}{2}(1+w)\mathcal{R}_k, \tag{4.171}$$

with Φ_k and $\mathcal{R}_k$ practically constant and $|\Phi_k| \ll |\mathcal{R}_k|$.

To see what happens after inflation, let us write Eq. (4.169) as

$$\frac{2}{3}\frac{1}{F}\frac{\partial}{\partial \ln a}(F\Phi_k) = -(1+w)\mathcal{R}_k, \tag{4.172}$$

where F satisfies

$$\frac{d \ln F}{d \ln a} = \frac{5+3w}{2}. \tag{4.173}$$

---

[11] One can be more explicit. The inflaton field ϕ is homogeneous on comoving slices because the momentum density given by Eq. (7.10) vanishes. As a result, the anisotropic stress vanishes (assumption 2 on page 59). The other perturbations are $\delta\rho = -\delta P = \delta(d\phi/dt_{pr})^2/2$. Using this result, we can obtain Eq. (4.171) by direct calculation (e.g., Lyth 1985).

Setting $\Phi_{\mathbf{k}}/\mathcal{R}_{\mathbf{k}} \simeq 0$ during inflation, the solution is

$$F\Phi_{\mathbf{k}} \simeq -\frac{3}{2} \int_{\ln a_*}^{\ln a} (1+w) F\mathcal{R}_{\mathbf{k}} \, d(\ln a). \tag{4.174}$$

For any reasonable $w(a)$, this indeed will give $|\Phi_{\mathbf{k}}/\mathcal{R}_{\mathbf{k}}| \sim 1$. Therefore, $\mathcal{R}_{\mathbf{k}}$ indeed will be constant outside the horizon.

During radiation domination $w = 1/3$, and Eq. (4.170) becomes

$$\Phi_{\mathbf{k}} = -\frac{2}{3}\mathcal{R}_{\mathbf{k}}. \tag{4.175}$$

Using Eq. (4.91), this gives Eq. (4.8) for $\delta_{\mathbf{k}}$. We see in Sections 4.8.4 and 15.6 that anisotropic stress multiplies this relation by a numerical coefficient of order 1.

4.8.3   Subtleties

It is important to recognize that the "separate Universe" assumption is not exact. All we are saying is that the errors incurred by the assumption decrease with k/aH.[12] Indeed, if the assumption were really exact, it would follow immediately that $\mathcal{R}_{\mathbf{k}}$ is constant (corresponding to a small nonzero K in the Friedmann equation). However, in its definition (4.167), $\mathcal{R}_{\mathbf{k}}$ appears in the combination $(k/aH)^2 \mathcal{R}_{\mathbf{k}}$. As a result, an error of order $(k/aH)^2$ in the separate-Universe assumption, in general, may give an error of order 1 in the statement that $\mathcal{R}_{\mathbf{k}}$ is constant. In other words, the fractional change in $\mathcal{R}_{\mathbf{k}}$ per Hubble time, in general, may be of order 1. To show that this change is small, we needed the *additional* assumption that there is a single degree of freedom. In Chapter 6 we see how another degree of freedom, corresponding to an isocurvature perturbation, can give a varying $\mathcal{R}_{\mathbf{k}}$.

Because the separate-Universe assumption is not exact, the difference between the two sides of the generalized adiabatic condition (4.160) actually will be some *nonzero* quantity, which, however, decreases with k/aH. Consider in particular the adiabatic condition (4.163). Because we are working on comoving slices, $\delta_{\mathbf{k}}$ decreases rapidly as we go outside the horizon $[\delta_{\mathbf{k}} \sim (k/aH)^2 \mathcal{R}_{\mathbf{k}}]$. As a result the *fractional* difference between the two sides of the adiabatic condition (4.163) need not decrease, but a careful examination of the way that the condition is used in Section 15.6.1 reveals that this is not a problem. What is important is that the absolute difference decreases or, in other words, that δ_i decreases. As we see in Chapter 6, the absolute difference remains constant for an isocurvature perturbation.

4.8.4   Including anisotropic stress

Pressure defines a purely isotropic stress, which is the only kind that exists in a gas if particle collisions are frequent. Anisotropic stress occurs in a gas if diffusion or freestreaming are significant, and also can occur in a nongaseous fluid. We have no certain knowledge of the

---

[12] The perturbations are functions of the two independent quantities k and t, the latter determining aH. We are saying that the errors decrease as k decreases at fixed t, and also as aH increases at fixed k.

cosmic fluid before nucleosynthesis, though it normally is assumed to be gaseous after inflation (apart from brief phase transitions). Let us see what happens when we include anisotropic stress.

The anisotropic stress tensor Σ_{ij} will be defined by Eq. (14.26). We see in Chapter 14 that, when it is included, the Euler equation (4.150) becomes

$$a_i = -\frac{1}{a}\frac{\partial_i P + \partial_j \Sigma_{ij}}{\rho + P} \qquad \text{(Euler)}. \tag{4.176}$$

Taking the divergence of Eq. (4.152) now gives

$$\frac{dt}{dt_{pr}} - 1 = \frac{\delta P - \frac{2}{3}P\Pi}{\rho + P}, \tag{4.177}$$

where $\frac{2}{3}P\Pi_{\mathbf{k}} \equiv -(k_i k_j/k^2)\Sigma_{ij\,\mathbf{k}}$.

In contrast to pressure, the anisotropic stress Σ_{ij} is absent in the unperturbed Universe, and so, its Fourier component $\Sigma_{ij\,\mathbf{k}}$ must decrease with k/aH. The same is true of $\Pi_{\mathbf{k}}$ because $\frac{2}{3}P\Pi_{\mathbf{k}}$ is smaller (in magnitude) than any component of the tensor $\Sigma_{ij\,\mathbf{k}}$. We are going to argue that, as a result, it will not affect the constancy of $\mathcal{R}_{\mathbf{k}}$.

In the presence of anisotropic stress, the Raychaudhuri equation (4.151) becomes

$$\frac{dH}{dt_{pr}} = -H^2 - \frac{4\pi G}{3}(\rho + 3P) - \frac{1}{3}\nabla^2\left[\frac{\delta P - \frac{2}{3}P\Pi}{\rho + P}\right]. \tag{4.178}$$

As before, we can use Eq. (4.177) to extract the perturbations from Eqs. (4.144) and (4.178). The result can be written

$$\dot{\delta}_{\mathbf{k}} - 3wH\delta_{\mathbf{k}} = -3(1+w)\delta H_{\mathbf{k}} - 2Hw\Pi_{\mathbf{k}}, \tag{4.179}$$

$$(\delta H_{\mathbf{k}})\dot{} + H\delta H_{\mathbf{k}} = \frac{1}{3}\left(\frac{k}{a}\right)^2\left(\frac{\delta P_{\mathbf{k}}}{\rho + P} - \frac{2}{3}\frac{w}{1+w}\Pi_{\mathbf{k}} + \Psi_{\mathbf{k}}\right), \tag{4.180}$$

where we have introduced another potential, Ψ, analogous to the potential Φ defined by Eq. (4.91). Repeating the former definition for convenience, the two of them are defined by

$$\Phi_{\mathbf{k}} \equiv -4\pi G(a/k)^2\rho\,\delta_{\mathbf{k}} = \frac{3}{2}\left(\frac{aH}{k}\right)^2\delta_{\mathbf{k}}, \tag{4.181}$$

$$(\Psi_{\mathbf{k}} - \Phi_{\mathbf{k}}) \equiv -8\pi G(a/k)^2 P\,\Pi_{\mathbf{k}} = -3\left(\frac{aH}{k}\right)^2 w\Pi_{\mathbf{k}}. \tag{4.182}$$

If anisotropic stress is negligible, $\Phi_{\mathbf{k}} = \Psi_{\mathbf{k}}$. This is the case for a gas in which there is negligible diffusion or freestreaming. Including these effects, we are still likely to have $|\Phi_{\mathbf{k}}| \sim |\Psi_{\mathbf{k}}|$ (Chapter 15).

Still defining the curvature perturbation $\mathcal{R}$ by Eq. (4.167), its time dependence is now

$$\dot{\mathcal{R}}_{\mathbf{k}} = -H\frac{\delta P_{\mathbf{k}} - \frac{2}{3}P\Pi_{\mathbf{k}}}{(\rho + P)}. \tag{4.183}$$

As we noted earlier, the anisotropic stress $\Pi_{\mathbf{k}}$ vanishes during inflation, and we need only consider what happens afterward. We have found that if $\Pi_{\mathbf{k}}$ vanishes, $\mathcal{R}_{\mathbf{k}}(t)$ is a constant outside the horizon, of the same order as $\Phi_{\mathbf{k}}$. From the above equation, we now see that this conclusion will not be affected by $\Pi_{\mathbf{k}}(t)$, provided that $|\Pi_{\mathbf{k}}(t)|$ is, at most, of order $|\Phi_{\mathbf{k}}|$ at the epoch of horizon entry, and decreases with k/aH as we go to earlier times. This certainly will be true if $|\Psi_{\mathbf{k}}| \sim |\Phi_{\mathbf{k}}|$, as we expect for a gas, but we can expect it to hold far more generally. We have argued for the decrease of $\Pi_{\mathbf{k}}$ with k/aH, and from Eqs. (4.181) and (4.182) the condition $|\Pi_{\mathbf{k}}| \lesssim |\Phi_{\mathbf{k}}|$ at $k = aH$ is equivalent to

$$|P\Pi_{\mathbf{k}}| \lesssim |\delta\rho_{\mathbf{k}}|. \tag{4.184}$$

However, as we noted on page 99, $\delta\rho_{\mathbf{k}}$ will be the same order of magnitude as $\delta P_{\mathbf{k}}$ well outside the horizon. We can expect this to be roughly true at horizon entry, and then Eq. (4.184) just says that the stress should not be dominated completely by its anisotropic part. This seems reasonable for any medium because the separation of the anisotropic part from the isotropic part (pressure) is more mathematical than physical.

This argument that the generalized adiabatic condition leads to the constancy of $\mathcal{R}_{\mathbf{k}}$ far outside the horizon is not absolutely conclusive, but neither is its earlier counterpart that ignored anisotropic stress. What is being said is that the result will hold, provided that the as-yet-unknown functions $\rho(t)$, $P(t)$, and $\Pi_{\mathbf{k}}(t)$ have reasonable behaviour. In particular, we expect it to hold in a gaseous Universe, and during single-component inflation.

4.9 Peculiar velocity in the relativistic domain

In this section we see how the concept of peculiar velocity can be extended to the domain of superhorizon scales. The extension has no direct observational significance in the foreseeable future because there is no prospect of measuring the peculiar motion of a region of the Universe whose size is comparable with the Hubble distance. Nor is it relevant in the usual description of cosmological perturbations employing a perturbed metric, which we describe in Chapters 14 and 15. However, in the main body of this book, we are using the alternative approach, which avoids an explicit consideration of the metric. In that setting, we need the peculiar velocity field to derive the famous Sachs–Wolfe effect in Section 5.2.6.

As with our earlier discussion, we need to quote without proof some results from general relativity, which are established in Chapter 14. At the moment, we aim to explain the concepts.

4.9.1 Velocity gradient

The discussion of velocity in Section 4.4.6 remains valid in the region around a free-falling observer equipped with a locally inertial frame. In the limit of zero perturbations, the observer is taken to be comoving, with the spatial orientation of their locally inertial frame coinciding with the global coordinate system $\mathbf{x}$. This uniquely defines the velocity gradient u_{ij} at each space-time point, to first order in the perturbations.

Through Eq. (4.107), the velocity gradient defines the vorticity ω_{ij}, the shear σ_{ij}, and the local Hubble parameter $H(\mathbf{x}, t)$. In the absence of perturbations, the vorticity and shear vanish and the Hubble parameter $H(t)$ is independent of position. Including the perturbations, we can write $H(\mathbf{x}, t) = H(t) + \delta H(\mathbf{x}, t)$ after choosing a slicing, which defines the perturbation δH and the unperturbed quantity $H(t)$.

The vorticity and shear are also perturbations, which are independent of both slicing and threading (gauge independent). We can split the velocity gradient into an unperturbed part and a perturbation,

$$u_{ij}(\mathbf{x}, t) = u_{ij}(t) + \delta u_{ij}(\mathbf{x}, t), \tag{4.185}$$

where $u_{ij}(t) = H(t)\delta_{ij}$ is the unperturbed part.

All of this is similar to the Newtonian case. The difference is that in the general-relativistic case, one cannot, in general, find a globally defined peculiar velocity field $\mathbf{v}(\mathbf{x}, t)$ such that $\delta u_{ij} = a^{-1} \partial v_i / \partial x_j$. However, as we now describe, this turns out to be possible for the scalar perturbations.

4.9.2 Threading and peculiar velocity

One can think of a threading as a population of the Universe by observers (not necessarily free falling), who make measurements only at their own position. In particular, they measure the fluid velocity $\mathbf{v}$. A change in the threading corresponds to a coordinate change

$$\tilde{\mathbf{x}} = \mathbf{x} + \delta\mathbf{x}. \tag{4.186}$$

The velocity of the new observers relative to the old is $(\delta\mathbf{x})\dot{}$, and the fluid velocity measured by the former is

$$\tilde{\mathbf{v}} = \mathbf{v} - (\delta\mathbf{x})\dot{}. \tag{4.187}$$

In Section 4.9.1, we consider the shear, vorticity, and local Hubble parameter of the comoving worldlines. In a similar way, we can define corresponding quantities for any threading, though we had better not use the same symbols. To prevent confusion, we avoid the term "local Hubble parameter" for a generic threading, using instead the term **expansion parameter**.

We show in Chapter 14 that, keeping only the scalar perturbations, there is a unique threading with zero shear (Section 14.6.3). We go on to show (in Section 14.6.4) that if $\mathbf{v}$ is taken to denote the velocity of the cosmic fluid with respect to this threading, then the perturbation in the velocity gradient defined with the comoving slicing is given by $\delta u_{ij} = a^{-1} \partial v_i / \partial x_j$. This means that the zero-shear threading is the general-relativistic version of the Newtonian "uniform expansion" threading, and that $\mathbf{v}$ is the general-relativistic version of peculiar velocity. In accordance with the fact that scalar perturbations do not contribute to the vorticity, $\mathbf{v}$ is the gradient of a scalar, and so can be written in the form $\mathbf{v_k} = -i\mathbf{k}\, V_k/k$. This is the definition of V that we used in the Newtonian context, Eq. (4.117), which is equivalent to the relation Eq. (4.118) between δH and V. During matter domination, V is related to the density perturbation by the Newtonian expression Eq. (4.121). We use this fact in Section 5.2.6 to give a simple derivation of the Sachs–Wolfe effect.

Examples

4.1 Use Eq. (4.26) to show that the variance of a generic perturbation $g(\mathbf{x})$ is given by Eq. (4.16), independently of $\mathbf{x}$. For $\mathbf{x} = 0$, check that the same result is obtained from the spherical expansion Eq. (4.30), with the definition Eq. (4.39) of the spectrum.

4.2 Verify Eqs. (4.54) and (4.55). For this and the following question, you need

$$\int_0^\infty x^n e^{-x^2/2}\, dx = 2^{(n-1)/2}\, \Gamma\left(\frac{n+1}{2}\right) \qquad (n > -1).$$

4.3 Suppose wavenumbers and scales are related by $k_R = \alpha/R$, where α is a constant. Assuming that $\mathcal{P}_g(k) \propto k^m$, and using a Gaussian filter, find α such that Eq. (4.57) is exact. For what range of m is your calculation valid? Over roughly what range of k must the power-law approximation be maintained to obtain an approximately correct result?

4.4 By considering a single Fourier component, show that the vector part of a velocity field is determined by its vorticity.

4.5 Derive the Newtonian multifluid equation (4.122) by mimicking the derivation of the corresponding equation for the total density contrast.

4.6 Verify Eq. (4.130), giving the Jeans mass as a function of the baryon and photon temperatures.

4.7 The sound speed in a fluid is defined by $c_s^2 = dP/d\rho$. Demonstrate, using their redshift evolution, that if one has a combination of radiation and matter components in the fluid, then

$$c_s^2 = \frac{1}{3}\left(\frac{3}{4}\frac{\rho_m}{\rho_r} + 1\right)^{-1}.$$

What is the sound speed at matter–radiation equality?

4.8 In the absence of anisotropic stress, the time dependence of the curvature perturbation is related to the pressure perturbation by Eq. (4.166). Derive this equation using the method suggested in the text.

5 Simplest model for the origin of structure II

5.1 From horizon entry to galaxy formation

At the end of the preceding chapter, we derived equations for the time dependence of the total density perturbation $\delta\rho_k$, Eqs. (4.179) and (4.180). They can be solved for $\delta\rho_k$ if the pressure perturbation and anisotropic stress are known. Well before horizon entry these are negligible, which allowed us to solve the equations and in particular to show that $\mathcal{R}_k$ is constant. Using the adiabatic condition (4.5), this gave us the density perturbations of each particle species.

After horizon entry, the equations for the total density perturbation remain valid, but they cannot be solved on their own because the pressure perturbation and the anisotropic stress become significant. Rather, they become a small part of a complete system of equations, involving the densities of each species and, for relativistic species, the distribution functions specifying the particle motion. We develop this system in Chapter 15, and go on to see how it can be solved given the single initial condition $\mathcal{R}_k$. The result can be encoded in a set of linear transfer functions, which form the main focus of the present chapter. We describe their origin in a mostly qualitative way, and give a brief overview of their physical implications, paving the way for the detailed discussion of Chapters 9 through 12. We begin with the matter density perturbation, which is the origin of large-scale structure, and go on to consider the cosmic microwave background (cmb) anisotropy.

5.1.1 The matter-density transfer function

We discuss the matter-density contrast during matter domination, assuming critical density and zero cosmological constant.

On very large scales that enter the horizon after matter domination, we can use the analysis of Section 4.8.2. During matter domination, Eq. (4.170) becomes

$$\Phi_k = -\frac{3}{5}\mathcal{R}_k. \tag{5.1}$$

The density contrast, given by Eq. (4.91), is

$$\delta_k(t) = -\frac{2}{3}\left(\frac{k}{aH}\right)^2 \Phi_k = \frac{2}{5}\left(\frac{k}{aH}\right)^2 \mathcal{R}_k. \tag{5.2}$$

For the relevant scales, these relations continue to hold after horizon entry because the pressure perturbation remains negligible.

On smaller scales, the pressure perturbation is still negligible during matter domination (except at early times on scales below the Jeans scale). As a result, the argument in Section 4.8.2 can again be invoked. The curvature perturbation has some constant value $\mathcal{R}^{(m)}$, and

$$\delta_{\mathbf{k}}(t) = -\frac{2}{3}\left(\frac{k}{aH}\right)^2 \Phi_{\mathbf{k}} = \frac{2}{5}\left(\frac{k}{aH}\right)^2 \mathcal{R}_{\mathbf{k}}^{(m)}. \tag{5.3}$$

We derived these expressions for $\delta_{\mathbf{k}}$ by appealing to the relativistic analysis of Section 4.6, and so, they are valid both before and after horizon entry. In the latter regime, however, they also may be derived from the Newtonian analysis of Section 4.4. This is because the relativistic continuity and Euler equations [Eqs. (4.144) and (4.145), respectively] coincide with the Newtonian ones during matter domination, *and* we are defining the density perturbation on comoving slices so that t_{pr} and t coincide even before horizon entry [Eq. (4.154) with $\delta P = 0$]. Had we used a different slicing, the evolution might have been non-Newtonian before horizon entry; for example, this happens in the commonly employed conformal Newtonian gauge, defined in Chapter 15. On the other hand, all of the usually considered slicings give the Newtonian equations after horizon entry.

The matter-density transfer function $T(k)$ is defined by

$$\mathcal{R}_{\mathbf{k}}^{(m)} = T(k)\mathcal{R}_{\mathbf{k}}. \tag{5.4}$$

It is equal to 1 on very large scales, where $\mathcal{R}_{\mathbf{k}}^{(m)}$ has its primordial value $\mathcal{R}_{\mathbf{k}}$. The matter-density contrast is

$$\delta_{\mathbf{k}}(t) = \frac{2}{5}\left(\frac{k}{aH}\right)^2 T(k)\mathcal{R}_{\mathbf{k}}. \tag{5.5}$$

Its spectrum is given by

$$\mathcal{P}_\delta(k, t) = \frac{4}{25}\left(\frac{k}{aH}\right)^4 T^2(k)\mathcal{P}_{\mathcal{R}}(k)$$

$$= \left(\frac{k}{aH}\right)^4 T^2(k)\delta_H^2(k), \tag{5.6}$$

and its time dependence, carried by the aH factor, is $\mathcal{P}_\delta \propto (1 + z)^{-2}$, corresponding to $\delta \propto a \propto 1/(1 + z)$. In the second line, we have introduced a quantity

$$\delta_H^2(k) \equiv \frac{4}{25}\mathcal{P}_{\mathcal{R}}(k). \tag{5.7}$$

The subscript H stands for horizon entry, and the idea is that δ_H is roughly the root mean square (rms) value of δ when it enters the horizon.

Standard practice is to normalize the spectrum to match the cosmic microwave background anisotropy measured by the Cosmic Background Explorer (COBE) satellite. We derive this result later, but it is useful to quote it now for use in this chapter. For critical-density Universes, and assuming that the primordial spectrum can be approximated by a power law, it is given by

Eq. (9.6), ignoring gravitational waves:

$$\delta_H(k) = 1.91 \times 10^{-5} \left(\frac{k}{k_{pivot}} \right)^{(n-1)/2},$$

(5.8)

where n is the spectral index and $k_{pivot} = 7.5 a_0 H_0$ is the scale at which the normalization is n-independent. This result is discussed fully in Section 9.1.2.

5.1.2 Calculating the transfer function

The matter-density transfer function $T(k)$ is the vital link between theory and observations of large-scale structure. Its detailed form depends on the cosmology under consideration, and to calculate it, we have to consider what happens to each of the particle species present. There is a competition between gravity, which tries to increase the density contrast by attracting more matter to overdense regions, and random particle motion, which inhibits this increase.

The cold dark matter (CDM) has negligible random motion by definition, but until matter domination, its gravitational interaction is too weak to increase its density very much. We can get a rough estimate of the transfer function by pretending that no growth at all occurs between horizon entry and matter domination. This gives

$$T(k) = 1 \quad (k < k_{eq} = 14\Omega_0^{-1} h^{-2} \, \text{Mpc}^{-1}),$$
$$T(k) = (k_{eq}/k)^2 \quad (k > k_{eq}).$$

(5.9)

The most direct route to this behaviour is from Eq. (4.91); normally, Φ_k is constant, but for scales inside the horizon during radiation domination ($k > k_{eq}$) this is replaced by the δ_k constant.

We have included Ω_0 in the formula for k_{eq}, even though $\Omega_0 = 1$ in the simplest model under consideration at the moment. The above estimate of T, and the more sophisticated ones that we are about to consider, continue to hold for $\Omega_0 < 1$.

On scales below the Silk scale (defined on page 120), the photons diffuse so that their density perturbation decays away. The photons carry the baryons with them, but the CDM is unaffected. The density contrast δ_c of the latter evolves under the action of its Newtonian self-gravity, even though it lives in a uniform sea of radiation. The evolution is given by Eq. (4.98), with ρ the matter-density contrast. Because radiation dominates, we have $H = 1/2t$. Defining $y = a/a_{eq}$, we obtain what is called the Meszaros equation:

$$2y(y+1)\frac{d^2\delta_{ck}}{dy^2} + (2+3y)\frac{d\delta_{ck}}{dy} - 3\delta_{ck} = 0.$$

(5.10)

The growing solution is

$$\delta_{ck} = A_k(1 + 1.5y),$$

(5.11)

and the decaying solution is

$$\delta_{ck} = B_k \left[(1+1.5y) \ln \frac{(1+y)^{1/2}+1}{(1+y)^{1/2}-1} - 3(1+y)^{1/2} \right],$$

(5.12)

where A_k and B_k are constants.

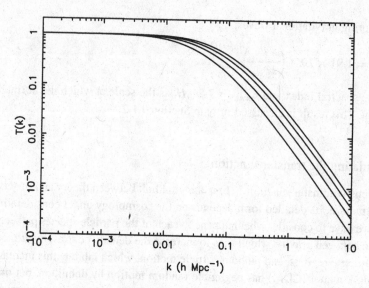

Fig. 5.1. The BBKS transfer function, showing (top to bottom) $\Gamma = 0.5, 0.4, 0.3$, and 0.2.

The growing solution increases only by a factor of order 2 between horizon entry [when Eq. (4.98) first becomes valid] and the epoch $a_{\rm eq}$ when matter starts to dominate. However, we should keep both modes, matching to the (approximately correct) time dependence $\delta_{\rm ck} \propto (aH)^{-2}$ at horizon entry. We then have a logarithmic growth, corresponding to a transfer function going like $k^{-2} \ln k$.

To obtain an accurate transfer function over the full range of scales, we have to solve the equations of Chapter 15 numerically. Transfer functions have been tabulated in the literature for a wide range of cosmological models. For the CDM model, which we are considering at the moment, a widely used parameterization of the result is that of Bardeen et al. (1986), the BBKS transfer function

$$T(q) = \frac{\ln(1 + 2.34q)}{2.34q} \left[1 + 3.89q + (16.1q)^2 + (5.46q)^3 + (6.71q)^4\right]^{-1/4}. \quad (5.13)$$

Here, $q = k/\Gamma h$ Mpc^{-1}, where the **shape parameter** is given by

$$\Gamma = \Omega_0 h \exp(-\Omega_{\rm b} - \sqrt{2h}\,\Omega_{\rm b}/\Omega_0). \quad (5.14)$$

The transfer function is plotted in Figure 5.1 for several Γ values, which simply corresponds to sliding the horizontal scale.

In the limit $\Omega_{\rm b} = 0$, q is a multiple of $k_{\rm eq}$, in accordance with the fact that the latter is the only relevant scale. The coefficients were given for this case by Bardeen et al. (1986). The correction for the baryons is an empirical one, given for $\Omega_0 = 1$ by Peacock and Dodds (1994) and for low-density models by Sugiyama (1995).

Although accurate enough for present data, the BBKS transfer function can give errors of several percent, especially at the crucial bend in the spectrum. Future data sets such as large

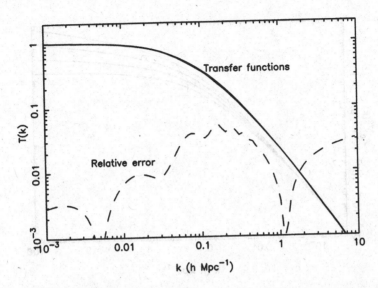

Fig. 5.2. The BBKS transfer function for $h = 0.5$ and $\Omega_b h^2 = 0.016$ is compared with that from an accurate calculation using CMBFAST. The dashed line shows the relative error of the BBKS transfer function, which can be several percent.

galaxy surveys will require more accurate determinations to allow comparison with theory. Accurate transfer functions can be computed using the CMBFAST code (Seljak and Zaldarriaga 1996), with which a comparison is made in Figure 5.2. Improved fitting functions have been constructed by Eisenstein and Hu (1999).

In structure-formation studies it is often useful to have a fiducial model with which others can be compared. For historical reasons, the model considered is often the **Standard Cold Dark Matter (SCDM) model**, even though this provides a poor fit to the whole data set. It has parameters $n = 1$, $h = 0.5$, Ω_b given by nucleosynthesis (we take $\Omega_b h^2 = 0.016$), and the spectrum normalized to COBE ($\delta_H = 1.91 \times 10^{-5}$). For short-scale work, the COBE normalization sometimes is replaced by a normalization to the abundance of galaxy clusters, which gives a better, though still unsatisfactory, fit to the short-scale observational data.

5.1.3 Bottom-up picture of structure formation

First-order cosmological perturbation theory has given us a prediction for $\mathcal{P}_\delta(k, z)$ in terms of n, h, and Ω_b: Eqs. (5.6) and (5.13). Because the Fourier components satisfy linear equations, first-order cosmological perturbation theory usually is called linear theory. As we now discuss, linear theory breaks down when gravitationally bound objects form on the relevant scale, and it then has to be replaced by a nonlinear theory. On scales $k^{-1} \gtrsim 10 \, h^{-1}$ Mpc, bound objects have yet to form and linear theory is still valid. On smaller scales, the regime of linear theory ends at an epoch $z_{nl}(k)$, after which nonlinearity sets in. We estimate $z_{nl}(k)$ in this subsection.

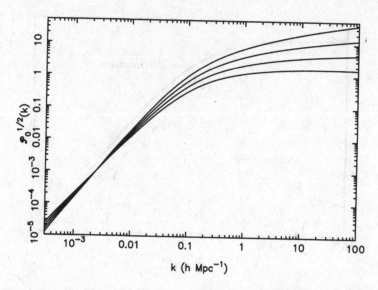

Fig. 5.3. Spectrum of the density contrast during matter domination, in the standard CDM model with $h = 0.5$ and $\Omega_b h^2 = 0.016$. Lines correspond to spectral indices $n = 0.6, 0.8, 1$, and 1.2, from bottom to top.

In the linear regime, the spectrum is given by

$$\mathcal{P}_\delta^{1/2}(z, k) = \frac{\mathcal{P}_0^{1/2}(k)}{1 + z}, \tag{5.15}$$

where $\mathcal{P}_0$ is the *linear prediction* for the present value.

In addition to $\mathcal{P}_\delta(k, z)$, it is useful to consider $\sigma(R, z)$, the *rms* of the density contrast smoothed on scale R (the filtered density contrast). In the linear regime,

$$\sigma(R, z) = \frac{\sigma_0(R)}{1 + z}, \tag{5.16}$$

where σ_0 is the linear prediction for the present value, given by Eqs. (4.56) and (5.6) as

$$\sigma_0^2(R) = \int_0^\infty W^2(kR) \left(\frac{k}{a_0 H_0}\right)^4 T^2(k)\, \delta_H^2(k) \frac{dk}{k}. \tag{5.17}$$

Figure 5.3 shows $\mathcal{P}_0^{1/2}(k)$, and Figure 5.4 shows $\sigma_0(R)$. The predictions are normalized to COBE, with $h = 0.5$ and $\Omega_b h^2 = 0.016$ and four choices of the spectral index n. Pure CDM with $h = 0.5$ requires $n \simeq 0.7$ to fit the observational data, shown later in Figure 12.1 on page 304. In the extensions of the model considered in the next chapter, we can have $n \simeq 1.0 \pm 0.3$ but, on scales $R \lesssim 10\,\mathrm{Mpc}$, the result for $\sigma(R)$ still resembles the case $n \simeq 0.7$ (when the parameters are chosen to provide a reasonable fit to the data). Accordingly, we focus on that case in what follows; it corresponds to the region between the lowest pair of lines in the next three figures.

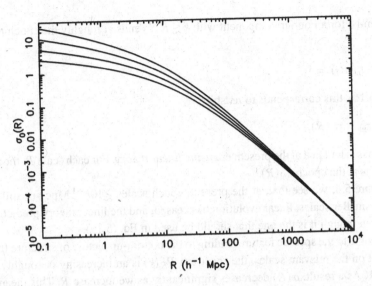

Fig. 5.4. Dispersion of the filtered density contrast for the models of Figure 5.3. Pure CDM with $h = 0.5$ requires $n \simeq 0.7$ to fit the data.

The predictions of linear theory, as well as direct observation, suggest a **bottom-up** picture of structure formation. In this picture, the first objects form at more or less the same time with a wide range of masses. Later, heavier objects form in succession. The bottom-up picture holds because the spectrum $\mathcal{P}_\delta(k)$ of the density contrast is an increasing function of k, corresponding to the fact that we are dealing with CDM. As a result, the spectrum $W^2(kR)\mathcal{P}_\delta(k)$ of the *filtered* density contrast peaks at $k \sim k_R \equiv 1/R$. It is dominated by Fourier components with this wavenumber, and therefore contains structures mostly with size of order R. One manifestation of this is the approximate equality of $\mathcal{P}_0^{1/2}(k)$ and $\sigma_R(1/k)$, evident from Figures 5.3 and 5.4.

Let us see in more detail how things work out. Without any smoothing, $\delta(\mathbf{x}, t)$ will be small almost everywhere in space, as long as the *rms* density contrast $\sigma(t)$ is significantly less than 1. This will be the case just after inflation ends, and for some time afterward, as we discuss shortly. Even while $\sigma(t)$ is small though, there will be rare regions of space where $\delta(\mathbf{x}, t)$ exceeds 1. In them, the matter has collapsed under its own weight and linear theory ceases to apply. This early formation of rare objects is important, but it will not have a significant effect on the evolution of the Fourier components. Thus, linear theory will remain valid for all of the Fourier components, and almost everywhere in space, as long as $\sigma(t)$ is significantly less than 1.

When $\sigma(t)$ becomes of order 1, the linear regime ends and a large fraction of the mass of the Universe collapses into gravitationally bound objects. Then, the linear equations that we have developed no longer apply to the unfiltered density contrast.

However, after smoothing on scale R, we may expect the linear evolution equations to apply until $\sigma(R, t) \sim 1$. Smoothing on scale R does not affect Fourier components with wavenumber $k \lesssim R^{-1}$, and so, they too will evolve linearly. The linear regime for smoothing scale

R (equivalently, for a Fourier component with $k \lesssim R^{-1}$) ends at roughly the epoch $t = t_{nl}(R)$, defined by

$$\sigma(R, t_{nl}(R)) = 1. \tag{5.18}$$

From Eq. (5.16), this corresponds to redshift

$$1 + z_{nl} = \sigma_0(R), \tag{5.19}$$

where $\sigma_0(R)$ is calculated at the present era *using linear theory*. For each scale R, the nonlinear regime begins at the epoch $z_{nl}(R)$.

From Figure 5.4, we see that, at the present, epoch scales $\gtrsim 10h^{-1}$ Mpc are still evolving linearly. On smaller scales, linear evolution has ceased, and the linear theory prediction for σ_0 is wrong, even though it is the one that should be used in Eq. (5.19).

Now we come to the specific feature leading to the bottom-up picture of structure formation, which is that on the relevant scales, the spectrum $\mathcal{P}_{\mathcal{R}}(k)$ is an increasing or roughly constant function of k. As a result, $\sigma(R)$ decreases significantly as we increase R. This means that, at the epoch $z_{nl}(R)$, when regions of the Universe with comoving size of order R are collapsing, the density contrast smoothed on significantly bigger scales is still evolving linearly. In other words, while regions of a given size are collapsing, significantly bigger ones are still expanding with the Universe. This is the the bottom-up picture of structure formation.

An alternative is that the spectrum cuts off below some scale, called the **coherence length**. Then, the bottom-up picture will apply only on scales bigger than the coherence length. In the models we are considering, the coherence length corresponds to the scale entering the horizon when the nonbaryonic dark matter first becomes nonrelativistic. For the case of CDM that we are considering in this and the preceding chapter, the coherence length is by definition negligibly small.

In view of the bottom-up picture, it is useful to specify a comoving scale R by the mass $M(R)$, which is enclosed by a sphere of radius a/k in the almost-homogeneous early Universe. This is given by Eq. (4.54), and shown in Figure 5.5. If one wants to associate a mass with a given k, one can take $R = 1/k$, though this is less well defined, with different conventions appearing in the literature, such as $1/k$ replaced by $\lambda/2$ or $\lambda/4$ (where $\lambda = 2\pi/k$ is the wavelength), which increases $M(k)$ by a factor π^3 or $(\pi/2)^3$.

Let us now be more quantitative and see how the bottom-up picture refers to the various objects in the Universe. As we discuss in Section 5.4.1, the CDM might collapse into very light objects but, for the moment, let us consider only baryonic objects, formed when CDM *and* baryons collapse. At $z \sim 10$, the Jeans mass equation (4.130) is of order $10^5 M_\odot$, and it is not wildly different if we consider the epoch $z \sim 20$ to 30. This provides a very crude estimate of the mass of the first baryonic objects to form, though arguably the true value might be an order of magnitude or two lower when nonlinear collapse is described properly. Looking at Figure 5.5, and focusing on the case $n \sim 0.7$, we see that a large fraction of the mass of the Universe collapses into baryonic objects at a relatively recent epoch, somewhat later than $z = 10$. However, we see in Chapter 11 that even at $z \sim 15$ to 20, baryonic objects may be sufficiently numerous to reionize the Universe in this model.

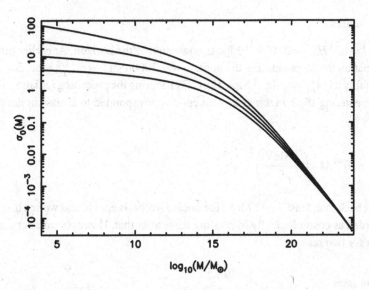

Fig. 5.5. Dispersion of the filtered density contrast for the models of Figure 5.4, but now showing σ_0 as a function of filter mass M.

We can see from Figure 5.5 that, after the first baryonic objects form, heavier ones form not much later, because $\sigma_0(M)$ is quite flat on small scales. Objects with mass $M \sim 10^{11} M_\odot$ or so will form at a redshift of a few, and these might correspond to the damped Lyman-alpha systems observed at redshift $z \sim 4$, which are widely supposed to be early galaxies. (In some of the models discussed in the next chapter, sufficiently early formation of such structures can be problematic.) After the first galaxies form, they can merge and split, leading to a complicated evolution, which prevents us from saying much about present-day galaxies on the basis of linear evolution.

The situation is different for the scales $M \gtrsim 10^{13} M_\odot$ to $10^{15} M_\odot$, which correspond to galaxy clusters. In this range, $\sigma(M)$ is decreasing quite strongly, and we can expect a more sharply defined bottom-up picture, with the heavier structures forming quite a bit later than the lighter ones. This is consistent with the rather sparse information that is available from high-redshift observations.

We can quantify these remarks by estimating the maximum mass M_{max} of structures forming at the epoch $z_{nl}(M)$, in other words, by estimating the width of the band M to M_{max} of the masses that are formed at that epoch. It is the mass such that $\sigma_0(M_{max})$ is significantly less than 1, so that the density contrast filtered on the scale M_{max} still is evolving linearly almost everywhere in the Universe. To be definite, let us require that it still is evolving linearly in 90 percent of the volume of the Universe. From the Gaussian distribution, this corresponds to $\sigma_0(M_{max}) = 1/1.63 = 0.61$. Let us illustrate this for the parameter choice $h = 0.5$ and $n = 0.7$. For $M = 10^6 M_\odot$ one has $M_{max} \simeq 10^{10} M_\odot$, and objects in this broad band of masses form around the epoch $z_{nl}(10^6 M_\odot)$. For $M = 10^{12} M_\odot$ in contrast, $M_{max} \simeq 3 \times 10^{13} M_\odot$, and so, a much narrower range of masses collapses at the epoch $z = z_{nl}(10^{12} M_\odot)$.

5.1.4   Tiny scales

The scale $k^{-1} = a_0^{-1} H_0^{-1} = 3000 h^{-1}$ Mpc is now entering the horizon. At earlier times during matter domination the scale entering the horizon was $3000 h^{-1} (1 + z)^{-1/2}$ Mpc, and at matter–radiation equality it is $k_{\text{eq}}^{-1} = 14 h^{-2}$ Mpc, Eq. (2.37). During the preceding radiation-dominated era, the scale entering the horizon at a given epoch corresponded to a mass in nonrelativistic particles of

$$M_{\text{nr}} \sim 10^{-2} M_\odot \left(\frac{1 \, \text{MeV}}{T} \right)^3 . \tag{5.20}$$

This era goes back to at least $T \sim 1$ MeV, the nucleosynthesis epoch, and we see that the scales usually regarded as cosmologically interesting leave after that. However, smaller scales might be important for two reasons:

(1) *Boson stars*

CDM has negligible pressure, and so, it will collapse in any region where the matter-density contrast exceeds 1. If it is bosonic, it may form what is sometimes called a boson star (Jetzer 1992; Liddle and Madsen 1992). If, as is quite likely, inflation is followed by a long era of matter domination, such collapse could occur on the very small scales entering the horizon during that era.

Alternatively, the collapse might occur during a phase transition, perhaps associated with topological defects. For example, it has been suggested that axion star formation might be triggered by domain walls at the epoch $T \sim 1$ GeV, which would correspond to a mass $\sim 10^{-11} M_\odot$ (assuming that radiation domination has begun by then). It is not known whether boson stars would survive to the present, nor whether they can be cosmologically significant.

(2) *Primordial black holes*

If n is significantly greater than 1, as predicted by some models of inflation, $\mathcal{P}_{\mathcal{R}}(k)$ can become comparable with unity (or bigger) on very small scales. In that case, there will be regions where the density contrast is roughly of order unity at horizon entry, leading to the formation of black holes (Carr 1975). Black holes formed this way decay by Hawking radiation, and depending on their mass may be observationally constrained either at nucleosynthesis or at the present epoch (MacGibbon and Carr 1991). It has also been conjectured (MacGibbon 1987) that stable Planck-mass relics may remain, again potentially leading to observational constraints.

Alternatively, a density contrast of order unity at horizon entry might be generated by a phase transition long after inflation. This might occur because the sound speed drops, allowing some modest growth of the radiation density contrast. It also might be caused directly by the phase transition, if it is of first order. Assuming that the final radiation-dominated era has begun, the mass of a black hole forming in this way at temperature T is of order of the total energy $M_{\text{tot}}(k)$ in a sphere of radius k^{-1}. This is bigger than the mass of the nonrelativistic matter $M_{\text{nr}}(k)$ by a factor T/T_{eq}, which means that Eq. (5.20) is replaced by

$$M_{total} \sim 1M_\odot \left(\frac{100\,\text{MeV}}{T} \right)^2. \qquad (5.21)$$

If the quark–hadron phase transition is first order (not considered to be very likely) it could generate black holes with mass of order $1M_\odot$.

5.2 The cosmic microwave background anisotropy

Future observations of the cosmic microwave background anisotropy will be very sensitive, achieving an accuracy roughly of order 1 percent on degree scales, and the extremely small predicted polarization probably will also be measured, or at least its cross correlation with the temperature anisotropy. Even at this level of sensitivity, the anisotropy and polarization for the most part are well described by linear cosmological perturbation theory, at least if the redshift of reionization is in the expected range $\sim$10 to 30. The only significant exception, already observed and well understood, is the Sunyaev–Zel'dovich effect (Sunyaev and Zel'dovich 1972, 1980; Rephaeli 1995), which occurs when a nearby galaxy cluster is in the line of sight. This effect comes from photons gaining energy by scattering off hot gas on their way through the cluster. It occurs on scales of order 1 arc-minute and is identified readily for nearby clusters (Jones et al. 1993; Myers et al. 1997). The cumulative effect of distant clusters on the cosmic microwave background spectrum is expected to be negligible.

In Chapter 15, we give the full details of the linear calculation; we devote the rest of this chapter to an overview. We also give the basic formalism for analyzing the anisotropy and polarization, which follows from general principles, independently of the detailed calculation.

5.2.1 Multipoles

The cosmic microwave background photons that we observe are characterized by the number of photons per quantum state (known as the occupation number). Ignoring the perturbations, the occupation number is a function $f(p)$ of the magnitude of the photon momentum (equivalently, of the photon energy). To high accuracy it is given by the blackbody distribution

$$f(p) = \frac{1}{e^{p/T} - 1}, \qquad (5.22)$$

with $T = 2.728$ K.

The perturbation $\delta f(\mathbf{p})$ depends on both the magnitude and the direction of the photon momentum. The total is then

$$f(\mathbf{p}) = f(p) + \delta f(\mathbf{p}). \qquad (5.23)$$

In Chapter 15, we show that, to first order in the perturbations, $f(\mathbf{p})$ has the blackbody form (5.22), with a fractional temperature perturbation $\delta T/T$ that is *independent of p*. It depends on the direction $\mathbf{n}$ of the photon momentum or, equivalently, on the direction $\mathbf{e} = -\mathbf{n}$ of observation.

The multipoles $a_{\ell m}$ of the cosmic microwave background anisotropy are defined by

$$\frac{\delta T(\mathbf{e})}{T} = \sum_{\ell m} a_{\ell m} Y_{\ell m}(\mathbf{e}).$$ (5.24)

The monopole a_{00} is unobservable; although it is well defined if we consider a position-dependent δT, there is no way of measuring it. The dipole would vanish if we were in the rest frame of the cosmic microwave background, defined as the frame in which the momentum density of the cosmic microwave background vanishes. Through the Doppler shift, it is equal to $\mathbf{e} \cdot \mathbf{v}$, where $\mathbf{v}$ is our velocity relative to the cosmic microwave background rest frame. If the polar axis is chosen along $\mathbf{v}$, $a_{10} = v$ and $a_{1,\pm 1} = 0$. The diurnal average of the observed dipole gives the velocity of the Sun as $v = 371$ km$\cdot$s$^{-1} = 1.2 \times 10^{-3}$, the final figure in our chosen unit $c = 1$.

The multipoles with $\ell \geq 2$ represent the intrinsic anisotropy of the cosmic microwave background, which is present even in the rest frame. To first order in v (measured in units where $c = 1$), they are independent of our motion.[1]

5.2.2   Transfer function and spectrum

Later in this section, and in much more detail in Chapter 15, we see how to calculate the cosmic microwave background multipoles in terms of the primeval curvature perturbation $\mathcal{R}_k$. Because the equations are linear, and are invariant under rotations, the result must be of the form

$$a_{\ell m} = \frac{4\pi}{(2\pi)^{3/2}} \int_0^\infty T_\Theta(k, \ell) \mathcal{R}_{\ell m}(k) k \, dk.$$ (5.25)

Indeed, under a rotation of the coordinate system, $a_{\ell m}$ and $\mathcal{R}_{\ell m}(k)$ have the same transformation law (4.43). The transfer function $T_\Theta(k, \ell)$ is independent of m because of the invariance under rotations, which mix different m according to Eq. (4.43). We have chosen its normalization and k-dependence to fit the usual conventions of Chapter 15.

We saw in Section 4.3.2 that the real and imaginary parts of the coefficients $\mathcal{R}_{\ell m}(k)$ have Gaussian probability distributions, with no correlation between them except for the reality condition. The ensemble mean squares (variances) are the same for the real and imaginary parts, and independent of both ℓ and m.

From Eq. (5.25), these properties are bequeathed to the cosmic microwave background multipoles $a_{\ell m}$, except that the variances $C_\ell \equiv \langle |a_{\ell m}|^2 \rangle$ now depend on ℓ. An alternative definition of C_ℓ, encoding the random phases, is

$$\langle a_{\ell m} a_{\ell' m'}^* \rangle = C_\ell \delta_{\ell\ell'} \delta_{mm'}.$$ (5.26)

We call C_ℓ the **spectrum of the cosmic microwave background**. From Eq. (4.39),

$$C_\ell = 4\pi \int_0^\infty T_\Theta^2(k, \ell) \mathcal{P}_\mathcal{R}(k) \frac{dk}{k}.$$ (5.27)

[1] At higher orders, the relativistic Doppler effect changes each multipole by a fractional amount of order v^ℓ. The COBE value of the quadrupole has taken this effect into account, but even there, the effect is insignificant compared with the cosmic variance defined later.

As usual, the hypothesis is that we observe a typical realization of the ensemble. This means that we expect the difference between the observed values $|a_{\ell m}|^2$ and the ensemble averages C_ℓ to be of order of the mean-square deviation of $|a_{\ell m}|^2$ from C_ℓ. The latter is called the **cosmic variance** and, because we are dealing with a Gaussian distribution, it is equal to $2C_\ell$ for each multipole. For a single ℓ, averaging over the $(2\ell + 1)$ values of m reduces the cosmic variance by a factor $(2\ell + 1)$, but it remains a serious limitation for low multipoles.[2]

The correlation function of the cosmic microwave background anisotropy can be defined as

$$C(\theta_{12}) \equiv \left\langle \frac{\delta T(\mathbf{e}_1)}{T} \frac{\delta T(\mathbf{e}_2)}{T} \right\rangle, \tag{5.28}$$

where θ_{12} is the angle between the directions $\mathbf{e}_1$ and $\mathbf{e}_2$. In this expression, the expectation value is an ensemble average. It can be regarded as an average over the possible observer positions, but not in general as an average over the single sky that we observe, because of cosmic variance. Using the addition theorem Eq. (4.45), we find

$$C(\theta) = \sum_\ell \frac{2\ell + 1}{4\pi} C_\ell P_\ell(\cos\theta). \tag{5.29}$$

5.2.3   A small patch of sky

To describe the cosmic microwave background anisotropy on small angular scales, there is no need to consider the whole sky simultaneously. Instead, we can focus on a small patch of sky. In such a patch, taken without loss of generality to be near the pole of spherical coordinates, we can lay down almost-Cartesian coordinates as in Figure 5.6:

$$\vec{\theta} = (x, y) = (\theta \cos\phi, \theta \sin\phi). \tag{5.30}$$

On angular scales much smaller than the patch, it makes sense to write

$$\frac{\delta T(\vec{\theta})}{T} = \frac{1}{2\pi} \int d^2\vec{\ell} \, a(\vec{\ell}) e^{i\vec{\theta} \cdot \vec{\ell}}. \tag{5.31}$$

The Fourier integral is supposed to go over the whole $\vec{\ell}$ plane so that the functions $\exp(i\vec{\theta} \cdot \vec{\ell})/2\pi$ are orthonormal. The transformation from this expansion to the multipole expansion is unitary (because both use orthonormal functions); therefore, the coefficients $a(\vec{\ell})$ have Gaussian distributions which are independent except for the reality condition.

To be more precise about the transformation, we need polar coordinates $(\ell, \phi_{\vec{\ell}})$ in the $\vec{\ell}$ plane, as shown in Figure 5.7. They are defined by $\vec{\ell} = (\ell_x, \ell_y) = (\ell \cos\phi_{\vec{\ell}}, \ell \sin\phi_{\vec{\ell}})$, so that $\ell = |\vec{\ell}|$. As the notation suggests, ℓ is the continuous limit of the integer labelling the multipoles $a_{\ell m}$. To see this, substitute into Eq. (5.24) the small-angle approximation

$$P_\ell^m(\theta) \simeq \ell^m J_{-m}(\ell\theta), \tag{5.32}$$

---

[2] In principle, we might indirectly detect at least the quadrupole at other places in the Universe, via polarization induced by scattering from gas in clusters (Kamionkowski and Loeb 1997), which would allow some further reduction of the cosmic variance.

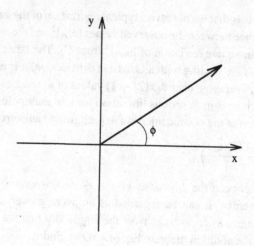

Fig. 5.6. In a small patch of sky near the pole, we can lay down almost-Cartesian coordinates $\vec{\theta} = (x, y) = (\theta \cos \phi, \theta \sin \phi)$. A typical wavevector **k** points well away from the patch, so that there is a unique Cartesian vector $\vec{\theta}_{\mathbf{k}}$ pointing toward it. The contribution of $\mathcal{R}(\mathbf{k})$ to the cosmic microwave background anisotropy $\delta T(\vec{\theta})/T$ is constant on the lines perpendicular to this vector.

(valid for $\theta \to 0$ with $\ell \theta$ constant) and the two-dimensional relation between the spherical and plane-wave expansions

$$J_{-m}(\ell \theta) = \frac{i^m}{2\pi} \int_0^{2\pi} e^{i\vec{\ell} \cdot \vec{\theta}} e^{im(\phi_{\vec{\ell}} - \phi)} d\phi_{\vec{\ell}}. \tag{5.33}$$

After taking the continuous limit $\sum_\ell \to \int d\ell$, this reproduces Eq. (5.31) with

$$a(\vec{\ell}) \equiv \left(\frac{1}{2\pi \ell}\right)^{1/2} \sum_m a_{\ell m} i^{-m} e^{im\phi_{\vec{\ell}}}. \tag{5.34}$$

Given this interpretation of ℓ, the unitary of the transformation from the multipole expansion to the small-patch expansion ensures that

$$\langle a^*(\vec{\ell}) a(\vec{\ell'}) \rangle = C(\ell) \delta^2(\vec{\ell} - \vec{\ell'}), \tag{5.35}$$

with $C(\ell) = C_\ell$ for large integer ℓ.

The correlation function can be calculated directly with one of the directions taken to be the pole, or it can be read off from Eq. (5.29) using the large-ℓ limit. We find

$$C(\theta_{12}) \equiv \left\langle \frac{\delta T(\vec{\theta}_1)}{T} \frac{\delta T(\vec{\theta}_2)}{T} \right\rangle = \frac{1}{2\pi} \int_0^\infty C(\ell) J_0(\ell \theta_{12}) \ell \, d\ell, \tag{5.36}$$

where $\theta_{12} = |\vec{\theta}_1 - \vec{\theta}_2|$. Mimicking the discussion of the three-dimensional Fourier transform in Section 4.3.3, we can effectively sample the ensemble (on small scales) by using a single patch of sky. In particular, the ensemble average in the definition of $C(\theta)$ can be replaced by

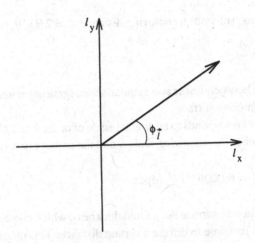

Fig. 5.7. The cosmic microwave background anisotropy within the patch may be written as a two-dimensional Fourier integral. The position vector in Fourier space is $\vec{\ell} = (\ell_x, \ell_y) = (\ell \cos \phi_{\vec{\ell}}, \ell \sin \phi_{\vec{\ell}})$. The contribution of $\mathcal{R}(\mathbf{k})$ to the anisotropy $a(\vec{\ell})$ is nonvanishing only if $\vec{\ell}$ points toward the direction of $\mathbf{k}$ in the sky.

an average over positions in the sky because cosmic variance is negligible. It therefore can be determined by observation, which provides one way of determining C_ℓ from observation, for high ℓ.

Rotational invariance requires that the contribution to $\delta T(\mathbf{e})/T$, from a given component $\mathcal{R}(\mathbf{k})$, depends on $\mathbf{e}$ only through the combination $\mathbf{e} \cdot \mathbf{k}$. Within a small patch, this means that $\delta T(\vec{\theta})$ is constant along lines in the sky perpendicular to the direction of $\mathbf{k}$. (The direction is unique, discounting the exceptional case that the direction of $\mathbf{k}$ is close to the direction of the small patch.) As a result, the Fourier transform $a(\vec{\ell})$ vanishes unless $\vec{\ell}$ points along this same direction.

5.2.4 Scales explored by the cosmic microwave background anisotropy

Between the epoch of decoupling and the epoch of reionization, the Universe is almost transparent to photons. As we discuss later, the probability that a given photon scatters after reionization is expected to be significantly less than 1, and so, to a useful approximation, we can say that all cosmic microwave background photons that we see originate on the surface of a precisely defined sphere around us, called the **last-scattering surface**.

The cosmic microwave background anisotropy comes partly from the anisotropy already present at last scattering, and partly from the additional anisotropy caused by gravity since then. (The separation between these effects is gauge dependent, but that does not matter for the present purpose.) With critical matter density, it turns out that the latter effect depends only on conditions at last scattering. As a result, each angular scale explores the linear scale that it subtends at the last-scattering surface, insofar as reionization can be ignored. This surface lies practically at the particle horizon, whose comoving distance is $2H_0^{-1} = 6{,}000h^{-1}$ Mpc, and so, an angle θ (in radians) subtends a comoving distance $x = 2H_0^{-1}\theta$, or

$$\frac{x}{100h^{-1}\,\mathrm{Mpc}} \simeq \frac{\theta}{1^0}. \tag{5.37}$$

This is the relationship between linear and angular scale, ignoring reionization and the finite thickness of the last-scattering surface.

A multipole of order ℓ corresponds to an angular scale of order $\theta \simeq 1/\ell$ [see Eq. (5.31)], and so, the ℓth multipole probes the comoving scale x and the comoving wavenumber k, given by

$$k^{-1} \simeq x \simeq \frac{2}{H_0\ell} = 6{,}000h^{-1}\ell^{-1}\,\mathrm{Mpc}. \tag{5.38}$$

In reality, the last-scattering surface has a finite thickness, which comes from the fact that, at any epoch, a photon has had time to diffuse a certain distance. This distance is called the **Silk scale**, and its value at decoupling is the thickness of the surface of last scattering. Let us estimate the Silk scale. Before decoupling, the photons perform a random walk, as they Thomson-scatter from one electron to another. The mean time between collisions is $t_c \sim (n_e\sigma_T)^{-1}$, where n_e is the electron number density and σ_T is the Thomson-scattering cross-section. The average number of steps in time t is $N = t/t_c$, and in that time a photon diffuses a distance $d \sim \sqrt{N}t_c \sim (tt_c)^{1/2}$. At decoupling, with $t \sim 1/H$, this is the thickness of the surface of last scattering. A careful estimate using the transport equations developed in Chapter 15 gives a thickness of order $7\Omega_0^{-1/2}h^{-1}\,\mathrm{Mpc}$, corresponding to $\ell \sim 10^3$. As a result, the anisotropy will be wiped out on much smaller scales (much larger ℓ).

5.2.5 COBE regime

The biggest scale on which causal processes can be relevant is the horizon scale at decoupling, which is $90h^{-1}\,\mathrm{Mpc}$, corresponding to $\ell \simeq 70$. On much larger scales (much smaller ℓ), the anisotropy is given by the Sachs–Wolfe effect, which we derive shortly. Comparing Eqs. (4.30) and (5.25), we find that the Sachs–Wolfe effect, Eq. (4.9), corresponds to the transfer function

$$T_\Theta(k,\ell) = -\frac{1}{5}j_\ell(2k/H_0). \tag{5.39}$$

From Eq. (5.27), this corresponds to the spectrum

$$C_\ell = \pi \int_0^\infty \frac{dk}{k}\, j_\ell^2\left(\frac{2k}{H_0}\right)\, \delta_H^2(k), \tag{5.40}$$

where δ_H is given by Eq. (5.7) and j_ℓ is the spherical Bessel function. If the spectrum is a power law, $\delta_H^2(k) \propto k^{n-1}$, the integration can be performed analytically [try Eq. (6.574) of Gradshteyn and Rhyzik (1994), and the Γ function duplication formula] to give

$$\ell(\ell+1)C_\ell = \frac{\pi}{2}\left\{\frac{\sqrt{\pi}}{2}\ell(\ell+1)\frac{\Gamma[(3-n)/2)]}{\Gamma[(4-n)/2]}\frac{\Gamma[\ell+(n-1)/2]}{\Gamma[\ell+(5-n)/2]}\right\}\delta_H^2(H_0/2). \tag{5.41}$$

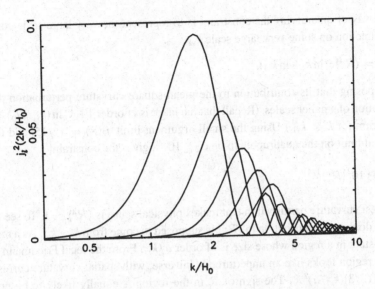

Fig. 5.8. Spherical Bessel functions $j_\ell^2(2k/H_0)$ for $\ell = 2, 3, 4, 5$, and 6, showing how the different C_ℓ sample scales k. For a scale-invariant $\delta_H(k)$, the C_ℓ is just the area under the curve.

The term within the curly braces is equal to 1 for $n = 1$. For $\ell \gg 1$ and $\ell \gg |n|$, it can be replaced by 1, if $\delta_H(k)$ is evaluated on the scale $k = \ell H_0/2$, which dominates the integral; that is,

$$\ell(\ell + 1)C_\ell \simeq \frac{\pi}{2} \delta_H^2 \left(\frac{\ell H_0}{2} \right). \tag{5.42}$$

If the spectrum $\delta_H(k)$ is approximately flat, so too will $\ell(\ell + 1)C_\ell$ be flat.

Grishchuk–Zel'dovich effect

With a flat spectrum the integrand in Eq. (5.40) is proportional to $k^{2\ell-1}$ for small k, so that scales far outside the horizon make practically no contribution to the predicted value of C_ℓ (see Figure 5.8). The possible effect of a sharp increase of the spectrum, on some very large scale, is called the Grishchuk–Zel'dovich effect (Grishchuk and Zel'dovich 1978). Because of the factor $k^{2\ell-1}$, the dominant effect is in the quadrupole, and because the observed quadrupole does not stick out above the others, we can say that no effect is observed.[3] Let us ask what constraint this places on the spectrum $\mathcal{P}_\mathcal{R}$ of the primordial curvature perturbation.

To get a significant contribution, $\mathcal{P}_\mathcal{R}$ has to increase rapidly on scales far outside the horizon. These scales left the horizon long before the observable Universe, and so, in principle, the absence of an effect represents a constraint on models of inflation at a very early epoch. However, for a power-law increase, one needs k^{-4} or faster, corresponding to the spectral index falling to $n \lesssim -3$ on scales far outside the horizon. Models of inflation so far proposed do not have such violent behaviour, though things can be rather different in the context of an open Universe, as discussed in Section 6.3.4.

---

[3] One can show that, for the dipole, a very large scale contribution to Eq. (5.40) would be cancelled by the Doppler shift caused by its contribution to our motion. In plain language, a very distant gravitational source pulls us and the cosmic microwave background equally.

Another way of quantifying the constraint is to represent the very large scale contribution by a delta function on some very large scale k_{vl}

$$\mathcal{P}_{\mathcal{R}} = \langle \mathcal{R}^2 \rangle \delta(\ln k - \ln k_{vl}). \tag{5.43}$$

We are supposing that its contribution to the mean-square curvature perturbation dominates the contribution of smaller scales. (Recall that the latter is of order $10^{-10} \ln (k_{vl}/H_0)$, assuming a flat spectrum for $k < k_{vl}$.) Using the small-argument limit $j_2(x) = x^2/15$, and taking the observational limit on the quadrupole to be $C_2 \lesssim 10^{-9}$, gives the constraint

$$\frac{H_0}{k_{vl}} \gtrsim 10^2 \langle \mathcal{R}^2 \rangle^{1/4}. \tag{5.44}$$

The biggest curvature perturbation that makes physical sense is $\langle \mathcal{R}^2 \rangle \sim 1$. To see this, note first that the dominant contribution to $\langle \mathcal{R}^2 \rangle$ is supposed to come from $k \sim k_{vl}$. As a result, $\mathcal{R}$ is roughly constant in a region whose size is of order a/k_{vl}. From the local Friedmann equation (4.165), the region looks like an unperturbed Universe, with spatial curvature corresponding to $K/a^2 = (2/3)(k_{vl}/a)^2 \mathcal{R}$. The sign of $\mathcal{R}$ in the region is equally likely to be positive or negative but, in the former case, it can be, at most, of order 1 because otherwise the region we are considering would close on itself. (From Section 14.5.2, the size of a Universe with positive K is of order $\sqrt{K}/a$, which would be less than the size of the patch we are considering if $\mathcal{R}$ were $\gtrsim 1$.) To avoid this catastrophe, at least over most of space, we indeed need $\langle \mathcal{R}^2 \rangle \lesssim 1$.

This bound will be roughly saturated in models in which slow-roll inflation begins after an era of eternal inflation, with k_{vl} the scale leaving the horizon at the transition era. The constraint (5.44) therefore suggests that eternal inflation must end more than a few e-folds ($\ln 100 \simeq 5$) before our Universe leaves the horizon.

More generally, the absence of an effect perhaps suggests that the observable Universe is part of a smooth patch whose size is $\gtrsim 100 H_0^{-1}$. However, as the original authors emphasized, the Grishchuk–Zel'dovich effect is a stochastic one that applies to our Universe on the assumption that it is a fairly typical realization of an ensemble whose stochastic properties are invariant under translations and rotations. The only known origin for such an ensemble is a vacuum fluctuation during inflation, which is why we emphasized the connection with models of inflation. The Grishchuk–Zel'dovich effect does *not* give model-independent knowledge of what lies beyond the edge of the observable Universe, which would in fact be a violation of causality.

5.2.6   A derivation of the Sachs–Wolfe effect

We now give a derivation of the Sachs–Wolfe effect, Eq. (4.10), using the comoving slicing and ascribing the effect to the accumulated redshift seen by a succession of comoving observers in the line of sight. In Chapter 15, we give a more conventional derivation using the metric perturbation.

Brightness function

To proceed, we need to consider the cosmic microwave background anisotropy measured at positions other than our own, and at earlier times. This is called the **brightness function**, and we denote it by $\Theta(t, \mathbf{x}, \mathbf{n})$:

$$\Theta \equiv \frac{\delta T(t, \mathbf{x}, \mathbf{n})}{T(t)}. \tag{5.45}$$

The photons with momentum $\mathbf{p}$ in a given range $d^3 p$ have intensity I (power per unit area perpendicular to $\mathbf{p}$) proportional to $T^4(t, \mathbf{x}, \mathbf{n})$. The fractional perturbation in this intensity is therefore

$$\frac{\delta I}{I} = 4\Theta. \tag{5.46}$$

Some authors include the factor 4 in the definition of the brightness function.

The brightness function depends on the direction $\mathbf{n}$ of the photon momentum or, equivalently, on the direction of observation $\mathbf{e} = -\mathbf{n}$. Using the latter, we expand it into multipoles at each point in space-time. From Eq. (5.46), the monopole is related to the photon energy density contrast $\delta_\gamma \equiv \delta\rho_\gamma/\rho_\gamma$ by

$$\Theta_{00}(t, \mathbf{x}) = \frac{1}{4}\delta_\gamma, \tag{5.47}$$

and one needs a slicing to define it. The dipole represents the motion of the observer relative to the cosmic microwave background rest frame, and we need a threading to define it. The higher multipoles are independent of both slicing and threading (gauge invariant).

Because the cosmic microwave background travels freely after last scattering, we can write the observed anisotropy as

$$\frac{\delta T}{T} = \Theta(t_{ls}, \mathbf{x}_{ls}, \mathbf{n}) + \left(\frac{\delta T}{T}\right)_{jour}. \tag{5.48}$$

Here, $\mathbf{x}_{ls} = -x_{ls}\mathbf{n}$ is the point of origin of the photon coming from direction $\mathbf{e} = -\mathbf{n}$. The comoving distance of the last-scattering surface is $x_{ls} = 2/H_0$. The first term corresponds to the anisotropy already existing at last scattering. The second term is the additional anisotropy acquired on the journey toward us, equal to minus the fractional perturbation in the redshift of the radiation. The separation between the two terms depends on the slicing, though the sum does not. We use the comoving slicing and find that the second term gives the Sachs–Wolfe effect, whereas the first term is negligible. In Chapter 15, we see how a different slicing leads to a different separation between the two terms, though of course the same final result.

Redshift perturbation

Consider the redshift perturbation first. To calculate it, we imagine that the Universe is populated by comoving observers along the line of sight. The relative velocity of adjacent comoving observers is equal to their distance, times the velocity gradient measured in the direction $\mathbf{n}$ of the photon. Assuming that gravitational waves are negligible, the velocity gradient is given by Eq. (4.111), and the corresponding Doppler shift is

$$\frac{d\lambda}{\lambda} = \frac{da}{a} + n_i n_j \frac{\partial v_i}{\partial x_j} \, dx, \tag{5.49}$$

where $\mathbf{v}(\mathbf{x}, t)$ is the peculiar velocity field of the cosmic fluid.

As we note in Section 5.1.1, the evolution of the peculiar velocity is Newtonian even on scales outside the horizon. It therefore is related to the peculiar gravitational potential by Eq. (4.120), giving

$$\left(\frac{\delta T}{T}\right)_{\text{jour}} = \int_0^{x_{\text{ls}}} \frac{t}{a} \frac{d^2\Phi(x)}{dx^2} \, dx. \tag{5.50}$$

The photon trajectory is $a d\mathbf{x}/dt = \mathbf{n}$, and using $a \propto t^{2/3}$ gives

$$x(t) = \int_t^{t_0} \frac{dt}{a} = 3\left(\frac{t_0}{a_0} - \frac{t}{a}\right). \tag{5.51}$$

Integrating Eq. (5.50) by parts, we find

$$\left(\frac{\delta T}{T}\right)_{\text{jour}} = \frac{1}{3}[\Phi(\mathbf{x}_{\text{ls}}) - \Phi(0)] + \mathbf{e} \cdot [\mathbf{v}(0, t_0) - \mathbf{v}(\mathbf{x}_{\text{ls}}, t_{\text{ls}})]. \tag{5.52}$$

The potential at our position contributes only to the unobservable monopole and can be dropped. On scales well outside the horizon at decoupling, Eq. (4.121) shows that we also can drop $\mathbf{v}$ at last scattering. The term $\mathbf{e} \cdot \mathbf{v}(0, t_0)$ represents the effect of our peculiar velocity calculated in linear theory, and it should be replaced by the actual peculiar velocity of Earth. The remaining term is the Sachs–Wolfe effect, Eq. (4.10),

$$\frac{\delta T(\mathbf{e})}{T} = \frac{1}{3}\Phi(\mathbf{x}_{\text{ls}}) = -\frac{1}{5}\mathcal{R}(\mathbf{x}_{\text{ls}}). \tag{5.53}$$

The last equality follows from Eq. (5.1).

Anisotropy at last scattering

Finally, we need to see that the initial perturbation $\Theta(t_{\text{ls}}, \mathbf{x}_{\text{ls}}, \mathbf{n})$ is negligible compared with $\Phi(\mathbf{x}_{\text{ls}})$. We are dealing only with low multipoles, which corresponds to smearing the direction of $\mathbf{x}_{\text{ls}}$ over a patch of sky that corresponds to a linear scale well outside the horizon at decoupling. This smearing will leave only the superhorizon modes of $\Phi(\mathbf{x}_{\text{ls}})$ [those with $k \ll (aH)_{\text{ls}}$], but that will not dramatically reduce its typical value because the spectrum $\mathcal{P}_\Phi = \frac{9}{25}\mathcal{P}_\mathcal{R}$ is practically flat. However, the same smearing will make the multipoles $\Theta_{\ell m}(t_{\text{ls}}, \mathbf{x}_{\text{ls}})$ of the brightness function negligibly small. For all multipoles except the monopole, this is an immediate consequence of our assumption that the Universe looks homogeneous when smoothed on a superhorizon scale (Section 4.8.1), and is confirmed by the explicit calculation of Chapter 15. The case of the monopole is more subtle. It is equal to $\delta_\gamma/4$, and from the adiabatic condition (4.161) (with $P = 0$ corresponding to matter domination), $\delta_\gamma/4$ is equal to $\delta/3$, which indeed decreases as we go outside the horizon [Eq. (5.3)].

We see in Section 6.6 that the monopole remains constant for an isocurvature perturbation, which means that it gives a contribution comparable to the Sachs–Wolfe effect.

5.2.7   Acoustic peaks

On angular scales smaller than the COBE regime, a range of physical effects contribute and the anisotropies can only be accurately calculated numerically, though there do exist semianalytic

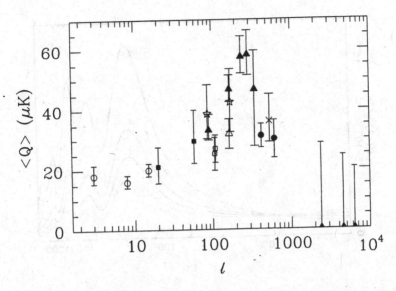

Fig. 5.9. Compilation of cosmic microwave background data, as of early 1998 (data courtesy of Martin White). The different experiments are identified in a replot of this figure on page 253. The quantity shown is $\langle Q \rangle^2 \equiv \ell(\ell + 1)C_\ell T_0^2/4\pi$. The COBE data have been combined into three bins (the leftmost points).

methods that give physical insight (Hu and Sugiyama 1995; Hu et al. 1997). In Figure 5.9, we summarize the existing data for C_ℓ. Some theoretical curves are shown in Figure 5.10, calculated using the publically available CMBFAST code of Seljak and Zaldarriaga (1996), which is based on the earlier COSMICS package described by Bertschinger (1995). As we discuss in detail in Chapter 9, the situation regarding the data will be revolutionized in a few years, when two satellites, Microwave Anisotropy Probe (MAP) and Planck, should provide very accurate data over the whole range of interest.

The data are seen to be in agreement with the prediction of approximate flatness for the low multipoles (the COBE regime). Going up in ℓ, the theoretical curves have a series of peaks. Historically, these have been called the Doppler peaks but more correctly should be called **acoustic peaks** because they come from the standing-wave oscillation of the baryon density. The observations show evidence of at least some peak structure.

5.3 Polarization

The cosmic microwave background is predicted to have linear polarization, which should be observable by the next generation of satellites. As shown in Section 15.5, the cosmic microwave background is practically unpolarized before decoupling. It acquires some polarization by Thomson scattering occurring as it emerges from the last-scattering surface, on linear scales less than the horizon size at that epoch, which correspond to $\ell \gtrsim 100$. After reionization, there is more Thomson scattering, producing polarization also in the low-ℓ multipoles. Both contributions are expected to be measurable.

Early studies of microwave background polarization were made by Polnarev (1985) and Bond and Efstathiou (1987).

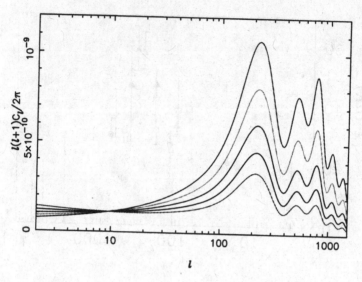

Fig. 5.10. Theoretical predictions for C_ℓ, taking $h = 0.5$ and $\Omega_b h^2 = 0.016$. Shown, from bottom to top, are $n = 0.8, 0.9, 1, 1.1$, and 1.2.

5.3.1 Stokes parameters

We first recall the classical description of polarized electromagnetic radiation (Chandrasekhar 1960). Consider a plane wave, arriving at the observer's position from the $+z$ direction. At this position, write the electric field as a Fourier series in the time variable, performed in some time interval much bigger than the one of interest. Each Fourier component is of the form

$$E_x(t) = E_x \cos(\omega t - \delta_1),$$
$$E_y(t) = E_y \cos(\omega t - \delta_2).$$

$$(5.54)$$

The amplitude depends on ω, and we have denoted it by the same symbol as the wave itself.

For angular frequencies in a given range $d\omega$, we can calculate the intensity of linearly polarized radiation in a plane making an angle $\phi_{\rm pol}$ with the x axis (clockwise around the z axis). It is $d\omega\langle E_{\rm pol}^2\rangle$, where $E_{\rm pol}$ is the component of $\mathbf{E}$ in the plane of polarization, and the average is over both time and frequency. The frequency average is taken over the interval $d\omega$, which is assumed to contain many Fourier components. When each component is a stochastic quantity, as is the case for the cosmic microwave background, the frequency average can be replaced by the ensemble average applied to each Fourier component. Performing the averages, we find

$$4\frac{\text{intensity}}{d\omega} = 4\langle E_{\rm pol}^2\rangle = I + Q \cos 2\phi_{\rm pol} + U \sin 2\phi_{\rm pol}.$$

$$(5.55)$$

The coefficients are

$$I = \langle E_x^2 + E_y^2\rangle,$$

$$(5.56)$$

$$Q = \langle E_x^2 - E_y^2\rangle,$$

$$(5.57)$$

$$U = \langle 2E_x E_y \cos(\delta_1 - \delta_2)\rangle,$$

$$(5.58)$$

where now E_x and E_y are the time-independent amplitudes, and the average is only over frequency. The total intensity is $I d\omega/4$, and Q and U are the **Stokes parameters** specifying the plane polarization.

By detecting circular polarization, we can measure a third Stokes parameter:

$$V = \langle 2E_x E_y \sin(\delta_1 - \delta_2) \rangle .\tag{5.59}$$

It vanishes if there is no circular polarization (equal intensity for right- and left-polarized waves).

Rotating the x and y axes gives new Stokes parameters. If the rotation angle is ψ (in the clockwise direction), then this will increase ϕ_{pol} by $-\psi$ while leaving E_{pol} unchanged, and so,

$$\begin{pmatrix} Q \\ U \end{pmatrix} \rightarrow \begin{pmatrix} \cos 2\psi & +\sin 2\psi \\ -\sin 2\psi & \cos 2\psi \end{pmatrix} \begin{pmatrix} Q \\ U \end{pmatrix} .\tag{5.60}$$

Also, a reflection $y \rightarrow -y$ or $x \rightarrow -x$ has the effect

$$\begin{pmatrix} Q \\ U \end{pmatrix} \rightarrow \begin{pmatrix} Q \\ -U \end{pmatrix} .\tag{5.61}$$

This reflection is a parity transformation, changing the usual right-handed coordinate system into a left-handed one. (The former was assumed when specifying the directions of ϕ_{pol} and ψ.)

There is a preferred choice for the orientation of the x-z plane that makes U vanish, and this defines the "plane of polarization" of the radiation. The quantity $(Q^2 + U^2)$ is independent of the orientation and satisfies

$$0 \leq \frac{\sqrt{Q^2 + U^2}}{I} \leq 1.\tag{5.62}$$

This is the degree of linear polarization.

We are interested in the cosmic microwave background, and the above description applies to the photons coming from a given direction $\mathbf{e}$, with frequency in a given range $d\omega$. We already considered the total intensity, noting that, even in the perturbed Universe, it is given by the blackbody distribution Eq. (5.22) with the temperature perturbation $\delta T(\mathbf{e})/T$ independent of frequency. Since the intensity in a given interval $d\omega$ is proportional to T^4, we have $\delta T(\mathbf{e})/T = \frac{1}{4}\delta I(\mathbf{e})/I$, where I is given by Eq. (5.56). The Stokes parameters vanish in the unperturbed Universe, since there can be no polarization in the absence of a preferred direction. It is therefore convenient to define "cosmological" Stokes parameters by

$$\begin{pmatrix} \delta T(\mathbf{e})/T \\ Q_{\text{cosm}}(\mathbf{e}) \\ U_{\text{cosm}}(\mathbf{e}) \end{pmatrix} \equiv \frac{1}{4} \begin{pmatrix} \delta I/I \\ Q/I \\ U/I \end{pmatrix} .\tag{5.63}$$

From now on we redefine Q and U to be the objects on the left, dropping the subscripts cosm. As we see in Section 15.5, the quantities Q and U thus defined are predicted to be independent of photon energy, just like $\delta T/T$.

For a given direction in the sky, the x axis can be any perpendicular direction. To define Q and U over the whole sky, it is convenient to choose some spherical polar coordinate system (θ, ϕ), and then define the Stokes parameters in each direction with reference to an x axis that points away from the pole $\theta = 0$.

As with $\delta T/T$, we can consider Q and U as functions of the observer's position $\mathbf{x}$. We then can consider just one Fourier component, $Q_{\mathbf{k}}$ and $U_{\mathbf{k}}$, and take the pole $\theta = 0$ to be along the $\mathbf{k}$ direction. With this special choice of the polar direction, we define

$$E_{\mathbf{k}} \equiv Q_{\mathbf{k}}, \tag{5.64}$$
$$B_{\mathbf{k}} \equiv U_{\mathbf{k}}. \tag{5.65}$$

We have introduced new symbols E and B because the definition holds only if the pole is along the $\mathbf{k}$ direction.[4]

Through the equations that we will develop in Chapter 15, the polarization with given $\mathbf{k}$ is caused by the perturbation in the electron density with the same $\mathbf{k}$. We are assuming that this perturbation comes from the primeval curvature perturbation $\mathcal{R}_{\mathbf{k}}$. In that case (and for the more general scalar perturbation that might be caused by an isocurvature density perturbation) the electron density is constant in each plane perpendicular to $\mathbf{k}$. As a result, the polarization $B_{\mathbf{k}}$ vanishes because it is odd under the parity transformation and there is nothing in the physics to distinguish right- and left-handedness. Also, $E_{\mathbf{k}}$ is independent of ϕ, depending only on $\mathbf{e} \cdot \mathbf{k}$. Gravitational waves, not present in the simplest model that we are considering at the moment, generate to both $E_{\mathbf{k}}$ and $B_{\mathbf{k}}$. Vector perturbations would do the same, but they are not generated by inflation and we are taking them to be absent.

5.3.2   A small patch of sky

We need to add up the contributions from different $\mathbf{k}$, and this is simple if we consider a small patch of sky, as in Section 5.2.3. Within the small patch, each Stokes parameter can be written as a Fourier integral,

$$Q(\vec{\theta}) = \frac{1}{2\pi} \int d^2\vec{\ell}\, Q(\vec{\ell})\, e^{i\vec{\theta} \cdot \vec{\ell}}, \tag{5.66}$$

$$U(\vec{\theta}) = \frac{1}{2\pi} \int d^2\vec{\ell}\, U(\vec{\ell})\, e^{i\vec{\theta} \cdot \vec{\ell}}. \tag{5.67}$$

Consider first the contribution from a given $\mathbf{k}$, and temporarily choose the x axis in the sky pointing towards the direction of $\mathbf{k}$. We have just seen that $U_{\mathbf{k}} \equiv B_{\mathbf{k}}$ vanishes. We have also seen that $Q_{\mathbf{k}}(\mathbf{e})$ is independent of the azimuthal angle about the $\mathbf{k}$ direction, which means that, within the small patch, $Q_{\mathbf{k}}(\vec{\theta})$ is constant along lines in the sky perpendicular to the direction of $\mathbf{k}$. (As always we discount the case that the direction of $\mathbf{k}$ is close to the direction of the small patch.) As a result, the Fourier transform in the sky $Q_{\mathbf{k}}(\vec{\ell}) \equiv E_{\mathbf{k}}(\vec{\ell})$ is nonvanishing only if ℓ points towards $\mathbf{k}$, which is the direction we have temporarily chosen for the x axis. Transforming to the $\mathbf{k}$-independent choice of x axis that is to be used to sum the Fourier components, we will have $Q_{\mathbf{k}}(\vec{\ell}) = E_{\mathbf{k}}(\vec{\ell})\cos(2\phi_{\vec{\ell}})$ and $U_{\mathbf{k}}(\vec{\ell}) = E_{\mathbf{k}}(\vec{\ell})\sin(2\phi_{\vec{\ell}})$. Adding up

---

[4] We stress that this notation of E and B for the polarization components has nothing to do with the electric and magnetic fields of the incident radiation. The notation is derived from an analogy with gravitational radiation.

all components we therefore find

$$Q(\vec{\ell}) = E(\vec{\ell})\cos(2\phi_{\vec{\ell}}),$$ (5.68)

$$U(\vec{\ell}) = E(\vec{\ell})\sin(2\phi_{\vec{\ell}}).$$ (5.69)

This remarkable relation between the two Stokes parameters is here seen to follow from general principles. Pedestrian derivations have been given in the literature, either working within a small patch (e.g., Seljak 1997) or by taking the small-angle limit of the all-sky formalism that we encounter in a moment (e.g., Zaldarriaga and Seljak 1997).

Within the small patch of sky, $E(\vec{\ell})$ is a two-dimensional Gaussian random field, depending linearly on $\mathcal{R}(\mathbf{k})$ just like the temperature perturbation $a(\vec{\ell})$. Including the latter for completeness, the correlations of temperature, polarization, and the cross correlation between temperature and polarization are

$$\langle a^*(\vec{\ell})a(\vec{\ell}')\rangle = C(\ell)\,\delta^2(\vec{\ell} - \vec{\ell}'),$$ (5.70)

$$\langle a^*(\vec{\ell})E(\vec{\ell}')\rangle = C_{\text{cross}}(\ell)\,\delta^2(\vec{\ell} - \vec{\ell}'),$$ (5.71)

$$\langle E^*(\vec{\ell})E(\vec{\ell}')\rangle = C_E(\ell)\,\delta^2(\vec{\ell} - \vec{\ell}'),$$ (5.72)

where

$$C(\ell) = 4\pi \int_0^\infty T_\Theta^2(k,\ell)\mathcal{P}_\mathcal{R}(k)\frac{dk}{k},$$ (5.73)

$$C_{\text{cross}}(\ell) = 4\pi \int_0^\infty T_\Theta(k,\ell)T_E(k,\ell)\mathcal{P}_\mathcal{R}(k)\frac{dk}{k},$$ (5.74)

$$C_E(\ell) = 4\pi \int_0^\infty T_E^2(k,\ell)\mathcal{P}_\mathcal{R}(k)\frac{dk}{k},$$ (5.75)

and $T_E(k,\ell)$ is some transfer function. From observation, we will be able to check that these very special expressions are valid and to determine the three $C(\ell)$. One route would be to use, for C_{cross} and C_E, expressions analogous to Eq. (5.36) for C. More simply, we could determine the $E(\vec{\ell})$ by performing the Fourier expansion in a finite patch of sky. All of this is analogous to the situation for the density contrast observed around us, discussed in Section 4.3.3.

5.3.3 An all-sky analysis

As noted earlier, the polarization is expected also to be detectable on large angular scales, in which case we cannot use a small patch of sky. For an all-sky analysis (Kamionkowski et al. 1997a,b; Seljak and Zaldarriaga 1997; Zaldarriaga and Seljak 1997; Hu and White 1997c), we use spherical polar angles (θ, ϕ), with some arbitrary choice for the pole and the azimuthal orientation. Then, to define the Stokes parameters $Q(\mathbf{e})$ and $U(\mathbf{e})$ for a given direction in the sky, we choose the x axis to point along the θ direction. However, for each Fourier component, we still take the polar axis to be in the direction of $\mathbf{k}$, and use this axis to define the contribution of that component to the Stokes parameters. We need a formalism that allows the Fourier components to be added over the whole sky.

From Eq. (5.60), a clockwise rotation by angle ψ, about a given direction **e** in the sky, has the effect

$$Q(\mathbf{e}) \pm iU(\mathbf{e}) \to \exp(\mp 2i\psi)\,[Q(\mathbf{e}) \pm iU(\mathbf{e})]. \tag{5.76}$$

The appropriate spherical expansion for these quantities is one in terms of the **spin-weighted harmonics** $_{\pm 2}Y_{\ell m}(\mathbf{e})$ (Sazhin and Shulga 1996). Under a rotation, they have the same transformation law (4.42) as the ordinary harmonics $Y_{\ell m}(\mathbf{e})$, except for an additional factor $\exp[\mp 2i\psi(\mathbf{e})]$, where $\psi(\mathbf{e})$ is the rotation of the coordinate lines at a given point in the sky. They also have the same orthonormality and completeness properties. They are defined in Section 15.5.

We use the notation $_{\pm 2}Y_{\ell m} \equiv Y_{\ell m}^{\pm}$. Then the multipoles of the polarization are defined by

$$Q(\mathbf{e}) \pm iU(\mathbf{e}) = \sum_{\ell m} a_{\ell m}^{\pm} Y_{\ell m}^{\pm}(\mathbf{e}). \tag{5.77}$$

Under a change in the coordinates (θ, ϕ), the multipoles $a_{\ell m}^{\pm}$ of $Q \pm iU$ have the same transformation (4.43) as the multipoles $a_{\ell m}$ of the temperature anisotropy. In particular there is no mixing between different ℓ, nor between $+$ and $-$.

It is useful to write

$$a_{\ell m}^{\pm} \equiv (E_{\ell m} \pm i B_{\ell m}). \tag{5.78}$$

From Eqs. (5.64) and (5.65) et seq., $B_{\ell m}$ vanishes[5] and we have only $E_{\ell m}$. Analogously with Eq. (5.25), there will be some transfer function $T_E(k, \ell)$ such that

$$E_{\ell m} = \frac{4\pi}{(2\pi)^{3/2}} \int T_E(k, \ell)\mathcal{R}_{\ell m}(k)k\,dk. \tag{5.79}$$

This formalism neatly reproduces the results that we gave for the small patch of sky (Zaldarriaga and Seljak 1997). As the notation implies, $T_E(\ell)$ for large ℓ becomes the transfer function that we considered for that case. However, we now can go to low ℓ as well. Each of the multipoles $E_{\ell m}$ has an independent Gaussian distribution, except for the reality condition, and the nonzero correlations are

$$\langle a_{\ell m}^* a_{\ell' m'} \rangle = C(\ell)\delta_{\ell\ell'}\delta_{mm'}, \tag{5.80}$$

$$\langle a_{\ell m}^* E_{\ell' m'} \rangle = C_{\text{cross}}(\ell)\delta_{\ell\ell'}\delta_{mm'}, \tag{5.81}$$

$$\langle E_{\ell m}^* E_{\ell' m'} \rangle = C_E(\ell)\delta_{\ell\ell'}\delta_{mm'}, \tag{5.82}$$

with C, C_{cross}, and C_E given by Eqs. (5.73), (5.74), and (5.75), respectively.

In Figure 5.11, we show predictions for these correlations, again using CMBFAST (Seljak and Zaldarriaga 1996).

5.4   Reionization

The Gunn–Peterson test, which we discuss in more detail in Section 11.6, shows that there is, at most, a very low level of absorption of quasar light by neutral hydrogen in the intergalactic

---

[5] For each Fourier component, this is true with the pole in the **k** direction, and it remains true for an arbitrary choice of the pole because each of $a_{\ell m}^{\pm}$ transforms in the same way as an ordinary multipole.

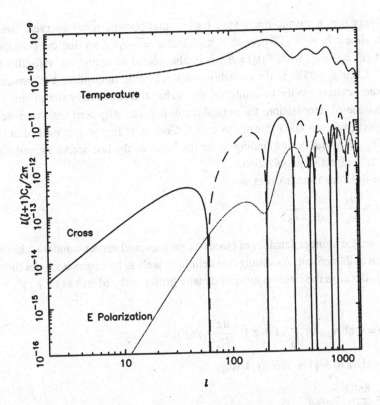

Fig. 5.11. Temperature and polarization spectra for standard CDM (no reionization); the parameters are $n = 1, h = 0.5$, and $\Omega_b h^2 = 0.016$. For the cross correlation, the dashed part of the line is anticorrelated.

medium, the conclusion being that already by redshift 5, and conceivably much earlier, the Universe was in a high state of ionization. The most plausible energy source for this reionization is early structure formation, and in Section 11.6 we examine how such reionization might have come about and estimate the redshift.

In this section, we take for granted that reionization happened at some redshift z_{ion}, and examine its effect on the cosmic microwave background anisotropies. The reason for the effect is that the cosmic microwave background photons are free to scatter off the free electrons thus created, whereas they lack the energy to interact with bound atoms. Despite this, the interaction cross-section is actually rather low for the electron densities at low redshift.

5.4.1 Optical depth

The probability per unit time for a photon to scatter is $n_e\sigma_T$, where n_e is the number density of free electrons and σ_T is the Thomson-scattering cross-section. The **optical depth** $\kappa(t)$ is defined as

$$\kappa(t) = \sigma_T \int_t^{t_0} n_e(t)\, dt. \tag{5.83}$$

The probability that a cosmic microwave background photon, now observed, has travelled freely since time t is $e^{-\kappa(t)}$. [To see this, note that $\dot{\kappa} = -\sigma_T n_e$, so that the probability $P(t)$ satisfies $dP/dt = -\dot{\kappa}P$, with $P(t_0) = 1$.] It is also useful to define the **visibility function**, $g = -\dot{\kappa}e^{-\kappa}$. Clearly, $g(t)\,dt$ is the probability that a cosmic microwave background photon, now observed, scattered in the time interval dt and has travelled freely since then.

In the absence of reionization, the optical depth is practically zero until we reach the decoupling epoch, when it rises sharply to $\kappa \gg 1$. Correspondingly, g is peaked at the epoch of decoupling, its width corresponding to the thickness of the last-scattering surface. We are interested in the effect of reionization.

We define the ionization fraction as

$$\chi(z) \equiv \frac{n_e}{n_p} \simeq \frac{n_e}{0.88\,n_b}, \tag{5.84}$$

where these are the number densities of electrons, protons, and baryons, and we take the helium mass fraction as 24 percent. Assuming that helium as well as hydrogen becomes fully ionized gives $\chi = 1$. Because the electron number density grows with redshift as $(1+z)^3$, the optical depth is

$$\kappa(z) = 0.88\,n_{b,0}\,\sigma_T \int_0^z (1+z')^2\,\frac{dz'}{H(z')}\,\chi(z'), \tag{5.85}$$

where we used $dz = -(1+z)H\,dt$. Using

$$\Omega_b = \frac{8\pi G}{3H_0^2}m_b n_{b,0} \tag{5.86}$$

and Eq. (2.34) gives the optical depth to redshift z as

$$\kappa(z) = \kappa^* \int_0^z (1+z')^{1/2}\sqrt{\frac{\Omega(z')}{\Omega_0}}\,\chi(z')\,dz', \tag{5.87}$$

where the redshift dependence of Ω is given by Eq. (2.48) or Eq. (2.51), depending on whether the Universe is open or flat, and

$$\kappa^* = 0.88\,\frac{3H_0\Omega_b\sigma_T}{8\pi Gm_b} \simeq 0.050h\,\Omega_b. \tag{5.88}$$

Let us assume instantaneous reionization at redshift z_{ion}, so that $\chi(z) = 1$ for $z < z_{ion}$ and zero otherwise. Then $\kappa(z)$ has a constant value between decoupling and reionization, equal to $\kappa(z_{ion})$. We study the critical-density case, though Eq. (5.87) also can be integrated in the low-density cases. With critical density, it is given by

$$\kappa(z_{ion}) \simeq 0.033h\,\Omega_b\,(1+z_{ion})^{3/2}. \tag{5.89}$$

[Equivalently, this expression gives $\kappa(z)$, assuming full reionization at all epochs.] If we insert the central nucleosynthesis value for Ω_b, this becomes

$$\kappa(z_{ion}) \simeq 5 \times 10^{-4}h^{-1}(1+z_{ion})^{3/2}. \tag{5.90}$$

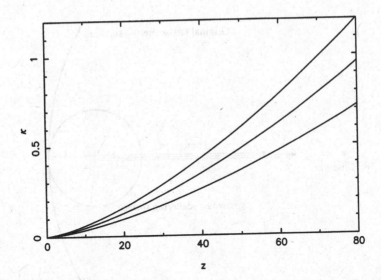

Fig. 5.12. Optical depth $\kappa(z)$, assuming full ionization, for $h = 0.5$ and (from bottom to top) $\Omega_b = 0.06$, 0.08, and 0.10.

This tells us that, to achieve an optical depth of unity, we would need to reionize the Universe completely by a redshift of around 100. We see in Chapter 11 that models of the type we are considering give $z_{ion} \sim 10$ to 40, when their parameters are constrained by existing data. We conclude that at least some reasonable fraction of the cosmic microwave background photons from the original last-scattering surface will make it to our location.

The Gunn–Peterson test tells us that the Universe must have been highly ionized by redshift 5. However, putting this into Eq. (5.90) gives an optical depth of only 2 percent or so, implying that, if reionization occurs as late as possible, there will be only a tiny effect on the cosmic microwave background. In particular, this also implies that even if the neutral hydrogen absorption systems contain a high fraction of (unseen) ionized gas, they have only a tiny effect on the cosmic microwave background photons. Figure 5.12 shows the optical depth as a function of reionization redshift, for different choices of baryon density.

5.4.2 Effect of reionization on the cosmic microwave background anisotropy

Let us turn to the effect of this on the cosmic microwave background anisotropy. Consider first the effect on the temperature anisotropy $\delta T / T$. The probability that a cosmic microwave background photon, now observed, scattered at least once after decoupling is given by

$$P_{scatt} = 1 - \exp(-\kappa),\qquad(5.91)$$

where κ is evaluated at decoupling. What we see looking out along a given direction is therefore a superposition, consisting of a fraction $1 - P_{scatt}$ of photons from the original last-scattering surface, plus a fraction P_{scatt} that have been scattered into the line of sight from a variety of different directions. The rescattered photons have originated from a surface whose radius is

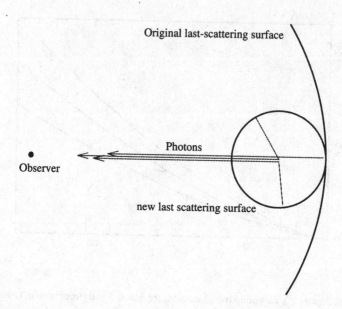

Fig. 5.13. Schematic representation of cosmic microwave background photons scattering from ionized electrons. Photons that scatter a single time after last scattering originate on the last-scattering sphere surrounding the rescattering point. As seen by a distant observer, that sphere just touches the original last-scattering surface. Photons that scatter more than once originate from inside the sphere. The plot is in comoving coordinates **x**.

the distance between the original last-scattering surface and the rescattering point, in the case of a single rescattering, or from within that sphere if multiple scatterings have occurred. This is shown in Figure 5.13.

The fraction of the photons that reach us directly from the last-scattering surface without rescattering clearly give the usual contribution to the anisotropy because they have been unaffected by reionization. Those that did scatter will not give the usual contribution because their point of origin was somewhere on the surface of the small circle shown in Figure 5.13 (or within it in the case of multiple rescattering), which has a different anisotropy pattern than the original last-scattering surface. When all rescattered photons are taken into account, we observe an averaging of the temperature over their possible points of origin.

The effect on the anisotropy depends on the scale being examined. On large angular scales, larger than the horizon size at the time of rescattering (the small circle in Figure 5.13), the anisotropies are unaffected by rescattering because the rescattering occurs among photons within a large region sharing the same large-scale temperature contrast. This is simply a statement of causality; rescattering cannot affect physics on scales larger than the horizon size at that time. On small scales the situation is different; the temperature on the original last-scattering surface is uncorrelated with the mean temperature on the sphere[6] from which the rescattered photons originated (the small sphere in Figure 5.13). This means that when

---

[6] In general, this has to be weighted both for multiple scatterings and for the anisotropy of Thomson scattering, but neither affect this argument.

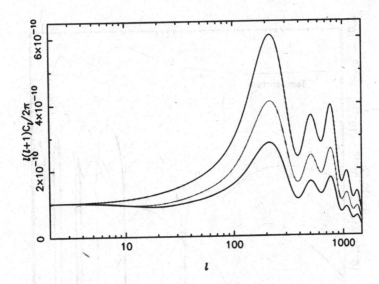

Fig. 5.14. The cosmic microwave background spectrum of standard CDM with reionization. The top curve has no reionization; the lower two have optical depths κ of 0.2 and 0.4. Each curve is normalized to the same matter spectrum normalization, rather than to COBE.

considering the small-scale anisotropy, the rescattered photons can be considered to have the *mean* cosmic microwave background temperature, so that they contribute no anisotropy at all.

We consequently have two limiting behaviours:

$$C_\ell^{obs} = C_\ell^{int} \qquad \text{(Small } \ell \text{)}, \tag{5.92}$$

$$C_\ell^{obs} = \exp(-2\kappa)\, C_\ell^{int} \qquad \text{(Large } \ell \text{)}, \tag{5.93}$$

where C_ℓ^{int} is the intrinsic cosmic microwave background spectrum on the last-scattering surface at $z \simeq 1,000$, and C_ℓ^{obs} is that which we actually see. Equation (5.90) gives κ, and the factor 2 is simply because the C_ℓ is given by the square of the temperature contrast. The break between these two behaviours can be estimated from the angular size of the region in which the rescattered photons originate, that is, from the horizon size at rescattering, as being at ℓ of a few tens.

The overall effect of reionization is therefore to leave the largest scales untouched, but to suppress the C_ℓ on small angular scales by a uniform multiplier.

A more detailed treatment has been given by Hu and White (1997a), who derived what they call the reionization damping envelope, which interpolates between the two regimes as a function of optical depth and horizon size at last scattering. This shows that indeed the large-ℓ regime with its constant suppression is attained, typically for $\ell \gtrsim 30$. The first ten or so multipoles are affected only marginally even if the total optical depth is high.

Finally, we can perform the full calculation of linear perturbation theory, as described in Chapter 15. This is necessary to include new anisotropies generated by the peculiar velocity of the electrons from which the photons rescatter, ignored in the preceding discussion. Figure 5.14 shows the C_ℓ curves for standard CDM with various different redshifts of reionization, produced

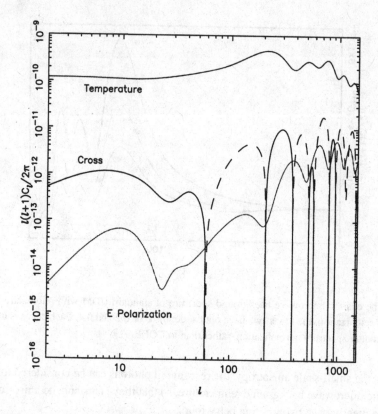

Fig. 5.15. Effect of reionization on polarization of the cosmic microwave background, for optical depth $\kappa = 0.2$. Note the large boost to the small-ℓ polarization compared to that of Figure 5.11.

by CMBFAST (Seljak and Zaldarriaga 1996). Note the limiting behaviours at small and large ℓ, as discussed above.

In the future, it should be possible to measure the polarization as well as the temperature anisotropy. In the absence of reionization, polarization is produced only as the radiation emerges from the last-scattering surface. Then it is significant only at $\ell \gtrsim 100$, corresponding to scales within the horizon at photon decoupling. However, more polarization is produced after reionization, which will produce significant polarization at low ℓ. Measuring the precise amount should allow z_{ion} to be determined from observation, giving information about the spectrum $\mathcal{P}_R$ on very small scales. Spectra for optical depth $\kappa(z_{ion}) = 0.2$ are shown in Figure 5.15.

Because reionization suppresses the original anisotropies, it can mean that second-order effects give the most important contribution. The largest of these is the Ostriker–Vishniac effect (Ostriker and Vishniac 1986; Vishniac 1987; Hu et al. 1994a; Dodelson and Jubas 1995), which is a coupling of long-wavelength velocities to short-wavelength density perturbations and comes in at large ℓ. The best calculations of this at present are those of Hu and White (1996a) and Jaffe and Kamionkowski (1998). However, detecting it will be a daunting observational challenge.

–All of the above assumes that reionization occurs homogeneously throughout the Universe. In reality, most existing models predict that it will occur through mergers of ionized bubbles over a range of redshifts. Several estimates of the extent to which inhomogeneous reionization can imprint anisotropies onto the cosmic microwave background have been made (Aghanim et al. 1996; Gruzinov and Hu 1998; Knox et al. 1998; Peebles and Juszkiewicz 1998).

Already, observation tells us that the Universe cannot have reionized very early, at $z \sim 100$ or so, because there is fairly clear evidence of the first acoustic peak (see Figure 5.9 on page 125). This is in accordance with the prediction of our models, and is of course enormously good news for satellite experiments such as MAP and Planck. Very early reionization would have erased almost all the cosmological information that they would aim to extract.

Examples

5.1 In Eq. (4.98), during radiation domination, it is convenient to replace t by $y \equiv a/a_{eq}$, to obtain what is called the Meszaros equation (5.10). Derive it and verify the growing and decaying solutions (5.11) and (5.12). Determine the ratio of the coefficients by imposing the time dependence $\delta_{ck} \propto (aH)^{-2}$ around horizon entry, and hence show that δ_{ck} grows logarithmically between horizon entry and matter domination.

5.2 Use Eq. (5.34) for $a(\vec{\ell})$, with the definition (5.26) of C_ℓ and the definition (4.34) of $\mathcal{P}_\mathcal{R}$, to show that, before taking the continuous limit for ℓ, one has the ensemble average $\ell \langle a(\vec{\ell}) a^*(\vec{\ell'}) \rangle = C_\ell \delta_{\ell\ell'} \delta(\phi_{\vec{\ell}} - \phi_{\vec{\ell'}})$. After taking the continuous limit $\sum_\ell \to \int d\ell$, this provides a pedestrian derivation of Eq. (5.35), derived in the text from general principles.

5.3 Derive Eq. (5.55), giving the intensity of polarized radiation in terms of the Stokes parameters.

5.4 Compute the present electron density, and hence the electron density at recombination. Taking care to distinguish comoving and physical coordinates, use this to obtain a numerical value for the thickness of the last-scattering surface, using $d = (t_c/H_{rec})^{1/2}$ quoted in Section 5.2.4. Compare this with the true value $d = 7\Omega_0^{-1/2} h^{-1}$ Mpc.

5.5 Obtain Eq. (5.52) giving the Sachs–Wolfe effect, by integrating Eq. (5.50) by parts.

5.6 Integrate Eq. (5.87) for the low-density case, both flat and open, to obtain the generalization of Eq. (5.89).

5.7 If we relax the assumption of critical density, what redshift of reionization would lead to an optical depth of unity, as a function of Ω_0?

6 Extensions to the simplest model

There is considerable evidence that the simplest model, discussed in the preceding two chapters, is not correct; in particular, observations favour a low-density Universe. In this chapter we investigate a series of possibilities for modifying the simplest model. We begin with modifications of the pure cold dark matter (CDM) hypothesis, in the interest of simplicity keeping the (probably unrealistic) assumption that the total matter density is critical. Then we discuss the case of a low-density Universe, both with and without a cosmological constant, now for simplicity assuming the dark matter is entirely cold. We go on to examine the possibility of significant gravitational waves. Finally, we entertain the possibility that the density perturbation is not purely adiabatic, but rather has an isocurvature component, again making the simplifying assumption of critical mass density.

6.1 Modifying the cold dark matter hypothesis

CDM is defined as having constituents with negligible random motion. This is the simplest possibility but not the only one and, in this section, we discuss alternatives. The opposite extreme is hot dark matter (HDM), normally assumed to be in the form of neutrinos. Pure HDM is not viable, but a mixture of up to 30 percent or so is allowed, and tends to improve the agreement between theory and observation (Bonometto and Valdarnini 1984; Fang et al. 1984; Shafi and Stecker 1984). This is the cold plus hot dark matter (CHDM) model.

6.1.1 CHDM model

The present density of a neutrino species with mass m_ν is given by a standard abundance calculation (2.69) as

$$\Omega_\nu h^2 = \frac{m_\nu}{94\,\text{eV}}.$$

(6.1)

The energy per neutrino is $\simeq 3k_B T_\nu$ (see Section 2.2.1), until the neutrinos become nonrelativistic at the epoch $3k_B T_\nu \simeq m_\nu$, corresponding to

$$z_{nr} \simeq 6 \times 10^4 h^2 \Omega_\nu.$$

(6.2)

For Ω_ν big enough to be of interest, this is about the same as the epoch of matter–radiation equality. This is no coincidence; because the temperatures of neutrinos and photons are similar,

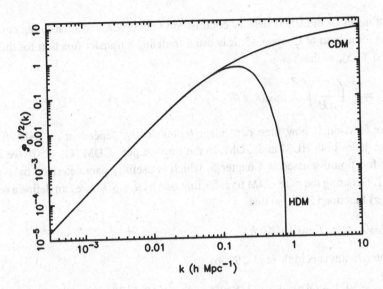

Fig. 6.1. The present spectrum $\mathcal{P}_0^{1/2}(k)$ for pure HDM compared to that of standard CDM, both computed using CMBFAST.

their densities are also similar, and we are requiring that the neutrinos have a density similar to that of the nonrelativistic matter once they become nonrelativistic. In this case, we have $z_{eq} = 2.8 \times 10^4 \, h^2$, slightly different from Eq. (2.32) because now we have only two massless neutrino species.

Consider first the case of pure HDM, $\Omega_\nu \simeq 1$. The scale entering the horizon at $z = z_{eq}$ is $k^{-1} \sim 13 \, h^{-2}$ Mpc, using Eq. (2.35) modified for two neutrino types. Because $z_{eq} \simeq z_{nr}$, the HDM freestreams on smaller scales (after horizon entry), which wipes out its density perturbation, as shown in Figure 6.1. As a result, pure HDM is completely unviable; with no surviving perturbations on galaxy scales, we cannot hope to match observations.

We therefore assume that CDM is also present. Then, the HDM can fall into the potential wells already created by the CDM, so that the HDM density contrast grows to match that of the CDM. At any epoch, this can occur on scales below a scale called the neutrino Jeans length. The root mean square (rms) velocity v_ν of the HDM particles is comparable to c at z_{nr}, and afterward falls like $1/a$. The comoving Jeans length $k_J = 2\pi a/\lambda_J$ therefore is given by Eq. (4.132) as

$$k_J^{-1} = \frac{2.3 \times 10^{-2}(1+z)^{1/2}}{h^3 \Omega_\nu} \text{ Mpc}. \tag{6.3}$$

On scales $\gtrsim (h^3 \Omega_\nu)^{-1}$ Mpc (e.g., $\gtrsim 100$ Mpc), the HDM falls into the potential wells straight away, at the epoch $z_{nr} \sim 10^4$. On such scales, the transfer function is the same as for pure CDM. As the scale is reduced, the transfer function falls because the growth of the HDM density contrast is delayed. Finally, on scales $\lesssim 1$ Mpc, it flattens out because the growth of the HDM density contrast still has not begun by the present epoch $z = 0$.

These estimates of the HDM density contrast are very crude; for an accurate calculation, we need the formalism of Chapter 15. It is usual to define a transfer function for the CHDM model by Eq. (5.5), so that

$$\delta_{\mathbf{k}}(t) = \frac{2}{5} \left(\frac{k}{aH} \right)^2 T_{\text{CHDM}}(k, z) \mathcal{R}_{\mathbf{k}}. \tag{6.4}$$

The transfer function is now time dependent (equivalently, dependent on redshift z), and $\delta \equiv \delta\rho/\rho$ includes both HDM and CDM. In the limit of pure CDM ($\Omega_\nu = 0$), we arrive at the transfer function discussed in Chapter 5, which is usefully parameterized by Eq. (5.13) with $\Omega_0 = 1$. Denoting the pure CDM transfer function by $T_{\text{CDM}}(k)$, we can define a reduction factor $D(q, z)$ function (5.13), so that

$$T_{\text{CHDM}}(k, z) = T_{\text{CDM}}(k)D(k, z). \tag{6.5}$$

One parameterization is (Liddle et al. 1996b)

$$D(q, z) = \left[\frac{1 + (Aq)^2 + a_{\text{eq}}(1 + z)(1 - \Omega_\nu)^{1/\beta}(Bq)^4}{1 + (Bq)^2 - (Bq)^3 + (Bq)^4} \right]^\beta, \tag{6.6}$$

where

$$\beta = \frac{5}{4} \left(1 - \sqrt{1 - 24\Omega_\nu/25} \right),$$

$$A = 17.266 \frac{(1 + 10.912\Omega_\nu)\sqrt{\Omega_\nu(1 - 0.9465\Omega_\nu)}}{1 + (9.259\Omega_\nu)^2},$$

$$B = 2.6823 \frac{1.1435}{\Omega_\nu + 0.1435},$$

$$a_{\text{eq}} = \frac{4.212 \times 10^{-5}}{h^2}, \tag{6.7}$$

and $q = (k/h^2)\exp(2\Omega_B)$ as before.[1] The form of $D(q)$ at the present epoch is illustrated in Figure 6.2.

A more accurate parameterization, along with numerical codes for its implementation, is given by Eisenstein and Hu (1999).

In Figure 6.3, we compare the CHDM and CDM predictions for $\sigma_0(M)$, taking a neutrino density of 0.3 for the former. We see that the bottom-up structure formation will proceed more or less as with pure CDM. The reduction of small-scale power means, though, that there is a lack of high-redshift objects, and we find later that a value Ω_ν significantly bigger than 0.3 is excluded for this reason.

In the preceding account, we assume that only one neutrino species has significant mass. An alternative is to suppose that two or even all three species have almost the same mass, and this affects the detailed prediction without changing the qualitative picture. At the time of writing,

---

[1] The epoch of matter–radiation equality here differs slightly from what we have been quoting because it was based on an older measurement of the microwave background temperature.

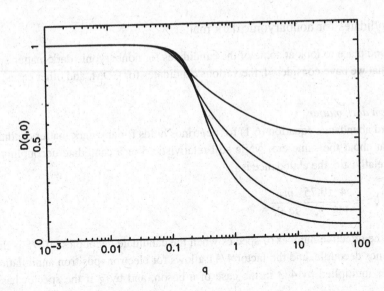

Fig. 6.2. The reduction factor $D(q, z)$ at the present epoch, showing (from top to bottom) $\Omega_\nu = 0.1, 0.2, 0.3$, and 0.4.

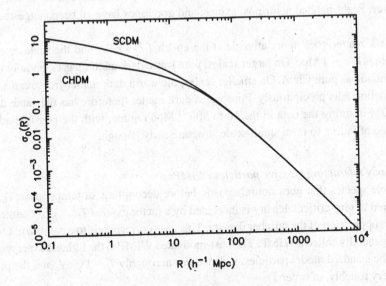

Fig. 6.3. The standard CDM model compared with a CHDM model with $\Omega_\nu = 0.3$.

a naïve acceptance of all currently available hints about the neutrino mass would demand this mass degeneracy, but there is no good theoretical reason to expect it.

Finally, let us note that the cosmic microwave background (cmb) anisotropy is the same as in the pure CDM model, except for the very highest multipoles corresponding to the scale of order $10h^{-1}$ Mpc.

6.1.2   Candidates for nonbaryonic dark matter

This is a good place to look at some of the candidates for nonbaryonic dark matter, including the HDM that we have considered, the various candidates for CDM, and other cases.

Warm dark matter

The standard abundance equation (6.1) for neutrinos holds for any dark matter candidate that decouples at about the same era, while still relativistic. For a candidate decoupling earlier, while still relativistic, the abundance is

$$\Omega_{dm} h^2 = \frac{4}{11} \frac{10.75}{g_*} \frac{m_s}{94 \, eV}. \tag{6.8}$$

Here, g_* is the effective number of species when the candidate decouples, 10.75 is the value when neutrinos decouple, and the factor 4/11 allows for electron–positron annihilation. This abundance is multiplied by 4/3 in the case of a boson, and by g if the species has g spin states.

From Figure 2.1 on page 19, we see that a candidate decoupling at $T \sim 1 \, GeV$, where $g_* \simeq 75$, would have critical density with mass $m_s \sim 1 \, keV$. This case usually is called **warm dark matter**. Right-handed neutrinos, axinos, and gravitinos have all been suggested as candidates.

Warm dark matter goes nonrelativistic at the epoch $T \sim 1 \, keV$, and the scale entering the horizon then is $k^{-1} \sim 1 \, Mpc$. On larger scales, pure warm dark matter has essentially the same transfer function as pure CDM. On smaller scales, the warm dark matter freestreams, and the transfer function falls precipitously. Pure warm dark matter therefore has the same difficulty as pure CDM in fitting the data in the 10 to $50h^{-1} \, Mpc$ regime, with the possible additional disadvantage of failing to form small-scale structure early enough.

Weakly interacting massive particles (WIMPs)

For a particle species that goes nonrelativistic before decoupling, at temperature T_{freeze}, the mass required to give critical density is increased by a factor $\exp(m/T_{freeze})$, to counteract the Boltzmann suppression of the number density. This case corresponds to a candidate for CDM, of the type generally called a WIMP. The best-motivated WIMP is the lightest supersymmetric partner of the standard model particles. It decouples at roughly $T \sim 1 \, GeV$, and the predicted Ω_{CDM} is, very roughly, of order 1.

No thermal equilibrium

The dark matter might not start out in thermal equilibrium. The best-motivated possibility here is the axion, generated through one of the mechanisms described in Section 7.8. It is a CDM candidate. Alternatively, right-handed neutrinos may be produced by the neutrino oscillation phenomenon, which would give something like warm dark matter.

Another possibility arises if the usual radiation-dominated era, containing the epoch of nucleosynthesis, is preceded by an era of matter domination. Such an era would end when the relevant particles decay, and the decay products could include nonbaryonic dark matter and/or baryons. If a nonbaryonic decay product is nonrelativistic, it is CDM; otherwise it is some version of hot or warm dark matter, with a nonthermal distribution function.

For a single species, we might end up with both thermal and nonthermal components to the distribution. This possibility has been explored both for a known neutrino species (Kaiser et al. 1993) and for the axion's superpartner, the axino (Bonometto et al. 1994).

Finally, we might consider dark matter that originates as a quantum fluctuation during inflation, or that is created during the reheating era after inflation.

Once we abandon the requirement of thermal equilibrium, almost any mass becomes possible for the dark matter particles, depending on the production mechanism. The axion must be very light, but candidates coming from decay or from quantum fluctuations might be very heavy. Then they are necessarily CDM particles. Being very rare compared with nuclei, they could be electrically charged, or even have colour (the quantum number carried by quarks). At least in the latter case they would be combined with quarks, to give a very heavy and rare type of atomic nucleus.

Finally, there is the possibility that CDM particles interact, at least in the early Universe, through a new long-range force.

6.1.3 Producing additional radiation

We end this section by mentioning a somewhat different modification of the simplest model, which has nothing to do with HDM. Going back to the idea that radiation domination is preceded by an era of matter domination, the decay products of the matter might include radiation that interacts too weakly to thermalize. (Photons, by contrast, thermalize to reach the blackbody distribution if they are produced before a redshift of a million or so, leaving no permanent effect.) Such radiation is very dangerous for nucleosynthesis, but a detailed study of the case of a decaying τ neutrino shows that it need not be fatal (Dodelson et al. 1994). We also can make it completely safe, by having the relevant particle decay after nucleosynthesis (White et al. 1995a), but a suitable candidate is hard to find.

How does such radiation affect structure formation? The point is that increasing the radiation density by a factor α delays the onset of matter domination, increasing k_{eq}^{-1} by a factor $\alpha^{1/2}$, and decreases the shape parameter Γ of Eq. (5.14) by the same factor. To obtain the favoured value $\Gamma \sim 0.3$ with $n = 1$ and $h = 0.5$ requires $\alpha \sim 3$. We therefore need a lot of radiation to get a significant effect. The effect normally is described in terms of $g_* = 3.36\alpha$.

6.2   **ΛCDM model**

Now we contemplate the possibility that the present total matter density Ω_0 is less than 1, with a cosmological-constant contribution $\Omega_{\Lambda 0}$. For simplicity, the latter is chosen so that $\Omega_0 + \Omega_{\Lambda 0} = 1$, corresponding to a flat spatial geometry, consistent with standard inflation models. At the completion of this book, the ΛCDM model (with $\Omega_0 \sim 0.3$) is widely regarded as the model best capable of matching the observational data.

In these models, it is useful to define the density transfer function $T(k)$ by applying Eq. (5.4) at early times, before Ω breaks away from 1 and hence before Ω_Λ becomes significant. That is because the *shape* of the spectrum already is determined before these things happen because the growth of δ is given by the scale-independent expression (4.104).

The transfer function is independent of the cosmological constant because the latter is negligible at early times. It does depend on Ω_0, which determines the matter density and, in particular, the epoch of matter–radiation equality. To good accuracy it is given by Eq. (5.13). It follows that lowering Ω_0 offers another way of decreasing Γ toward the observed range. However, the normalization of the spectrum is changed when we go to a low-density Universe because the interpretation of the microwave anisotropies changes.

6.2.1   Matter-density contrast

For nonzero cosmological constant, with or without critical matter density, the time dependence of the matter-density contrast is given by Eq. (4.104). Until Ω breaks away from 1, δ grows like a, and then it grows more slowly. Equivalently, $\Phi_{\mathbf{k}}$ initially has some constant value Φ^{early}, and then is suppressed by some factor

$$g \equiv \frac{\Phi_{\mathbf{k}}(t)}{\Phi_{\mathbf{k}}^{\text{early}}}.$$

(6.9)

From Eq. (4.92), the corresponding suppression factor of δ is g/Ω. The suppression is given by Eq. (4.104), normalized so that $D_1 = a$ at early times; that is,

$$a\frac{g}{\Omega} = D_1 \equiv \frac{5}{2}\Omega_0 H_0^2 H \int_0^a \frac{da}{(aH)^3},$$

(6.10)

with

$$H^2 = H_0^2 \left(\frac{\Omega_0}{a^3} + \Omega_{\Lambda 0} \right).$$

(6.11)

The integral can be written in terms of elliptic functions (Eisenstein 1997), but is more commonly performed numerically or parameterized. An accurate parameterization, shown in Figure 6.4, is (Carroll et al. 1992):

$$g(\Omega) = \frac{5}{2}\Omega \left(\frac{1}{70} + \frac{209\Omega}{140} - \frac{\Omega^2}{140} + \Omega^{4/7} \right)^{-1}.$$

(6.12)

Note that Ω is related to $a^{-1} = 1 + z$ by Eq. (2.51).

Combining these results, the matter-density contrast is

$$\delta_{\mathbf{k}}(t) = \frac{2}{5}\left(\frac{k}{aH} \right)^2 \frac{g(\Omega)}{\Omega} T(k, \Omega_0)\mathcal{R}_{\mathbf{k}},$$

(6.13)

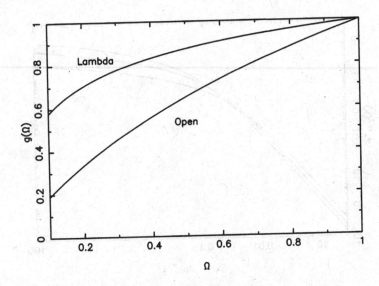

Fig. 6.4. Growth suppression factor in ΛCDM and open CDM models.

and its spectrum is

$$\mathcal{P}_\delta = \frac{4}{25}\left(\frac{k}{aH}\right)^4\left[\frac{g(\Omega)}{\Omega}\right]^2 T^2(k,\Omega_0)\mathcal{P}_\mathcal{R}(k). \tag{6.14}$$

We need a generalization of the definition (5.7) of $\delta_H(k)$ for low-density Universes. We choose to include the growth suppression factor, so that

$$\delta_H^2(k) = \frac{4}{25}\left[\frac{g(\Omega)}{\Omega}\right]^2 \mathcal{P}_\mathcal{R}(k). \tag{6.15}$$

With this definition, formulae (5.6) and (5.17) remain valid in the low-density case.

As we see in Section 6.2.2, the Cosmic Background Explorer (COBE) normalization of $\mathcal{P}_\mathcal{R}$ in the ΛCDM model is practically independent of Ω_0. This means that, on large scales, $\mathcal{P}_\delta$ is increased relative to the pure CDM model by a factor $(g/\Omega)^2$. Predictions for the present value of $\mathcal{P}_\delta$ are given in Figure 6.5.

For the peculiar velocity, it is useful to define a suppression factor f at fixed density contrast. From Eq. (4.119),

$$f = \frac{a}{D_1}\frac{dD_1}{da}. \tag{6.16}$$

A good approximation at the present epoch is

$$f(\Omega_0) \simeq \Omega_0^{0.6}. \tag{6.17}$$

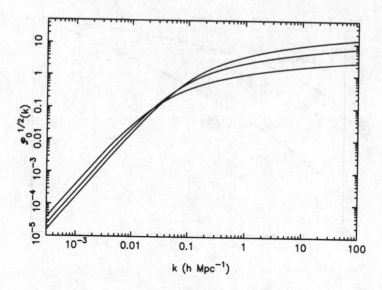

Fig. 6.5. Three spectra for the ΛCDM model, each normalized to COBE and with $h = 0.5$ and $\Omega_b h^2 = 0.016$. From top to bottom at the right, we show $\Omega_0 = 1$ (standard CDM, or SCDM), 0.6, and 0.3. Note the higher COBE normalization at low density, offset on short scales by the larger-scale turnaround of the spectrum. In models with the same-shape Γ, the low-density ones would be higher at all k.

6.2.2   The cosmic microwave background anisotropy

As in Eq. (5.48), the cmb anisotropy can be separated into that which is present at last scattering, and that which is acquired on the journey toward us. Using the comoving slicing, the former is negligible for low ℓ, corresponding to scales far outside the horizon at last scattering, whereas the latter gives the Sachs–Wolfe effect. Taking into account the time dependence of Φ, we show in Section 15.3.1 that the Sachs–Wolfe effect is

$$\left(\frac{\delta T(e)}{T}\right)_{\text{SW}} = \frac{1}{3}\Phi(ex_{\text{ls}}, \tau_{\text{ls}}) + 2\int_{\tau_{\text{ls}}}^{\tau_0} \frac{\partial \Phi(ex, \tau)}{\partial \tau}\, d\tau. \tag{6.18}$$

The second term is called the **integrated Sachs–Wolfe effect** (Sachs and Wolfe 1967). The integral goes over the photon trajectory, so that $dx = -d\tau \ (\equiv -dt/a)$, leading to $x(\tau) = \tau_0 - \tau$. Using the spherical expansion (4.30), this becomes

$$a_{\ell m} = \int_0^\infty dk\, \Phi_{\ell m}^{\text{early}}(k)\left[\frac{1}{3}\Pi_{k\ell}(x_{\text{ls}}) + 2\int_0^{\tau_0} d\tau\, \Pi_{k\ell}(x)\frac{dg(\tau)}{d\tau}\right], \tag{6.19}$$

where $\Pi_{k\ell}(x) = k\sqrt{2/\pi}\, j_\ell(kx)$ is the radial function. We have set the lower limit of the τ integral equal to 0, an excellent approximation. To find C_ℓ, we take the mean square of this expression, and use the definition (4.39) of the spectrum of $\mathcal{R} = -(5/3)\Phi^{\text{early}}$. This gives

$$C_\ell = 2\pi^2 \int_0^\infty \frac{dk}{k}\mathcal{P}_\mathcal{R}(k)I_{k\ell}^2, \tag{6.20}$$

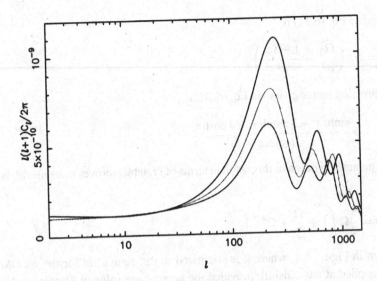

Fig. 6.6. The ΛCDM and SCDM models are compared, for $h = 0.5$, $\Omega_b h^2 = 0.016$, and spectral index $n = 1$, with all curves normalized to COBE. From bottom to top, the lines are $\Omega_0 = 1$ (SCDM), $\Omega_0 = 0.6$, and $\Omega_0 = 0.3$.

where

$$kI_{k\ell} = \frac{1}{5}\Pi_{k\ell}(x_{ls}) + \frac{6}{5}\int_0^{\tau_0} \Pi_{k\ell}(x(\tau))\frac{dg(\tau)}{d\tau}\,d\tau. \tag{6.21}$$

Unless Ω_0 is unfeasibly small, the first term dominates (Kofman and Starobinsky 1985), corresponding to the first term in Eq. (6.18), which is the ordinary Sachs–Wolfe effect.

Going back to Eq. (5.48), the perturbation in the cmb already present at last scattering becomes dominant as we move up to higher ℓ. In Figure 6.6, we give a comparison between the SCDM and ΛCDM predictions for the cmb anisotropy, normalized to the COBE data.

6.3 Open CDM model

The open CDM model reduces the matter density Ω_0, without introducing a cosmological constant. The spatial geometry is then non-Euclidean, with a negative curvature scale of order H_0^{-1}. It is convenient to normalize the scale factor so that $K = -1$, and to use conformal time τ.

6.3.1 Matter-density contrast

We can ignore spatial curvature when considering the matter-density contrast, because it is observed only on scales much less than the curvature scale. For zero cosmological constant, the equivalent of Eq. (6.11) is

$$H^2 = H_0^2 \left(\frac{\Omega_0}{a^3} - \frac{1 - \Omega_0}{a^2} \right),$$

(6.22)

and the suppression factor given via Eq. (6.10) is

$$g(\tau) = 5 \frac{\sinh^2 \tau - 3\tau \sinh \tau + 4 \cosh \tau - 4}{(\cosh \tau - 1)^3}.$$

(6.23)

An accurate parameterization of this result in terms of Ω, which proves more useful, is (Carroll et al. 1992)

$$g(\Omega) = \frac{5}{2} \Omega \left(1 + \frac{\Omega}{2} + \Omega^{4/7} \right)^{-1}.$$

(6.24)

This is shown in Figure 6.4, where it is compared to the equivalent for the ΛCDM model. This can be applied at any redshift, provided the appropriate value of $\Omega(z)$ is used, given by Eq. (2.48).

As in the ΛCDM case, we generalize the definition of δ_H to the low-density case, including the growth suppression factor (6.15). As long as we do not venture near the curvature scale, the dispersion $\sigma(R)$ still can be calculated using the flat-space formula (5.17).

The velocity suppression factor (6.17) is a good approximation in the open model as well as the cosmological-constant one.

6.3.2   Radial functions

To handle the cmb anisotropy, we have to modify the spherical expansion (4.30) to take account of the non-Euclidean geometry. We continue to denote the eigenfunctions of the spatial Laplacian by $Z_{k\ell m}$, so that $\nabla^2 Z_{k\ell m} = -(k/a)^2 Z_{k\ell m}$. With $k \geq 1$, these form a complete orthonormal set, and the multipole expansion of a generic function g is

$$g(x, \theta, \phi) = \int_1^\infty dk \sum_{\ell=0}^\infty \sum_{m=-\ell}^\ell g_{k\ell m} Z_{k\ell m}(x, \theta, \phi).$$

(6.25)

The mode functions are

$$Z_{k\ell m}(x, \theta, \phi) = \Pi_{k\ell}(x) Y_{\ell m}(\theta, \phi).$$

(6.26)

The radial functions $\Pi_{k\ell}(x)$ satisfy

$$\left[\frac{1}{\sinh^2 x} \frac{d}{dx} \left(\sinh^2 x \frac{d}{dx} \right) + \frac{\ell(\ell+1)}{\sinh^2 x} \right] \Pi_{k\ell}(x) = -k^2 \Pi_{k\ell}(x).$$

(6.27)

They are given by

$$\Pi_{k\ell}(x) = N_{k\ell} \, \tilde{\Pi}_{k\ell}(x),$$

(6.28)

where

$$\tilde{\Pi}_{k\ell}(x) = q^{-2}(\sinh x)^{\ell} \left(\frac{-1}{\sinh x}\frac{d}{dx}\right)^{\ell+1} \cos(qx), \qquad (6.29)$$

and

$$N_{k\ell} = \sqrt{\frac{2}{\pi}} q^2 \prod_{s=0}^{\ell}(s^2+q^2)^{-1/2}. \qquad (6.30)$$

In the last two expressions, $q^2 \equiv k^2 - 1$. In the flat-space limit, $\Pi_{k\ell}(x)$ reduces to $\sqrt{2/\pi}\, k\, j_{\ell}(kx)$.

The normalization $N_{k\ell}$ is chosen to correspond to the orthonormality relation

$$\int dV Z_{k\ell m}^*(\mathbf{x}) Z_{k'\ell'm'}(\mathbf{x}) = \delta(q - q')\delta_{\ell\ell'}\delta_{mm'}, \qquad (6.31)$$

where $dV = \sinh^2 x \sin\theta\, dx\, d\theta\, d\phi$.

Just as in the Euclidean case, the inflationary vacuum fluctuation will give each $\mathcal{R}_{k\ell m}$ an independent Gaussian probability distribution. This distribution is not affected by a change in the origin or orientation of the coordinates (x, θ, ϕ) because such a change induces a unitary transformation on the functions $Z_{k\ell m}$ at fixed k. The unitarity of the transformation is guaranteed by the orthonormality relation.

The spectrum is defined by

$$\langle g_{k\ell m} g_{k'\ell'm'}^* \rangle = \frac{2\pi^2}{qk^2}\mathcal{P}_g(k)\delta(q - q')\delta_{\ell\ell'}\delta_{mm'}, \qquad (6.32)$$

leading to the position-independent ensemble average

$$\langle g^2(x, \theta, \phi) \rangle = \int_1^{\infty} \frac{dk}{k}\mathcal{P}_f(k). \qquad (6.33)$$

Given the position independence, this relation is derived easily by choosing $x = 0$.

6.3.3 The cmb anisotropy

The low multipoles are dominated by the Sachs–Wolfe effect, given by Eqs. (6.20) and (6.21), with g now given by Eq. (6.23). In contrast to the ΛCDM model, the integral now gives a large contribution, which means that we are probing a wide range of relatively small linear scales, in contrast to the narrow range of large scales that is probed by the other term.

As we go to larger ℓ, the perturbation $\Theta(t_{ls}, \mathbf{x}_{ls}, \mathbf{n})$ at last scattering becomes significant [first term of Eq. (5.48)], but it relates only to a single linear scale that is small compared with the curvature scale. It is therefore a reasonable approximation to ignore the curvature when evaluating $\Theta(t_{ls}, \mathbf{x}_{ls}, \mathbf{n})$, except insofar as it increases the angle subtended by a given distance at last scattering. As shown in Section 2.4, the increase is approximately a factor $\Omega_0^{-1/2}$, and this shifts the first acoustic peak to higher ℓ by the same factor.

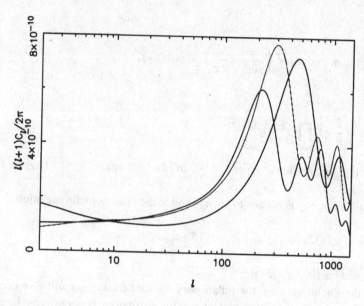

Fig. 6.7. Spectrum of the temperature anisotropy, comparing the open model with the SCDM model, both normalized to COBE. The parameters are $h = 0.5$, $\Omega_b h^2 = 0.016$, and $n = 1$, corresponding to $\mathcal{P}_\mathcal{R}$ being scale independent. SCDM has its first peak at $\ell \simeq 220$; the others, with the peak location moving rightward, are $\Omega_0 = 0.6$ and $\Omega_0 = 0.3$.

In Figure 6.7, the anisotropies in the open model are compared to those of the SCDM model, taking $\mathcal{P}_\mathcal{R}$ to be scale independent in both cases.

Similarly to the ΛCDM model, the matter-spectrum normalization is higher in the open Universe case, at least as long as Ω_0 is not too small. The enhancement is not quite as much as $g^2(\Omega_0)/\Omega_0^2$ because the line-of-sight term in the Sachs–Wolfe integral is not negligible unless we are very close to the flat case. Note also that the growth suppression factor is larger than in the cosmological-constant case. The COBE normalization of open models is given in detail in Chapter 9. Some spectra are shown in Figure 6.8.

6.3.4   Subtleties

Super-curvature modes

According to single-bubble models of inflation, which give an open Universe as discussed in Section 8.7, the preceding analysis needs two types of modification. The first is relatively trivial. When we calculate the vacuum fluctuation, we find that the usual formula for $\mathcal{P}_\mathcal{R}(k)$ is modified in the regime $1 < k \lesssim 2$, so that a flat potential does not lead to a spectrum $\mathcal{P}_\mathcal{R}(k)$ that is independent of k. The change in the predicted C_ℓ turns out to be insignificant, essentially because the integrated Sachs–Wolfe effect is dominant, so that the scales probed by even the low multipoles are quite small.

The second point is more subtle. The mode functions discussed earlier form a complete orthonormal set, in the sense that any square-integrable function has a unique expansion in

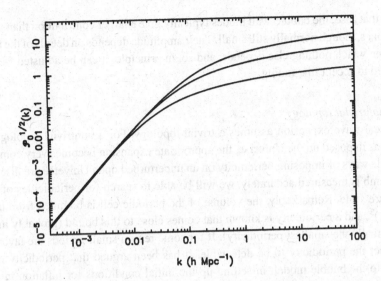

Fig. 6.8. Three spectral for the open model, each normalized to COBE and with $h = 0.5$ and $\Omega_0 h = 0.016$. From top to bottom at the right, we show $\Omega_0 = 1$, 0.6, and 0.3. The change in COBE normalization is much less dramatic than that in Figure 6.5.

terms of them. However, to produce the most general random Gaussian scalar field, we need to include the range $0 < k < 1$, using appropriate analytic continuations of the radial functions (Lyth and Woszczyna 1995). These are called **super-curvature modes** because the inverse eigenvalue k^{-1} is bigger than the curvature scale.

The single-bubble inflation models produce a super-curvature mode with a discrete value of k, the precise value depending on the details of the tunneling that creates the open Universe (Bucher and Turok 1995; Yamamoto et al. 1996). It also may be absent, though it is present in all the favourite models. Often, it can be approximated by a mode with $k = 0$.

There is more yet. To generate the most general random Gaussian *tensor* field, which can be obtained by acting on a scalar field with an nth order differential operator, we need also discrete modes with $k^2 = 1 - n^2$ (García-Bellido et al. 1997). The open inflation models generate a set of these modes with $n = 2$, originating as perturbations of the nucleated bubble from perfect sphericity. We can call these the **bubble-wall modes** (Cohn 1996; García-Bellido 1996; Garriga 1996; Hamazaki et al. 1996; Yamamoto et al. 1996). In fact, they should be interpreted directly as tensor modes, though the cmb anisotropy can be computed either way (Garriga et al. 1998; Sasaki et al. 1997; Tanaka and Sasaki 1997).

The super-curvature modes give the open-Universe version of the Grishchuk–Zel'dovich effect. It now may significantly affect several of the lower multipoles, but is still excluded by the data because it gives the wrong ℓ-dependence for the low multipoles (García-Bellido et al. 1995; Lyth and Woszczyna 1995). In contrast to the flat case, the absence of the effect *is* a significant constraint on some models, in particular, excluding a model introduced by Linde as we describe in Section 8.7, though it is still small in most of them. The problematic models are ones in which the vacuum energy after tunneling is much less than before.

Barring that case, the bubble-wall modes typically give a larger contribution than the super-curvature ones, though normally still small. Their amplitude depends on details of the tunneling potential, on which nothing else depends, and so, in principle, it can be adjusted freely if we are prepared to accept fine-tuning.

A nontrivial topology?

The spherical wave expansion assumes a trivial topology. For a nontrivial topology, where periodicity is imposed on the Universe, the appropriate expansion becomes very complicated. The possible ways of imposing periodicity on an unperturbed open Universe are discrete and, when the cmb is measured accurately, we will be able to search for periodicities of order of the curvature scale. Remarkably, the volume of the periodic cell is bounded from below by $(0.550\, d_{curv})^3$, and a periodicity is known that comes close to this bound (probably indicating that it is actually the smallest periodicity). If for some reason small periods are favoured, we might expect the periodicity to be detectable. It has been argued that periodicity offers an alternative to the bubble model, in setting up the initial conditions for inflation in an open Universe (Cornish et al. 1996).

We also can impose a nontrivial topology in the flat case, for example, by keeping a finite box size L in the Fourier expansion. From the observed cmb anisotropy, we can deduce a lower limit on L, somewhat larger than the size of the observable Universe (Stevens et al. 1993; de Oliveira-Costa et al. 1996). This is less interesting than in the open case because there is no reason to expect the period to be of order H_0^{-1}.

6.4   Fine-tuning issues

We have now considered three options for changing the simplest model. All of them involve fine-tuning, but the type of fine-tuning is different in each of the three cases.

In the CHDM model, the ratio of HDM to CDM is fine-tuned. Assuming that the HDM and CDM have independent origins, this fine-tuning is in addition to what already is required to give critical matter density (or, stated more correctly, to give the Hubble parameter and the cmb temperature simultaneously their observed values). The CDM and HDM might have a common origin, which would alleviate the problem. In any case, the question is why Nature has chosen particular values for certain couplings and masses, and this same question already presents itself in several other cosmological contexts.

In the open CDM model, we have to ask why we live soon after the epoch when Ω breaks away from 1. In the context of a bubble model of inflation, our Universe must leave the horizon not too long after the bubble nucleates. This involves a fine-tuning of particle physics parameters, over and above what is needed in the usual inflation models.

In the ΛCDM model, a particle physics explanation would require an understanding of why the vacuum energy density has the tiny value $\rho_{vac}^{1/4} \sim 10^{-3}$ eV. At present, no reasonable proposal exists.

Instead of appealing to particle physics, we might invoke the anthropic principle. In its most concrete setting, this supposes that the observable Universe is only one of many. The others might be encountered as one moves around in space, for instance, in models giving an early epoch of eternal inflation, or they might exist only as possible quantum states. The other possible Universes might have different particle physics, meaning either different parameters or completely different forms for the action. Then we note that most of the other possible Universes would be inhospitable to life, and the possibility arises that the observable Universe is the way it is because we are here to observe it.

The anthropic approach is particularly attractive for the cosmological constant (Efstathiou 1995), whose small value is proving so difficult to understand on the basis of particle physics. Let us suppose that, at least in the cosmologically interesting regime $\rho_{vac}^{1/4} \lesssim 10^{-3}$ eV, the probability distribution of ρ_{vac} is flat. (This distribution refers to a random choice among the possible Universes, with some sensible choice for the probability measure.) Then the probability $\mathbf{P}(\rho_{vac}) \, d\rho_{vac}$ of us observing ρ_{vac} in a given range will be something like

$$\mathbf{P}(\rho_{vac}) \, d\rho_{vac} = \frac{N(\rho_{vac}) \, d\rho_{vac}}{\int_0^\infty N(\rho_{vac}) \, d\rho_{vac}}, \tag{6.34}$$

where $N(\rho_{vac})$ is the number of galaxies that ever form in a given comoving region. The function $\mathbf{P}$ defined by this expression can be estimated using the Press–Schechter approximation, discussed in Chapter 11. Present estimates indicate that it is fairly flat in the range corresponding to $0 \lesssim \Omega_{\Lambda 0} \lesssim 0.8$, falling off steeply only at higher values. The anthropic principle therefore might explain why the cosmological constant is so small, and at the same time lead us to expect a value significantly different from zero.

6.5   Gravitational waves

For the rest of this chapter we turn to modifications of the simplest model which are motivated by theory rather than by observation.

Gravitational waves are an inevitable consequence of all inflationary models (Grishchuk 1974; Rubakov et al. 1982; Starobinsky 1982, 1983; Abbott and Wise 1984), being created as a vacuum fluctuation in exactly the manner of the density perturbations. In the classification of Section 4.6.2, the gravitational waves are a tensor perturbation, and they cause anisotropy and polarization of the cmb at some level.

6.5.1   The gravitational-wave spectrum

In the Robertson–Walker Universe, a gravitational wave corresponds to the spatial metric perturbation h_{ij}, defined by Eqs. (14.133) and (14.135). It is traceless, $\delta^{ij} h_{ij} = 0$, and transverse, $\partial_i h_{ij} = 0$. This means that each Fourier component is of the form

$$h_{ij} = h_+ e_{ij}^+ + h_\times e_{ij}^\times. \tag{6.35}$$

In a coordinate system where $\mathbf{k}$ points along the z axis, the nonzero components of the polarization tensors are defined by $e_{xx}^+ = -e_{yy}^+ = 1$ and $e_{xy}^\times = e_{yx}^\times = 1$.

Like any perturbation, $h_{ij}(\mathbf{x}, t)$ can be taken to live in unperturbed space-time because it would be a second-order effect to include the effect of metric perturbations on the coordinates $\mathbf{x}$ and t.

From the Einstein action Eq. (14.82), we find that each of the amplitudes h_+ and $h_\times$ in Eq. (6.35) has the same action as a free, massless, scalar field. To be precise, the canonically normalized scalar fields corresponding to these amplitudes are

$$\psi_{+,\times} \equiv \frac{M_{\mathrm{Pl}}}{\sqrt{2}} h_{+,\times}. \tag{6.36}$$

This result for the action is valid to second order, corresponding to a result for the field equation that is valid to first order. One needs to go one order higher in the action than in the field equations.

As we note on page 75, slow-roll inflation gives any free, massless, scalar field ψ a vacuum fluctuation, whose spectrum well after horizon exit is

$$\mathcal{P}_\psi = \left(\frac{H}{2\pi}\right)^2 \bigg|_{k=aH}. \tag{6.37}$$

Each of $h_{+,\times}$ therefore has a spectrum $\mathcal{P}_{\mathrm{grav}}$, given by

$$\mathcal{P}_{\mathrm{grav}}(k) = \frac{2}{M_{\mathrm{Pl}}^2} \left(\frac{H}{2\pi}\right)^2 \bigg|_{k=aH}. \tag{6.38}$$

Because H is slowly varying, $\mathcal{P}_{\mathrm{grav}}$ is nearly scale independent, just like the spectrum $\mathcal{P}_\mathcal{R}$ of the primordial curvature perturbation. Its scale dependence normally can be taken to be a power law,

$$\mathcal{P}_{\mathrm{grav}}(k) \propto k^{n_{\mathrm{grav}}}. \tag{6.39}$$

On each scale, the primordial gravitational wave amplitudes $h_{+,\times}$ remain constant until the approach of horizon entry. Then they start to vary, and well after horizon entry we have a decaying oscillation, corresponding to redshifting gravitational waves.

The variation of the gravitational-wave amplitudes gives rise to anisotropy and polarization of the cmb, which we now consider.

6.5.2   The cmb anisotropy

The dominant contribution to the cmb anisotropy is in the low multipoles $\ell \ll 100$ that correspond to scales well outside the horizon at decoupling. For higher multipoles, corresponding to smaller scales, the amplitude of the gravitational waves has been reduced from its primordial value by the redshift. As in the case of the scalar perturbation, practically all of the anisotropy on the large scales is acquired by the cmb on its journey toward us.[2]

---

[2] Gravitational waves do not contribute to the monopole and dipole of the brightness function discussed in Section 5.2.6. As we noted there, all multipoles except the monopole are expected to be small on scales well outside the horizon.

As in Section 5.2.6, this anisotropy may be ascribed to the successive redshifts seen by a sequence of observers. The fractional perturbation in the distance between adjacent observers, with separation $\delta \mathbf{r}$, is $h_{ij} e_i e_j$, where $\mathbf{e} = \delta \mathbf{r}/|\delta \mathbf{r}|$. (This is the definition, Eqs. (14.133) and (14.135), of the gravitational-wave amplitude.) The perturbation in the velocity gradient is therefore $\partial h_{ij}/\partial t$, and the gravitational Sachs–Wolfe effect is

$$\frac{\delta T(\mathbf{e})}{T} = -\int_{\tau_{ls}}^{\tau_0} e_i e_j \frac{\partial h_{ij}(x, \tau)}{\partial \tau} \, d\tau, \tag{6.40}$$

where the photon trajectory is $x(\tau) = \tau_0 - \tau$. A more standard derivation of this formula, using the geodesic equation, is given in Chapter 15.

The evolution of the gravitational-wave amplitude during matter domination is given by the massless field equation (14.136), whose solution is

$$h_{+,\times}(\tau) = \left[3\sqrt{\frac{\pi}{2}} \frac{J_{3/2}(k\tau)}{(k\tau)^{3/2}} \right] h_{+,\times}. \tag{6.41}$$

Here, $\tau = 2/(aH)$ is conformal time, and $h_{+,\times}$ on the right-hand side denotes the primordial amplitude well outside the horizon whose spectrum is given by Eq. (6.38).

We can calculate the spectrum C_ℓ corresponding to Eq. (6.40) by projecting out the multipole $a_{\ell m}$ and using the definition (4.39) of $\mathcal{P}_g$ (applied with $g = h_{+,\times}$ at the epoch t_*). This is best done by using the spin-2 spherical harmonics considered in Section 5.3.3. For $\ell \ll 100$, the time dependence of $h_{+,\times}$ is given by Eq. (6.41), and we can take $\tau_{ls} = 0$ under the pretense that matter domination extends into the infinite past. This gives a result of the form (Starobinsky 1985a)

$$\ell(\ell+1)C_\ell = \frac{\pi}{9} \left(1 + \frac{48\pi^2}{385} \right) \mathcal{P}_{\text{grav}} c_\ell, \tag{6.42}$$

where $c_2 = 1.118$, $c_3 = 0.878$, and $c_4 = 0.819$ with $c_\infty = 1$.

A full calculation takes into account both photon scattering and the departure from matter domination. We find that the above form works for $\ell \lesssim 10$, with c_ℓ close to 1 at $\ell \sim 10$. In this regime, the above result is good also if $\mathcal{P}_{\text{grav}}(k)$ has moderate scale dependence, provided that it is evaluated at the scale $k \simeq \ell H_0/2$, which dominates the ℓth multipole.

Figure 6.9 shows the result of a full calculation of the C_ℓ spectrum for power-law gravitational-wave spectra, using the CMBFAST code (Seljak and Zaldarriaga 1996). The cutoff is seen clearly, as is the shape correction [represented by c_ℓ in Eq. (6.42)] at low ℓ.

In a given inflationary model, the crucial question is how this contribution to the cmb anisotropies compares with that of the density perturbations, and we investigate this in detail in Chapters 7 and 8. Obviously, there being only a single microwave sky, we would observe the sum of the two contributions, rather than each separately. An illustrative example, not tied to any particular inflationary model, is shown in Figure 6.10. As we see, in almost all current inflationary models the contribution of gravitational waves is in fact negligible.

In the open model, the preceding results are modified. The cutoff will be moved to higher ℓ because a given scale at decoupling corresponds to a smaller angle, roughly as $\ell \propto \Omega_0^{-1/2}$.

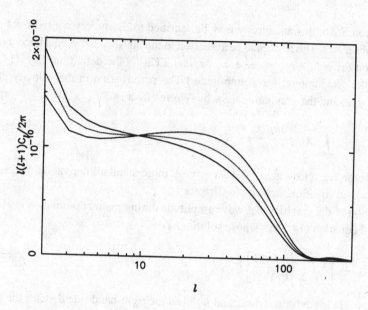

Fig. 6.9. The C_ℓ for gravitational waves with $n_{\text{grav}} = 0, -0.1$, and -0.2 (reading upward at the left-hand edge).

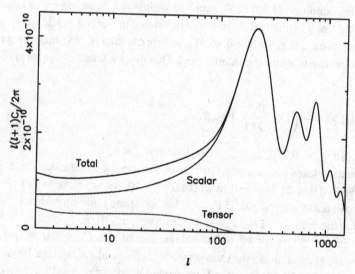

Fig. 6.10. The C_ℓ for scale-invariant scalars and tensors, plus the total, normalized to be 70 percent scalars at the COBE scale.

Also, on very large angular scales, we have to include the effect of spatial curvature when calculating the spectrum $\mathcal{P}_{\text{grav}}$. Considerable care is needed in calculating the latter effect but, as in the case of the adiabatic density perturbation, it has little observational significance when cosmic variance is taken into account.

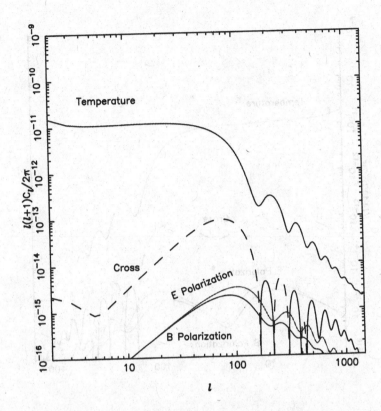

Fig. 6.11. The four temperature and polarization spectra for a scale-invariant tensor spectrum. We normalize so that the gravitational waves contribute 10 percent of the COBE temperature anisotropy signal.

6.5.3   Polarization of the cmb

Extending the analysis of Section 5.3.3, gravitational waves contribute to both $E_{\ell m}$ and $B_{\ell m}$. There are now four observable correlations:

$$\langle a^*_{\ell m} a_{\ell' m'} \rangle = C(\ell)\, \delta_{\ell \ell'} \delta_{m m'}, \tag{6.43}$$

$$\langle a^*_{\ell m} E_{\ell' m'} \rangle = C_{\text{cross}}(\ell)\, \delta_{\ell \ell'} \delta_{m m'}, \tag{6.44}$$

$$\langle E^*_{\ell m} E_{\ell' m'} \rangle = C_E(\ell)\, \delta_{\ell \ell'} \delta_{m m'}, \tag{6.45}$$

$$\langle B^*_{\ell m} B_{\ell' m'} \rangle = C_B(\ell)\, \delta_{\ell \ell'} \delta_{m m'}. \tag{6.46}$$

Each of the C is the sum of the scalar contribution given by Eqs. (5.73)–(5.75) (which vanishes for C_B) and the gravitational-wave contribution given by expressions of the same form with $\mathcal{P}_\mathcal{R}$ replaced by $\mathcal{P}_{\text{grav}}$, and different transfer functions. There is no cross correlation between B and the other quantities because it has odd parity.

In Figures 6.11 and 6.12, we show some theoretical predictions for the shape of the gravitational-wave contribution to the cmb spectra, generated using the code CMBFAST.

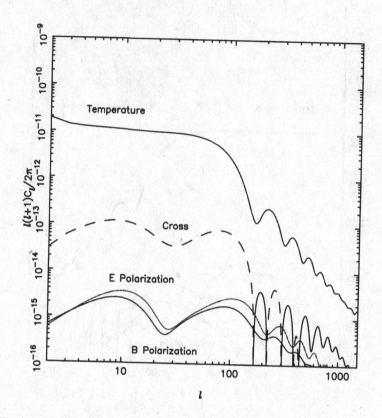

Fig. 6.12. The four temperature and polarization spectra for a scale-invariant tensor spectrum, and with an optical depth $\kappa = 0.2$. We normalize so that the gravitational waves contribute 10 percent of the COBE temperature anisotropy signal.

6.6   Isocurvature perturbations

Instead of the adiabatic initial condition (4.5), we could consider an *isocurvature* initial condition, which specifies perturbations in the energy densities of two (or more) species that add up to zero. It is called an isocurvature initial condition because there is no perturbation in the spatial curvature of comoving slices (the primeval curvature perturbation $\mathcal{R}_\mathbf{k}$ vanishes). The specific reference to comoving slices usually is not spelled out in the literature, and one might even form the impression that it is the curvature of space-time that is supposed to be unaffected. Of course, that is not so because, in general, isocurvature perturbations affect the pressure, which is related through the field equation to the space-time geometry. In particular, the curvature scalar of space-time is $R = -8\pi G(\rho - 3P)$.

Each particle species constitutes either radiation or matter (except for a possible brief epoch of transition between the two regimes, occurring, for example, in the case of neutrino dark matter). It is useful to define the "entropy perturbation" of a species i by

$$S_i = \frac{\delta(n_i/n_\gamma)}{(n_i/n_\gamma)} = \frac{\delta n_i}{n_i} - \frac{\delta n_\gamma}{n_\gamma} \tag{6.47}$$

$$= \begin{cases} \delta_i - \frac{3}{4}\delta_\gamma & \text{(matter)}, \\ \frac{3}{4}\delta_i - \frac{3}{4}\delta_\gamma & \text{(radiation)}. \end{cases} \qquad (6.48)$$

This term is indeed appropriate for a matter species because, then, $-S_i$ is the perturbation in the entropy per matter particle (recall that the entropy is dominated by the radiation, with the photons carrying a fixed fraction of it). Note that S_i is gauge independent because n_i/n_γ is time independent in the absence of perturbations. An isocurvature perturbation is specified by giving the value of S_i for each species, at the same initial epoch that we use to specify an adiabatic density perturbation.[3]

The most general density perturbation is a superposition of an isocurvature density perturbation (specified by one or more initial entropy perturbations S_i with initial $\mathcal{R} = 0$) and an adiabatic density perturbation (specified by the initial curvature perturbation $\mathcal{R}$ with all the $S_i = 0$). We see that, according to observation, the isocurvature component is subdominant if it exists at all. In all probability it is zero or completely negligible, but its study is still important from the viewpoint of particle physics. The reason is that the particle theory may be constrained significantly by the requirement that the isocurvature perturbation be no bigger than what is observed.

We focus exclusively on an isocurvature perturbation between the CDM density and the rest. Thus, we assume that, for the isocurvature perturbation, there is initially a common radiation-density contrast δ_r, a baryon-density contrast $\delta_b = 3\delta_r/4$, and a CDM density contrast specified by the initial value of

$$S = \delta_c - \frac{3}{4}\delta_r = \frac{\rho_r\delta\rho_c - \frac{3}{4}\rho_c\delta\rho_r}{\rho_r\rho_c} = \frac{\rho_r + \frac{3}{4}\rho_c}{\rho_r\rho_c}\delta\rho_c \simeq \delta_c. \qquad (6.49)$$

We have used the fact that $\delta\rho_r + \delta\rho_c = 0$, and the last equality holds because the initial epoch is during radiation domination.

6.6.1 Observational consequences of an isocurvature perturbation

To calculate the effect of an isocurvature perturbation on large-scale structure, we need to know the scale dependence of its spectrum. Motivated by the particle theory modelling described in Section 7.8, we assume that the dependence is of power-law form

$$\mathcal{P}_S(k) \propto k^{n_{\text{iso}}}, \qquad (6.50)$$

where scale invariance means $n_{\text{iso}} = 0$. Note that some of the literature defines n_{iso} so that it has the value -3 for a scale-invariant entropy fluctuation, coming from the use of Eq. (4.17) as the definition of the spectrum; another part of the literature, including CMBFAST (1999 version), uses

---

[3] Recall that this epoch is a few Hubble times before cosmological scales start to enter the horizon, and that it occurs during radiation domination but well after nucleosynthesis. Recall also that along each comoving worldline, the Universe on scales well outside the horizon is supposed to evolve in the same way as an unperturbed Universe. This means that, on each scale, S_i is time independent until the approach of horizon entry.

$n_{\mathrm{iso}} = 1$, presumably to mimic the adiabatic case. Keeping to our convention, the spectral index n_{iso} is expected to be slightly below 0, though a value significantly above 0 is also possible.

In contrast to the case of an adiabatic perturbation, inflation can generate a non-Gaussian isocurvature perturbation almost as easily as a Gaussian one. In what follows, we use the formalism that we developed in the context of Gaussian perturbations, noting that much of it applies also to the non-Gaussian case.

Isocurvature matter-density transfer function

For an isocurvature perturbation, $\mathcal{R}$ vanishes during the radiation-dominated era preceding the present matter-dominated era. However, a nonzero $\mathcal{R}^{(m)}$ has appeared by the time of matter domination. We can define a transfer function as

$$\mathcal{R}_{\mathbf{k}}^{(m)} = \frac{1}{3} T_{\mathrm{iso}}(k) S_{\mathbf{k}}. \tag{6.51}$$

The right-hand side is evaluated well before horizon entry, which for most scales means during matter domination.

We chose the normalization so that $T_{\mathrm{iso}} = 1$ on the very large scales entering the horizon well after matter domination. To check this normalization, we can integrate Eq. (4.166), between the initial epoch and one well after matter domination. The integrand is significant only around the epoch $\rho_{\mathrm{r}} \sim \rho_{\mathrm{c}}$, and using Eq. (6.49) with $\delta\rho_{\mathrm{c}} = -\delta\rho_{\mathrm{r}}$ gives

$$\mathcal{R}_{\mathbf{k}}^{(m)} = -\frac{1}{3} \int_{t_1}^{t_2} dt\, H \frac{\delta\rho_{\mathrm{rk}}}{\rho + P}, \tag{6.52}$$

$$\simeq \frac{1}{3} S_{\mathbf{k}} \int_0^{\infty} \frac{dx}{\left(1 + \frac{3}{4}x\right)\left(\frac{4}{3} + x\right)}, \tag{6.53}$$

$$= \frac{1}{3} S_{\mathbf{k}}. \tag{6.54}$$

On the second line, $x = \rho_{\mathrm{c}}/\rho_{\mathrm{r}}$.

The situation on smaller scales is similar to the one that we described for the adiabatic case. To estimate the transfer function roughly, we note that $\delta_{\mathrm{c}} \sim S$ at horizon entry and that there is little subsequent growth of δ_{c} until matter domination. As a result, the transfer function again is given roughly by Eq. (5.9). For a more accurate calculation, we should use the formalism of Chapter 15. In Figure 6.13, we compare this with the adiabatic transfer function.

The cosmic microwave background anisotropy

The Sachs–Wolfe effect, Eq. (5.53), gives the anisotropy acquired by the radiation on its way toward us, whether the initial condition is adiabatic or isocurvature. For the isocurvature case, Eq. (6.54) allows us to write it as

$$\left[\frac{\delta T(\mathbf{e})}{T}\right]_{\mathrm{jour}} = -\frac{S(\mathbf{e} x_{\mathrm{ls}})}{15}. \tag{6.55}$$

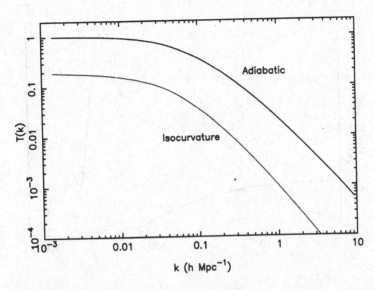

Fig. 6.13. The isocurvature and adiabatic transfer functions are compared, with $h = 0.5$ and $\Omega_b h^2 = 0.016$, both computed from CMBFAST. The adiabatic one is normalized to unity at large scales, and the isocurvature one shown at the relative amplitude assuming both spectra COBE normalized (rather than to one on large scales as in the text). As estimated in the text, the isocurvature spectrum is about a factor 6 smaller on large scales, and even more so on small ones.

However, the initial anisotropy is not now negligible. Because we are in matter domination, Eq. (6.49) gives $S = -3\delta_r/4$, and therefore,

$$\Theta = \frac{1}{4}\delta_r = -\frac{1}{3}S. \tag{6.56}$$

This is 5 times bigger than the Sachs–Wolfe effect (Starobinsky and Sahni 1984; Efstathiou and Bond 1986; Kodama and Sasaki 1987), with the same sign, giving $\delta T/T = -6S(e\mathbf{x}_{ls})/15$.

Normalizing to COBE, assuming flat spectra, and making the rough approximation that the transfer functions are equal, we conclude that during matter domination a pure isocurvature density perturbation is only one-sixth as big as a pure adiabatic density perturbation. The full calculation gives a similar result. Because the pure adiabatic density perturbation more or less agrees with the data, the pure isocurvature one is ruled out. A detailed study (Stompor et al. 1996) shows that an isocurvature contribution of up to 50 percent or so is allowed to the cmb anisotropy, but it would do nothing to improve the observational situation in the 10 to $50h^{-1}$ Mpc regime.

A tilted Universe?

We noted in Chapter 5 that a very large scale contribution to the spectrum of an adiabatic density perturbation would contribute mainly to the quadrupole cmb anisotropy. In particular, the dipole contribution to the Sachs–Wolfe effect is suppressed because a very large scale

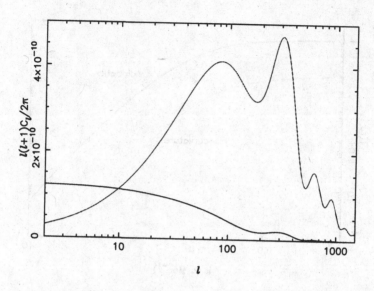

Fig. 6.14. The cmb spectrum for isocurvature models, for spectra with $n_{iso} = 0$ and 1, the latter being the one with several oscillations. The axion isocurvature model in fact gives $n_{iso} \leq 0$, whereas the observational data require n_{iso} significantly greater than 0.

adiabatic perturbation pulls us and the cmb equally. The suppression of the Sachs–Wolfe effect still holds for an isocurvature perturbation, but now there is the anisotropy at last scattering, Θ. It gives a contribution to the dipole, which is the dominant effect (Turner 1991). In words, a very large scale isocurvature perturbation affects the motion of the cmb, but not our motion. This possibility is sometimes called a **tilted Universe**, because the slice orthogonal to our worldline is tilted, relative to that of the cmb worldline. Note that this has nothing to do with the word "tilt" when used to imply a non-scale-invariant spectrum.

An observational bound on the degree of tilt is provided by the velocity v of the galaxies within around $100h^{-1}$ Mpc of our position, relative to the cmb rest frame. It seems clear that this velocity is significantly smaller than that of the Sun, but how much smaller remains somewhat controversial, and so, to be conservative, we assume that at least it is no bigger. With this crude estimate we can ignore cosmic variance, yielding $C_{1VL}^{1/2} \lesssim v/c = 1.2 \times 10^{-3}$ (Bennett et al. 1996; Fixsen et al. 1996).

As with the Grishchuk–Zel'dovich effect, the spectrum of S would have to rise very steeply on scales bigger than the horizon, to give an observable effect. Let us represent the very large scale contribution by a delta function $\mathcal{P}_S = \langle S_{VL}^2 \rangle \delta(\ln k - \ln k_{VL})$. Because we deal with a scale far outside the horizon, $\Theta(\mathbf{ex}_{ls}) = -S(\mathbf{ex}_{ls})/3$, and using Eqs. (4.30) and (4.39), we find that

$$C_{1VL}^{1/2} = \sqrt{\frac{2}{3}} \frac{2\pi}{3} \frac{k_{VL}}{H_0} \langle S_{VL}^2 \rangle^{1/2} . \tag{6.57}$$

Because S is defined as a fractional perturbation, the biggest value that makes sense is $\langle S_{VL}^2 \rangle \sim 1$. If this bound is saturated, the observational constraint on C_{1VL} requires $k_{VL}^{-1} \gtrsim 10^3 H_0$.

Examples

6.1 For any gas, Eq. (4.132) gives an estimate of the Jeans length. Show that this leads to Eq. (6.3) for neutrino HDM.

6.2 In an open Universe, a generic perturbation has the spherical expansion (6.25), and its spectrum is defined by Eq. (6.32). Show that, at the origin $x = 0$, the ensemble mean square of the perturbation is given by Eq. (6.33).

6.3 In a flat Universe, the expansion and spectrum of a generic perturbation $g(\mathbf{x})$ are defined by Eqs. (4.30) and (4.39). The cmb anisotropy due to a primordial isocurvature perturbation $S(\mathbf{x})$ is given by Eq. (6.55), and the spectrum of the cmb anisotropy is defined by Eq. (5.26). Putting all this together, show that the cmb dipole due to a very large scale isocurvature perturbation is given by Eq. (6.57). Explain the physical origin of this contribution to the dipole and contrast it with the physical origin that usually is ascribed to the observed dipole.

7 Scalar fields and the vacuum fluctuation

The analysis in the preceding three chapters used only the general features of the vacuum fluctuation. There is a Gaussian primordial curvature perturbation with a nearly scale-invariant spectrum, which gives rise to an adiabatic density perturbation. There is also, at some level, a primordial gravitational-wave amplitude. Finally, there may be a primordial isocurvature density perturbation, coming from the vacuum fluctuation of a non-inflaton field.

In this chapter, we calculate the vacuum fluctuations from first principles. We first derive the slow-roll formula quoted in Section 4.3.4 for the spectrum of the primordial curvature perturbation, and we calculate the corresponding spectral index n. We go on to consider a more accurate calculation of this quantity, which takes into account the effect of departures from the slow-roll approximation. Then we consider the prediction for gravitational waves, and for an isocurvature perturbation, and finally consider the possibility of a multicomponent inflaton.

In this chapter, we consider only models of inflation that give a spatially flat Universe. A descriptive account of open-Universe inflation models is given in the next chapter.

Although we work in the settings of special and general relativity, we have little need at this stage of the detailed formalism of either theory, which is left to Chapter 14. Results derived there occasionally are quoted.

7.1 Classical scalar field

7.1.1 Action principle

As far as we know, the properties of any system in fundamental physics are specified by its **action**

$$S = \int_{-\infty}^{\infty} L \, dt. \tag{7.1}$$

The Lagrangian L depends on the "coordinates" (also called "degrees of freedom") needed to specify the system. It has dimensions of energy, corresponding to a dimensionless action, and this fixes the dimensions of the "coordinates" that appear in it. (Recall that we are setting $\hbar = c = 1$.) Different actions may be physically equivalent. Actions differing by a total divergence are physically equivalent at the classical level, and at least perturbatively at the quantum level also. In gauge theories and gravity theories, there are additional possibilities for equivalence. From the equivalence class, there is usually a simplest Lagrangian, and from now on, we take it for granted that this has been chosen.

In Newtonian mechanics, and in ordinary quantum mechanics, we are dealing with N particles and the number of degrees of freedom is finite (corresponding to the position coordinates of each particle, plus relevant internal degrees of freedom such as the two angles needed to specify the orientation of a diatomic molecule). In classical or quantum field theory, we are dealing with fields that are functions of position. The number of degrees of freedom is now infinite because we need an infinite set of numbers to specify the field, namely, its value at each point in space or, equivalently, the values of its Fourier components.

To have a Lorentz-invariant action, the Lagrangian for the fields must be of the form $L = \int \mathcal{L} \, d^3 r$, where the **Lagrangian density** $\mathcal{L}$ is Lorentz invariant and has dimensions [energy]^4. It is a function of the fields and their first derivatives with respect to space and time. Higher derivatives, or no derivatives at all, would not lead to sensible physics. Fields can be classified according to the spin of the corresponding particles; in the Standard Model, we have spin-0 (Higgs boson), spin-$\frac{1}{2}$ (quarks and leptons), and spin-1 (gauge bosons). Gravitational waves correspond to the graviton with spin-2.

The spin-0 fields are called **scalar fields**, and they are what we need for inflation. Happily, there are lots of scalar fields in supersymmetric extensions of the Standard Model. This is because every spin-$\frac{1}{2}$ field is accompanied by either a spin-0 or a spin-1 field (respectively, a **chiral** or a **gauge** multiplet), with the first case ubiquitous.

7.1.2 Scalar-field Lagrangian

In what follows, we consider the part of the action containing *only* scalar fields, though a full description of them requires a knowledge of their coupling to other fields and to gravity. The main effect of the other fields, as far as inflation is concerned, is likely to be in the loop correction mentioned later.

For the moment, we consider only a single, real field ϕ, which is presumed to be decoupled from the others. Then $\mathcal{L}$ is of the form

$$\mathcal{L} = \frac{1}{2}(\dot{\phi}^2 - \nabla\phi \cdot \nabla\phi) - V(\phi). \tag{7.2}$$

In this expression, $V(\phi)$ is some function, called the potential, and the other term is called the kinetic term, with the overdot denoting $\partial/\partial t$. Note that V has dimension [energy]^4, while ϕ has dimension [energy]^1. The names "potential" and "kinetic terms" are chosen because of the analogy with the Lagrangian of a single particle moving in one dimension; if the position coordinate is q, this is $L = \frac{1}{2}m^2\dot{q}^2 - V(q)$, where the first term is the kinetic energy and $V(q)$ is the potential.

Using relativistic index notation, so that the coordinates are $x^\mu \equiv (t, \mathbf{r}) \equiv (x^0, x^i)$, we have

$$\mathcal{L} = -\frac{1}{2}\eta^{\mu\nu}\partial_\mu\phi \, \partial_\nu\phi - V(\phi), \tag{7.3}$$

where ∂_μ denotes $\partial/\partial x^\mu$. Up to a field redefinition, this is the only Lorentz-invariant expression containing first derivatives but no higher.

The evolution of a Lagrangian system is given by the action principle. This states that $\delta S = 0$, where δS is the change in S resulting from a small change in the time dependence of the "coordinates," in this case ϕ and $\partial_\mu\phi$. The change is arbitrary, except for a boundary condition that depends on the system. In our case,

$$\delta S = \int d^4x \left[\frac{\partial \mathcal{L}}{\partial \phi} \delta \phi + \frac{\partial \mathcal{L}}{\partial(\partial_\mu\phi)} \delta(\partial_\mu\phi) \right]. \tag{7.4}$$

Integrating the second term by parts, in each of the four variables, x^μ, gives

$$\delta S = \int d^4x \left\{ \frac{\partial \mathcal{L}}{\partial \phi} - \frac{\partial}{\partial x^\mu} \left[\frac{\partial \mathcal{L}}{\partial(\partial_\mu\phi)} \right] \right\} \delta\phi + \int d^4x \frac{\partial}{\partial x^\mu} \left[\frac{\partial \mathcal{L}}{\partial(\partial_\mu\phi)} \delta\phi \right]. \tag{7.5}$$

We impose the boundary condition $\delta\phi = 0$ at infinity, so that the last term vanishes. Then $\delta S = 0$ is equivalent to

$$\frac{\partial \mathcal{L}}{\partial \phi} - \frac{\partial}{\partial x^\mu} \left(\frac{\partial \mathcal{L}}{\partial(\partial_\mu\phi)} \right) = 0. \tag{7.6}$$

This is the equation of motion of the field, called the **field equation**. Taking $\mathcal{L}$ from Eq. (7.3) gives

$$\ddot{\phi} - \nabla^2\phi + V'(\phi) = 0, \tag{7.7}$$

where, as before, the prime denotes $d/d\phi$. For a spatially homogeneous field, continuing to ignore gravity for the time being, this becomes

$$\ddot{\phi} + V'(\phi) = 0. \tag{7.8}$$

7.1.3　Energy and momentum

According to Newton's theory, the source of gravity is mass density. According to general relativity, the sources are energy density, momentum density, pressure, and anisotropic stress. These quantities appear as sources in Einstein's field equation, and ultimately are defined by that equation.

In Chapter 14, we show that the energy density ρ is

$$\rho = \frac{1}{2}\dot{\phi}^2 + \frac{1}{2}\nabla\phi \cdot \nabla\phi + V(\phi), \tag{7.9}$$

and the momentum density is

$$\mathbf{N} = -\dot{\phi}\nabla\phi. \tag{7.10}$$

The total energy of an isolated system is

$$E^{\text{total}} = \int d^3r \, \rho(\mathbf{r}, t), \tag{7.11}$$

and the total momentum is

$$\mathbf{p}^{\text{total}} = \int d^3r \, \mathbf{N}(\mathbf{r}, t). \tag{7.12}$$

As described in Chapter 14, these quantities are time independent (conserved) by virtue of Einstein's field equation.

An alternative route to the *total* energy and momentum, which has nothing to do with gravity, is through the Noether procedure. As described in the standard texts, an isolated system possesses a conserved quantity corresponding to each invariance of the action, and we can define energy and momentum as the quantities associated with invariance under time and space translations, respectively. However, this determines the *densities* of energy and pressure only up to a divergence, which is not good enough in the context of cosmology.

For a spatially homogeneous field, the momentum density vanishes, and the stress is purely isotropic, corresponding to pressure

$$P = \frac{1}{2}\dot{\phi}^2 - V(\phi). \tag{7.13}$$

The energy density in that case is

$$\rho = \frac{1}{2}\dot{\phi}^2 + V(\phi). \tag{7.14}$$

The equation of motion is the same as for a unit-mass particle moving in one dimension, with coordinate ϕ, under the influence of a potential $V(\phi)$. Also, the expression for ρ is the same as the expression for the energy of such a particle. That is why $V(\phi)$ is called the potential.

7.1.4 Free field

A noninteracting, or **free**, field has potential $V = m^2\phi^2/2$, where m is the mass of the corresponding particle.[1] The field equation (7.7) is then linear:

$$\ddot{\phi} - \nabla^2\phi + m^2\phi = 0. \tag{7.15}$$

This is the **Klein–Gordon** equation. Writing the Fourier series

$$\phi(\mathbf{r}, t) = \sum_{\mathbf{p}} \phi_{\mathbf{p}}(t)e^{i\mathbf{p}\cdot\mathbf{r}}, \tag{7.16}$$

it becomes

$$\ddot{\phi}_{\mathbf{p}} + E_p^2\phi_{\mathbf{p}} = 0, \tag{7.17}$$

where $E_p = \sqrt{p^2 + m^2}$, with p denoting the magnitude of $\mathbf{p}$. This is the harmonic-oscillator equation, with solutions $\phi_{\mathbf{p}} = \exp(\pm i E_p t)$. The most general solution of the original equation

---

[1] The following development also goes through if m is time dependent. This case occurs if m is a function of other fields with known time dependence. See also Section 7.4.6 for another occurrence.

is a superposition of plane waves, of the form

$$\phi(\mathbf{r}, t) = \mathcal{V}^{-1/2} \sum_{\mathbf{p}} \sqrt{\frac{1}{2E_p}} \{a_{\mathbf{p}} \exp[i(\mathbf{p} \cdot \mathbf{r} - E_p t)] + a_{\mathbf{p}}^* \exp[-i(\mathbf{p} \cdot \mathbf{r} - E_p t)]\},$$

(7.18)

where $a_{\mathbf{p}}$ is a complex amplitude and $\mathcal{V}$ is the volume of the box used to make the Fourier expansion. The factor $\mathcal{V}^{-1/2}/\sqrt{2E_p}$ is pulled out to make later expressions simpler.

Taking the integrals (7.11) and (7.12) over the box, and using the orthonormality relation (4.58), we find that

$$E^{\text{total}} = \sum_{\mathbf{p}} N_{\mathbf{p}} E_p,$$

(7.19)

$$\mathbf{p}^{\text{total}} = \sum_{\mathbf{p}} N_{\mathbf{p}} \mathbf{p},$$

(7.20)

where $N_{\mathbf{p}} \equiv |a_{\mathbf{p}}|^2$. In the quantum theory, which we look at next, these expressions have a very simple interpretation. Each plane wave is equivalent to $N_{\mathbf{p}}$ particles, each particle carrying momentum $\mathbf{p}$ and energy E_p.

If the potential contains higher-order terms, the equation of motion becomes nonlinear. For instance, one could consider the potential

$$V = \frac{1}{2} m^2 \phi^2 + \frac{1}{4} \lambda \phi^4.$$

(7.21)

This adds a term $\lambda \phi^3$ to the left-hand side of the field equation (7.15). To obtain the same expressions for the energy and momentum, we can still make the plane-wave expansion, but the amplitude $a_{\mathbf{p}}$ becomes time dependent. The plane waves are interacting with each other and, after quantization, so are the particles.

It turns out that the interaction typically has a small effect if $\lambda \ll 1$, whereas if $\lambda \gg 1$, it is so strong that the free-field theory is not useful. We therefore only consider values $\lambda \lesssim 1$.

7.2   Quantized free scalar field in flat space-time

To arrive at the quantum theory of the free field, let us go back to the Lagrangian,

$$L = \frac{1}{2} \int d^3 r (\dot{\phi}^2 - \nabla \phi \cdot \nabla \phi - m^2 \phi^2).$$

(7.22)

For each term, we insert the plane-wave expansion, and integrate using the orthonormality relation (4.58). This gives

$$L = \frac{1}{2} \mathcal{V} \sum_{\mathbf{p}} (|\dot{\phi}_{\mathbf{p}}|^2 - E_p^2 |\phi_{\mathbf{p}}^2|).$$

(7.23)

In the second expression, the contribution of a given $\mathbf{p}$ is the same as that of $-\mathbf{p}$ because of the reality condition $\phi_{\mathbf{p}}^* = \phi_{-\mathbf{p}}$. Let us replace $\sum$ by $\sum'$, where the prime means that we keep only one of each pair $\mathbf{p}$ and $-\mathbf{p}$. Also, we introduce the real and imaginary parts by writing

$$\phi_{\mathbf{p}} = R_{\mathbf{p}} + i I_{\mathbf{p}}. \tag{7.24}$$

Then the Lagrangian is

$$L = \mathcal{V} \sum_{\mathbf{p}}{}' [(\dot{R}_{\mathbf{p}}^2 - E_p^2 R_{\mathbf{p}}^2) + (\dot{I}_{\mathbf{p}}^2 - E_p^2 I_{\mathbf{p}}^2)]. \tag{7.25}$$

Focussing on $R_{\mathbf{p}}$, the Lagrangian is $(\dot{q}^2 - E_p^2 q^2)/2$, where $q \equiv \sqrt{2}\,\mathcal{V}^{1/2} R_{\mathbf{p}}$. This is the same as for a unit-mass particle moving in one dimension, with coordinate q and potential $E_p^2 q^2/2$. The equation of motion is

$$\ddot{q} = -E_p^2 q, \tag{7.26}$$

which reproduces Eq. (7.17).

In quantum theory, q becomes a Hermitian operator. In ordinary quantum mechanics, we take operators such as q to be time independent, whereas the state vectors satisfy the Schrödinger equation

$$i\frac{d}{dt}|t\rangle = H|t\rangle, \tag{7.27}$$

where H is the Hamiltonian operator. However, the only physically significant expressions are the *expectation values* of Hermitian operators, which correspond to physical observables, and they are unaltered if we make the replacement

$$|t\rangle \to |0\rangle = e^{iHt}|t\rangle, \tag{7.28}$$

while making, for every Hermitian operator A, the replacement

$$A \to e^{iHt} A e^{-iHt}. \tag{7.29}$$

The original scheme is called the **Schrödinger picture**, whereas the new one is called the **Heisenberg picture**. The latter is more convenient for our purpose.

From Eq. (7.29), we learn that, in the Heisenberg picture,

$$\frac{dA}{dt} = i[H, A]. \tag{7.30}$$

The equivalence of this to the classical equation of motion $\dot{q} = \partial H/\partial p$ is ensured by imposing the canonical commutation relation $[q, p] = i$, where the conjugate momentum p is defined in terms of the Lagrangian by $p \equiv (\partial L/\partial \dot{q}) = \dot{q}$, and H is defined by $H = p\dot{q} - L$. In our case, $p = \dot{q} = \sqrt{2}\,\mathcal{V}^{3/2} \dot{R}_{\mathbf{p}}$.

The expectation value of each observable also satisfies the classical equation of motion. In the Heisenberg picture, this happens because the state vector is time independent, so that the time dependence of the expectation value is simply that of the operator. (In the Schrödinger picture, the operator is time independent, and all of the time dependence comes from the state vector.)

We focussed on the part of the Lagrangian involving $R_{\mathbf{p}}$, but the considerations apply quite generally and, in particular, they apply also to $I_{\mathbf{p}}$. We conclude that, in quantum theory, $R_{\mathbf{p}}$ and $I_{\mathbf{p}}$ become Hermitian operators, with the commutation relations

$$[R_{\mathbf{p}}, \dot{R}_{\mathbf{p}'}] = [I_{\mathbf{p}}, \dot{I}_{\mathbf{p}'}] = (2\mathcal{V})^{-1} i \delta_{\mathbf{p}\mathbf{p}'}, \tag{7.31}$$

$$[R_{\mathbf{p}}, I_{\mathbf{p}'}] = [R_{\mathbf{p}}, \dot{I}_{\mathbf{p}'}] = [I_{\mathbf{p}}, \dot{R}_{\mathbf{p}'}] = 0. \tag{7.32}$$

The Fourier component $\phi_{\mathbf{p}}$ is not itself Hermitian, but it satisfies the condition $\phi_{\mathbf{p}}^{\dagger} = \phi_{-\mathbf{p}}$, which ensures that $\phi(\mathbf{r}, t)$ is Hermitian. In the plane-wave expansion (7.18), $a_{\mathbf{p}}$ is an operator, and $a_{\mathbf{p}}^{*}$ becomes $a_{\mathbf{p}}^{\dagger}$, ensuring that ϕ is Hermitian. This expansion is equivalent to

$$\phi_{\mathbf{p}} = w_p(t) a_{\mathbf{p}} + w_p^{*}(t) a_{-\mathbf{p}}^{\dagger}, \tag{7.33}$$

where

$$w_p \equiv \mathcal{V}^{-1/2} \sqrt{\frac{1}{2E_p}} \, \exp(-i E_p t). \tag{7.34}$$

Using this expression, we can verify that Eqs. (7.31) and (7.32) are equivalent to

$$[a_{\mathbf{p}}, a_{\mathbf{p}'}^{\dagger}] = \delta_{\mathbf{p}\mathbf{p}'}, \qquad [a_{\mathbf{p}}, a_{\mathbf{p}'}] = 0. \tag{7.35}$$

To completely define the quantum theory, we need state vectors on which the operators will act, and a correspondence between Hermitian operators and observables. At this point, remember that the Fourier expansion in the box gives discrete momenta $\mathbf{p}_1, \mathbf{p}_2, \ldots$. We therefore can write $a_n \equiv a_{\mathbf{p}_n}$ and so on. Inspired by Eqs. (7.19) and (7.20), let us define the **number operator**

$$\widehat{N}_i \equiv a_i^{\dagger} a_i. \tag{7.36}$$

The plan is that $\widehat{N}_i$ will be the operator corresponding to the number of particles with momentum $\mathbf{p}_i$. Then there will be states $|N_1, N_2, \ldots\rangle$, such that

$$\widehat{N}_i |N_1, N_2, \ldots\rangle = N_i |N_1, N_2, \ldots\rangle, \tag{7.37}$$

where N_i is the number of particles with momentum $\mathbf{p}_i$. We can construct these states, assuming that there is a vacuum state $|0\rangle \equiv |0, 0, \ldots\rangle$. They are obtained by the repeated action of the **creation operators** $a_i^{\dagger}$, according to the formula

$$a_1^{\dagger} |N_1, N_2, \ldots\rangle = (N_1 + 1)^{1/2} |N_1 + 1, N_2, \ldots\rangle, \tag{7.38}$$

and similarly for the action of $a_2^{\dagger}, a_3^{\dagger}, \ldots$.

Let us check that these states have the required properties. Using the commutation relation $[N_i, a_j^{\dagger}] = \delta_{ij} a_i^{\dagger}$, we see that $a_i^{\dagger}$ does indeed raise N_i by 1. What about the normalization? We take the vacuum to be normalized:

$$\langle 0|0 \rangle = 1. \tag{7.39}$$

Taking the inner product of both sides of Eq. (7.37) with a state $|N_1' + 1, N_2', \ldots\rangle$ gives zero unless $N_i = N_i'$ for all i. In the latter case, we can use $a_i a_i^\dagger = N_i + 1$ (equivalent to the commutation relation) to find that all states have the same norm. In summary, the states are orthonormal. Acting on both sides with a_1 gives, for $N_1 > 0$,

$$a_1|N_1, N_2, \ldots\rangle = N_1^{1/2}|N_1 - 1, N_2, \ldots\rangle. \tag{7.40}$$

This shows that the a_i are **annihilation operators**. Finally, we *define* $a_i|0\rangle$ to be the zero vector, so that there are no states with negative N_i. We say that a_i "annihilates the vacuum."

The orthonormal states $|N_1, N_2, \ldots\rangle$ are the basis of a vector space called the Fock space. We assume that every state vector lives in this space.

Repeating the derivation of Eqs. (7.19) and (7.20), we find that the basis states have definite energy:

$$E^{\text{total}} = \sum_{\mathbf{p}} \left(N_{\mathbf{p}} + \frac{1}{2} \right) E_{\mathbf{p}}, \tag{7.41}$$

$$\mathbf{p}^{\text{total}} = \sum_{\mathbf{p}} N_{\mathbf{p}} \, \mathbf{p}. \tag{7.42}$$

Increasing $N_{\mathbf{p}}$ by 1 increases the energy by E_p and the momentum by $\mathbf{p}$, which justifies the particle interpretation.

7.2.1 Vacuum fluctuation

In the vacuum state, the field components $\phi_{\mathbf{p}}$ do not have definite values. The real and imaginary parts of each $\phi_{\mathbf{p}}$ has a probability distribution, which is the modulus-squared of the ground-state wavefunction of a simple harmonic oscillator. Using Eqs. (7.33) and (7.35), and remembering that a_i annihilates the vacuum, we find the expectation value

$$\langle 0 | \, |\phi_{\mathbf{p}}|^2 \, |0\rangle \equiv \langle |\phi_{\mathbf{p}}|^2 \rangle = |w_p|^2. \tag{7.43}$$

Just as in Eq. (4.15), we can define the spectrum

$$\mathcal{P}_\phi(p) = \mathcal{V}\frac{p^3}{2\pi^2}|w_p|^2 \tag{7.44}$$

$$= \frac{p^3}{4\pi^2 E_p}, \tag{7.45}$$

such that the mean-square field is

$$\langle \phi^2(\mathbf{r}) \rangle = \int_0^\infty \mathcal{P}_\phi(p) \frac{dp}{p}. \tag{7.46}$$

It diverges, but becomes finite after smoothing the field on a scale R because this filters out the Fourier components with $p^{-1} \lesssim R$. In the massless case $E_p = p$, we then find $\langle \phi^2 \rangle \sim R^{-2}$.

In the present context of special relativity, the energy due to the vacuum fluctuation normally can be ignored; indeed, according to the standard interpretation of quantum physics it has absolutely no meaning until someone measures it, and there is no need to suppose that such a measurement is made. However, if we take the extreme case of a smoothing scale $R \sim 1/M_{\text{Pl}}$, the typical magnitude of a positive fluctuation in the energy is enough to form a black hole, which can hardly be ignored! The conclusion is presumably that distances, and therefore times, below the Planck scale make no sense. This situation sometimes is described by saying that space-time looks like a foam on such scales.

In the next section, we see that the situation is very different during inflation because the quantum fluctuation becomes a classical quantity after it leaves the horizon.

7.2.2   Fourier integral

Some authors take the limit of large box size, writing

$$\phi(\mathbf{r}) = \frac{1}{(2\pi)^{3/2}} \int \phi(\mathbf{p}) e^{i \mathbf{p} \cdot \mathbf{r}} d^3 p. \tag{7.47}$$

The creation operators are defined by

$$\phi(\mathbf{p}) = w(p) a(\mathbf{p}) + w^*(p) a^\dagger(\mathbf{p}), \tag{7.48}$$

with

$$w(p) = \sqrt{\frac{1}{2E_p}} \, \exp(-i E_p t). \tag{7.49}$$

The commutation relations are

$$[a(\mathbf{p}), a^\dagger(\mathbf{p}')] = \delta^3(\mathbf{p} - \mathbf{p}'), \tag{7.50}$$
$$[a(\mathbf{p}), a(\mathbf{p}')] = 0, \tag{7.51}$$

and the spectrum is

$$\mathcal{P}_\phi(p) = \frac{p^3}{2\pi^2} |w(p)|^2. \tag{7.52}$$

Using Eqs. (4.27), (4.28), and (4.29), we can check that these formulae are consistent with the ones we gave. The last formula reproduces our result for $\mathcal{P}_\phi$.

7.2.3   Vacuum energy and the cosmological constant

In contrast to the classical expression (7.19), the quantum expression (7.41) does not give zero energy in the vacuum state. The number of momentum states per unit volume is $d^3 p/(2\pi)^3$, and so, the vacuum energy due to quantum states with $p < p_{\text{max}}$ is

$$\rho_{\text{vac}} = \frac{1}{(2\pi)^2} \int_0^{p_{\text{max}}} p^3 \, dp \sim p_{\text{max}}^4. \tag{7.53}$$

Nowadays, we regard a quantum field theory as an effective theory, valid below some cutoff energy scale. In that case, we can identify p_{max} with this scale, and the predicted energy density is finite. If we assume the biggest possible range of validity for our theory, p_{max} is of order M_{Pl}. At the other extreme, we might take $p_{max} \sim 100\,\text{GeV}$, the limit to which the Standard Model has been tested.

In the context of special relativity, a constant energy density has no physical significance, but in the context of general relativity, it *is* significant, being the **cosmological constant** that we studied earlier. Do we conclude that the cosmological constant is predicted to be of order of the cutoff scale? If so, we have a disaster because the value required by observation is $\rho_{vac}^{1/4} \lesssim 10^{-3}\,\text{eV}$. However, in the absence of any other theoretical constraints, we are free to add a constant V_0 to the potential without altering the dynamics of the free field, in which case there is actually *no* prediction for the cosmological constant. On the other hand, the presence of large known contributions to it means that the small observed value requires cancellations, which in our present state of knowledge seem accidental. A possible way out of this impasse was mentioned in Section 6.4.

7.3 Several scalar fields

We often need to consider two or more real fields. In particular, complex fields, corresponding to a pair of real fields, are mandatory in supersymmetry.

7.3.1 Several fields

With two real fields, the simplest Lagrangian density is

$$\mathcal{L} = \frac{1}{2}(\dot{\phi}^2 - \nabla\phi \cdot \nabla\phi) + \frac{1}{2}(\dot{\psi}^2 - \nabla\psi \cdot \nabla\psi) - V(\phi, \psi). \tag{7.54}$$

The field equations, continuing for the time being to ignore gravity, are

$$\ddot{\phi} - \nabla^2\phi + \frac{\partial V(\phi, \psi)}{\partial \phi} = 0, \tag{7.55}$$

$$\ddot{\psi} - \nabla^2\psi + \frac{\partial V(\phi, \psi)}{\partial \psi} = 0. \tag{7.56}$$

The extension to further fields is similar.

It is often appropriate to combine two real fields ϕ_1 and ϕ_2 into a single complex field, defined by convention as

$$\phi = \frac{1}{\sqrt{2}}(\phi_1 + i\phi_2). \tag{7.57}$$

Its kinetic term is

$$\mathcal{L}_{kin} = (\dot{\phi}^*\dot{\phi} - \nabla\phi^* \cdot \nabla\phi). \tag{7.58}$$

This form is particularly appropriate if the action is invariant under a $U(1)$ transformation corresponding to $\phi \to e^{i\alpha}\phi$, which will be true if V depends only on $|\phi|$.

With more than one field, it is no longer true that the most general Lorentz-invariant Lagrangian density $\mathcal{L}$ can be reduced to the form of Eq. (7.54) by a field redefinition. For several real fields ϕ_n, the most general kinetic term involving derivatives is

$$\mathcal{L}_{\text{kin}} = \sum_{m,n} G_{mn}\eta^{\mu\nu}\partial_\mu\phi_m\,\partial_\nu\phi_n, \tag{7.59}$$

where G_{mn} is an arbitrary function of the fields. Equation (7.54) is recovered only if $G_{mn} = \delta_{mn}$. With more than one field, it generally is not possible to recover this form by a field redefinition. If it is impossible, the space of the fields is said to be curved. At least for fields with only gravitational-strength interaction, we expect (see Section 8.2) that the curvature scale will be of order M_{Pl}, allowing us to choose $G_{mn} = \delta_{mn}$ to high accuracy in the regime $|\phi_n| \ll M_{\text{Pl}}$.

7.3.2   Internal symmetry

In addition to Lorentz invariance, the action usually will be invariant under a group of transformations acting exclusively on the fields, with no effect on the space-time indices. This is called an **internal symmetry**. For a single real field, V is often an even function and then there is invariance under $\phi \to -\phi$ (the Z_1 group). For a single complex field, V often depends only on $|\phi|$ and then there is invariance under

$$\phi \to e^{i\alpha}\phi, \tag{7.60}$$

known as the $U(1)$ group. If there are only scalar fields, α has to be independent of space-time because of the derivatives in the kinetic term, and we say that there is a **global symmetry**. In Nature though, there are also gauge fields such as the electromagnetic potential A^μ, and they transform in such a way that the full action is invariant with a space-time–dependent α. (In other words, the change in the kinetic term is cancelled by a change in the part of the action involving the gauge fields.) This is a **local symmetry**, also called a **gauge symmetry**. With more fields, there can be higher symmetries, including the $SU(2)$ and $SU(3)$ gauge symmetries present in the Standard Model of particle physics.

As in the preceding cases, the symmetry is defined conveniently by taking the origin $\phi = 0$ to be a point that is fixed under the action of the group, called a **fixed point.**

7.3.3   Spontaneously broken symmetry

Now we look at spontaneously broken symmetry. It is important for inflation model building, and for many other aspects of cosmology.

The concept of spontaneously broken symmetry
We remarked earlier that the action will be invariant under some group of internal symmetries. For instance, if there is a single real field and $V(\phi)$ is an even function, the action is invariant

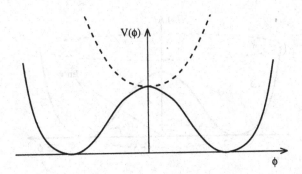

Fig. 7.1. The solid line illustrates schematically the potential equation (7.61), and the dashed line shows the potential equation (7.63) with $\psi^2 > m^2/\lambda'$.

under $\phi \to -\phi$. The potential we just wrote down is one example of this case, but now, consider instead

$$V(\phi) = V_0 - \frac{1}{2}m^2\phi^2 + \frac{1}{4}\lambda\phi^4. \qquad (7.61)$$

As shown in Figure 7.1, this potential has minima at $\phi = \pm(m/\sqrt{\lambda})$ and, taking for example the positive one, we can define a new field $\tilde{\phi} = \phi - (m/\sqrt{\lambda})$. Then, if the constant V_0 is chosen to make the potential vanish at the minimum, near the minimum we have

$$V = \frac{1}{2}\tilde{m}^2\tilde{\phi}^2 + A\tilde{\phi}^3 + B\tilde{\phi}^4 + \cdots, \qquad (7.62)$$

where $\tilde{m} = \sqrt{2}m$ and we are not interested in the precise values of A and B. The minima represent possible vacuum values of the field, which in the quantum theory are called **vacuum expectation values**. Oscillations around a given vacuum correspond to an almost-free field if the higher terms are small. (It turns out that the criterion for this is $\lambda \lesssim 1$, which, as in the previous case, we assume to be valid.) On the other hand, the original symmetry will not be evident in this almost-free field theory, and we say that it has been **spontaneously broken**.

A slightly more complicated case is the $U(1)$ symmetry present if V is a function only of the modulus of a complex field. The preceding discussion goes through if ϕ is replaced by $|\phi|$. Considered as a function defined in the complex ϕ plane, V has a "Mexican-hat" shape, shown in Figure 7.2. The vacuum now consists of the circle $|\phi| = m/\sqrt{\lambda}$ at the bottom of the rim. About any point in the vacuum, there is a radial mode of oscillation corresponding to the one we have already have considered, plus an angular mode with zero frequency.

For a global symmetry, the zero-mass particle corresponding to the angular mode is called the **Goldstone boson** of the symmetry; the particle corresponding to the radial mode has no particular name. For a gauge symmetry the radial mode corresponds to the Higgs particle, whereas the Goldstone boson loses its identity to become one of the degrees of freedom of the gauge boson. This latter case, generalized to the $SU(2)$ group, occurs in the electroweak sector of the Standard Model.

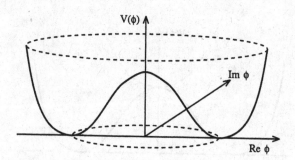

Fig. 7.2. Mexican-hat potential.

A global symmetry may be only approximate, in which case the Goldstone boson acquires mass to become a pseudo-Goldstone boson. A gauge symmetry must be exact.

Symmetry restoration and topological defects

Symmetries that are spontaneously broken in the present Universe may be present in the early Universe because of field interactions. This goes under the somewhat misleading name of symmetry restoration. As the simplest example, consider real fields ϕ and ψ, with a potential

$$V(\phi, \psi) = V_0 - \frac{1}{2}m^2\phi^2 + \frac{1}{4}\lambda\phi^4 + \frac{1}{2}\lambda'\phi^2\psi^2. \tag{7.63}$$

With $\psi = 0$, this becomes Eq. (7.61), and the symmetry $\phi \to -\phi$ is spontaneously broken. However, if ψ^2 is spatially homogeneous and time independent, the effective mass-squared of ϕ (defined as the second derivative of the potential at the origin) is $-m^2 + \lambda'\psi^2$. It is positive for $\psi^2 > m^2/\lambda'$, in which case the symmetry is restored as shown by the dashed line in Figure 7.1

The most common case is when ψ is in thermal equilibrium, in which case $\psi^2 \sim T^2$ and symmetry is restored above some critical temperature $T_c \sim m$. (To be precise, ψ is fluctuating rapidly with the preceding as its mean-square value.) In that case, all fields in thermal equilibrium (including ϕ itself) should be considered using finite-temperature field theory. Another case is when ψ is the inflaton field, leading to the hybrid inflation model described in Section 8.4.1. In that case, we need to add an extra piece to the potential, so that the inflaton field is rolling slowly toward the origin, inflation ending when ψ falls below the critical value $\psi_c^2 = m^2/\lambda'$.[2]

The epoch at which symmetry restoration fails is given by $T = T_c$ for thermal equilibrium, or by $\psi^2 = \psi_c^2$ for hybrid inflation. Symmetry restoration also occurs with $U(1)$ or a higher symmetry.

When a symmetry is restored in the early Universe, and then spontaneously broken, stable field configurations called **topological defects** may form (Vilenkin and Shellard 1994), which have $\phi = 0$ at their centre. To see this, consider the simplest case of a real field. Just before symmetry breaking, the spatial average of ϕ is 0, but at any particular point, ϕ will be nonzero

---

[2] When we actually discuss inflation models, we exchange the labels ϕ and ψ to conform with our convention that ϕ denotes the inflaton field.

because of thermal or quantum fluctuations. Then, according to its sign, ϕ rolls down to one of the two vacua $\phi = \pm(m/\sqrt{\lambda})$. Separating the regions of positive and negative ϕ are two-dimensional objects called **domain walls**, with $\phi = 0$ at the centre, and a thickness typically of order m^{-1}.

Topological defects also form in the case of a complex field possessing $U(1)$ symmetry. Then they are one-dimensional **cosmic strings**. For a higher symmetry, there may be domain walls or strings, or else **monopoles** or **textures** that have zero dimension. The possible cosmological significance of topological defects is considered briefly in Sections 3.1.3 and 12.4.

7.4 Vacuum fluctuation of the inflaton field

The central result of this section is Eq. (7.87) for the spectrum of the inflaton field perturbation, which we quoted without proof in Chapter 4.

In our development of field theory, we took the usual attitude that gravity is negligible, so that the discussion takes place in the arena of special, as opposed to general, relativity. In the cosmological setting, this means that the treatment applies on scales well within the horizon, where space-time curvature is negligible. Outside the horizon, the equations undergo a simple modification, which we quote without proof pending the full discussion of Chapter 14.

7.4.1 Scalar field equation in an expanding Universe

In Chapter 14, we show that the scalar field equation, in the absence of cosmological perturbations, is

$$\ddot{\phi} + 3H\dot{\phi} + \nabla^2\phi + \frac{dV}{d\phi} = 0. \tag{7.64}$$

This neglects the perturbation of the space-time geometry, caused by the inhomogeneity of the field, but we see that such neglect introduces only a small correction. The overdot now indicates differentiation with respect to t at fixed $\mathbf{x}$, and

$$\nabla^2 = a^{-2} \sum \frac{\partial^2}{\partial x_i^2}. \tag{7.65}$$

This equation includes, through $V(\phi)$, the effect of any interaction of ϕ with itself or other scalar fields. In addition, there could be an interaction with the space-time curvature, for example, a linear term $-\xi R\phi^2$, where $R = -(\rho - 3P)/M_{\text{Pl}}^2$ is the curvature scalar. During inflation $R = -12H^2$ is practically constant. We therefore can take this term to be included in V if it is present, but note that we must have $|\xi| \ll 1$ if the term is not to spoil inflation, assuming that it is not cancelled. Also, we should have $|\xi|\phi^2 \ll M_{\text{Pl}}^2$ if the corresponding term in the Lagrangian is not to change the effective value of M_{Pl}.

Some quantum effects can be taken into account by taking $V(\phi)$ to be an effective potential that includes loop corrections, and depending on the inflation model, these may or may not be significant. If they are, we assume that they have been included.

We now split the field into an unperturbed part and a perturbation:

$$\phi(\mathbf{x}, t) = \phi(t) + \delta\phi(\mathbf{x}, t). \tag{7.66}$$

For definiteness, we choose the spatially flat slicing, though the exact choice is unimportant at this stage. Perturbing Eq. (7.64) and linearizing, we find

$$(\delta\phi)\ddot{} + 3H(\delta\phi)\dot{} - \nabla^2\delta\phi + m^2\delta\phi = 0, \tag{7.67}$$

where $m^2(t) = V''$. For a given Fourier component, this becomes

$$(\delta\phi_{\mathbf{k}})\ddot{} + 3H(\delta\phi_{\mathbf{k}})\dot{} + \left(\frac{k}{a}\right)^2 \delta\phi_{\mathbf{k}} + \frac{1}{2}m^2\delta\phi_{\mathbf{k}} = 0. \tag{7.68}$$

Until a few Hubble times after horizon exit, the last term is negligible and we are dealing with a massless field. To see this, compare the last term with the one before it. At the epoch of horizon exit $k = aH$, it is negligible because the slow-roll condition $M_{\rm Pl}^2 V'' \ll V$ is equivalent to $m^2 \ll H^2$. However, after horizon exit, aH increases like $a \propto \exp(Ht)$, and so, after a few Hubble times, the last term generally will no longer be negligible.

As long as the inflaton field has negligible mass,

$$(\delta\phi_{\mathbf{k}})\ddot{} + 3H(\delta\phi_{\mathbf{k}})\dot{} + \left(\frac{k}{a}\right)^2 \delta\phi_{\mathbf{k}} = 0. \tag{7.69}$$

Well after horizon exit, the last term is negligible and $\delta\phi_{\mathbf{k}}$ settles down to a constant value.

Well before horizon exit, we recover the flat space-time equation

$$(\delta\phi_{\mathbf{k}})\ddot{} + \left(\frac{k}{a}\right)^2 \delta\phi_{\mathbf{k}} = 0. \tag{7.70}$$

7.4.2   Some subtleties

Effect of space-time curvature

In Chapter 14, we show that the field equation in a generic coordinate system is

$$\Box\phi + V' = 0, \tag{7.71}$$

where $\Box$ is given terms of the metric components by

$$\Box \equiv -\frac{1}{\sqrt{-g}}\, \partial_\mu(\sqrt{-g}\, g^{\mu\nu}\partial_\nu). \tag{7.72}$$

Inserting Eq. (7.66) into this equation, and dropping the second-order term $(\delta\Box)(\delta\phi)$, we find

$$(\ddot\phi + 3H\dot\phi - \nabla^2\phi) + (\delta\Box)\phi(t) + V' = 0, \tag{7.73}$$

where the full $\phi(\mathbf{x}, t)$ appears in the first three terms. The part of the differential operator $\delta\Box$ that acts on the space coordinates has no effect on $\phi(t)$. Thus, we can ignore the perturbation $\delta\Box$ if $\phi(t)$ is sufficiently slowly varying.

By explicit calculation (Section 7.6), we find that, for the inflaton field, the slow-roll conditions ensure sufficiently slow variation, when $\delta\phi$ is defined on spatially flat slices. For a massless field, V vanishes and $\phi(t)$ is indeed time independent, apart from a decaying mode. These are the only two cases of interest, and so, we can indeed ignore the perturbation in the space-time curvature.

Almost any gauge will do!

Although we have stated that $\delta\phi$ is to be defined using the spatially flat slicing, this is not essential in the present context. If we choose a different slicing, corresponding to a coordinate $\tilde{t} = t + \delta t(t, \mathbf{x})$, the new perturbation is given by Eq. (4.141) as

$$\widetilde{\delta\phi} - \delta\phi = -\dot\phi\,\delta t. \tag{7.74}$$

If the time displacement δt remains finite in the slow-roll limit, the difference vanishes in that limit. We can choose any slicing that remains nonsingular, and obtain the same result in the slow-roll limit. Corrections to the slow-roll approximation, where the spatially flat slicing becomes crucial, are considered later.

However, a very important point is that the comoving slicing could *not* have been chosen because that would lead to a zero perturbation. The reason is that the momentum density seen by any observer, given by Eq. (7.10), vanishes by definition for a comoving observer. This means that the spatial gradient vanishes for such an observer, making ϕ homogeneous on comoving slices.

In fact, the comoving slicing becomes singular in the slow-roll limit $\dot\phi \to 0$. To see this, consider the time displacement $\delta t(\mathbf{x}, t)$ required to switch between the flat and comoving slicings. It is given by $\delta t = \delta\phi/\dot\phi$ [Eq. (4.142)], where $\delta\phi$ is the perturbation on the flat slicing. The typical value of $\delta\phi$ is of order H (because $\mathcal{P}_\phi^{1/2} \sim H$), which is independent of $\dot\phi$. The typical value of δt therefore goes infinite in the slow-roll limit $\dot\phi \to 0$, indicating that the comoving slicing indeed becomes singular. For instance, the spatial curvature of comoving slices is specified [Eq. (4.143)] by $\mathcal{R} = H\delta t$, which becomes singular.

Is $\delta\phi$ a free field?

We derived the field equation for $\delta\phi$ by *assuming* that it is linear. At first sight, this assumption seems harmless because it is equivalent to the validity of cosmological perturbation theory. However, it is also equivalent to the neglect of any interaction between $\delta\phi$ and other fields. It might happen that the interaction is not negligible, in which case cosmological perturbation theory breaks down during inflation, to become valid only later. In that case, the vacuum fluctuations of different Fourier modes will couple, leading to a non-Gaussianity of the perturbations, which might be observable. This point has been investigated by a few authors, who include the *self*-interaction of $\delta\phi$ and find that it is small because of the flatness of the potential. [Indeed, the higher-order terms in the expansion (7.67) are $3V'''(\delta\phi)^2 + \cdots$.] It seems reasonable that the same conclusion holds also for the interaction with other fields; either they have mass much less than H, in which case the potential is very flat, making their interaction negligible, or they have mass much greater than H, in which case their vacuum fluctuation is negligible.

However, as far as we know, this has not been verified in explicit models. Here, we simply assume that higher-order terms are negligible.

7.4.3   Classical homogeneity before horizon exit

While it is well within the horizon, each Fourier component $\delta\phi_\mathbf{k}$ is described by flat space-time field theory.[3] Our fundamental assumption is that the inflaton field is in the vacuum state, corresponding to no inflaton particles. Let us ask about the status of this assumption.

The idea behind it is that, at any epoch during inflation, there is a cutoff k_{max} in the energy spectrum of inflaton particles. Such a cutoff obviously is present in order to avoid infinite energy density, but the point at issue is its magnitude, in Hubble units. If inflation is preceded by an epoch of thermal equilibrium, the equilibrium occupation number $2[\exp(k/aT) \pm 1]^{-1}$ (Kolb and Turner 1990) implies $k_{max}/a \sim T$. In that case, inflation begins when the radiation energy $\sim T^4$ falls below the inflaton-field energy $\sim V$, and so, the cutoff is independent of the epoch and is given by $(k_{max}/a)^4 \sim V_{begin}$. However, because V decreases during inflation, and a increases rapidly, this implies that, once inflation is under way,

$$\left(\frac{k_{max}}{a}\right)^4 \ll V. \tag{7.75}$$

Irrespective of what happens before inflation, this condition is in any case necessary for consistency, just to ensure that the radiation energy density due to the inflaton particles is much less than the potential V, which is supposed to be dominant. Indeed, because the density of states in $\mathbf{k}$ space is $(2\pi)^{-3}$, one inflaton per state up to k_{max} implies a radiation energy density given by

$$\rho_{rad} = \frac{1}{2\pi^2} \int_0^{k_{max}} \left(\frac{k}{a}\right)^4 \frac{dk}{k} = \frac{1}{8\pi^2} \left(\frac{k_{max}}{a}\right)^4. \tag{7.76}$$

Apart from the numerical factor this leads to the desired condition, Eq. (7.75).

This condition is enough to justify the vacuum assumption. To see this, rewrite it in the form

$$\left(\frac{k_{max}}{aH}\right) \ll \left(\frac{M_{Pl}}{H}\right)^{1/2}. \tag{7.77}$$

We see that, typically, the right-hand side is of order 10^3, and so, assume this value for definiteness. Now suppose that inflation begins some number N_{before} of Hubble times before the observable Universe leaves the horizon. Because the scale factor a grows like $\exp(Ht)$, with H slowly varying, Eq. (7.77) applied at the beginning of inflation leads to

$$\frac{k_{max}}{k_1} \lesssim 10^3 \exp(-N_{before}). \tag{7.78}$$

---

[3]   One can verify this explicitly from the general-relativistic Lagrangian (14.85), but there is hardly any more reason to do that in the present context than there is for terrestrial applications of flat space-time field theory. Working in a locally inertial frame, gravity can be ignored on time and distance scales much less than the space-time curvature scale, which in cosmology is of order H^{-1}.

The vacuum assumption therefore is ensured if inflation begins at least 7 or so Hubble times before the observable Universe leaves the horizon. Clearly, the result is not very sensitive to the assumed value of M_{Pl}/H.

7.4.4 Vacuum fluctuation after horizon exit

Well before horizon exit, the vacuum fluctuation of the inflaton field is given by the flat space-time field theory of Section 7.2. We need to evolve it forward to the epoch a few Hubble times after horizon exit. Generalizing Eq. (7.33), it will be of the form

$$\delta\phi_{\mathbf{k}}(t) = w_k(t)a_{\mathbf{k}} + w_k^*(t)a_{-\mathbf{k}}^\dagger,$$
(7.79)

where $w_k(t)$ is a solution of the field equation (7.69):

$$\ddot{w}_k + 3H\dot{w}_k + \left(\frac{k}{a}\right)^2 w_k = 0.$$
(7.80)

This is the most general operator satisfying both the reality condition and the field equation.

To recover flat space-time field theory well before horizon exit, we adopt the same commutation relation (7.35) for the annihilation operator $a_{\mathbf{k}}$, and look for a solution w_k that matches Eq. (7.34) up to a phase factor that varies slowly on the Hubble timescale. The solution need only be valid for a few Hubble times either side of horizon exit. During that era, we ignore the variation of H, and then the appropriate solution is

$$w_k(t) = L^{-3/2}\frac{H}{(2k^3)^{1/2}}\left(i + \frac{k}{aH}\right)\exp\left(\frac{ik}{aH}\right).$$
(7.81)

We easily verify that it is a solution. Also, around some epoch $t = T$ well before horizon exit, we have

$$\frac{k}{aH} = \left.\frac{k}{aH}\right|_{t=T} - \frac{k}{a}(t - T) + \cdots,$$
(7.82)

where the remaining terms are negligible for $|t - T| \ll H^{-1}$. Thus, apart from a slowly varying phase factor, we have, well before horizon exit,

$$w_k = \left(\frac{1}{aL}\right)^{3/2}\sqrt{\frac{1}{2E_k}}\exp(-iE_k t),$$
(7.83)

where $E_k = k/a$. This is precisely Eq. (7.34), remembering that aL is the physical box size, k/a is the physical momentum, and the mass m is negligible. Thus, we have the correct flat space-time limit at early times.

We are assuming that the quantum state is the vacuum state, annihilated by $a_{\mathbf{k}}$. The mean-square vacuum fluctuation is therefore

$$\langle|\delta\phi_{\mathbf{k}}|^2\rangle = |w_k|^2.$$
(7.84)

At an epoch $t = t_*$, a few Hubble times after horizon exit, Eq. (7.81) gives it as

$$\langle |\delta\phi_{\mathbf{k}}|^2\rangle = \frac{H^2(t_*)}{2L^3k^3},$$

(7.85)

and the spectrum, from equation (7.44), is

$$\mathcal{P}_\phi(k, t_*) = \left[\frac{H(t_*)}{2\pi}\right]^2.$$

(7.86)

Because we are dealing with slow-roll inflation, H has negligible variation in a few Hubble times, and we can replace this expression by

$$\mathcal{P}_\phi(k, t_*) = \left(\frac{H}{2\pi}\right)^2\bigg|_{k=aH},$$

(7.87)

where H is evaluated simply at the epoch of horizon exit $k = aH$.

Except for the reality conditions, the phases of $\phi_{\mathbf{k}}$ are random, so that inflation has generated a Gaussian perturbation.

7.4.5   Conformal-time formalism

An equivalent formalism uses conformal time, and works with $u \equiv a\delta\phi$.

Conformal time τ is often convenient in theoretical cosmology. It is defined, up to a constant, by $d\tau = dt/a$. We can make the following choices during inflation, radiation domination, and matter domination:

$$\tau = -(aH)^{-1} \quad \text{(inflation)},$$

(7.88)

$$\tau = (aH)^{-1} \quad \text{(radiation domination)},$$

(7.89)

$$\tau = 2(aH)^{-1} \quad \text{(matter domination)}.$$

(7.90)

The first expression is the one we need for the moment. It ignores the variation of H, and so may be valid only for a few Hubble times, but that will be enough.

To be in line with most of the literature, we use a Fourier integral instead of a sum. The Fourier expansion is

$$u(\mathbf{x}, \tau) = \frac{1}{(2\pi)^{3/2}} \int d^3k\, u(\mathbf{k}, \tau)e^{i\mathbf{k}\cdot\mathbf{r}}.$$

(7.91)

The creation operators are given by

$$u(\mathbf{k}, t) = w(k, \tau)a(\mathbf{k}) + w^*(k, \tau)a^\dagger(-\mathbf{k}).$$

(7.92)

The equation of motion for $u(\mathbf{k}, \tau)$, also satisfied by w, is

$$\frac{\partial^2 u(\mathbf{k}, \tau)}{\partial\tau^2} + \left(k^2 - \frac{2}{\tau^2}\right)u(\mathbf{k}, \tau) = 0.$$

(7.93)

The commutation relation is taken to be

$$[a(\mathbf{k}), a^\dagger(\mathbf{k}')] = \delta^3(\mathbf{k} - \mathbf{k}'). \tag{7.94}$$

Well before horizon exit, we want to reproduce the flat space-time formulae of Section 7.2. Because $u = a\delta\phi$ and $p = k/a$, equating Eqs. (7.47) and (7.91) gives $u(\mathbf{k}) = a^{-2}u(\mathbf{p})$. Remembering that $\delta(\mathbf{k} - \mathbf{k}') = a^{-3}\delta(\mathbf{p} - \mathbf{p}')$, we also have $a(\mathbf{k}) = a^{-3/2}a(\mathbf{p})$. Combining these gives $w(\mathbf{k}) = a^{-1/2}w(\mathbf{p})$. Because $E_p = p$ for a massless fluctuation, Eq. (7.49) becomes

$$w(k) = \frac{1}{\sqrt{2k}} \exp\left(-\frac{ikt}{a}\right) \tag{7.95}$$

up to a slowly-varying phase factor.

The appropriate solution is

$$w(k, \tau) = -\frac{1}{\sqrt{2k^3}} (i - k\tau) \frac{\exp(-ik\tau)}{\tau}. \tag{7.96}$$

Well before horizon exit, it becomes

$$w(k, \tau) \to \frac{1}{\sqrt{2k}} \exp(-ik\tau), \tag{7.97}$$

which indeed has the right behaviour because $d\tau = dt/a$.

From Eq. (7.52), the spectrum is

$$\mathcal{P}_u = a^2 \mathcal{P}_\phi = \frac{k^3}{2\pi^2} |w^2(k)|. \tag{7.98}$$

Well after horizon exit, $|w|^2 = (2k^3\tau^2)^{-1} = a^2 H^2 (2k^3)^{-1}$, and $\mathcal{P}_\phi = (H/2\pi)^2$, in agreement with Eq. (7.87).

7.4.6 Formalism using the Lagrangian in curved space-time

The preceding approaches calculate the vacuum fluctuation well before the horizon exit, and then propagate the result forward in time using the perturbed classical field equation. The latter was obtained from the full field equation, which came ultimately from the full action, which we display as Eq. (14.85). Schematically,

action → field equation → perturbed field equation.

A different approach is to perturb the full action, to obtain the action for $\delta\phi$. We find that the perturbed action, in terms of $u = a\delta\phi$ and conformal time, is simply (Mukhanov 1989; Mukhanov et al. 1992)

$$S = \frac{1}{2} \int \left[(\partial_\tau u)^2 - (\partial_i u)^2 + 2\frac{u^2}{\tau^2} \right] d\tau \, d^3x. \tag{7.99}$$

This has the same form as the action of a free field in flat space-time (with position coordinate **x**, time coordinate τ, and a time-dependent mass-squared $-2\tau^{-2}$). Using the flat space-time formulae with the replacements $\mathbf{p} \to \mathbf{k}$ and $t \to \tau$, we then deduce the equations of the preceding section. Note that this is a true derivation, not just "working by analogy"; the action determines the quantum theory, and any time parameter can appear in the integral.

This approach may be represented schematically by

action $\to$ perturbed action $\to$ perturbed field equation.

It looks more elegant, but the calculation of the perturbed action is very long, especially when we go beyond the slow-roll approximation in the next section. Also, the approach rather loses sight of the fact that we need only invoke flat space-time quantum theory, plus the classical equation of motion.

7.4.7   Classicality

We noted earlier that the vacuum state is a superposition of states with well-defined field values, of which our Universe is supposed to be a typical one. At that point, we took it for granted that these field values can be regarded as classical quantities, but is this really correct? In other words, do there exist states that are the analogue of the "wave packets" that one considers in ordinary quantum mechanics, such that the fields have sharply defined values over a long period of time?

Well before horizon exit, we have flat space-time field theory, and the answer to the question is "no." The flat space-time vacuum fluctuation is an essentially quantum object, because it is not possible to put the field into a condition in which it remains sharply defined over an extended period of time, unless the field strength is far out on the tail of the probability distribution. This is clear from the fact that the real and imaginary parts of each Fourier component have the same dynamics as a quantum oscillator, the vacuum state being the ground state of the oscillator. It is well known that it is not possible to make a wave packet that does not spread significantly, unless its displacement from the origin is far in excess of the root mean square (rms) value for the ground state.

The situation is quite different well after horizon exit. Then, we can check that it is possible to give the field a sharply defined value, which is around the quantum expectation value (Guth and Pi 1985; Lyth 1985; Grishchuk 1993; Albrecht et al. 1994). We can see this immediately from the formalism of Section 7.4.5, where the quantum theory is that of a nonrelativistic particle in a harmonic oscillator potential $k^2 - 2(aH)^2$. After horizon exit, the second term dominates and the particle moves away from the origin into the classical regime.

The underlying reason for the classical behaviour seems to be the fact that w_k becomes almost real after horizon exit (Starobinsky 1982a, 1986), so that $\phi_{\mathbf{k}}$ has only trivial time dependence:

$$\phi_{\mathbf{k}} \simeq w_k(t)(a_{\mathbf{k}} + a^{\dagger}_{-\mathbf{k}}). \tag{7.100}$$

This quantum-to-classical behaviour is a tremendous success for the theory, which does not seem to have received adequate publicity. If it had failed, the prediction for the spectrum of the density perturbation would have had nothing to do with reality.

Of course there remains the usual interpretation problem, arising whenever we talk about measurement in quantum theory. Even after horizon exit, we are not allowed to say that the ϕ_k have particular values before they are measured, because, in principle, we are supposed to be able to have chosen instead to measure observables whose operators do not commute with the ϕ_k. A similar situation arises when we observe an electron created during, for example, beta decay. Even if its observed position is a metre away from the source, we are not allowed to say that it arrived at that position on the classical trajectory. The reason is that we might, in principle, have chosen to measure, for example, its angular momentum, which involves the full outgoing spherical wavefunction. The problem with the inflaton field perturbation is, however, more severe because it defines the density perturbation. Opinions differ about an electron, but probably nobody is willing to say that the density perturbation of the Universe is not there until it is measured. The problem has been discussed by many authors using the concept of decoherence (Sakagami 1988; Halliwell 1989; Padmanabhan 1989), but it seems far from a satisfactory solution.

7.4.8   Stochastic approach

We have so far focused on a single Fourier component $\delta\phi_k$, without paying much attention to the total perturbation $\delta\phi(\mathbf{x}, t)$ or, more or less equivalently, the total field $\phi(\mathbf{x}, t) = \phi(t) + \delta\phi(\mathbf{x}, t)$, where the first term is the spatial average.

To have a classical quantity, we need to smooth $\phi(\mathbf{x}, t)$ on a scale much bigger than the horizon. If one chooses the scale to be comoving, $\phi(\mathbf{x}, t)$ will obey the classical equation of motion. During inflation though, it is of interest to choose instead an almost fixed physical scale, equal to $(\varepsilon H)^{-1}$, where $\varepsilon \ll 1$ is a constant.

Smoothing on the fixed scale filters out Fourier components with $aH/k < \varepsilon^{-1}$. As time goes by, modes are constantly leaving the horizon, to become part of the smoothed $\phi(\mathbf{x}, t)$. Along a given worldline, specified by some value of $\mathbf{x}$, they give the smoothed ϕ random kicks, which have to be added to the classical motion. During one Hubble time, scales in an interval $\Delta \ln k = 1$ leave the horizon, and the rms size of the kick is $\Delta_{\text{vac}} = H/2\pi$. During the same time, the classical evolution changes ϕ by an amount $\Delta_{\text{class}} = \dot\phi H^{-1}$. The ratio is

$$\frac{\Delta_{\text{vac}}}{\Delta_{\text{class}}} = \frac{H}{2\pi}\frac{H}{\dot\phi}. \tag{7.101}$$

The formalism applies to any field satisfying the slow-roll conditions during inflation. Specializing to the inflaton field, the ratio in Eq. (7.101) is just the predicted value of the spectrum of the curvature perturbation $\mathcal{P}_{\mathcal{R}}^{1/2}(k)$, with $k = aH$. While astrophysically observable scales are leaving the horizon, it must be small, but at much earlier or later times, it can become bigger than unity.[4] In that case, using the stochastic formalism, we can show that there is a regime called **eternal inflation** (Linde 1986; Linde et al. 1994). The study of eternal inflation

---

[4] In that case the identification of the ratio with $\mathcal{P}_{\mathcal{R}}^{1/2}$ ceases to make sense because positive curvature in a region of given size can never exceed the corresponding scale. However, this does not affect the validity of the stochastic formalism.

for very large scales is important in principle because it may set the initial conditions (see Section 3.4.4), but we do not pursue it here.

On very small scales, it may be that we can never get deep into the regime of eternal inflation, for the following reason. Such scales leave the horizon toward the end of inflation, and even a value $\mathcal{P_R}(k)$ of order 1 will be enough to cause a good fraction of the energy density of the Universe to collapse into black holes, soon after the scale $1/k$ reenters the horizon. Objects with mass below 10^{10} g evaporate quickly by Hawking radiation, but if a good fraction of their mass is converted into stable (presumably Planck-mass) objects, the abundance of the black holes is constrained by the requirement that they contribute less than the critical density at present. More interestingly, black holes of masses above 10^{10} g evaporate at nucleosynthesis or later and are highly constrained by a number of observations (Carr 1985; MacGibbon and Carr 1991). Because black holes normally form with the horizon mass, which for a radiation-dominated Universe is given roughly by

$$M_{\text{hor}} \sim 10^{18}\,\text{g}\,\left(\frac{10^7\,\text{GeV}}{T}\right)^2,$$
(7.102)

the relevant mass scale depends on the inflationary energy scale. The conclusion is that $\mathcal{R}$ is constrained to be a factor of a few below unity even on very short scales. The use of these results in constraining inflationary perturbations was investigated by Carr et al. (1994) and Green and Liddle (1997b).

7.5   Spectrum of the primordial curvature perturbation

7.5.1   Spectrum

The primordial curvature perturbation $\mathcal{R}_k$ is equal to $-(H/\dot\phi)\delta\phi_k$ evaluated at $t = t_*$ [Eq. (4.6)]. Using Eq. (7.87) for $\mathcal{P}_\phi$, this gives

$$\mathcal{P_R}(k) = \left(\frac{H}{\dot\phi}\right)^2\left(\frac{H}{2\pi}\right)^2,$$
(7.103)

where the right-hand side simply can be evaluated at the epoch of horizon exit $k = aH$. Using the slow-roll formula $3H\dot\phi = -V'$, and the critical-density relation $3H^2 M_{\text{Pl}}^2 = V$, this becomes

$$\mathcal{P_R}(k) = \frac{1}{12\pi^2 M_{\text{Pl}}^6}\frac{V^3}{V'^2} = \frac{1}{24\pi^2 M_{\text{Pl}}^4}\frac{V}{\epsilon},$$
(7.104)

where $\epsilon \equiv \frac{1}{2}M_{\text{Pl}}^2(V'/V)^2$ is the slow-roll parameter introduced on page 42.

In terms of $\delta_H^2 = (4/25)\mathcal{P_R}$ [Eq. (5.7)], this is

$$\delta_H^2(k) \simeq \frac{1}{75\pi^2 M_{\text{Pl}}^6}\frac{V^3}{V'^2} = \frac{1}{150\pi^2 M_{\text{Pl}}^4}\frac{V}{\epsilon}.$$
(7.105)

As we see in Chapter 9, the cosmic microwave background (cmb) anisotropy measured by the Cosmic Background Explorer (COBE) satellite provides a good normalization of $\delta_H(k)$ on a very large scale: $k \simeq k_{\text{pivot}} \equiv 7.5 a_0 H_0$. This is given by Eq. (9.6):

$$\delta_H(k_{\text{pivot}}) = 1.91 \times 10^{-5}, \tag{7.106}$$

assuming that adiabatic density perturbations give the only contribution (otherwise equalities become upper limits). It gives

$$\frac{V^{3/2}}{M_{\text{Pl}}^3 V'} = 5.2 \times 10^{-4} \tag{7.107}$$

or, equivalently,

$$\frac{V^{1/4}}{\epsilon^{1/4}} = 0.027 M_{\text{Pl}} = 6.6 \times 10^{16} \, \text{GeV}, \tag{7.108}$$

with a 1-sigma uncertainty of 90 percent; see Eq. (9.10) for a generalization. The left-hand sides are to be evaluated at the epoch of horizon exit, $aH = k_{\text{pivot}}$, which can be calculated via an equation such as Eq. (3.19). This relation provides a crucial constraint on models of inflation. In practice, it is used to eliminate one of the parameters in the potential, usually the overall normalization. In principle, though, we would like to predict it from first principles in terms of some fundamental parameters of particle physics. Suggestions as to how this might come about are as yet very tentative and, unfortunately, provide at best an order-of-magnitude explanation for what is a very accurate observational result.

Because ϵ must be less than unity for inflation to occur, the inflationary energy scale $V^{1/4}$ is at least a couple of orders of magnitude below the Planck scale (Lyth 1984).

7.5.2   Spectral index

Now we discuss the scale dependence of the spectrum. No matter what its form, we can define an "effective spectral index" $n(k)$ as

$$n(k) - 1 \equiv \frac{d \ln \mathcal{P}_\mathcal{R}}{d \ln k}. \tag{7.109}$$

Over an interval of k, where $n(k)$ is constant, this is equivalent to the power-law behaviour that is assumed when defining the spectral index in the normal way,

$$\mathcal{P}_\mathcal{R}(k) \propto k^{n-1}. \tag{7.110}$$

We can work out $n(k)$ and its derivatives by using the slow-roll conditions. Because the right-hand side of Eq. (7.104) is evaluated when $k = aH$, and because the rate of change of H is negligible compared with that of a, we have $d \ln k = H \, dt$. From the slow-roll condition $dt = -(3H/V') d\phi$, we find

$$\frac{d}{d \ln k} = -M_{\text{Pl}}^2 \frac{V'}{V} \frac{d}{d\phi}. \tag{7.111}$$

Derivatives of the slow-roll parameters are given by

$$\frac{d\epsilon}{d\ln k} = 2\epsilon\eta - 4\epsilon^2, \tag{7.112}$$

$$\frac{d\eta}{d\ln k} = -2\epsilon\eta + \xi^2, \tag{7.113}$$

$$\frac{d\xi^2}{d\ln k} = -2\epsilon\xi^2 + \eta\xi^2 + \sigma^3, \tag{7.114}$$

where

$$\xi^2 \equiv M_{\rm Pl}^4 \frac{V'(d^3V/d\phi^3)}{V^2}, \tag{7.115}$$

$$\sigma^3 \equiv M_{\rm Pl}^6 \frac{V'^2(d^4V/d\phi^4)}{V^3}. \tag{7.116}$$

Following, for instance, Liddle et al. (1994), we have defined these as a square and a cube, respectively, even though the quantity inside the root never itself appears in an equation. This convenient device indicates their order in the slow-roll expansion. Also, in the case $V' \propto \phi^p$, with $p \neq 1$ or 2, we have $|\eta| \sim |\xi| \sim |\sigma|$. The hierarchy can be continued, each new equation introducing a new quantity $M_{\rm Pl}^{2n} V^{m-1}(d^{n+1}V/d\phi^{n+1})$.

Using Eqs. (7.112) and (7.104), we find (Liddle and Lyth 1992)

$$n - 1 = -6\epsilon + 2\eta. \tag{7.117}$$

As usual, the right-hand side is to be evaluated as the interesting scales cross outside the horizon. We see in Chapter 8 that ϵ is negligible in most models of inflation.[5]

Because slow-roll requires $\epsilon \ll 1$ and $|\eta| \ll 1$, we draw an important conclusion: *Inflation predicts that the variation of the spectrum is small in an interval $\Delta \ln k \sim 1$*. This prediction is amply verified by observation, which conservatively requires something like $|n-1| < 0.3$.

In a similar fashion, we can calculate the rate of change of n, to find (Kosowsky and Turner 1995)

$$\frac{dn}{d\ln k} = -16\epsilon\eta + 24\epsilon^2 + 2\xi^2. \tag{7.118}$$

Cosmologically observable scales correspond to a fairly narrow range of $\ln k$, about $\Delta \ln k \sim 10$. As we see in Section 9.3.4, cmb satellite observations can potentially measure n to order 0.01. For the Planck satellite, $\Delta \ln k$ is about 3 on either side of the central point, suggesting that $dn/d\ln k$ may be observable if it exceeds a few thousandths, and this is confirmed by a detailed analysis (Jungman et al. 1996c; Copeland et al. 1997). Such a variation is predicted by some models of inflation, whereas in others the variation is negligible. So-called designer inflation models (Salopek et al. 1989; Hodges and Blumenthal 1990; Hodges et al. 1990) have

---

[5] This equation also can be derived using the Hamilton–Jacobi version of the slow-roll approximation described in Section 3.6, yielding $n - 1 = -4\epsilon_{\rm H} + 2\eta_{\rm H}$. The different coefficient arises from the relation $\eta_{\rm H} \simeq \eta - \epsilon$ in the slow-roll limit.

long been known to be given a dramatic scale dependence, but the running mass model of Stewart (1997b) gives the same result with better motivation.

If n has negligible variation, we can write $\mathcal{P_R} \propto k^{n-1}$, the usual definition of n. When n is not equal to 1, it often is referred to as a **tilted spectrum** (Cen et al. 1992). When the tilt is to $n > 1$, the terminology **blue spectrum** (Mollerach et al. 1994) often is used.

7.5.3 Errors in the slow-roll predictions

We end this section by estimating the uncertainty in the slow-roll predictions.

In deriving the prediction for $\mathcal{P_R}$, we use the slow-roll conditions $\epsilon \ll 1$ and $|\eta| \ll 1$, as well as the slow-roll approximation $3H\dot{\phi} = -V'$. Differentiating the last expression, we find

$$\frac{\ddot{\phi}}{H\dot{\phi}} = \epsilon - \eta. \tag{7.119}$$

Comparing with the exact equation,

$$\ddot{\phi} + 3H\dot{\phi} + V' = 0, \tag{7.120}$$

we see that this is the fractional error in the slow-roll approximation. As a result, we expect $\mathcal{P_R}$ to pick up fractional errors of order ϵ and η,

$$\frac{\Delta \mathcal{P_R}}{\mathcal{P_R}} = \mathcal{O}(\epsilon, \eta). \tag{7.121}$$

Using Eqs. (7.112), (7.113), and (7.114), we therefore expect

$$n - 1 = 2\eta - 6\epsilon + \mathcal{O}(\xi^2), \tag{7.122}$$

$$\frac{dn}{d\ln k} = -16\epsilon\eta + 24\epsilon^2 + 2\xi^2 + \mathcal{O}(\sigma^3). \tag{7.123}$$

In the first expression, we ignore errors that are quadratic in ϵ and η because, barring cancellations, the corresponding fractional errors are small by virtue of the conditions $\epsilon \ll 1$ and $|\eta| \ll 1$. In the second expression, we ignore errors that are cubic in ϵ, η, and ξ. Barring cancellations, the accuracy of the prediction for $n - 1$ requires

$$|\xi^2| \ll \max(\epsilon, |\eta|), \tag{7.124}$$

and the accuracy of the prediction for its derivative requires, in addition,

$$|\sigma^3| \ll \max(\epsilon^2, \epsilon|\eta|, |\xi^2|). \tag{7.125}$$

7.6 Beyond the slow-roll approximation

The slow-roll predictions given earlier in this chapter are very convenient because they involve only V and its first few derivatives evaluated at the epoch of horizon exit. The use of slow-roll is not mandatory, however; on the contrary, one can obtain predictions using essentially no assumptions beyond linear perturbation theory.

In linear perturbation theory, the quantity $u = a\delta\phi$ satisfies the following exact equation:

$$\frac{\partial^2 u}{\partial \tau^2} + \left(k^2 - \frac{1}{z}\frac{d^2 z}{d\tau^2}\right)u = 0. \tag{7.126}$$

Here, τ is conformal time and

$$z \equiv \frac{a\dot{\phi}}{H}, \tag{7.127}$$

from which we find

$$\frac{d^2 z}{d\tau^2} = 2a^2 H^2\left(1 + \epsilon_{\mathrm{H}} - \frac{3}{2}\eta_{\mathrm{H}} + \frac{1}{2}\eta_{\mathrm{H}}^2 - \frac{1}{2}\epsilon_{\mathrm{H}}\eta_{\mathrm{H}} + \frac{1}{2H}\frac{d\epsilon_{\mathrm{H}}}{dt} - \frac{1}{2H}\frac{d\eta_{\mathrm{H}}}{dt}\right), \tag{7.128}$$

where

$$\epsilon_{\mathrm{H}} \equiv \frac{1}{2}\frac{\dot{\phi}^2}{H^2} = -\frac{\dot{H}}{H^2}, \qquad \eta_{\mathrm{H}} \equiv -\frac{\ddot{\phi}}{H\dot{\phi}} \tag{7.129}$$

are the Hamilton–Jacobi slow-roll parameters of Section 3.6 and an overdot denotes d/dt. To derive Eq. (7.126), note first that $u = -z\mathcal{R}$, so that the equation is equivalent to an expression for $\ddot{\mathcal{R}}$ in terms of $\dot{\mathcal{R}}$ and $\mathcal{R}$. This expression may be obtained by differentiating Eq. (4.166) with the aid of Eqs. (4.156) and (4.167) because on comoving slices ϕ, and therefore $V(\phi)$, is constant, which from Eq. (3.3) means that $\delta P = \delta\rho$.

This relatively straightforward derivation was given (in slightly different forms) by Mukhanov (1985) and Sasaki (1986). Later it was noted (Mukhanov 1989) that the equation also follows from the action of u:

$$S = \frac{1}{2}\int d\tau\, d^3 x\left[(\partial_\tau u)^2 - (\partial_i u)^2 + \frac{d^2 z}{d\tau^2}\frac{u^2}{z}\right]. \tag{7.130}$$

This action is obtained by perturbing the full action equation (14.85) and dropping various total derivatives, but the calculation is extremely long, and unnecessary in view of what has been said.

These equations can be solved numerically (e.g., Grivell and Liddle 1996), but here we explore improved analytical calculations. We assume that inflation is near enough slow-roll that $k|\tau| \gg 1$ a few Hubble times before horizon exit, and $k|\tau| \ll 1$ a few Hubble times after. Then, there is a solution $u = w$ of Eq. (7.126) that satisfies Eq. (7.97) a few Hubble times before horizon exit. A few Hubble times after horizon exit, this solution has the behaviour

$$\frac{w}{z} \to \text{constant}. \tag{7.131}$$

From Eq. (7.98), the spectrum of $\mathcal{R}$ then is given by

$$\mathcal{P}_{\mathcal{R}}(k) = \frac{k^3}{2\pi^2 z^2}|w(k)|^2. \tag{7.132}$$

Given an inflationary trajectory defined by $a(\tau)$ and $\dot{\phi}(\tau)$, this method gives a practically unique, and accurate, result in all reasonable cases. The trajectory in turn follows from the potential practically independently of the initial conditions, if slow-roll becomes very accurate at some early epoch.

We noted earlier that in the regime where the slow-roll predictions for $\mathcal{P}_{\mathcal{R}}$, $n - 1$, *and* $dn/d \ln k$ are approximately valid, the slow-roll conditions (3.9), (7.124), and (7.125) are also valid. In that case, the "exact" solution yields an improved version of the slow-roll predictions for $\mathcal{P}_{\mathcal{R}}$ and $n - 1$ (Stewart and Lyth 1993). Let us see how this goes.

Equations (7.119), (7.112), and (7.113) and the slow-roll conditions give the approximation

$$\frac{d^2 z}{d\tau^2} = 2a^2 H^2 \left(1 + \epsilon_{\mathrm{H}} - \frac{3}{2} \eta_{\mathrm{H}} \right), \tag{7.133}$$

with ϵ_{H} and η_{H} slowly varying on the Hubble timescale. Solving Eq. (7.126) with the appropriate vacuum initial conditions leads to the approximation (Stewart and Lyth 1993)

$$\mathcal{P}_{\mathcal{R}}^{1/2}(k) = [1 - (2C + 1) \epsilon_{\mathrm{H}} + C \eta_{\mathrm{H}}] \frac{H^2}{2\pi |\dot{\phi}|}, \tag{7.134}$$

where $C = -2 + \ln 2 + b \simeq -0.73$, with b the Euler–Mascheroni constant. As always, the right-hand side is evaluated at $k = aH$.

We want an expression involving V and its derivatives. Substituting Eq. (3.8) into $\rho = V + \frac{1}{2}\dot{\phi}^2$ gives

$$\frac{3M_{\mathrm{Pl}}^2 H^2}{V} = 1 + \frac{1}{3} \epsilon, \tag{7.135}$$

and substituting Eq. (7.119) into Eq. (7.120) gives

$$-\frac{3H\dot{\phi}}{V'} = 1 - \frac{1}{3} \epsilon + \frac{1}{3} \eta. \tag{7.136}$$

These are improvements in the slow-roll formulae, valid to linear order in ϵ and η. Squaring the last equation gives

$$\frac{\epsilon_{\mathrm{H}}}{\epsilon} = 1 - \frac{2}{3} \epsilon + \frac{1}{3} \eta, \tag{7.137}$$

and Eq. (7.119) is

$$\eta_{\mathrm{H}} = \epsilon - \eta. \tag{7.138}$$

Inserting these four expressions into Eq. (7.134) gives

$$\mathcal{P}_{\mathcal{R}}^{1/2}(k) = \frac{1}{\sqrt{12}\,\pi\, M_{\mathrm{Pl}}^3} \frac{V^{3/2}}{|V'|} \left[1 - \left(2C + \frac{1}{6} \right) \epsilon + \left(C - \frac{1}{3} \right) \eta + \mathcal{O}(\xi^2) \right]. \tag{7.139}$$

The fractional error in this improved expression for $\mathcal{P}_{\mathcal{R}}$ is expected to be of order ξ^2, plus terms quadratic in ϵ and η that we did not display. The ξ^2 term will be present because it contributes to the variation of η per Hubble time, Eq. (7.113).

Using $k = aH$ with Eqs. (7.135) and (7.136), and the exact expression

$$\dot{H} = -\frac{1}{2}\frac{\dot{\phi}^2}{M_{\rm Pl}^2},$$
(7.140)

one finds the improved formula

$$\frac{d\phi}{d\ln k} = \sqrt{2\epsilon}\left(1 + \frac{1}{3}\epsilon + \frac{1}{3}\eta\right).$$
(7.141)

This leads to

$$\frac{n-1}{2} = -3\epsilon + \eta - \frac{5+36C}{3}\epsilon^2 + (8C-1)\epsilon\eta + \frac{1}{3}\eta^2 - \frac{3C-1}{3}\xi^2 + \mathcal{O}(\sigma^3).$$
(7.142)

The fractional error of order σ^3 comes from differentiating the error of order ξ^2 in Eq. (7.139) [$d\xi^2/d\ln k$ is given by Eq. (7.114)]. There also will be error terms cubic in ϵ, η, and ξ, which we do not display. The improved formula becomes exact in the case of $V = V_0 \pm \frac{1}{2}m^2\phi^2$, in the limit $\phi \to 0$ when $\epsilon \to 0$ and $\eta_{\rm H}$ becomes constant.

In the case of power-law inflation, for which $\epsilon_{\rm H} = \eta_{\rm H}$ is constant, there is an exact solution to Eq. (7.126) (Lyth and Stewart 1992a), which becomes a Bessel equation. The solution satisfying the appropriate boundary conditions features a fractional-order Hankel function, being

$$w(k,\tau) = \frac{\sqrt{\pi}}{2}\exp\left[i\left(\nu+\frac{1}{2}\right)\frac{\pi}{2}\right]\sqrt{-\tau}\,H_\nu^{(1)}(-k\tau)$$
(7.143)

$$\to \exp\left[i\left(\nu-\frac{1}{2}\right)\frac{\pi}{2}\right]2^{\nu-3/2}\frac{\Gamma(\nu)}{\Gamma(3/2)}\frac{1}{\sqrt{2k}}(-k\tau)^{1/2-\nu}$$

$$\text{as } \frac{aH}{k} \to \infty,$$
(7.144)

where

$$\nu = \frac{3}{2} + \frac{\epsilon_{\rm H}}{1-\epsilon_{\rm H}}.$$
(7.145)

From this, the spectrum of $\mathcal{R}$ is obtained as

$$\mathcal{P}_{\mathcal{R}}^{1/2}(k) = \sqrt{\frac{k^3}{2\pi^2}}\left|\frac{w(k)}{z}\right| = 2^{\nu-3/2}\frac{\Gamma(\nu)}{\Gamma(3/2)}\left(\nu-\frac{1}{2}\right)^{1/2-\nu}\frac{H^2}{2\pi|\dot{H}|}\bigg|_{k=aH},$$
(7.146)

which is valid for any constant $\epsilon_{\rm H} < 1$.

In some models, the improvement to the slow-roll result is big enough to measure *with fixed values of the parameters in the potential*. However, in the cases that have been examined to date, this change can be practically cancelled by varying the parameters. As a result, the improvement is probably going to be useful only if gravitational waves are detected.

7.7   Gravitational waves

7.7.1   Gravitational-wave spectrum

We saw in Section 6.5 that the gravitational-wave amplitudes $h_{+,\times}$ have the same action as massless scalar fields, up to a numerical factor. The analysis leading to Eq. (7.87) remains valid for any massless scalar field and, as a result, the spectrum $\mathcal{P}_{grav}$ of the amplitudes is equal to H^2 evaluated at horizon exit, up to a numerical factor, the precise result being Eq. (6.38), namely

$$\mathcal{P}_{grav}(k) = \frac{2}{M_{Pl}^2}\left(\frac{H}{2\pi}\right)^2\Bigg|_{k=aH} \tag{7.147}$$

The spectral index is defined by

$$n_{grav} = \frac{d\ln\mathcal{P}_{grav}(k)}{d\ln k}, \tag{7.148}$$

and using Eq. (7.111), we find (Liddle and Lyth 1992) that

$$n_{grav} = -2\epsilon. \tag{7.149}$$

7.7.2   Relative amplitude of gravitational waves

We now have seen that inflation generates spectra of both density perturbations and gravitational waves, and that, in particular, both spectra can generate microwave background anisotropies. A crucial question therefore is how important the two contributions are, for a given inflationary model. This comes from comparing Eqs. (5.41) and (6.42). Assuming that we look at an ℓ value large enough to drop the correction coefficient c_ℓ in the latter, we see that the ratio of the contributions from the gravitational waves and the adiabatic density perturbation is given by (Liddle and Lyth 1992)

$$r \equiv \frac{C_\ell(grav)}{C_\ell(ad)} \simeq 12.4\epsilon. \tag{7.150}$$

We use r to indicate the relative amplitude of the gravitational waves and density perturbations.

The preceding treatment is approximate in several ways. First, it refers to scale-invariant spectra, to give an ℓ-independent result. In general, the spectral indices of the two spectra will be different, and r has to be evaluated at the scale corresponding to the ℓ value under consideration. Second, both expressions used assume that only the Sachs–Wolfe effect applies, and that the Universe is perfectly matter-dominated. In fact, both of these approximations break down at the 10-percent level or so if ℓ is big enough for the c_ℓ correction factors to be negligible. In Chapter 9, we see that this has to be included in making precision comparisons with the COBE data.

Often, r is defined as the ratio at the quadrupole and, including the correction factor, that would give $\tilde{r} = 13.8\epsilon$.

Even if the gravitational waves have a large enough amplitude to be detectable in the cmb, their amplitude on the much shorter scales probed by interferometer experiments such as LIGO, VIRGO, and GEO will be far too small to allow their direct detection. This is a firm result for the models we have been discussing, because the gravitational-wave spectral index is necessarily negative.

7.7.3　Consistency relation

The results we have obtained show that the spectra of gravitational waves and density perturbations are not independent, but rather are related, from Eqs. (7.149) and (7.150), by

$$r = -6.2\,n_{\mathrm{grav}}. \tag{7.151}$$

Within the paradigm of single-field slow-roll inflation, this holds regardless of the scalar-field potential, and has become known as the consistency equation. Let us stress that it is the *spectra*, giving the typical size of perturbations on a given scale, that are related. There is no correlation between the *phases* of the two perturbations in a given realization.

The idea of a consistency equation in single-field models is very general; it simply states that, because both the density perturbation and the gravitational wave spectra are generated during the same inflationary period from the potential $V(\phi)$, they must be connected because we can solve to eliminate $V(\phi)$. However, we see later that even the simplest form as quoted above is probably impossible to test. Further, if we generalize to multifield inflation models (Section 7.9), the consistency equation is replaced by an inequality, even if slow-roll holds.

7.7.4　Why r is negligible in most models of inflation

In Section 8.2, we discuss some general expectations for the form of the inflaton potential. One important idea is that, in supergravity and superstring theories, we presently understand the form of the potential only for field values at most of order of the Planck mass. As we now explain, any model with an inflaton field variation $\ll M_{\mathrm{Pl}}$ will give undetectable gravitational waves, and in practice, we expect that this will be true even if the variation is of order M_{Pl} (Lyth 1997).

We first need to decide what will be detectable. Presently, there is only a very weak limit, perhaps $r \lesssim 1$, from the sorts of observations discussed in Chapter 12. The only way of significantly improving this is from the cmb anisotropy, and we must cope with the uncertainties in all the cosmological parameters. We study this in detail in Section 9.3.4. Using the temperature anisotropies alone is very unpromising (Jungman et al. 1996c; Bond et al. 1997), but things improve somewhat once polarization is included as well, because it breaks a degeneracy between the effects of reionization and gravitational-wave modes. Zaldarriaga et al. (1997) estimate that the Planck satellite with polarized detectors can measure r with an error $\Delta r \simeq 0.05$, and so, we assume that a 95-percent confidence detection requires $r > 0.1$. Note that, from Eqs. (7.108) and (7.150), a detection of r at this level would give a value $V^{1/4} = (2 \text{ to } 6) \times 10^{16}$ GeV.

In Eq. (7.150), the right-hand side is evaluated when $k = aH$. Because the ℓth multipole of the cmb anisotropy corresponds to a scale $k^{-1} \simeq 2/(H_0\ell)$, and gravitational waves only contribute up to $\ell \sim 100$, the relevant range corresponds to only $\Delta \ln k \simeq 4$. Through Eq. (7.111), we can relate r to the change in the inflaton field per Hubble time, at the time the relevant scales are leaving the horizon,

$$\frac{1}{M_{\text{Pl}}}\left|\frac{d\phi}{dN}\right| = M_{\text{Pl}}\left|\frac{V'}{V}\right| = \left(\frac{r}{6.2}\right)^{1/2}. \tag{7.152}$$

While the scales corresponding to $2 \leq \ell \lesssim 100$ are leaving the horizon, the amount of expansion is $\Delta N \simeq 4$, and so, the corresponding field variation is

$$\frac{\Delta\phi}{M_{\text{Pl}}} \simeq 4\left(\frac{r}{6.2}\right)^{1/2} = 0.5\left(\frac{r}{0.1}\right)^{1/2}. \tag{7.153}$$

We see that a detectable r requires $\Delta\phi \gtrsim 0.5M_{\text{Pl}}$; that is, the field has to move of order of a Planck mass while the relevant perturbations are generated. This is a minimum estimate for the total field variation because inflation must continue afterward for many more e-foldings. We see in the next chapter that almost all of the particle physics–motivated models of inflation proposed so far give negligible gravitational waves.

7.8 Generating an isocurvature perturbation

Now we see how an isocurvature perturbation might be generated, focussing particularly on the axion, which is the best-motivated possibility. As we saw in Section 6.6, we do not expect that such a perturbation is present in Nature at a significant level. However, even the requirement that it be no *bigger* than the total density perturbation provides a significant constraint on axion physics, for a given model of inflation.

7.8.1 The general picture

We begin by describing the general situation, which includes the axion as a special case. The isocurvature perturbation comes from the vacuum fluctuation of a non-inflaton field χ, and we lose nothing essential by taking its potential during inflation to be simply $V(\chi) = \frac{1}{2}m_\chi^2\chi^2$. For the fluctuation to be significant (and to become classical after horizon exit), we need $m_\chi \lesssim H$. In the context of supergravity, the mass of a generic field is expected to be at least this big (Section 8.2.5). However, we take for the moment the usual view that χ is a special field, the mass of which is so small that it has negligible classical motion during a few e-folds of inflation, even though it is not sitting at the minimum of its potential. This corresponds to

$$m_\chi^2 \ll V''(\ll H^2), \tag{7.154}$$

where, as usual, V'' is the second derivative of V in the direction of the inflaton field ϕ.

If χ is a pseudo-Goldstone boson, as is the case for the axion, the smallness of m_χ during inflation might be ensured by the relevant symmetry. At the end of this section we consider what happens in the opposite case of $V'' \lesssim m_\chi^2 \lesssim H^2$.

Because m_χ is negligible, the vacuum fluctuation $\chi_\mathbf{k}$ satisfies the massless field equation (7.69). It has a constant value after horizon exit, the spectrum of which is $\mathcal{P}_\chi(k) = (H/2\pi)^2$ with H evaluated at the epoch of horizon exit. Its spectral index is the same as for gravitational waves, $n_\chi = -2\epsilon$ [Eqs. (7.148) and (7.149)]. However, there is a big difference between the two cases. To get significant gravitational waves, we need $\epsilon \gtrsim 0.1$, and their spectral index will not then be completely negligible. However, this happens only in special models of inflation, which have no particular status in the context of isocurvature perturbations. Accordingly, we are entitled to assume that n_χ is practically 0 (under our current assumption that m_χ is negligible).

If χ survives after inflation, it might generate an isocurvature density perturbation in a variety of ways. The simplest case, and the only one that we consider, is if the χ particles constitute at least some mass fraction f_χ of the nonbaryonic dark matter.

The χ particles are created when χ starts to oscillate about the minimum of the potential. This happens when H falls below m_χ, which is supposed to occur after inflation.[6] In the case of axion physics, this type of particle creation usually is referred to as **misalignment** (of the field value with the value $\chi = 0$ corresponding to the minimum of the potential).

The initial isocurvature perturbation is given by Eq. (6.49) as

$$S = \frac{\delta\rho_c}{\rho_c} = f_\chi \frac{\delta\rho_\chi}{\rho_\chi}. \tag{7.155}$$

(In this definition the adiabatic perturbation is ignored, so that $\delta\rho_c = \delta\rho_\chi$.) At each point in space, the density ρ_χ of the χ particles is proportional to the square of the initial oscillation amplitude $\chi + \delta\chi$, and two cases arise.

If $|\delta\chi| \ll |\chi|$, we have a Gaussian isocurvature perturbation $S = 2\delta\chi/\chi$. Its spectrum is

$$\mathcal{P}_S = \frac{4f_\chi}{\chi} \left(\frac{H}{2\pi}\right)^2, \tag{7.156}$$

with H evaluated at horizon exit during inflation. The spectral index n_{iso} [defined by Eq. (6.50)] is then equal to n_χ and presumably very close to 0. In this case, the observational requirement $\mathcal{P}_S \lesssim 10^{-5}$ can be satisfied with $f_\chi \sim 1$, which is what we generally assume in the case of the axion.

In the opposite case of $|\delta\chi| \gg |\chi|$, we have

$$S \sim f_\chi \frac{(\delta\chi)^2}{\langle(\delta\chi)^2\rangle}. \tag{7.157}$$

This is a non-Gaussian perturbation of a particular type, namely, chi-squared distributed. Assuming that $n_\chi = 0$, we again find $n_{\text{iso}} \simeq 0$ (Lyth 1992). Also, $\mathcal{P}_S \sim f_\chi^2$; for this case to be viable, we need $f_\psi \lesssim 10^{-5}$.

---

[6] Here, m_χ still denotes the time-varying effective mass.

Finally, we mention the possibility $V'' \lesssim m_\chi^2 \lesssim H^2$. In that case, we should assume that the classical value of χ is 0 because, otherwise, χ becomes the inflaton (or at least, part of a multicomponent inflaton as discussed in Section 7.9). Then Eq. (7.157) still applies, but we now have (Linde and Mukhanov 1997)

$$n_{\text{iso}} = \frac{2m^2}{3H^2}, \qquad (7.158)$$

which is *bigger* than 0 and may be significant.

7.8.2 The axion

The axion is the only well-motivated candidate for generating an isocurvature perturbation, and is also one of the best candidates for cold dark matter. It is the pseudo-Goldstone boson of a spontaneously broken global $U(1)$ symmetry, called Peccei–Quinn symmetry. This symmetry usually is supposed to be present in extensions of the standard model because it guarantees the charge-parity (CP) invariance of the strong interaction. Let us assume for simplicity that only one field ψ is charged under Peccei–Quinn symmetry, though in the context of supersymmetry there actually will be at least two. When the symmetry is spontaneously broken, $\psi = \sqrt{2} f_{\text{a}} \exp(i\theta)$, where f_{a} is some constant. There is then an axion, with canonically normalized field $\chi = f_{\text{a}}\theta$.

In the early Universe, Peccei–Quinn symmetry usually is taken to be exact, making the axion massless. However, at $T \sim 1$ GeV, nonperturbative quantum chromodynamic (instanton) effects generate an axion mass $m(t)$. The mass increases with time, levelling out to its true value m by $T \sim 100$ MeV. One can show that $m f_{\text{a}} \sim m_\pi^2$, where m_π is the pion mass, or, to be precise,

$$\frac{m}{6.2 \times 10^{-4} \text{ eV}} = \frac{10^{10} \text{ GeV}}{f_{\text{a}}/N}, \qquad (7.159)$$

where N is an integer that we take to be 1. Astrophysics and collider physics combine to give the constraint $m \lesssim 10^{-2}$ eV, corresponding to $f_{\text{a}} \gtrsim 10^9$ GeV.

The axion field exists only when the Peccei–Quinn symmetry is spontaneously broken. To obtain an isocurvature perturbation, the symmetry needs to be spontaneously broken at all epochs after our Universe leaves the horizon during inflation, so that the axion field has a continuous existence.

This is the case only in a limited regime of parameter space. During inflation, the quantum fluctuation will restore the symmetry if $H \gtrsim f_{\text{a}}$ (Lyth and Stewart 1992b,c). Afterward, it will, in any case, be restored if the reheat temperature is bigger than f_{a}. If the symmetry is restored, axions are produced mostly from radiation by axionic cosmic strings forming when it breaks again.

We are interested in the case in which the symmetry is not restored, so that the axion can give an isocurvature perturbation. In that case, axions are produced only by the misalignment mechanism described earlier. We discount here the possibility of bubbles of domain wall, which occur only in a very limited regime of parameter space (Linde and Lyth 1990).

In both cases, axion number is conserved after the mass starts to become significant. This allows us to calculate the present axion density Ω_a in terms of the relevant parameters. Assuming the standard cosmology after 1 GeV, with production either from cosmic strings or from the misalignment mechanism with θ not too small, the requirement that the present axion density be $\Omega_a \lesssim 1$ gives the constraint $m \gtrsim 10^{-3}$ to 10^{-4} eV. Combined with the astrophysics/collider-physics constraint $m \lesssim 10^{-2}$ eV, this leaves an allowed "window" of, at most, a couple of orders of magnitude.

A late-decaying particle can widen the window by producing significant entropy after the massive axions first appear at 1 GeV. Alternatively, if there are no cosmic strings, we can postulate that we live in a part of the Universe where the misalignment θ is very small, which reduces Ω and again widens the window. (This is possible only if the vacuum fluctuation of θ is also very small.) In the end, f_a might be of order 10^{15} GeV, which may allow us to identify the Peccei–Quinn axion with a similar field arising in superstring theory.

Finally, in the regime where there are no cosmic strings, a further constraint is placed by the requirement that the isocurvature density perturbation be no bigger than the total observed density perturbation (Lyth 1990, 1992).

7.9   A multicomponent inflaton?

So far we have assumed that there is only one possible inflationary trajectory in the space of the scalar fields, which is automatically true if there is only a single scalar field. However, this is unlikely to be strictly correct; rather, there will be a whole family of possible inflaton trajectories. On the other hand, the trajectories may all be physically equivalent, in which case only one need be considered.

Let us see how this might work, taking the relevant fields to be canonically normalized. A trivial example would be if, in addition to the inflaton field ϕ, there are fields making practically no contribution to the potential during inflation (massless fields). The massless fields then label an infinity of inflationary trajectories, but unless the fields survive and become significant after inflation, all trajectories are physically equivalent. A more interesting example is if a symmetry ensures the equivalence of the trajectories. For instance, the inflaton field might be the modulus of a complex field charged under a $U(1)$ symmetry. Then the possible inflaton trajectories are the radial lines, but the $U(1)$ symmetry ensures that they are all equivalent. More generally, the trajectories will be equivalent if they are practically straight, and the transition from inflation to a matter- or radiation-dominated Universe is the same for each trajectory.

In this section we briefly discuss models in which the trajectories are not equivalent. In such models, the inflaton field is a multicomponent object with components ϕ_a. In the context of Einstein gravity, they often are called double inflation models because there are typically two components, with the trajectory turning a more or less sharp corner as in Figure 7.3. There are also models involving modified gravity, and both kinds are mentioned briefly in the next chapter.

Let us see how to calculate the spectrum of $\mathcal{R}$ in these models. It is assumed that, while cosmological scales are leaving the horizon, all components of the inflaton have the slow-roll

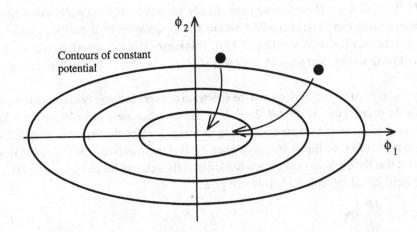

Fig. 7.3. A two-field potential of the form of an asymmetric bowl, seen from above, with the minimum at the origin and contours of constant V illustrated. The thick curves indicate different trajectories via which the fields may reach the minimum. Different trajectories in this potential may lead to different predictions for large-scale structure.

behaviour

$$3H\dot{\phi}_a = -\frac{\partial V}{\partial \phi_a}. \tag{7.160}$$

Differentiating this and comparing it with the exact expression $\ddot{\phi}_a + 3H\dot{\phi}_a = -\partial V/\partial \phi_a$ gives consistency, provided that

$$M_{\text{Pl}}^2 \left(\frac{\partial V/\partial \phi_a}{V}\right)^2 \ll 1, \qquad M_{\text{Pl}}^2 \left|\frac{\partial^2 V/\partial \phi_a \partial \phi_b}{V}\right| \ll 1. \tag{7.161}$$

The second condition actually could be replaced by a weaker one but let us retain it for simplicity. We expect slow-roll to hold if these conditions are satisfied. This implies that H (and therefore ρ) is slowly varying, giving quasiexponential inflation. The second condition ensures that $\dot{\phi}_a$ is slowly varying.

It is not necessary to assume that all of the fields continue to slow-roll after cosmological scales leave the horizon. For instance, one or more of the fields might start to oscillate, while the others continue to support quasiexponential inflation, which ends only when slow-roll fails for all of them. Alternatively, the oscillation of some field might briefly interrupt inflation, which resumes when its amplitude becomes small enough. These things might happen while cosmological scales leave the horizon too, but that case is not considered here.

We now calculate the adiabatic density perturbation generated by the vacuum fluctuation of a multicomponent inflaton, using the method of Sasaki and Stewart (1996) as summarized by Lyth and Riotto (1999).

For a multicomponent inflaton, the formula $\mathcal{R} = -H\delta\phi/\dot{\phi}$ remains valid if we take ϕ to denote the component along the direction of slow roll. To see this, we note first that the formula

$\mathcal{R} = H\delta t$ from Chapter 14, which we have already advertised, is universally valid. Next, the momentum density can be written $\dot{\phi}\nabla\delta\phi$, where ϕ is the component of the field parallel to the direction of the classical trajectory [Eq. (7.10)]. This means that the perturbation of ϕ vanishes in the comoving slicing, and so, with any other slicing, it is $\delta\phi = -\dot{\phi}\,\delta t$, where δt is the time displacement.

A few Hubble times after horizon exit the spectrum of every inflaton field component, and in particular the parallel one, is still $(H/2\pi)^2$. If $\mathcal{R}$ had no subsequent variation, this would lead to the usual prediction, but we are considering the case in which the variation is significant. It is given in terms of δP by Eq. (4.166), and when δP is significant, it can be calculated from the assumption that the evolution along each worldline is the same as for an unperturbed Universe with the same initial inflaton field. This will give

$$\delta P = \frac{\partial P}{\partial\phi_a}\,\delta\phi_a, \tag{7.162}$$

where $\delta\phi_a$ is evaluated at the initial epoch and $P(\phi_1, \phi_2, \ldots, t)$ represents the evolution of P in an unperturbed Universe. Choosing the basis so that one of the components is the parallel one, and remembering that all components have spectrum $(H/2\pi)^2$, we can calculate the final spectrum of $\mathcal{R}$. The only input is the evolution of P in the unperturbed Universe, corresponding to a generic initial inflaton field (close to the classical initial field).

Using this expression, we can track the evolution of $\mathcal{R}$ until it stops varying significantly. This will happen after inflation, assuming that the Universe is dominated by matter, radiation, or by a *single* homogeneous field, because in those cases, $\delta P/\delta\rho$ has a definite ratio (respectively 0, 1/3, and -1) and $\delta\rho$ is negligible. The crucial point is that, with a multicomponent inflaton field, it need not happen *before* the end of inflation, whereas with a single component, it happens soon after horizon exit.

In this discussion, we started with Eq. (4.6) for the initial $\mathcal{R}$, and then invoked Eq. (4.166) to evolve it. The equations actually can be combined to give

$$\mathcal{R} = \delta N, \tag{7.163}$$

where N is the number of Hubble times measured by a comoving observer passing from the initial (flat) slice that defines $\delta\phi$ to the final (comoving) slice that defines $\mathcal{R}$. This remarkable expression was given by Starobinsky (1985b) and proved by Salopek (1995) and Sasaki and Stewart (1996). The approach we are using is close to that of Sasaki and Stewart (1996).

To see that this expression is indeed equivalent to Eqs. (4.6) and (4.166), note first that from Eq. (4.143) it certainly gives the correct result at the initial epoch. Thus we need only convince ourselves that δN gives the *change* in $\mathcal{R}$, if N is redefined to be the number of Hubble times between the initial *comoving* slice and the final slice. Representing the initial time as usual by t_*, we find that the redefined N has a perturbation

$$\delta N(\mathbf{x}, t) = \delta\left[\int_{t_*}^{t} H(\mathbf{x}, t)\frac{dt_{\text{pr}}}{dt}\,dt\right] \simeq \int_{t_*}^{t} H(\mathbf{x}, t)\delta\frac{dt_{\text{pr}}}{dt}\,dt, \tag{7.164}$$

where we used the fact that δH is negligible on scales far outside the horizon (Section 4.8).

However, Eq. (4.154), evaluated to first order, shows that this is precisely the change in $\mathcal{R}$ given by Eq. (4.166).[7]

Because the evolution of H along a comoving worldline will be the same as for a homogeneous Universe with the same initial inflaton field, N is a function only of this field and we have

$$\mathcal{R} = \frac{\partial N}{\partial \phi_a} \delta \phi_a, \tag{7.165}$$

where repeated indices are summed over. The perturbations $\delta \phi_a$ are Gaussian random fields generated by the vacuum fluctuation, and have a common spectrum $(H/2\pi)^2$. The spectrum is therefore

$$\delta_H^2 = \frac{V}{75\pi^2 M_{\text{Pl}}^2} \frac{\partial N}{\partial \phi_a} \frac{\partial N}{\partial \phi_a}, \tag{7.166}$$

again with summation over a.

In the single-component case, $N' = M_{\text{Pl}}^{-2} V/V'$, and we recover the usual expression. In the multicomponent case, we can always choose the basis fields so that, while cosmological scales are leaving the horizon, one of them points along the inflaton trajectory, and then its contribution gives the standard result, with the orthogonal directions giving an additional contribution. Because the spectrum of gravitational waves is independent of the number of components (being equal to a numerical constant times V) the relative contribution r of gravitational waves to the cmb is always *smaller* in the multicomponent case, and so, the consistency condition (7.151) is replaced by an inequality.

Using the slow-roll conditions, we can show that

$$n - 1 = -\frac{M_{\text{Pl}}^2 V_{,a} V_{,a}}{V^2} - \frac{2}{M_{\text{Pl}}^2 N_{,a} N_{,a}} + 2\frac{M_{\text{Pl}}^2 N_{,a} N_{,b} V_{,ab}}{V N_{,d} N_{,d}}, \tag{7.167}$$

where ",a" indicates partial derivative with respect to ϕ_a. Again, we recover the single-field case using $N' = M_{\text{Pl}}^{-2} V/V'$. Corrections to these formulae, along the lines of the single-field analysis of Section 7.6, are given by Nakamura and Stewart (1996).

These formulae are quite different from the usual ones that hold for a single-component inflaton. To evaluate them, we have to know the evolution of the homogeneous Universe corresponding both to the classical inflaton trajectory and to nearby trajectories. An important difference from the single-component case is that the classical trajectory is not uniquely specified by the potential, but rather has to be given as a separate piece of information.

This treatment can be generalized straightforwardly (Sasaki and Stewart 1996) to the case of noncanonical kinetic terms of the form of Eq. (7.59), which is expected in supergravity. However, in the small-field regime, we expect the curvature associated with the metric G_{ab} to

---

[7] This last step ignores anisotropic stress, which would contribute to dt/dt_{pr} without affecting $\dot{\mathcal{R}}$, but as we discuss in Section 4.8.4, anisotropic stress is indeed negligible, because after smoothing on a scale far outside the horizon, the Universe looks locally like an unperturbed one.

be negligible, and then we can recover the canonical normalization $G_{ab} = \delta_{ab}$ by redefining the fields.

Examples

7.1 Use Eqs. (7.9) and (7.10) to calculate the energy density ρ and the momentum density **N** of a scalar field $\phi = A\cos(\omega t - kz)$. Then specialize to the case of a massive free field and take the time average. Work out $\rho^2 - N^2$ and comment on the physical significance of this result.

7.2 Derive the expressions for the total energy and momentum of a free scalar field, Eqs. (7.19) and (7.20).

7.3 Derive the general form of the potential that, within the slow-roll approximation, gives scale-invariant density perturbations.

8 Building and testing models of inflation

In the preceding chapter we were on firm ground. The calculation of the vacuum fluctuation uses basic field theory, and is unlikely ever to be invalidated. Now we go to the more fluid topic of inflation model building. The subject has seen a renaissance in recent years, with an increasing emphasis on the particle-theory basis. After sketching the theoretical framework, we give a rapid survey of currently favoured models, corresponding to a snapshot of the present situation. More detail, with an extensive bibliography, is provided in a review by Lyth and Riotto (1999). The models give different predictions for the spectral index, which means that Microwave Anisotropy Probe and Planck satellite observations will be able to reject most of them. We end by discussing two special classes of model, not covered in the rest of the chapter; these are models invoking non-Einstein gravity, and models leading to an open Universe.

8.1 Overview

Although inflation most likely begins when V is at the Planck scale, we consider only the era that begins when cosmologically interesting scales start to leave the horizon. By that time, $V^{1/4}$ is at least 2 orders of magnitude below the Planck scale, from Eq. (7.108). Any memory of what happened earlier has been wiped out, except insofar as it sets the initial conditions.

At its most minimal, a model of inflation simply gives the form of the effective potential V during inflation (along with the kinetic term if it is nontrivial, and any modification of Einstein gravity). However, we also would like it to be sensible in the context of particle physics, the desire being that, in the end, it will belong to a complete field theory that describes all of the particles and interactions occurring in Nature. It also should look as if the inflaton field can reasonably find itself on the inflating portion, in the manner discussed in Section 3.4.4, though that is not a very strong constraint.

It is useful to distinguish two broad classes of model. In **single-field** models, the potential $V(\phi)$ is dominated by the slow-rolling inflaton field ϕ. It has a minimum corresponding to the vacuum expectation value (vev) of ϕ, at which V vanishes (or anyhow is much smaller than during inflation). Inflation ends when the minimum is approached, leading to a failure of the slow-roll conditions; ϕ then oscillates about its vev and the reheating process begins.

In **hybrid models**, we are dealing with at least one additional field, ψ, in addition to the slow-rolling inflaton ϕ. The potential $V(\phi, \psi)$ normally is dominated by ψ, which is held in place by its interaction with ϕ and gives a constant (or practically constant) contribution to V. Inflation ends only when the ψ field is destabilized and relaxes to its true minimum, when ϕ falls below some critical value ϕ_c.

Future measurements of n will distinguish strongly between different models. The prediction for n, as well as the Cosmic Background Explorer (COBE) constraint, usually depends on the number N of e-foldings of inflation occurring after the cosmologically interesting scales leave the horizon. The reason is that the model itself will only give the value ϕ_{end} of the inflaton field at the end of inflation, whereas we need the value when cosmological scales leave the horizon, which depends on all the subsequent evolution. This is given by Eq. (3.17), as

$$N(\phi) = \frac{1}{M_{Pl}^2} \int_{\phi_{end}}^{\phi} \frac{V}{V'} d\phi. \tag{8.1}$$

Taking cosmologically interesting scales to go from $H_0^{-1} \sim 10^4$ Mpc to 1 Mpc gives a range $\Delta N = \ln(10^4) \sim 9$. As we saw in Chapter 3, a working hypothesis is that the central value of N lies in the range 50 to 25, corresponding, respectively, to an uninterrupted Hot Big Bang, and to a delayed Hot Big Bang preceded by one bout of thermal inflation.

8.2   Form of the scalar field potential

We begin by discussing some of the particle physics motivations, which readers may choose to avoid if they prefer. More material can be found in the review by Lyth and Riotto (1999).

8.2.1   Supersymmetry and supergravity

Although there is no unique prediction, particle theory does give guidance about the likely form of the scalar field potential. The theory relevant during inflation will be some extension of the standard model, and practically all proposed extensions introduce supersymmetry (susy), because they invoke fundamental scalar fields. Without susy, quantum effects (Section 8.2.4) generically give each fundamental scalar field a mass of order M_{Pl}, and it requires a delicate cancellation to obtain sensible masses. This in general requires fine-tuning, but with susy, the cancellation is automatic. In about ten years, the Large Hadron Collider (LHC) at CERN will either discover susy, if it has not been discovered before then, or practically kill it.[1] In the latter eventuality the task of understanding whatever *is* observed at the LHC will take precedence over such relatively trivial matters as inflation model building, and so, let us suppose optimistically that susy is valid.

A brief account of susy can be found, for instance, in the review by Lyth and Riotto (1999), and it is the subject of several texts such as those by Bailin and Love (1994) and by Wess and Bagger (1983). Let us note a few important points. Supersymmetry is an extension of Lorentz invariance. Its outstanding prediction is that each fermion should have a bosonic superpartner,

---

[1] An alternative, which may become attractive if susy is not observed at the LHC, is to invoke the anthropic principle (Agrawal et al. 1998), arguing that such fine-tuning is required for our existence.

and vice versa, with identical mass and couplings in the limit of unbroken susy. In particular, each Standard Model particle has an undiscovered partner; there are squarks and sleptons with spin-0, Higgsinos with spin-$\frac{1}{2}$, and gauginos with spin-$\frac{1}{2}$. It is expected to be a local symmetry, and local susy is called **supergravity** because it automatically incorporates gravity, the graviton (spin-2) being accompanied by the gravitino (spin-$\frac{3}{2}$).

In a supersymmetric theory, all scalar fields are complex because they are the partners of left- or right-handed fermion fields with two components. Of course, we can still think in terms of real fields, corresponding to, for example, the real and imaginary parts of the complex fields, and the inflaton will be one of these, assuming that it has only one component.

Unbroken susy would require that each particle has the same mass as its partner. This is not observed, and so, susy must be broken in the present vacuum. Presumably this is achieved through the spontaneous breaking of supergravity, which at low energies looks like the explicit breaking of global susy, through certain so-called soft susy breaking terms in the Lagrangian. Soft susy breaking gives the sleptons and squarks masses very roughly of order 100 GeV, much bigger than those of their supersymmetric partners, the leptons and quarks (except the top quark, which is around that mass). The superpartners cannot be much lighter or they would have been observed, and they cannot be much heavier if we are to understand the existence of an elementary Higgs field without fine-tuning.

In the early Universe, additional susy breaking occurs. During inflation, it typically dominates soft susy breaking if $V^{1/4} \gg M_S$, where the scale M_S depends on the mechanism of susy breaking in the vacuum. In the usual "gravity-mediated" scheme, $M_S \sim (100 \, \text{GeV} \times M_{\text{Pl}})^{1/2} \sim 10^{10}$ GeV, but in "gauge-mediated" schemes it can be as low as 10^6 GeV.

Being merely a symmetry, susy does not fix the form of the field theory uniquely. For theoretical guidance about the likely form, we can look to superstring theory, which gives field theory as an approximation valid after the energy density falls significantly below the Planck scale.

8.2.2 Tree-level potential

In quantum field theory, the effective potential in the perturbative regime is given by a sum of Feynman diagrams. They can be classified according to the number of loops, the no-loop (tree-level) diagrams giving, roughly speaking, the classical object that appears in the Lagrangian. In many cases the tree-level potential provides a good approximation.

It will be built out of powers of the fields, and at least in the context of supergravity, there will be an infinite number of terms. We are interested in the potential as a function of the real inflaton field ϕ, during inflation. In most models, all other fields are fixed at the origin, and some symmetry usually forbids odd terms in the potential.[2] Then the inflaton potential is of the form

$$V(\phi) = V_0 \pm \frac{1}{2}m^2\phi^2 + \frac{1}{4}\lambda\phi^4 + \lambda' M_{\text{Pl}}^{-2}\phi^6 + \cdots. \tag{8.2}$$

---

[2] Typically, ϕ is the modulus of a complex field charged under a $U(1)$ (or higher) symmetry, which indeed forbids odd terms. In general, odd terms will appear, though the linear term is still forbidden if the origin is the fixed point of a subgroup Z_n of a $U(1)$.

As indicated, the quadratic term can be of either sign (so in principle can the others, but they are usually positive). In accordance with what was said on page 203, the positive sign corresponds to hybrid inflation, and the negative one to single-field inflation. A similar form holds for the potential as a function of any field, with all other fields fixed.

The lower-order terms of Eq. (8.2), which do not involve M_{Pl}, are called **renormalizable**. The higher-order terms, which disappear in the limit $M_{Pl} \to \infty$, are called **nonrenormalizable**.

Typically, one expects the dimensionless couplings λ, $\lambda' \cdots$ to be roughly of order 1, but this need not be so in a supersymmetric theory. In particular, λ is practically 0 in flat directions (Section 8.2.3). Also, superstring moduli fields are expected to have a potential of the form $V = Af(\phi/M_{Pl})$, with f and its low-order derivatives of order 1, which corresponds to dimensionless couplings of order $V/M_{Pl}^4 \ll 1$.

In the usual applications of particle theory, to such things as the Standard Model, neutrino masses, Peccei–Quinn symmetry, or a Grand Unified Theory, the relevant fields have values much less than M_{Pl}. In this regime, we expect the series to be dominated by a few low-order terms. As we consider bigger field values, the justification for keeping only a few terms becomes weaker. For values of order M_{Pl} we may still argue for keeping only a few terms, giving for instance a vev of order M_{Pl}. For field values orders of magnitude bigger than M_{Pl}, there is no theoretical argument for such an assumption, and in general it is not known what form the potential will take, or even if field theory is valid. The only case in which we have some guidance is that of the moduli fields of superstring theory, which parameterize the vacuum in the limit of unbroken susy. However, for moduli fields $\gg M_{Pl}$, the potential becomes either periodic in the fields, or is thought to be too steep to support inflation. Accordingly, models of inflation based on particle physics are constructed in the regime where all fields are, at most, of order M_{Pl}.

The appearance of M_{Pl} in the preceding discussion reflects the expectation that field theory will be valid down to distance scales of order M_{Pl}^{-1}. It may be that it breaks down already at some larger scale Λ^{-1}, in which case M_{Pl} should be replaced by Λ in the preceding discussion. Even if this does not happen for a complete theory, it may happen for a partial ("effective") theory containing only relatively light fields. We proceed on the assumption that M_{Pl} is the relevant scale for inflation model building.

8.2.3 Inflating with the tree-level potential

Let us see what constraint is imposed by the slow-roll conditions $|\eta| \ll 1$ and $\epsilon \ll 1$, if inflation is generated by a tree-level potential.

During inflation, the nonrenormalizable terms are usually negligible, and so, assuming that there are no odd terms, we have

$$V(\phi) \simeq V_0 \pm \frac{1}{2}m^2\phi^2 + \frac{1}{4}\lambda\phi^4. \tag{8.3}$$

Because the slow-roll conditions have to hold over a range of ϕ, they can hardly come from a cancellation between the two terms. Barring such a cancellation, the requirement $|\phi| \lesssim M_{Pl}$ leads to $\epsilon \lesssim |\eta|$, and so, we need only impose $|\eta| \ll 1$. Applied to each term separately, this gives

$$\lambda \frac{M_{\text{Pl}}^2 \phi^2}{V_0} \ll 1, \tag{8.4}$$

$$\frac{M_{\text{Pl}}^2 m^2}{V_0} \ll 1. \tag{8.5}$$

The first inequality depends on ϕ, and it must hold while cosmological scales are leaving the horizon. Using Eq. (8.1) and the COBE normalization (7.107), this implies $\lambda < 10^{-8}$ (Lyth and Riotto 1999). The second inequality can be written as

$$\frac{m^2}{(100\,\text{GeV})^2} \ll \left(\frac{V_0^{1/4}}{10^{10}\,\text{GeV}} \right)^4. \tag{8.6}$$

The constraint on λ looks very strong, whereas the one on m looks quite mild, but supersymmetric theories have special features. In a globally supersymmetric theory with only renormalizable terms, there typically are several fields in whose direction the tree-level potential is exactly flat when all other fields are fixed at the origin. Such fields are referred to as **flat directions** (in field space). So, in global susy we have a simple strategy for inflation model building; choose the inflaton to be a flat direction in the above sense, and then give the potential a small slope by invoking nonrenormalizable terms or quantum effects, or perhaps by explicitly breaking susy.

However, susy is expected to be a local symmetry (i.e., supergravity) with renormalizable global susy obtained only in the limiting case $M_{\text{Pl}} \to \infty$. We see in Section 8.2.5 that, when a globally supersymmetric theory is promoted to a supergravity theory, flat directions are lifted during inflation. Although λ remains small, we typically find $m^2 \sim V_0/M_{\text{Pl}}^2$, which makes $\eta \sim 1$ and spoils slow-roll inflation. We discuss possible solutions of this problem in Section 8.2.5.

8.2.4 The one-loop correction

In the rest of this section, we consider the one-loop correction to the potential, which typically dominates its slope unless the couplings are suppressed. Then we look at the problem of keeping the inflaton mass sufficiently small, in the context of supergravity. In both cases, the discussion involves more particle theory than appears elsewhere in the book, and the reader may prefer to skip it on a first reading.

For the one-loop correction, global susy is supposed to be an adequate approximation. We quote without comment a number of results. We are considering the correction to a tree-level potential $V(\phi)$, all scalar fields except the real field ϕ being adjusted to minimize the potential. The one-loop correction is[3]

$$\Delta V(\phi) = \sum_i \frac{\pm \mathcal{N}_i}{64\pi^2} M_i^4(\phi) \ln \left[\frac{M_i^2(\phi)}{M_{\text{Pl}}^2} \right]. \tag{8.7}$$

---

[3] In a nonsupersymmetric theory, there is the additional term of order $M_{\text{Pl}}^2 \phi^2$ mentioned on page 204.

The sum goes over all particle species, with the plus/minus sign for bosons/fermions, and $\mathcal{N}_i$ the number of spin states. The quantity $M_i^2(\phi)$ is the mass-squared of the species, in the presence of the constant ϕ field. For a scalar, $M_i^2 = \partial^2 V/\partial\phi_i^2$, which is valid for ϕ itself as well as other scalars.

Let us consider the case in which the loop correction comes from a fermion field and its superpartner. The superpartner is a complex field $\psi = (\psi_1 + i\psi_2)/\sqrt{2}$, whose real components ψ_i have true masses m_i. If there is an interaction $\frac{1}{2}\lambda\phi^2|\psi^2|$, this gives $M_i^2 = m_i^2 + \frac{1}{2}\lambda\phi^2$ ($i = 1, 2$). The fermionic partner typically has true mass $m_f = 0$, and we can show that its interaction with ϕ generates an effective mass-squared $M_f^2(\phi) = \frac{1}{2}\lambda_f\phi^2$, with $\lambda_f = \lambda$ by virtue of the susy. (This result is not affected by either spontaneous or soft symmetry breaking.) When ϕ is much bigger than m_i, the loop correction is therefore

$$\Delta V \simeq \frac{1}{32\pi^2}\left[\sum_{i=1,2}\left(m_i^2 + \frac{1}{2}\lambda\phi^2\right)^2 - 2\left(\frac{1}{2}\lambda\phi^2\right)^2\right]\ln\frac{\phi}{M_{\mathrm{Pl}}}. \tag{8.8}$$

The coefficient of ϕ^4 vanishes by virtue of the susy. Two cases arise for the other terms.

In the case $V_0^{1/4} \gg 10^{10}\,\mathrm{GeV}$, susy breaking during inflation will not be directly related to susy in our vacuum, and we can expect it to be spontaneous. Then typically $m_1^2 + m_2^2 = 0$, implying one of the fields has a negative mass-squared, and that the coefficient of ϕ^2 in Eq. (8.8) vanishes as well, leaving

$$\Delta V \simeq \frac{m_1^2 m_2^2}{32\pi^2}\ln\frac{\phi}{M_{\mathrm{Pl}}}. \tag{8.9}$$

We will see how this correction might dominate the slope of the potential, leading to the dramatic prediction $n = 0.97 \pm 0.01$.

In the opposite case, we may have soft susy breaking during inflation. Then one expects $|m_i|^2 \sim V_0/M_{\mathrm{Pl}}^2$. The quadratic term now dominates the loop correction, and adding it to the tree-level potential $V_0 + \frac{1}{2}m^2\phi^2$ gives an expression of the form

$$V = V_0 + \frac{1}{2}\left[m^2 + c\tilde{m}^2\ln\frac{\phi}{M_{\mathrm{Pl}}}\right]\phi^2, \tag{8.10}$$

where $\tilde{m}^2 = m_1^2 + m_2^2$ and $c = \lambda^2/16\pi^2$.

These expressions hold only if the loop correction is small. The correction represented by Eq. (8.10) becomes large when ϕ falls to a value of order $\exp(-1/c)M_{\mathrm{Pl}}$. In that regime, Eq. (7.118) should be replaced by

$$V = V_0 + \frac{1}{2}m^2(\phi)\phi^2. \tag{8.11}$$

The form of the "running mass" $m^2(\phi)$ is calculated from what are called renormalization group equations (RGEs).

Expressions such as Eqs. (8.10) and (8.11) are also obtained if the loop correction comes from a gauge field and its superpartner, with $c \sim g^2/16\pi^2$ and g the gauge coupling. In contrast with the case for fermion couplings such as λ, we expect that gauge couplings derived from superstring theory will not be extremely small, which means that we actually expect Eq. (8.11) to be needed. We will see how this leads to an attractive model of inflation, the observational signature of which might be a spectral index n with significant scale dependence.

Finally, we mention that actually all masses, and couplings, will have a dependence on the scale ϕ determined by the RGEs. As the scale is reduced, a gauge coupling can become of order 1 at some value ϕ_{strong}. We then reach the strong-coupling regime, where the analysis that we just presented makes no sense. In the weak-coupling regime $\phi > \phi_{strong}$, the analysis makes sense, but the tree-level potential may acquire additional terms proportional to ϕ^{-p}, with p an integer ≥ 1. They are associated with an internal symmetry whose spontaneous breaking comes from quantum effects rather than from the tree-level potential (dynamically broken symmetry).

8.2.5 Keeping the inflaton mass small in supergravity

The opinion of the research community concerning almost every aspect of physics beyond the Standard Model is constantly shifting, and in most respects supergravity is no exception. Remarkably though, there is a definite prediction concerning the form of the tree-level potential, and of the kinetic terms, which is true in any supergravity theory.[4] As we now explain, this generically lifts flat directions of global susy, giving $|\eta| \sim 1$ in contradiction to the slow-roll requirement $|\eta| \ll 1$.

Renormalizable global susy
In a globally supersymmetric theory without renormalizable terms, the potential is of the form $V = V_F + V_D$, where the terms are called, respectively, the F and D terms. The D term vanishes during inflation except in special models. The F term is of the form

$$V_F = \sum_i |W_i|^2. \tag{8.12}$$

The superpotential W is a holomorphic function of the complex fields ϕ_i, and a subscript i denotes $\partial/\partial\phi_i$.

We typically find that there are fields which appear in neither V_F nor V_D; these are the flat directions referred to earlier. In building a supersymmetric model of inflation, one assumes that the inflaton corresponds to a flat direction, making its tree-level potential exactly flat in the limit of renormalizable global susy.

Form of V in supergravity
To simplify the discussion of supergravity, we ignore the gaugino condensate that often is supposed to be responsible for susy breaking in our vacuum. This should be a good approximation in models of inflation where $V^{1/4} \gg 10^{10}$ GeV, and in any case, inclusion of the gaugino condensate essentially does not alter the situation.

---

[4] To be precise, it is true in any supergravity theory of the type known as $N = 1$. This is the type needed to construct extensions of the Standard Model.

To specify V in supergravity, we need the superpotential W, and also a further potential known as the Kahler potential K. The latter is a nonholomorphic function of the complex fields ϕ_n, taken to be a function of these fields and their complex conjugates $\bar{\phi}_n$.

The Kahler function specifies the kinetic terms, which are now nontrivial:

$$\mathcal{L}_{\text{kin}} = \sum_{n,m} (\partial_\mu \bar{\phi}_n) K_{\bar{n}m} (\partial^\mu \phi_m). \tag{8.13}$$

Here and in the following expressions, a subscript n denotes the derivative with respect to ϕ_n, whereas $\bar{n}$ denotes the derivative with respect to $\bar{\phi}_n$.

The potential V again consists of a D term and an F term, and we ignore the former. The latter gives

$$V = e^{K/M_{\text{Pl}}^2} \tilde{V}, \tag{8.14}$$

where

$$\tilde{V} = \left(W_n + M_{\text{Pl}}^{-2} W K_n\right) K^{n\bar{m}} \left(\bar{W}_{\bar{m}} + M_{\text{Pl}}^{-2} \bar{W} K_{\bar{m}}\right) - 3 M_{\text{Pl}}^{-2} |W|^2. \tag{8.15}$$

The matrix $K^{n\bar{m}}$ is the inverse of $K_{\bar{n}m}$ and a sum over repeated indices is understood. Global susy corresponds to the limit $M_{\text{Pl}} \to \infty$.

Near the origin of field space, we may choose the fields to be canonically normalized, corresponding to

$$K = \sum_n |\phi_n|^2 + \cdots, \tag{8.16}$$

giving

$$K_{\bar{n}m} = \delta_{nm} + \cdots. \tag{8.17}$$

If the equation looks strange, remember that the bar just indicates a derivative with respect to $\bar{\phi}_n$ instead of ϕ_n.

On the basis of superstring theory, we expect that the coefficients of the higher-order terms in Eq. (8.16) are of order 1 in Planck units. In the regime $|\phi_n| \lesssim M_{\text{Pl}}$, where the field theory is under control, we therefore can expect $K \simeq \sum |\phi_n|^2$. This gives roughly canonical kinetic terms and the global susy expression (8.12) for the *magnitude* of V.[5] As we now see though, this is *not* an adequate approximation for the slope of V.

Keeping the inflaton mass small

The inflaton field ϕ is supposed to be $\sqrt{2}$ times the real part of one of the ϕ_n, with exactly canonical normalization at the origin. Then, Eq. (8.14) gives its mass m as

$$\frac{M_{\text{Pl}}^2 m^2}{V} = 1 + \frac{M_{\text{Pl}}^2 \tilde{V}''}{\tilde{V}}, \tag{8.18}$$

---

[5] During inflation, there is no reason why the term $-3M_{\text{Pl}}^{-2}|W|^2$ should accurately cancel the other term, though that is supposed to happen in our vacuum to ensure that V is practically zero (page 172).

with the potential and its derivative evaluated at the origin. The first term comes from the factor $\exp(K/M_{\text{Pl}}^2)$. To have inflation, it must be cancelled by the second term, but this cancellation will not occur in the generic case (Lyth and Riotto 1999).

The degree of cancellation required depends on the model in question. It must reduce the generic prediction $M_{\text{Pl}}^2|m^2|/V_0 \sim 1$ by at least an order of magnitude because the contribution of the mass to η is $2M_{\text{Pl}}^2|m^2|/V_0$, and this is also the contribution to $\frac{1}{2}|n-1|$, which observations conservatively constrain below 0.2. However, in some models of inflation, η is one or more orders of magnitude smaller (n very close to 1), and so, a much higher degree of cancellation is required.

The generic prediction $M_{\text{Pl}}^2|m^2|/V_0 \sim 1$ clearly applies to any real field during inflation. For cosmologically dangerous noninflaton fields, such as certain moduli, it may be welcome and indeed was first noted in that connection. The difficulty that it poses for inflation model building was first emphasized by Copeland et al. (1994b).

Assuming that supergravity is indeed present in Nature, we can imagine four ways of keeping the inflaton mass small:

(1) Nature has chosen one of the special class of inflation models that have the D term dominating.
(2) The tree-level potential is not valid, because the one-loop correction or some other quantum effect dominates.
(3) The forms of K and/or W are such that the mass automatically vanishes in some well-defined limit.
(4) Some degree of cancellation occurs by accident. This has not been invoked much explicitly, but we can charitably apply it in cases in which the supergravity problem is ignored, and it may even be correct.

We encounter examples of all of these mechanisms, and see how a future measurement of the spectral index might distinguish between them.

8.3   Single-field models

In the next two sections we give a rapid survey, which includes most of the Einstein-gravity models proposed so far. In each case we give the prediction for the spectral index using Eq. (7.117), namely $n = 1 + 2\eta - 6\epsilon$, where ϵ and η are the slow-roll parameters. In most of the models, the parameter ϵ is actually negligible, so that we can use the simpler formula $n \simeq 1 + 2\eta$. We also give the constraint on the parameters corresponding to the COBE normalization (7.107), which usually boils down to a value for $V_0^{1/4}$.

Until recently, few model builders paid attention to the η problem of supergravity, and many did not even work within the framework of global susy. In our survey, we consider different forms for the inflationary potential that have been proposed, while paying comparatively little attention to any particle theory background.

We begin with single-field models. Two kinds have been proposed so far. In one of them, the inflaton field is of order $10M_{\text{Pl}}$ when cosmological scales leave the horizon, moving toward the origin, for example, under a monomial potential $V \propto \phi^\alpha$. A variant is the exponential potential of power-law inflation, perhaps with modifications to generate a minimum. In the other kind of model, the field is moving away from the origin under the influence of a potential $V \simeq V_0(1 - \mu\phi^p)$, ending up at a value at most of order M_{Pl}. The first case often is called "chaotic" inflation because it provides a way of descending from the Planck scale with chaotic initial field conditions (Linde 1983), and the second case is often called "new" inflation because that was the name given to the first example of it (Albrecht and Steinhardt 1982; Linde 1982).

In single-field models, inflation typically ends when one of the slow-roll parameters ϵ or η becomes of order 1. Then, Eq. (8.1) determines ϕ when cosmological scales leave the horizon.

8.3.1 Monomial potentials

The simplest potential giving inflation is (Linde 1983, 1990a)

$$V = \frac{1}{2}m^2\phi^2. \tag{8.19}$$

An alternative is $V = \lambda\phi^4/4$, and the obvious generalization is to consider an arbitrary monomial (i.e., a potential consisting of a single power of ϕ). It is of the form $V = \lambda M_{\text{Pl}}^{4-\alpha}\phi^\alpha$ with α a positive (and usually even) integer, and λ dimensionless.

The slow-roll parameters of Section 3.4 are

$$\epsilon = \frac{\alpha^2}{2}\frac{M_{\text{Pl}}^2}{\phi^2}, \qquad \eta = \alpha(\alpha - 1)\frac{M_{\text{Pl}}^2}{\phi^2}. \tag{8.20}$$

Inflation ends with the violation of the slow-roll conditions at $\phi_{\text{end}} \simeq \alpha M_{\text{Pl}}$, after which ϕ starts to oscillate about its vev $\phi = 0$. When cosmological scales leave the horizon, we find from Eq. (8.1) that $\phi = \sqrt{2N\alpha}\, M_{\text{Pl}}$. Utilizing the standard results (7.117) and (7.150) gives

$$n = 1 - \frac{2+\alpha}{2N}, \tag{8.21}$$

$$r = \frac{3.1\alpha}{N} = 6.2(1 - n) - \frac{6.2}{N}. \tag{8.22}$$

In this model the gravitational waves are big enough to be eventually observable ($r \gtrsim 0.1$; see Section 9.3), especially if N turns out to be toward the low end of its range. The COBE normalization (7.107) corresponds to $m = 1.8 \times 10^{13}\,\text{GeV}$ for the quadratic case, and to $\lambda = 7 \times 10^{-14}$ for the quartic case, taking $N = 50$ in each case.

Because the inflaton field is of order $10M_{\text{Pl}}$ when cosmologically interesting perturbations are generated, there is no particle physics motivation for these potentials. We could reasonably modify their shape, though of course the monomial is the simplest.

Finally, we note the proposal that the potential involves an inflaton with two or more components, for example, $V = A\phi_1^p + B\phi_2^q$. In such a model, slow-roll inflation can fail briefly, producing a shoulder or spike in the spectrum.

8.3.2 Power-law inflation

In the limit of a high power, the monomial potential gives the same predictions as the exponential potential:

$$V(\phi) = V_0 \exp\left(-\sqrt{\frac{2}{p}}\frac{\phi}{M_{\text{Pl}}}\right). \tag{8.23}$$

This is the power-law inflation model, which we discussed in Section 3.5.1 as a rare case of a simple potential for which an exact solution is available. The slow-roll parameters are given by

$$\epsilon = \frac{1}{p}, \qquad \eta = \frac{2}{p}, \tag{8.24}$$

and both are constants. The condition for inflation is $p > 1$. Because ϵ is constant, inflation never ends in the power-law inflation model, and so, clearly, this basic model requires some modification. The only theoretically viable proposal is the extended inflation model discussed later in this chapter, where power-law inflation arises after a conformal transformation of an apparently different theory, but unfortunately, as we see later, such models now are excluded by observations.

Because the slow-roll parameters are constant, the predictions for this model do not depend on how inflation comes to an end (at least insofar as any modifications to the potential are negligible when cosmologically interesting scales leave the horizon). The predictions are

$$n = 1 - \frac{2}{p}, \tag{8.25}$$

$$r = \frac{12.4}{p} = 6.2\,(1 - n). \tag{8.26}$$

That last relationship between n and r has received a lot of attention, and leads to observable gravitational waves unless n is very close to 1. It has even been advertised as a generic relationship, but as the rest of our summary shows, that is far from the case. On the contrary, the generic behaviour in modern model building is that r is negligible but $n - 1$ is not.

Although these are slow-roll results, power-law inflation has been a popular focus for studies of perturbation generation because the perturbation equations can be solved without making the slow-roll approximation, giving Eq. (7.146).

8.3.3 Inverted quadratic potential

Another possibility is the inverted quadratic potential:

$$V = V_0 - \frac{1}{2}m^2\phi^2 + \cdots. \tag{8.27}$$

Inflation with such a potential was discussed first by Binetruy and Gaillard (1986), and subsequently by many authors. The dots indicate the effect of higher powers, which are supposed to come in after cosmologically interesting scales leave the horizon. This potential gives

$$n \simeq 1 + 2\eta = 1 - \frac{2M_{\text{Pl}}^2 m^2}{V_0}. \tag{8.28}$$

In this subsection, we suppose that the contribution of higher powers is positive, giving ϕ a vev roughly of order

$$\phi_{\text{min}} \sim \frac{V_0^{1/2}}{m} = \left(\frac{2}{1-n}\right)^{1/2} M_{\text{Pl}}. \tag{8.29}$$

The end of inflation will occur at $\phi_{\text{end}} \sim \phi_{\text{min}}$. The opposite case in which they steepen the potential is considered briefly in Section 8.3.4.

If we are to understand the potential within the context of supergravity, the vev should not be much bigger than M_{Pl}, which requires $n - 1$ to be of order 0.1 as opposed to a much smaller value. (Recall that slow-roll inflation requires $|n - 1| \ll 1$, and observation requires $|n - 1| \lesssim 0.3$.) In that case, $|m^2|$ is only a factor 10 below the value V_0/M_{Pl}^2 associated with a generic supergravity theory, and it may not be unreasonable to ascribe the factor 10 to an accident.

The COBE normalization (7.107) gives

$$\frac{V_0^{1/2}}{M_{\text{Pl}}^2} = 5.3 \times 10^{-4} \left(\frac{1-n}{2}\right) \phi(N) \simeq 5 \times 10^{-4} \sqrt{\frac{1-n}{2}} e^{-x}, \tag{8.30}$$

where $x \equiv N(1 - n)/2$. Because $|1 - n| \lesssim 0.3$ with tiny values excluded, the COBE normalization requires $V_0^{1/4} \gtrsim 4 \times 10^{14}$ GeV.

In contrast to the other forms for the potential that we consider, the origin of ϕ for this one is not necessarily taken to be the natural origin. The latter, appearing for example in the discussion in Section 8.2.5, is defined as the fixed point of the internal symmetries. Instead of that choice, we may take the origin to be at the bottom of a Mexican-hat potential; this is the proposal of Freese et al. (1990), who called it natural inflation. From the discussion in Section 8.2.5, we might think that this proposal keeps the inflaton mass small in the context of supergravity, but that is not the case, because of nonrenormalizable terms in the potential (Lyth and Riotto 1999).

8.3.4   Cubic and higher potentials

If the quadratic term is heavily suppressed or absent, we will have

$$V \simeq V_0(1 - \mu\phi^p + \cdots), \tag{8.31}$$

with $p \geq 3$. For this potential, we expect that the integral (8.1) for N is dominated by the limit ϕ, as opposed to the limit ϕ_{end}. This leads to

$$\phi^{p-2} = \left[p(p-2)\mu N M_{\text{Pl}}^2\right]^{-1} \tag{8.32}$$

and

$$n \simeq 1 - 2\left(\frac{p-1}{p-2}\right)\frac{1}{N}. \tag{8.33}$$

The COBE normalization is

$$5.3 \times 10^{-4} = \left(p\mu M_{\mathrm{Pl}}^{p}\right)^{1/(p-2)}[N(p-2)]^{(p-1)/(p-2)}\frac{V_0^{1/2}}{M_{\mathrm{Pl}}^2}. \tag{8.34}$$

For $p = 4$ the dimensionless coupling is $\lambda \equiv 4V_0\mu = 3 \times 10^{-13}(50/N)^3$. Such a tiny value is problematical, as was recognized in the new inflation model in which a quartic term was first proposed. To be precise, the latter model invoked a loop-corrected potential roughly of the form

$$V = V_0 + \frac{1}{4}\lambda\phi^4\ln(\phi/Q). \tag{8.35}$$

As we note in Section 8.2.4, such a potential, however, will not occur in a supersymmetric theory because the massless scalar field ϕ will be accompanied by a massless fermion field, the loop correction of which will cancel the one being invoked.

We also could consider combinations of powers. In particular, we could allow drastic steepening of the quadratic potential, soon after cosmological scales leave the horizon. The simplest example is (Dine and Riotto 1997)

$$V = V_0 - \frac{1}{2}m^2\phi^2 - \frac{1}{4}\lambda\phi^4 + \cdots. \tag{8.36}$$

The steepening can dramatically reduce the COBE normalization of V_0. This may be desirable from a particle physics viewpoint, provided that we can understand the smallness of $|m^2|M_{\mathrm{Pl}}^2/V_0$.

8.3.5 Another form for the potential

Another potential that has been proposed is

$$V \simeq V_0\left[1 - \exp\left(-q\frac{\phi}{M_{\mathrm{Pl}}}\right)\right], \tag{8.37}$$

with q of order 1. This form is supposed to apply in the regime where V_0 dominates, which is $\phi \gtrsim M_{\mathrm{Pl}}$. Inflation ends at $\phi_{\mathrm{end}} \sim M_{\mathrm{Pl}}$, and when cosmological scales leave the horizon, we have $\phi \simeq \ln(q^2N)M_{\mathrm{Pl}}/q$ and

$$n \simeq 1 + 2\eta = 1 - \frac{2}{N}. \tag{8.38}$$

The potential is mimicked by $V = V_0(1 - \mu\phi^{-p})$ with $p \to \infty$. It occurs if inflation takes place near a singularity of the kinetic function, irrespective of the form of the potential (Stewart 1995a), with model-dependent values of q such as $q = 1$ or $\sqrt{2}$. We see later that it also occurs with R^2 gravity and scalar-tensor gravity, giving $q = \sqrt{2/3}$.

8.4　Hybrid inflation models

In **hybrid inflation** models (Linde 1990b, 1991), the slowly rolling inflaton field ϕ is not the one responsible for most of the energy density. That role is played by another field, ψ, which is held in place by its interaction with ϕ until the latter falls below a critical value ϕ_c. When that happens, ψ is destabilized, rolling to its true vacuum, and inflation ends.

The end of inflation can correspond either to a first-order phase transition (Linde 1990b), where the field tunnels through a potential barrier, or to a second-order phase transition (Linde 1991), where there is no barrier. In the context of Einstein gravity, which we are adopting in this section and the next, the second-order case is the usual one, and the only one considered here. In Section 8.6, we encounter the non-Einstein gravity model called extended inflation (La and Steinhardt 1989; Kolb 1991); this is, in effect, a first-order hybrid model, insofar as it fits the description of the preceding paragraph, though the term hybrid usually is not applied to it. Extended inflation actually predates Einstein-gravity hybrid inflation.

Because hybrid inflation ends with a phase transition, there is always the possibility of forming topological defects. In a sufficiently complete model, we could check whether this is the case, and whether the defects are cosmologically forbidden, but the answer to these questions may lie in a sector of the theory that is irrelevant for inflation itself.

8.4.1　Hybrid inflation with a quadratic potential

The potential for the original model of second-order hybrid inflation is (Linde 1991)

$$V = \frac{1}{2}m^2\phi^2 + \frac{1}{4}\lambda(\psi^2 - M^2)^2 + \frac{1}{2}\lambda'\psi^2\phi^2, \tag{8.39}$$

which we also can write as

$$V = V_0 + \frac{1}{2}m^2\phi^2 - \frac{1}{2}m_\psi^2\psi^2 + \frac{1}{4}\lambda\psi^4 + \frac{1}{2}\lambda'\psi^2\phi^2. \tag{8.40}$$

Comparing the two expressions, we learn that $V_0 = m_\psi^2 M^2/4$ and $m_\psi^2 = \lambda M^2$.

This potential is illustrated in Figure 8.1. The field ψ is fixed at the origin if $\phi^2 > \phi_c^2$, where

$$\phi_c^2 = \frac{m_\psi^2}{\lambda'} = \frac{\lambda}{\lambda'}M^2. \tag{8.41}$$

In this regime, slow-roll inflation can take place, with the quadratic potential

$$V = V_0 + \frac{1}{2}m^2\phi^2. \tag{8.42}$$

We suppose that the first term dominates, because in the opposite limit we have a single-field model with a monomial potential. There is a fine-tuned intermediate regime in which both terms are significant during inflation (Copeland et al. 1994b), and a less fine-tuned one in which the first term dominates inflation while the second term produces topological defects,

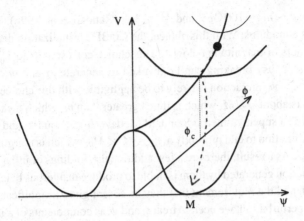

Fig. 8.1. The hybrid inflation potential. The field rolls down the $\psi = 0$ channel from large ϕ, until it encounters the instability point, after which the $\psi = 0$ solution becomes unstable and the fields roll to their true minimum at $\phi = 0$, $\psi = \pm M$.

but we do not consider either of them. The spectral index is then

$$n \simeq 1 + 2\eta = 1 + \frac{2M_{\mathrm{Pl}}^2 m^2}{V_0}. \tag{8.43}$$

The same inflationary potential is obtained with other forms for the last two terms of Eq. (8.40). For example (Randall et al. 1996), we can change the last term to $\lambda' M_{\mathrm{Pl}}^{-2} \psi^2 \phi^4$, leading to

$$\phi_c^2 = \frac{1}{\lambda'} M_{\mathrm{Pl}} m_\psi = \frac{\lambda^{1/2}}{\lambda'} M_{\mathrm{Pl}} M. \tag{8.44}$$

Another prescription for ϕ_c is provided by mutated hybrid inflation, which is described later. We also can replace $\lambda \phi^4 / 4$ by a nonrenormalizable term. The following discussion continues to apply if we make such changes.

When cosmological scales leave the horizon, we have

$$\frac{\phi}{\phi_c} = \exp\left(\frac{n-1}{2} N\right). \tag{8.45}$$

This typically gives $\phi \ll M_{\mathrm{Pl}}$ because the exponential is more than compensated by the smallness of ϕ_c. The COBE normalization (7.107) is

$$5.3 \times 10^{-4} = M_{\mathrm{Pl}}^{-3} \frac{V_0^{3/2}}{m^2 \phi_c} \exp\left(-\frac{n-1}{2} N\right). \tag{8.46}$$

With ϕ_c given by either of the above prescriptions, this imposes (Copeland et al. 1994b) a limit $n \lesssim 1.3$, assuming that V_0 dominates the potential and that $M \lesssim M_{\mathrm{Pl}}$.

This model has three independent parameters, for example, m_ψ, m, and λ'. It is possible to choose values that are reasonable from the viewpoint of particle theory, and which at the same time satisfy the COBE normalization, with no significant fine-tuning. One reasonable proposal for the masses is $m_\psi \sim m \sim 100\,\text{GeV}$ and $M \sim M_{\text{Pl}}$ (Randall et al. 1996). The big value of M is natural if ϕ is a modulus. With this choice, the COBE normalization determines a value λ' not too many orders of magnitude below 1. This choice corresponds to $V_0^{1/4} \sim 10^{10}\,\text{GeV}$, which means that soft susy breaking can be invoked to generate m and m_ψ. Also, the loop correction from the $\phi - \psi$ interaction is likely to be negligible with this choice. Note that mass of ϕ in the vacuum is about $\sqrt{\lambda'}M$, which is much greater than m, which means that ϕ cannot be one of the fields in a supersymmetric Standard Model (Higgs, squark, and slepton fields).

The criterion for inflation to end promptly at ϕ_c is $M \ll M_{\text{Pl}}$, which is marginally violated by the choice $M \sim M_{\text{Pl}}$. As a result, there are a few additional e-foldings of inflation after $\phi = \phi_c$. The density perturbation generated by them is liable to produce dangerous black holes (García-Bellido et al. 1996), and the requirement that they be absent gives a significant upper limit on n. To calculate this perturbation, we need to treat ϕ and ψ as components of a multicomponent inflation.

Another natural-looking choice of parameters (Linde 1991, 1994) is $m_\psi \sim M \sim V_0^{1/4}$, corresponding to coupling $\lambda = 1$. To avoid the loop correction dominating, this probably requires $V^{1/4} \gtrsim 10^{16}\,\text{GeV}$ and $\phi \sim M_{\text{Pl}}$ (Lyth and Riotto 1999).

The only unusual feature of the potential $V(\phi, \psi)$ is the absence of a term $\lambda_\phi \phi^4$. Such a term spoils the model, unless the coupling is extraordinarily small. In the context of susy this is no problem because there are many directions in field space where the term is indeed absent, the so-called flat directions.

Finally, we note that we can construct inverted hybrid models (Lyth and Stewart 1996b; Bastero-Gil and King 1998), where ϕ rolls *away* from the origin until ψ is destabilized by a term $-\lambda\phi^2\psi^2$.

8.4.2 Hybrid inflation with a cubic or higher potential

Instead of the quadratic potential (8.42), we might consider a potential

$$V = V_0(1 + c\phi^p), \tag{8.47}$$

with $p \geq 3$ (and $c > 0$). Then,

$$\eta = cM_{\text{Pl}}^2 p(p-1)\phi^{p-2}, \tag{8.48}$$

and inflation is possible only in the regime $\eta \ll 1$.[6] It is not clear how the inflaton is supposed to get into this regime.

The number of e-folds to the end of inflation is

$$N(\phi) \simeq \left(\frac{p-1}{p-2}\right)\left[\frac{1}{\eta(\phi_c)} - \frac{1}{\eta(\phi)}\right]. \tag{8.49}$$

---

[6] We are assuming that $V \simeq V_0$, when the expression for η is $\ll 1$. As usual, we consider only the case $\phi \lesssim M_{\text{Pl}}$.

For $\phi \gg \phi_c$, $N(\phi)$ approaches a constant

$$N_{max} \equiv \left(\frac{p-1}{p-2}\right)\frac{1}{\eta(\phi_c)}. \tag{8.50}$$

The spectral index is given by

$$\frac{n-1}{2} = \left(\frac{p-1}{p-2}\right)\frac{1}{N_{max} - N}. \tag{8.51}$$

The quartic case has been considered in some detail (Roberts et al. 1995), including the regime $\phi \gg M_{Pl}$ that we are ignoring here.

One also may consider the case where, for example, quadratic, cubic, and quartic terms are all important during observable inflation, but that clearly will involve considerable fine-tuning.

8.4.3   Mutated hybrid inflation

In both ordinary and inverted hybrid inflation, the other field ψ is precisely fixed during inflation. If it varies, an effective potential $V(\phi)$ can be generated even if the original potential contains no piece that depends only on ϕ. This mechanism was first proposed by Stewart (1995b), who called it mutated hybrid inflation. The potential considered was

$$V = V_0\left(1 - \frac{\psi}{M}\right) + \frac{1}{4}\lambda\phi^2\psi^2 + \cdots. \tag{8.52}$$

The dots represent one or more additional terms, which give V a minimum at which it vanishes but play no role during inflation. All of the other terms are significant, with V_0 dominating. For suitable choices of the parameters, inflation takes place with ψ held at the instantaneous minimum, leading to a potential

$$V = V_0\left(1 - \frac{V_0}{\lambda^2 M^2 \phi^2}\right). \tag{8.53}$$

This gives

$$n - 1 = -\frac{3}{2N}, \tag{8.54}$$

and the COBE normalization (7.107) is

$$5.2 \times 10^{-4} = (2N)^{3/4}\sqrt{\lambda}\,\frac{V_0^{1/4}\sqrt{M}}{M_{Pl}^{3/2}}. \tag{8.55}$$

A different version of mutated hybrid inflation (Lazarides et al. 1996) was called smooth hybrid inflation, emphasizing that any topological defects associated with ψ will never be produced. In that version, the potential is $V = V_0 - A\psi^4 + B\psi^6\phi^2 + \cdots$. It leads to $V = V_0(1 - \mu\phi^{-4})$.

Retaining the original name, the most general mutated hybrid inflation model with only two significant terms is (Lyth and Stewart 1996b)

$$V = V_0 - \frac{\sigma}{p} M_{\text{Pl}}^{4-p} \psi^p + \frac{\lambda}{q} M_{\text{Pl}}^{4-q-r} \psi^q \phi^r + \cdots. \tag{8.56}$$

In a suitable regime of parameter space, ψ adjusts itself to minimize V at fixed ϕ, and $\psi \ll \phi$ so that the slight curvature of the inflaton trajectory does not affect the field dynamics. Then, provided that V_0 dominates the energy density, the effective potential during inflation is

$$V = V_0(1 - \mu\phi^{-\alpha}), \tag{8.57}$$

where

$$\mu = M_{\text{Pl}}^{4+\alpha} \left(\frac{q-p}{pq}\right) \frac{\sigma^{q/(q-p)} \lambda^{-p/(q-p)}}{V_0} > 0, \tag{8.58}$$

$$\alpha = \frac{pr}{q-p}. \tag{8.59}$$

For $q > p$, the exponent α is positive as in the examples already mentioned, but for $p > q$, it is negative with $\alpha < -1$. In both cases it can be nonintegral, though integer values are the most common for low choices of the integers p and q. This potential is supposed to hold until V_0 ceases to dominate at

$$\phi_{\text{end}} \sim \mu^{1/\alpha}, \tag{8.60}$$

after which slow-roll inflation ends.

The situation in the regime $-2 < \alpha < -1$ is similar to the one that we discussed already for the case $\alpha = -2$; the prediction for n covers a continuous range below 1 because it depends on the parameters, but to have a model with $\phi \ll M_{\text{Pl}}$ the potential has to be steepened after cosmological scales leave the horizon. The COBE normalization in this case is

$$5.3 \times 10^{-4} = \frac{M_{\text{Pl}}^{\alpha-2} V_0^{1/2}}{|\alpha|\mu} \left[M_{\text{Pl}}^{|\alpha|-2} \phi_c^{2-|\alpha|} - |\alpha| (2-|\alpha|) M_{\text{Pl}}^{\alpha} \mu N \right]^{-(|\alpha|-1)/(2-|\alpha|)}. \tag{8.61}$$

In the cases $\alpha < -2$ and $\alpha > -1$, the situation is similar to the one that we encountered in Section 8.3.4 (except for the special cases $\alpha \simeq -2$ and $\alpha \simeq -1$, which we do not consider). In the case $\alpha < -2$, the integral (8.1) is dominated by the limit ϕ, provided that $\phi_{\text{end}} \ll \sqrt{N} M_{\text{Pl}}$, which we assume. In the case $\alpha > -1$, we have $\phi_{\text{end}} < \phi$, and assuming $\phi \ll M_{\text{Pl}}$ while cosmological scales leave the horizon again means that Eq. (8.1) is dominated by the limit ϕ. In all of these cases, the COBE normalization (8.34) and the prediction (8.33) are valid, with p replaced by $-\alpha$.

A different example of mutated hybrid inflation has been given by García-Bellido et al. (1996), in which ψ is a pseudo-Goldstone boson.

In this account, we assume that the original potential has no piece that depends only on ϕ. If there is such a piece, it has to be added to the inflationary potential (8.57). If it dominates while cosmological scales leave the horizon, the only effect that the ψ variation has on the inflationary prediction is to determine ϕ_c through Eq. (8.60).

8.4.4 Hybrid inflation from dynamical susy breaking

At the end of Section 8.2.4, we noted that dynamical symmetry breaking could give a potential such as

$$V = V_0 \left[1 + \alpha \left(\frac{M_{\mathrm{Pl}}}{\phi} \right)^p + \cdots \right], \tag{8.62}$$

where p is a positive integer, and the dots represent terms that are negligible during inflation. This potential has been proposed (Kinney and Riotto 1998) as a model of inflation. It gives

$$\eta = \alpha p(p+1) \left(\frac{M_{\mathrm{Pl}}}{\phi} \right)^{p-2}. \tag{8.63}$$

The potential satisfies the slow-roll conditions in the regime $\eta \ll 1$.[7] As with the model of Section 8.4.2, it is not clear how the inflaton is supposed to arrive on this part of the potential. Inflation is supposed to end when ϕ reaches a critical value ϕ_c, through some unspecified hybrid inflation mechanism.

The spectral index in this model is given by Eq. (8.51), with $p \to -p$.

8.4.5 Hybrid inflation with a loop correction: Spontaneous susy breaking

The models considered so far work at tree level. This is valid only if the couplings of the inflaton to other fields are suppressed strongly. In particular, the inflaton presumably has to be a gauge singlet (no coupling to gauge fields), because gauge couplings are not supposed to be suppressed.

In the absence of susy, the couplings indeed should be suppressed. The reason is that the loop correction is then $\Delta V \sim \phi^4 \ln(\phi/Q)$ [as in Eq. (8.35) with $\lambda \sim 1$], which would spoil inflation. However with susy, there is no reason to suppose that the inflaton couplings are suppressed.

As we saw in Section 8.2.4, the one-loop correction in a supersymmetric theory typically has one of two forms: $\Delta V \propto \ln(\phi/Q)$ or $\Delta V \propto \phi^2 \ln(\phi/Q)$. We discuss the first form in this subsection, and the second form in Section 8.4.6.

The first form typically arises if susy is broken spontaneously. Inflation with this form was considered first by Dvali et al. (1994) in an F-term inflation model proposed at tree level by Copeland et al. (1994b). It was first considered in a D-term inflation model by Binetruy and Dvali (1996), proposed at tree level by Stewart (1995a). The D-term model avoids the generic prediction $m^2 \sim V_0/M_{\mathrm{Pl}}^2$ for the inflaton mass in the context of supergravity. The F-term model does not have that feature (Copeland et al. 1994b; Lyth and Riotto 1999), which at first sight appears to favour the D-term model. However, we see in a moment that the inflaton field in the latter model has to be large. As a result, we have gained control over the mass term, only to be in danger of losing it for the nonrenormalizable terms.

---

[7] We are assuming that $V \simeq V_0$ as long as the right-hand side of expression (8.63) is much less than 1.

Assuming that tree-level terms are negligible during inflation, we find that the potential in both models is of the form

$$V = V_0 \left(1 + \frac{Cg^2}{8\pi^2} \ln \frac{\phi}{Q}\right). \tag{8.64}$$

In this expression, C may be taken to be the number of possible one-loop diagrams, in other words the number of fields that have significant coupling to the inflaton. The other factor g is a typical coupling of these fields (times a numerical factor of order 1). In the D-term model, it is a gauge coupling; in the F-term model, it is a Yukawa coupling. In the former case, $g \sim 1$ and C might be of order 100.

In both cases, this potential occurs as part of a hybrid inflation model. Depending on the parameters, inflation ends when either slow-roll fails ($\eta \sim 1$) or the critical value is reached, whichever is earlier.[8] However, the precise value of ϕ_{end} is irrelevant because the integral (8.1) is dominated by the limit ϕ. It gives

$$\phi \simeq \sqrt{\frac{NCg^2}{4\pi^2}} M_{\mathrm{Pl}} = 11 \sqrt{\frac{N}{50} \frac{C}{100}} g^2 M_{\mathrm{Pl}} = 0.2 \sqrt{\frac{N}{20} C \frac{g^2}{0.1}} M_{\mathrm{Pl}}, \tag{8.65}$$

where the different expressions indicate the effect of different choices of N, C, and g^2. This makes ϕ comparable to the Planck scale, and maybe bigger. We need $\phi \lesssim M_{\mathrm{Pl}}$ and preferably $\phi \ll M_{\mathrm{Pl}}$, to keep the theory under control and in particular to justify the assumption of canonical normalization for the fields. Let us proceed on the assumption that ϕ is not too big.

If we assume that the loop correction dominates the slope, and use Eq. (8.1), the slow-roll parameters are

$$\eta = -\frac{1}{2N}, \qquad \epsilon = C \frac{g^2}{8\pi^2} |\eta|. \tag{8.66}$$

The COBE normalization (7.107) is

$$V^{1/4} = 6.0 \left(\frac{50}{N}\right)^{1/4} C^{1/4} g \times 10^{15} \, \mathrm{GeV}. \tag{8.67}$$

The spectral index is given by

$$1 - n = \frac{1}{N} \left(1 + \frac{3Cg^2}{16\pi^2}\right). \tag{8.68}$$

Taking the parentheses to be close to 1, and N to be in the range 25 to 50, we obtain the distinctive prediction $n = 0.96$ to 0.98. With $g = 1$ and $C = 100$, $1 - n$ is increased by a factor $\simeq 2$, but it is clear anyhow that n is close to 1. This prediction eventually will be tested.

---

[8] If slow-roll fails at a value $\phi_{\mathrm{end}} > \phi_{\mathrm{c}}$, inflation will continue until the amplitude of the oscillation becomes of order ϕ_{c}. The number of e-folds of this type of inflation is $\Delta N \sim \ln(\phi_{\mathrm{end}}/\phi_{\mathrm{c}})$, which is typically negligible.

8.4.6 Hybrid inflation with a running mass

Now we turn to the case in which the loop correction is of the form $\phi^2 \ln(\phi/Q)$, which typically arises when susy is softly broken. Models of inflation invoking such a correction have been proposed by Stewart (1997a, b).

As we note in Section 8.2.4, this type of loop correction is equivalent to replacing the inflaton mass by a slowly varying mass $m^2(\phi)$, often called a running mass. At $\phi = M_{\text{Pl}}$, the running mass is supposed to have the magnitude $|m^2| \sim V_0/M_{\text{Pl}}^2$, which is the minimum one in a generic supergravity theory. The inflaton is supposed to have couplings (gauge, or maybe Yukawa) that are not too small, and for the most part we assume that $m^2(\phi)$ passes through zero before it stops running.[9] Because the couplings are small compared with unity, V' then vanishes at some relatively nearby point, which we denote as ϕ_*.

It is useful to write Eq. (8.11) in the form

$$V(\phi) = V_0\left[1 - \frac{1}{2}\mu^2(\phi)\frac{\phi^2}{M_{\text{Pl}}^2}\right], \tag{8.69}$$

where

$$\mu^2(\phi) \equiv -\frac{M_{\text{Pl}}^2 m^2(\phi)}{V_0}. \tag{8.70}$$

We are supposing that V_0 dominates because this is necessary for inflation in the regime $\phi \lesssim M_{\text{Pl}}$, where the field theory is under control. Then,

$$M_{\text{Pl}}\frac{V'}{V_0} = -\phi\left(\mu^2 + \frac{1}{2}\frac{d\mu^2}{dt}\right), \tag{8.71}$$

$$\eta \equiv M_{\text{Pl}}^2\frac{V''}{V_0} = -\left(\mu^2 + \frac{3}{2}\frac{d\mu^2}{dt} + \frac{1}{2}\frac{d^2\mu^2}{dt^2}\right), \tag{8.72}$$

where $t \equiv \ln(\phi/M_{\text{Pl}})$.

While observable scales are leaving the horizon, μ^2 is usually linear in $\ln\phi$:

$$\mu^2 \simeq \mu_*^2 + c\ln\frac{\phi}{\phi_*}, \tag{8.73}$$

where $|c| \ll 1$ is related to the couplings involved. This gives

$$M_{\text{Pl}}\frac{V'}{V_0} = c\phi\ln\frac{\phi_*}{\phi}, \tag{8.74}$$

$$\eta \equiv M_{\text{Pl}}^2\frac{V''}{V_0} = c\left(\ln\frac{\phi_*}{\phi} - 1\right). \tag{8.75}$$

Note that $\mu_*^2 = -c/2$ and that $\mu^2 = 0$ at $\ln(\phi_*/\phi) = -1/2$, whereas $V'' = 0$ at $\ln(\phi_*/\phi) = 1$.

---

[9] The running associated with a given loop will stop when ϕ falls below the mass of the particle in the loop.

The number $N(\phi)$ of e-folds to the end of slow-roll inflation is given by

$$N(\phi) = M_{\text{Pl}}^{-2} \int_{\phi_{\text{end}}}^{\phi} \frac{V}{V'} d\phi. \tag{8.76}$$

Using the linear approximation near ϕ_*, we obtain

$$N(\phi) = -\frac{1}{c} \ln \left(\frac{c}{\sigma} \ln \frac{\phi_*}{\phi} \right) \tag{8.77}$$

or

$$(\sigma/c) e^{-cN} = \ln \frac{\phi_*}{\phi}. \tag{8.78}$$

Knowing the functional form of $m^2(\phi)$, and the value of ϕ_{end}, we can evaluate the constant σ by taking the limit $\phi \to \phi_*$ in the full expression (8.76). On theoretical grounds, we expect (Covi and Lyth 1999)

$$|c| \lesssim |\sigma| \lesssim 1. \tag{8.79}$$

The spectral index $n = 1 + 2\eta$ is given in terms of c and σ by

$$\frac{n-1}{2} = \sigma e^{-cN} - c. \tag{8.80}$$

The COBE normalization is

$$\frac{V_0^{1/2}}{M_{\text{Pl}}^2} = 5.3 \times 10^{-4} M_{\text{Pl}} \frac{|V'|}{V_0}. \tag{8.81}$$

In our case, it is convenient to define a constant τ as

$$\ln \frac{M_{\text{Pl}}}{\phi_*} \equiv \frac{\tau}{|c|}. \tag{8.82}$$

Assuming that $|m^2|$ has the typical value V_0/M_{Pl}^2 at the Planck scale, the linear approximation (8.73) applied at that scale would give $\tau \simeq 1$. Will the linear approximation apply at that scale? If *all* relevant masses at the Planck scale are of order V_0/M_{Pl}^2, we expect on dimensional grounds that the linear approximation will be valid in the regime $|c \ln(\phi/\phi^*)| \ll 1$. Then the approximation will be just beginning to fail at the Planck scale. At least in this case, we expect τ to be very roughly of order 1.

Using the definition of τ, Eqs. (8.74) and (8.78) give

$$\frac{V_0^{1/2}}{M_{\text{Pl}}^2} = e^{-\tau/|c|} \exp\left(-\frac{\sigma}{c} e^{-cN_{\text{COBE}}}\right) |\sigma| e^{-cN_{\text{COBE}}} \times 5.3 \times 10^{-4}. \tag{8.83}$$

Four types of inflation model are possible, corresponding to whether ϕ_* is a maximum or a minimum, and whether ϕ during inflation is smaller or bigger than ϕ_*. In all cases (Covi and Lyth 1999), there is a regime of parameter space that is allowed by the observational constraint $|n - 1| \lesssim 0.2$, and it satisfies the theoretical expectation Eq. (8.79). In much of this regime, $|dn/d \ln k|$ is big enough eventually to be observable.

Table 8.1. *Predictions for the spectral index n and its variation*
$dn/d \ln k$ *displayed for some potentials of the form* $V_0(1 - c\phi^p)$

	$1 - n$		$-10^3 dn/d \ln k$	
p	$N = 50$	$N = 20$	$N = 50$	$N = 20$
$p \to 0$	0.02	0.05	(0.4)	2.6
$p = -2$	0.03	0.075	(0.6)	3.8
$p \to \pm\infty$	0.04	0.10	(0.8)	5.0
$p = 4$	0.06	0.15	(1.2)	5.4
$p = 3$	0.08	0.20	(1.6)	10.0

Note: The value of N depends on the history of the Universe after slow-roll
inflation ends, but $N < 60$ in all reasonable cases.

Table 8.2. *Predictions for various potentials*

Comments	$V(\phi)/V_0$	$\frac{1}{2}(n-1)$	$\frac{1}{2}\frac{dn}{d\ln k}$
Mass term	$1 \pm \frac{1}{2}c\frac{\phi^2}{M_{\rm Pl}^2}$	$\pm c$	0
Softly broken susy	$1 \pm \frac{1}{2}c\frac{\phi^2}{M_{\rm Pl}^2}\ln\frac{\phi}{Q}$	$\pm c + \sigma e^{\pm cN}$	$\mp c\sigma e^{\pm cN}$
Spontaneously broken susy	$1 + c\ln\frac{\phi}{Q}$	$-\frac{1}{2N}$	$-\frac{1}{2}\frac{1}{N^2}$
$p > 2$ or $-\infty < p < 1$ (self-coupling or hybrid)	$1 - c\phi^p$	$-\left(\frac{p-1}{p-2}\right)\frac{1}{N}$	$-\left(\frac{p-1}{p-2}\right)\frac{1}{N^2}$
Various models	$1 - e^{-q\phi}$	$-\frac{1}{N}$	$-\frac{1}{N^2}$
p an integer ≤ -1 (dynamical symmetry breaking) or ≥ 3 (self-coupling)	$1 + c\phi^p$	$\frac{p-1}{p-2}\frac{1}{N_{\rm max}-N}$	$-\left(\frac{p-2}{p-1}\right)\left(\frac{n-1}{2}\right)^2$

Note: Constants c, q, and Q are positive whereas σ and p can have either sign. In the first
three cases, there is a theoretical constraint $|c| \ll 1$. In the second case, we expect $|\sigma| \gtrsim |c|$.

8.5 The spectral index as a discriminator

In most models of inflation, the potential is of the form $V(\phi) = V_0 + \cdots$, with the constant
first term dominating and $\phi \lesssim M_{\rm Pl}$. With certain qualifications stated in the text (notably a
requirement $\phi \ll M_{\rm Pl}$ that needs to be imposed in certain cases), the amplitude of the spectrum
of the gravitational waves is too small ever to observe. With similar qualifications, the prediction
for the spectral index of the primordial curvature perturbation is the one shown in Tables 8.1
and 8.2. We also show the prediction for the $dn/d \ln k$. The latter quantity eventually will be
observed if it is bigger than a few times 10^{-3} (Copeland et al. 1998), and in Table 8.1 we have
placed values less than 2×10^{-3} in parentheses.

The simplest cases are $V = V_0 \pm \frac{1}{2}m^2\phi^2$, which give a scale-independent spectral index
that may or may not be close to 1.

Next in simplicity come the cases $V = V_0(1 - c\phi^p)$. Here, p can be an integer ≥ 3, corresponding to self-coupling of the inflaton at tree level, or it can be in the ranges $2 < p < \infty$ or $-\infty < p < 1$ (not necessarily an integer), corresponding to mutated hybrid inflation. Related to these, as far as the prediction is concerned, are the cases $V = V_0(1 - e^{-q\phi})$ (Section 8.3.5), which corresponds to $p \to -\infty$, and $V = V_0[1 + c\ln(\phi/Q)]$ (Section 8.4.5), which corresponds to $p \to 0$. In all these cases, the predictions are

$$\frac{1}{2}(n - 1) = -\left(\frac{p-1}{p-2}\right)\frac{1}{N}, \tag{8.84}$$

$$\frac{1}{2}\frac{dn}{d\ln k} = -\left(\frac{p-1}{p-2}\right)\frac{1}{N^2}. \tag{8.85}$$

The second expression can be written

$$\frac{1}{2}\frac{dn}{d\ln k} = -\left(\frac{p-2}{p-1}\right)\left(\frac{n-1}{2}\right)^2. \tag{8.86}$$

Excluding the cases $p \simeq 1$ and $p \simeq 2$, the factor $(p-1)/(p-2)$ is of order 1. As a result, $(n-1)$ is far enough below 0 to be eventually observable, because $N < 60$. Its scale dependence may or may not be observable, as we see from Table 8.1. If it is observable, Eq. (8.86) may allow discrimination between different values of p, without our knowing N.

Next consider the case $V = V_0(1 + c\phi^p)$ with p an integer ≥ 3 (tree-level self-coupling) or ≤ -1 (dynamical symmetry breaking). In these cases, there is a maximum possible number of e-folds of inflation, the value of which is unknown. If it is not too big, $n - 1$ may be far enough above 0 to be detected eventually. The scale dependence is given by Eq. (8.86), and will be observable if $|n - 1|$ is more than a few times 0.01. Note that, in these models, it is (more than usually) unclear how the inflaton is supposed to arrive at the inflaton part of the potential.

Finally, we come to the case of a running inflaton mass (Section 8.4.6). This gives a distinctive prediction for the scale dependence of $n(k)$, which is big enough to observe in a large region of parameter space (Copeland et al. 1998; Covi and Lyth 1999).

It cannot be emphasized too strongly that this is a snapshot of an evolving subject. The situation presented in Tables 8.1 and 8.2 therefore might be quite ephemeral, but it has one remarkable feature that is likely to persist in any updated version, which is that an observational accuracy $\Delta n \sim 0.01$, which we see in Chapter 9 is expected from future satellite measurements of the cosmic microwave background (cmb), is more or less what is required to distinguish different models! At the most extravagant, we might have ordered an accuracy $\Delta n \sim 10^{-3}$ for this purpose.

8.6   Models from extended theories of gravity

An entirely different strand of inflationary model building is based on the possibility that general relativity fails to be a correct description of gravity in the early Universe. As we discuss in Section 14.4.6, such a breakdown is practically inevitable at the earliest moments, as the Universe emerges from the Planck epoch. The models we now discuss suppose that general relativity still has not been restored when cosmological scales start to leave the horizon. Recall that $V^{1/4}$ is then at least 2 orders of magnitude below the Planck scale.

Models of this type typically have been constructed without reference to particle theories such as supergravity and superstrings. As in single-field inflation models, the question of whether there is any problem in considering field values much bigger than of order of the Planck mass is seldom asked.

In this section, we invoke the description of gravity based on an action S, which is described in Sections 14.4.5 and 14.4.6. The evolution of the space-time metric, and of the matter fields,[10] is determined from the action S by the action principle. Einstein gravity corresponds to the Einstein–Hilbert action,

$$S = \int d^4x \sqrt{-g} \left(\frac{M_{Pl}^2}{2} R + \mathcal{L}_{matt} \right).$$ (8.87)

Here, $\mathcal{L}_{matt}$ is the Lagrangian of the matter fields. Gravity is described by the first term, where R is the space-time curvature scalar, whose coefficient is $M_{Pl}^2 = (8\pi G)^{-1}$, where G is Newton's gravitational constant. We are talking about the possibility of replacing R by something more complicated.

There have been primarily three separate approaches to looking at extended theories of gravity. The first adds extra geometric terms to the Einstein–Hilbert action (terms constructed entirely from the curvature tensor). The first model of inflation was devised in precisely this form by Starobinsky (1980). The second approach, giving what are called scalar–tensor theories, amends gravity by making the gravitational "constant" a scalar field, thus permitting both spatial and temporal variations of the strength of the gravitational interaction. A special case of this second class of theories is the much investigated Jordan–Brans–Dicke theory. Finally, there are approaches based on extra dimensions, which we do not consider in this book.

All these theories have the advantage that, at the present epoch, general relativity is known to hold very accurately, thus providing constraints on the theory that have nothing to do with the way inflation proceeds (Will 1993). Particularly in the case of scalar–tensor theories, this can be a vital constraint. There are also constraints that can be applied on the time variation of the gravitational constant from several sources, such as nucleosynthesis, which provides a strong temporal "lever arm" on the variation of the gravitational constant stretching back to when the Universe was only 1 s old.

8.6.1 Conformal transformation

In analyzing these types of theories, there is an extremely powerful technique that can be brought into play, known as the **conformal transformation** (Whitt 1984; Barrow and Cotsakis 1988; Maeda 1989; Wands 1994). This is a position-dependent transformation mapping the

---

[10] The term "matter field" in general relativity is synonymous with the term "field" in particle physics. The point is that, in the former case, the space-time metric also is regarded as a field.

original metric $g_{\mu\nu}$ into a new metric $\tilde{g}_{\mu\nu}$, according to

$$\tilde{g}_{\mu\nu} = \Omega^2 g_{\mu\nu}. \tag{8.88}$$

This transformation yields a new Ricci scalar,

$$\tilde{R} = \frac{1}{\Omega^2}\left[R + \frac{6}{\Omega}\Box\Omega\right], \tag{8.89}$$

where

$$\Box \equiv -\frac{1}{\sqrt{-g}}\,\partial_\mu\left[\sqrt{-g}\,\partial^\mu\right] \tag{8.90}$$

is evaluated with the original metric. The function Ω may depend both on the space-time curvature and on the matter fields. The trick is that, by a careful choice of the function Ω, this mapping can restore general relativity. In this way, a nonstandard theory of gravity has been mapped into a standard theory of gravity, with a more complicated behaviour in the matter sector. Because general relativity, even with quite complicated matter contents, is a very well explored theory, this is often a great aid to progress in analyzing these theories.

It should be made completely clear that the conformal transformation is not, in general, a coordinate redefinition; such a redefinition would be trivial because general relativity is a covariant theory. Rather, the conformal transformation is a field redefinition that mixes up the gravitational and matter degrees of freedom. Each choice of field definitions is commonly called a **frame**.[11] The frame in which gravity takes the form of Einstein's theory is known (when it exists) as the **Einstein frame**.

From the viewpoint of quantum field theory, the statement that Einstein gravity is valid at the present epoch amounts to saying that there is an Einstein frame, *and* that in this frame $\mathcal{L}_{mat}$ is given by the Standard Model. We could use a different frame, but then $\mathcal{L}_{mat}$ would look more complicated.

As in the preceding example, the change from one frame to another does not correspond to a change in the physics. True, it changes the space-time curvature, but it also changes the Lagrangian of the nongravitational (matter) fields and the latter provide the definition of the ideal clocks that are to be used in measuring the space-time curvature. Phenomena that appear to be due to gravity in one frame may appear to have their origin in the matter sector in another.

8.6.2 R^2 inflation

The original inflation model studied by Starobinsky (1980) contained an action with both powers and derivatives of the curvature, motivated by one-loop quantum corrections to the Einstein–Hilbert action. We study a slightly simpler possibility, where the action is simply

$$S = \frac{M_{\mathrm{Pl}}^2}{2}\int d^4x \,\sqrt{-g}\,\left(R + \frac{R^2}{6M^2}\right), \tag{8.91}$$

---

[11] Note that this use of the word frame has nothing to do with the usage in Chapters 4 and 14.

the first term in the parentheses being the usual Einstein–Hilbert action describing general relativity. This action has been derived in the context of supergravity and, because of its simplicity, has received a lot of attention over the years. A nice feature of the model is that inflation is entirely a property of the gravitational sector, unlike in the case of scalar–tensor theories, examined shortly, in which it is the interaction between the modified gravity sector and the matter sector which leads to inflation.

In contrast to the other models we are considering, it is not possible to bring Eq. (8.91) into the Einstein–Hilbert form with a conformal transformation. However, the presence of the R^2 term means that the field equation for R is of higher order, and that there are effectively additional degrees of freedom. We can introduce a scalar field into the action, which takes care of this problem (Wands 1994). Then, after a conformal transformation

$$\Omega^2 = 1 + \frac{R}{3M^2}, \qquad (8.92)$$

the action is

$$S = \int d^4x \sqrt{-g} \left[\frac{M_{\text{Pl}}^2}{2} \tilde{R} - \frac{1}{2} \tilde{g}_{\mu\nu} \partial_\mu \phi \partial_\nu \phi - V(\phi) \right], \qquad (8.93)$$

with

$$V(\phi) = \frac{3M_{\text{Pl}}^2 M^2}{4} \left[1 - \exp\left(-\sqrt{\frac{2}{3}} \frac{\phi}{M_{\text{Pl}}} \right) \right]^2, \qquad (8.94)$$

where a tilde indicates the quantities in the transformed frame (Barrow 1988). The potential is shown in Figure 8.2. In the slow-roll regime, the potential is

$$V(\phi) = \frac{3M_{\text{Pl}}^2 M^2}{4} \left[1 - 2\exp\left(-\sqrt{\frac{2}{3}} \frac{\phi}{M_{\text{Pl}}} \right) \right], \qquad (8.95)$$

which reproduces Eq. (8.23) after redefining the origin of ϕ.

8.6.3 Variable Planck mass model

A similar result arises in a model known as the **variable Planck mass model** (Salopek et al. 1989). This is a half-way house between Einstein gravity and a scalar–tensor theory, in that the Einstein term remains in the action, but there also is a term $-\xi\phi^2 R/2$ coupling the scalar field and gravity. Here, ξ is a coupling constant, taken to be large and negative in this model. A conformal transformation to Einstein gravity, and a field redefinition to obtain a

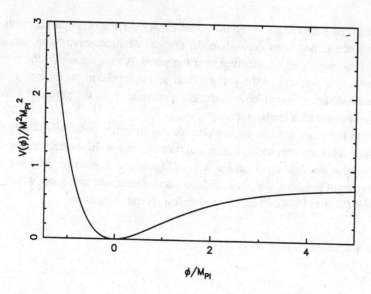

Fig. 8.2. The potential of Eq. (8.94).

canonical kinetic term, yields the potential (Salopek et al. 1989)

$$V(\psi) \propto \left[1 - \exp\left(-2\alpha \frac{\psi}{M_{Pl}} \right)^2 \right],$$

(8.96)

with

$$\alpha = \left(\frac{|\xi|}{1 + 6|\xi|} \right)^{1/2},$$

(8.97)

so that α approaches $\sqrt{1/6}$ in the large $|\xi|$ limit. This again can be treated as a version of Eq. (8.37), with $q \simeq \sqrt{2/3}$.

A variant of this model has been proposed by Futamase and Tanaka (1999), who invoke a loop correction along the lines of Eq. (8.11).

8.6.4 Extended inflation

Dynamics

Extended inflation is an attempt to reintroduce the concepts behind the model that Alan Guth used to introduce inflationary cosmology, a model now known as **old inflation**, which already was recognized as unviable even in Guth's original paper (Guth 1981). The old inflation model assumed general relativity plus a single scalar field trapped in a metastable vacuum state of nonzero potential energy, as illustrated in Figure 8.3. The initial conditions supposedly were provided by thermal equilibrium, which at high temperature localizes the field in the vicinity of the origin. Once inflation gets under way, all other matter redshifts away, cooling the Universe and leaving the scalar field trapped at the origin. In such a situation, inflation

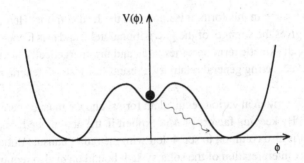

Fig. 8.3. Potential with a metastable vacuum state at the origin. Inflation ends when the field quantum tunnels through to the true minimum.

proceeds exponentially, with

$$a(t) \propto \exp\left(\sqrt{\frac{V_0}{3M_{\mathrm{Pl}}^2}}\, t\right). \tag{8.98}$$

Although classically stable, inflation can end by quantum tunnelling of the field into the true vacuum, nucleating a bubble of true vacuum in the "sea" of false vacuum. Provided that enough bubbles are nucleating, they will expand and meet, bringing the inflationary epoch to a close.

The problem is that the Universe is expanding at an exponential rate while the bubbles are nucleating. Even though the bubbles nucleate at a constant rate per volume per time, they will be unable to meet and percolate unless the nucleation rate Γ is high enough. However, if it is, then the phase transition will be completed almost immediately, failing to provide sufficient inflation to solve the cosmological problems. If inflation lasts long enough, it will last forever.

The crucial parameter is not the nucleation rate itself, but the nucleation rate per Hubble volume per Hubble time (not to be confused with the slow-roll parameter ϵ):

$$\varepsilon = \frac{\Gamma}{H^4}. \tag{8.99}$$

During old inflation, the perfect de Sitter expansion means that this remains constant. For the phase transition to be completed, ε must be of order unity; to obtain sufficient inflation, it must be much less.

The extended inflation model came about when La and Steinhardt (1989) realized that the fundamental problem, the constancy of ε, could be addressed by implementing the idea of a trapped field in the context of an extended gravity theory, for which the solution need not be perfect de Sitter. They chose to implement the theory in the simplest scalar–tensor theory, known as the Jordan–Brans–Dicke theory.

The action for Jordan–Brans–Dicke theory is given by

$$S = \int d^4x \sqrt{-g}\left[\frac{M_{\mathrm{Pl}}^2}{2}\left(\Phi R - \frac{\omega}{\Phi}\partial_\mu \Phi \partial^\mu \Phi\right) + \mathcal{L}_{\mathrm{matter}}\right]. \tag{8.100}$$

When the action is written in this form, it is known as the Jordan frame. Here, Φ is the Brans–Dicke field, which gives the strength of the gravitational field, and ω is a coupling constant. In the limit of large ω, the kinetic term becomes large and the energetically favoured situation is that Φ be constant, recovering general relativity. Present-day tests of general relativity require $\omega > 500$.

This action can be written in various equivalent forms, and we must be careful to keep track of which is chosen. By keeping factors of M_{Pl} explicit in the action, we have chosen to make Φ dimensionless; it is also common to see Φ left with the rather unusual dimensions of mass-squared. The correct interpretation of the solar system bound on ω also requires care with the form of the action. One way of rewriting this action is simply to redefine the Brans–Dicke field in order to have the normal dimensions and kinetic term; this is brought about by defining a new field and coupling parameter as

$$\phi^2 = 4\omega\,\Phi\,M_{\mathrm{Pl}}^2, \qquad \xi = \frac{1}{8\omega}, \tag{8.101}$$

to give

$$S = \int d^4x\sqrt{-g}\left(\frac{1}{2}\xi\phi^2 R - \frac{1}{2}\partial_\mu\phi\partial^\mu\phi + \mathcal{L}_{\mathrm{matter}}\right). \tag{8.102}$$

In the extended inflation model, the matter is provided by a scalar field in a metastable vacuum state, of the type shown in Figure 8.3; in that limit, the matter Lagrangian is simply a constant that we denote $V(0) = M^4$. In that situation, the appropriate homogeneous solution is well known:

$$a(t) = a_0\,(1 + Bt)^{\omega+1/2}, \tag{8.103}$$

$$\Phi(t) = \Phi_0\,(1 + Bt)^2, \tag{8.104}$$

where $B^2 = M^4/6\alpha^2\Phi_0$ and $\alpha = (2\omega + 3)(6\omega + 5)/12$. This is not the most general solution, but it is known to be a late-time attractor. After an initial transient, the solutions are power laws.

The advantage of the power-law behaviour is clear from studying the behaviour of the nucleation rate ε, which in this theory grows as t^4 at late times. Consequently, ε can start sufficiently small so as to allow enough inflation to occur, but will inevitably grow to reach order unity and allow the phase transition to be completed. That is, the exponential production of bubbles is always able eventually to overpower the power-law expansion of the volume. This scenario bears many resemblances to a first-order hybrid inflation scenario.

An interesting new view of extended inflation comes from carrying out a conformal transformation to Einstein gravity. The required conformal factor Ω in this case is easy to spot from Eq. (8.89), being $\Omega = \sqrt{\Phi}$. Along with this, we define a new scalar field, in order to obtain a canonical kinetic term in the Einstein frame, as

$$\frac{\psi}{M_{\text{Pl}}} = \sqrt{\omega + \tfrac{3}{2}} \ln \Phi, \tag{8.105}$$

which puts the action in the form

$$\tilde{S} = \int d^4\tilde{x}\sqrt{-\tilde{g}} \left[\frac{M_{\text{Pl}}^2}{2}\tilde{R} - \frac{1}{2}\partial_\mu\psi\partial^\mu\psi + V(\psi) \right], \tag{8.106}$$

where $V(\psi) = M^4/\Omega^4$ becomes

$$V(\psi) = M^4 \exp\left(-\sqrt{\frac{2}{p}}\frac{\psi}{M_{\text{Pl}}}\right), \qquad 2p = \omega + \frac{3}{2}. \tag{8.107}$$

This is exactly the power-law inflation potential of Eq. (8.23)! So, extended inflation is actually a new way of obtaining the power-law inflation model, with one huge advantage – power-law inflation as discussed in Section 8.3.2 would go on forever, but in extended inflation the decay of the metastable vacuum state will bring inflation to an end. When the tunnelling happens, the false vacuum energy M^4 disappears and the ψ potential vanishes.

Big-bubble problem

It was quickly realized (Weinberg 1989; La et al. 1989) that ensuring percolation of the bubbles was not sufficient to bring about a successful exit to the inflationary era. Because the nucleation rate is increasing slowly (in comparison to the rate of expansion of the Universe), there is still a reasonable probability of some bubbles nucleating long before the end of inflation. These bubbles have an interesting evolution. After nucleating, they are much smaller than the Hubble length and expand at the speed of light. Quickly though, they approach the Hubble radius and from then on are swept up by the expansion of the Universe, which becomes the dominant effect in the evolution of their size. Bubbles that nucleate very early, for example, 50 e-foldings before the end of inflation, can be swept up to very large sizes, comparable indeed to our present observable Universe. Such bubbles are unable to thermalize properly, and can lead to substantial inhomogeneities, which would be visible, for example, in the microwave sky.

This constraint clearly favours low values of ω because, then, ε increases more quickly relative to the expansion of the Universe (the former going as t^4 and the latter going as $t^{\omega+1/2}$ in the Jordan frame). The question is how low ω must be. Early calculations imposed a fairly crude criterion that a certain volume of the Universe (normally taken as 10^{-4}) is the maximum allowed in bubbles over a certain comoving size (normally taken as the horizon size at decoupling). Liddle and Wands (1991) made a more precise comparison by computing the bubble spectrum and calculating the Sachs–Wolfe perturbations, but in fact finished with more or less the same result. The bubble spectrum is only compatible with observations if

$$\omega \lesssim 20. \tag{8.108}$$

The conclusion is clear. In this version of extended inflation, the constraint of avoiding big bubbles is incompatible with the present-day constraints on ω from our knowledge that general relativity is an accurate descriptor of the physics within our solar system.

8.6.5   Extended inflation variants

The simplest way to try and resolve this conflict is to evade the solar system bound on ω. That is, we take advantage of the fact that these two conflicting constraints are applied at wildly different epochs: one during inflation and one at the present day. In fact, this is quite easy to do, by assuming that we can add a potential for the Brans–Dicke field Φ. This potential is assumed to be unimportant during inflation, but long after inflation, when the energy scale is reduced, it is responsible for anchoring the scalar field at its minimum. Such a potential will enforce a constant Φ at late times, and the solar system bounds, which rely on a free Φ field, will no longer apply. Notice that if this step is taken, the present-day strength of the gravitational interaction is fixed by the location of the minimum of the potential, rather than dynamically as it is in the massless Φ version of the theory.

This "fix" allows us to implement extended inflation with a low enough ω to satisfy the bubble constraint, without having to worry about the present-day bound. However, this model has a further, fatal, flaw; it solves the big-bubble problem by breaking the scale invariance of the bubble distribution, but in doing so, it destroys the scale invariance of the density perturbation spectrum produced from quantum fluctuations, to an extent that is incompatible with observations. We see this from the power-law inflation predictions

$$n = 1 - \frac{2}{p} = 1 - \frac{4}{\omega + 3/2}, \tag{8.109}$$

$$r = \frac{12.4}{p} = \frac{24.8}{\omega + 3/2}. \tag{8.110}$$

From Eq. (8.108), we find $n < 0.81$ and $r > 1.15$, which we see in Chapter 12 are too extreme to fit current data.

An alternative strategy is to have a genuinely large ω, but make it look small during inflation; this can be achieved (Holman et al. 1990) by postulating that the Brans–Dicke coupling to the invisible sector, taken to include the inflaton, differs from that to the visible-matter sector. Again, this evades the bubble constraint but will fall foul of the need to produce adequately scale-invariant density perturbations.

The best strategy therefore is to abandon completely the idea of a constant ω. The most general scalar–tensor theory automatically includes this possibility; its action is

$$S = \int d^4x \sqrt{-g} \left\{ \frac{M_{\text{Pl}}^2}{2} \left[\Phi R - \frac{\omega(\Phi)}{\Phi} \partial_\mu \Phi \partial^\mu \Phi \right] + \mathcal{L}_{\text{matter}} \right\}, \tag{8.111}$$

where $\omega(\Phi)$ is an arbitrary function. Such an action was studied by Barrow and Maeda (1990). This theory can be written equivalently with a canonical scalar field, analogously to Eqs. (8.101) and (8.102), as

$$S = \int d^4x \sqrt{-g} \left[\frac{f(\phi)}{2} R - \frac{1}{2} \partial_\mu \phi \partial^\mu \phi + \mathcal{L}_{\text{matter}} \right], \qquad (8.112)$$

with the transformation

$$\Phi = \frac{f(\phi)}{M_{\text{Pl}}^2}, \qquad \omega(\Phi) = \frac{f(\phi)}{(df/d\phi)^2}. \qquad (8.113)$$

This form of the action was discussed by Steinhardt and Accetta (1990).

Although formally these are equivalent, they have led to rather different model building, for the reason that, for example, $\omega(\Phi)$ can diverge while $f(\phi)$ remains perfectly well behaved. The different possibilities have been analyzed by Liddle and Wands (1992), who analyzed the bubble constraint as well as the inflationary dynamics. In the Barrow–Maeda (1990) version, ω is allowed to grow from a small value during inflation (compatible with the bubble constraint) to a large present value. In practice, this seems difficult to arrange because the transition between these regimes must be rapid (Liddle and Wands 1992). The Steinhardt–Accetta (1990) strategy is almost the opposite; they rely on *reducing* the effective ω, until it becomes so low that inflation stops. Bubbles then can safely nucleate in the postinflationary era to remove the false vacuum, without any danger of creating large ones. Then, further dynamics are used to drive ω back up to an acceptably large value. This is a perfectly workable scenario [see, e.g., the extensive analysis by Liddle and Wands (1992)], with the drawback that some of the original motivation is lost because inflation is not ended by the first-order transition itself but must end through dynamical reasons first.

It is clear that it is extremely difficult to design viable models based on the extended inflation paradigm in which inflation actually is ended by the first-order phase transition (Liddle and Lyth 1992, 1993a; Green and Liddle 1996).

8.6.6 Multifield inflation in scalar–tensor gravity

An alternative possibility is to implement a standard slow-roll inflation model in a scalar–tensor theory. Assuming there is nothing to prevent it, the Brans–Dicke field also will evolve, and so, we have a multicomponent inflaton as discussed in Section 7.9. Indeed, this is probably the most natural implementation of the multicomponent scenario. The standard perturbation formulae cease to apply in the multicomponent case and, in particular, $\mathcal{R}$ is no longer constant on scales well above the horizon. A more complicated perturbation calculation needs to be carried out, and this was done by Starobinsky and Yokoyama (1995) for Brans–Dicke theory, and by García-Bellido and Wands (1995) for general scalar–tensor theories. Using the conformal transformation to the Einstein frame, we use these calculations as part of the general formalism set out in Section 7.9.

8.7   Open inflation models

Among the most recent additions to the pantheon of inflationary models are predictive models that lead to open Universes, based on a combination of previous ideas. In these models there are two inflationary periods, separated by a tunnelling event that does not end inflation.

Historically, it long had been realized that it is possible to obtain a Universe of any present density from the standard inflationary paradigm, provided that suitable initial conditions are chosen. Because we are perfectly entitled to run the present conditions back in time through an inflationary cosmology to obtain the appropriate initial conditions, any present Universe can be obtained. However, the initial conditions required to achieve a homogeneous open present Universe by this route seem very contrived.

In an open Universe, the value of Ω is given directly by the ratio of the Hubble length to the curvature scale (Section 2.3.1). Because the curvature scale is a fixed comoving scale, we write

$$\sqrt{1 - \Omega} = \frac{r_{\text{hub}}}{r_{\text{curv}}} = \frac{H^{-1}/a}{|K|^{-1/2}}. \tag{8.114}$$

Because the purpose of inflation is to make the present comoving Hubble length much less than the one at some early stage in inflation, in order to solve the horizon problem, this equation implies that the initial value of Ω_0 has to be further from unity than the present one, and can only be similar if a minimal amount of inflation takes place. On the face of it, it seems rather unlikely that exactly enough inflation might have occurred to solve the horizon and flatness problems at the present epoch, but not any subsequent one.[12]

Were such an inflationary epoch to have occurred, it would represent a serious blow to the predictiveness of the inflationary scenario, which is supposed to have the advantage of removing any dependence on initial conditions from our present observable Universe. Although there is certainly no logical obstacle to present conditions depending on some unknowable initial condition, it would be a major setback to the hopes of discerning information about physics at high energies from cosmological observations.

8.7.1   Dynamics of the open inflation model

In the light of preceding discussion, it is extremely interesting that there now exist models of inflation that give $\Omega_0 < 1$, but in which Ω_0 is determined from parameters of the physical theory rather than from the initial conditions. We call these **open inflation** models. The idea on which they are based has a surprisingly long history. In 1980, Coleman and de Luccia realized that the inside of a nucleated bubble could look like an open Universe, and Gott (1982) was quick to make a connection to the inflationary cosmology. However, its incorporation into a working inflationary model was carried out much more recently, through papers by Sasaki et al. (1993) and Bucher et al. (1995). The idea is to consider a tunnelling from a metastable

---

[12] On the other hand, it also has been argued (Ellis et al. 1991) that if we did observe the Universe to be open, the initial conditions required, given a period of inflation, are less fine-tuned than had no inflation occurred. In the absence of any guidance toward a measure on the space of possible initial conditions, it seems hard to assess this argument.

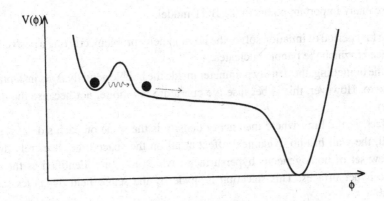

Fig. 8.4. Schematic of the potential for the BGT model. A first period of inflation occurs with the field trapped in the metastable state, and then a second period occurs after quantum tunnelling, when the field slow-rolls down the potential toward the true minimum.

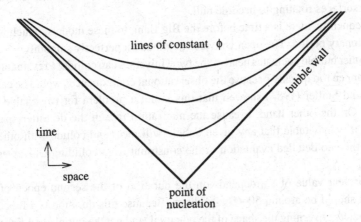

Fig. 8.5. Space-time diagram of the nucleation. The bubble walls expand at the speed of light, while inside the bubble the surfaces of constant scalar field value form hyperbolas that will be identified as the constant time slices.

vacuum state into a new state that is not the true vacuum, as illustrated in Figure 8.4. The field then slow-rolls down toward the true minimum, driving a second epoch of inflation within the nucleated bubble. Remarkably, the resulting Universe inside this bubble has the appearance of an open Universe, the present density of which is governed by the length of the second inflationary epoch, which in turn is determined by the shape of the potential. Figure 8.5 shows a space-time diagram of this process.

We describe the model of Bucher et al. (1995) (which we call the BGT model), based on a potential of the form shown in Figure 8.4. The shape of the potential is rather contrived, but more appealing models have been devised by Linde (1995) and Linde and Mezhlumian (1995) and now there are several others. We return to those models at the end of this section.

There are many important points to the BGT model.

- The first period of inflation solves the homogeneity problem, creating a perfect de Sitter space in which the bubble nucleates.
- At the tunnelling, the density parameter inside the bubble actually is set instantaneously to zero. However, this is because the curvature is infinite, not because the density is zero.
- In fact, in the limit where the energy density is the same on each side of the bubble wall, the wall has no dynamical effect at all on the space-time. It merely draws out a new set of homogeneous hypersurfaces, which later are identified as the constant time hypersurfaces. This selection is made by the scalar field dynamics inside the bubble.
- The bubble wall forms the initial singularity (zero time) surface of the open Universe.
- The open de Sitter singularity is merely a coordinate singularity, not a real one; as we see from the space-time diagram, it corresponds to the constant time (i.e., constant ϕ) hypersurfaces rotating to become null.
- Consequently, there is a time before the Big Bang in these models, which is the initial inflationary period. The open Universe is joined on perfectly smoothly.
- If another bubble nucleates nearby, the two Universes can collide! Presumably this must not happen too quickly because the observational consequences would be catastrophic. Gott and Statler (1984) showed that this is not a problem for reasonable tunnelling rates. On the other hand, because the nucleation rate in the de Sitter space is constant, it is inevitable that any given bubble will meet and collide with others. As we write this, no detailed evaluation of the constraint from colliding Universes has been made!
- The present value of Ω_0 depends on the duration of the second epoch of inflation, which should be around 50–60 e-foldings. Because this duration is a function of the parameters governing the shape of the potential, and not the initial conditions, the BGT model gives a predictive theory of Ω_0.
- The tuning of the length of the second period of inflation to obtain a plausible low-density Universe is of order of a few percent. This is over and above any tuning needed to ensure the right level of density perturbations, but does not seem especially severe.

The unnatural appearance of the BGT model detracts from its appeal as an inflationary model. Further, it turns out that unless the minimum, within which the first inflationary phase happens, is very deep, tunnelling proceeds not by bubble formation but rather by tunnelling to the top of the potential, the Hawking–Moss instanton. Hence, although influential, the BGT model is not considered a very viable scenario. However, shortly after their paper appeared, more elegant models were introduced by Linde (1995) and were analyzed further by Linde and Mezhlumian (1995); these models rely on the dynamics of two separate scalar fields.

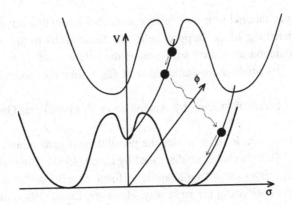

Fig. 8.6. Potential for Linde's open inflation model. The field rolls down the $\sigma = 0$ channel; then, at some point, σ tunnels across to its true minimum. After the tunnelling, ϕ continues to roll toward 0, driving a second phase of inflation.

The first of these is, in many ways, even more natural than the simplest single-field models, relying on the potential

$$V(\phi, \sigma) = \frac{1}{2}m^2\phi^2 + V(\sigma), \qquad (8.115)$$

where $V(\sigma)$ is a potential giving a first-order transition. Notice that the σ field is completely uncoupled from the inflaton ϕ; because we certainly believe in symmetry breakings other than those directly connected to inflation, including this extra term requires almost no assumption whatsoever. The potential is illustrated in Figure 8.6. The scenario is that ϕ, as usual, starts well up its potential, and σ rests in its false vacuum state. At some point during the ϕ field's travels toward its minimum, the σ field tunnels, nucleating a bubble as in the BGT model. Inside the bubble, the inflaton continues to evolve, though at a different rate because the vacuum energy of the σ field has been removed, generating more inflation. If the appropriate amount of inflation occurs inside the bubble, the present density can again fall anywhere between 0 and the critical density.

This model is more complicated than the BGT model because of the second field. In particular, the inflaton field continues to evolve outside the bubble, albeit more slowly than inside, and this means that the bubble wall, forming the initial hypersurface for the open Universe, is not perfectly homogeneous. We shortly see that this is a fatal problem for this model. However, it can be solved in a variant, where, instead of the mass term, we include a $\phi^2\sigma^2$ coupling, *a la* hybrid inflation (Linde 1995). Then the potential in the false vacuum is completely flat (as $\psi = 0$), and the coupling term generates the slope down which the potential runs inside the bubble after tunnelling.

Yet another implementation has been given by Green and Liddle (1997a), who used a scalar–tensor theory in which the Brans–Dicke field has a potential, along with a trapped scalar field. While the scalar field is in the false vacuum, the Brans–Dicke field is held displaced from its minimum by its gravitational influence, giving a static solution that drives the first period of inflation. After tunnelling of the trapped field, the dilaton rolls to the minimum, giving a second period of inflation as it does so. Because the solution before tunnelling is static, in this model all bubbles produce the same value of Ω_0, unlike the model of the preceding paragraph.

Other models have been constructed by Amendola et al. (1996) and García-Bellido and Linde (1997).

Before ending this section, let us consider the possibility of generating closed Universes rather than open ones. This also has been discussed by Linde (1995). The trick detailed above for obtaining open Universes cannot be adapted to form closed Universes because bubble walls cannot be made to behave in the right way. However, Linde (1995) discusses a rather less attractive, but perhaps still viable, model in which a closed Universe can be obtained via tunnelling from nothing. The idea is that the tunnelling is supposed to be so suppressed that the nucleated closed Universe is already very symmetric (i.e., homogeneous), and then we again have a minimal amount of inflation designed to drive the Universe close, but not too close, to flatness at the present epoch. This strategy seems considerably more dubious than the open-Universe strategy, and it is not clear how it can be formed into a complete scenario (including, for example, a theory of the production of density perturbations). Given the lack of any observational support for closed Universes, we can hope that it is a road we shall not be forced down!

8.7.2   Perturbations in the open inflation model

As we already mentioned in Section 6.3, there are many new issues that have to be taken into account when considering perturbations in an open Universe, as well as the extra algebraic complexity of mode functions in a hyperbolic geometry. These do now all seem to have been solved, however, allowing detailed predictions for the various open inflation models. The perturbations we have to worry about are

(1) sub-curvature modes,
(2) super-curvature modes,
(3) perturbations to the sphericity of the bubble wall,
(4) gravitational-wave modes, and
(5) classical inhomogeneity from evolution outside the bubble,

Let us consider these in turn before analyzing their combined effect.

Sub-curvature modes

These are the only modes present in the flat-space limit. The original calculation of the spectrum of these modes by Lyth and Stewart (1990a) [see also Ratra and Peebles (1995)] assumed an initial vacuum state now known as the conformal vacuum, and concluded that $\mathcal{P}_{\mathcal{R}}$ still would be given by the standard formula (7.104). However, this vacuum state is not exactly that appropriate to the single-bubble models, and that formula receives a correction at very low

k, which, in the simplest models, takes the form of a $\coth(\pi q)$ multiplier, where $q^2 = k^2 - 1$ (Bucher et al. 1995; Sasaki et al. 1995; Yamamoto et al. 1996). In practice, the effect of this tends to be hidden within the cosmic variance (Section 5.2.2) of the large-angle anisotropies.

Super-curvature modes

As discussed in Chapter 6, the most general Gaussian random scalar field requires the inclusion of super-curvature modes, with $0 < k^2 < 1$ (Lyth and Woszczyna 1995). In general, there is no reason why there should not be a continuous spectrum of these, but it turns out that open inflation models either do not excite them at all, or only excite the modes with a single discrete k value (Bucher and Turok 1995; Yamamoto et al. 1996). This mode only appears if the mass of a scalar field satisfies $m^2 < 2H_{\text{false}}^2$, where H_{false} is the expansion rate in the false vacuum, and its wavenumber approaches $k = 0$ in the massless limit. This is exactly the type of mode that gives the open-Universe Grishchuk–Zel'dovich effect discussed in Chapter 6. It affects only the lowest few multipoles of the microwave background anisotropy.

Normally, its amplitude is negligibly small, the exception being when the vacuum energy is much smaller after tunnelling than before (Sasaki and Tanaka 1996), whereupon it gets enhanced by the ratio of the vacuum energies. In the model of Eq. (8.115), this enhancement may be large, but in models such as the Green–Liddle model, it is small (García-Bellido and Liddle 1997; Green and Liddle 1997a).

Bubble-wall modes

Although the nucleation of a perfectly spherical bubble has the smallest action and hence is most likely, neighbouring configurations may not be greatly suppressed, and so, the bubble may nucleate with asphericities (Hamazaki et al. 1996). The extent to which this happens is extremely dependent on the precise details of the tunnelling potential. Under idealized circumstances, the bubble-wall modes give rise to a scalar perturbation with a negative eigenvalue $k^2 = -3$ (Cohn 1996; García-Bellido 1996; Garriga 1996; Yamamoto et al. 1996), which can influence the lowest few multipoles of the microwave background anisotropy.

Scalar perturbations with this eigenvalue are actually equivalent to tensor modes (Garriga 1996; Yamamoto et al. 1996) because the usual scalar–vector–tensor decomposition fails (Stewart 1990; García-Bellido et al. 1997). A detailed analysis (Sasaki et al. 1997) shows that the bubble-wall modes in realistic circumstances should be interpreted as a collection of gravitational-wave modes with wavenumbers at the low wavenumber end of the sub-curvature spectrum. The cmb anisotropies from the bubble-wall fluctuations are reproduced by the long-wavelength gravitational wave modes, at least in the weak gravity limit (Tanaka and Sasaki 1997; Sasaki et al. 1997).

Gravitational-wave modes

The gravitational wave production is a subtle problem, finally solved by Sasaki et al. (1997) and Bucher and Cohn (1997). A comprehensive analysis has been given by Garriga et al. (1998). As we just said, the bubble-wall modes should be interpreted strictly as low-wavenumber gravitational-wave modes, though in fact the calculation of cmb anisotropies using the bubble-wall mode remains valid. The microwave anisotropy spectrum from the gravitational waves has been computed by Hu and White (1997b). There are no super-curvature gravitational-wave modes, which correspond to pure gauge modes (Tanaka and Sasaki 1997; Garriga et al. 1998).

Classical inhomogeneity

In models such as Eq. (8.115), the inflaton field is not static outside the bubble, but rather continues to roll. This breaks the homogeneity of the bubble wall, which can give a classical contribution to the inhomogeneities over and above the quantum ones we have so far discussed. This source of perturbations was estimated by Linde (1995), Linde and Mezhlumian (1995), and Sasaki and Tanaka (1996), and finally calculated precisely by Garriga and Mukhanov (1997).

Putting it all together

From all of the above, it is clear that the phenomenology of the open models is much more complicated than the usual ones. However, all the extra features only come in at very large scales, affecting the first few multipoles of the microwave anisotropy. For anything other than COBE then, we can ignore these subtleties. Indeed, it turns out that they are already negligible by the COBE pivot point at $\ell \simeq 15$, and so, even the COBE normalization is unaffected by the precise form of these extra modes (Liddle et al. 1996a; Yamamoto and Bunn 1996).

What is affected though is the shape of the spectrum at very low multipoles. Although the effect of cosmic variance is large, it still would be possible to see features above it were they sufficiently prominent, for example, a particularly strong Grishchuk–Zel'dovich effect (García-Bellido et al. 1995; Lyth and Woszczyna 1995).

The most interesting conflict is between the classical inhomogeneity and the super-curvature modes. We mentioned that the super-curvature modes are enhanced if the difference between the energy density before and after tunnelling is large. However, such a large difference is exactly what is needed to suppress the classical rolling of the field outside the bubble (by increasing the friction term in the scalar field equation). Linde and Mezhlumian (1995) noted that these needs may be conflicting in the model of Eq. (8.115) unless Ω_0 is uninterestingly close to 1, and this indeed has been confirmed through more precise calculations of each effect by Sasaki and Tanaka (1996) and Garriga and Mukhanov (1997). Particular open inflation models therefore can be predictive enough to find themselves in conflict with the data and be excluded. However, the other models we have mentioned remain viable at present, and cosmic variance will prevent the observational situation on the large-angle anisotropies from improving very much. On the other hand, the geometry of the Universe is very easy to measure with high-resolution microwave background observations, which may render this whole class of models uninteresting in the near future.

8.7.3   Open inflation from instantons

Just as we were completing this book, another idea for creating an open Universe was introduced, by Hawking and Turok (1998). Rather than tunnelling from an initial de Sitter phase, they propose a tunnelling from nothing, using ideas from quantum cosmology. The tunnelling is mediated by the Hawking–Turok instanton. Using the Hartle–Hawking no-boundary proposal, they conclude that the favoured tunnelling is to a Universe with the lowest possible density, that is, a completely empty Universe. Because that obviously is unacceptable, they then use anthropic-principle arguments to say that at least some galaxies must form. Unfortunately, this still gives a very low density Universe, $\Omega \sim 0.01$, in contradiction to observations.

Further, the instanton is unusual in being a singular one, albeit an integrable singularity giving a finite action and tunnelling rate. The interpretation of the instanton, and the ways in which it can be used, is highly controversial (Garriga 1998; Linde 1998; Turok and Hawking 1998; Vilenkin 1998).

Examples

8.1 For a nonhybrid model of inflation, the minimum of the potential corresponds to the true vacuum. Use this fact to determine the height V_0 of the potential, in terms of the mass and the self-coupling, on the assumption that m^2 is negative. What happens if it is positive?

8.2 In many models of inflation, the inflaton potential is of the form $V(\phi) = V_0 + \cdots$, with the constant term V_0 dominating. Because ϕ varies while cosmological scales leave the horizon, it is reasonable to impose the slow-roll condition $|V''| \ll V_0/M_{\text{Pl}}^2$ separately on each term of V. Use Eq. (8.1), and the COBE normalization (7.107), to show that, as a result, the coefficient of the term $\frac{1}{4}\lambda\phi^4$ can be, at most, of order 10^{-8}.

8.3 In the text, hybrid inflation with a quadratic potential is worked out for the usual case that m^2 is positive. Another possibility is to have "inverted hybrid inflation" in which m^2 is negative, and one way of achieving that is to reverse the signs of both m^2 and λ' in Eq. (8.40). Work out the prediction for n and the COBE normalization, in that case.

8.4 In the model of Eq. (8.115), estimate the ϕ value at which tunnelling must occur to give an open but significantly non-empty Universe.

9 The cosmic microwave background

The remainder of the main body of the book concerns observations and the ways in which they can be used to constrain the theoretical development we have made so far. The level of technical detail will be considerably less, because the observational situation doubtless will change. Roughly speaking, we shall be working from the largest scales to the smallest, beginning here with the cosmic microwave background (cmb).

9.1 Large angles and the COBE satellite

When we study cmb anisotropies on large angular scales, we are as close as one can get to directly studying the initial perturbations. In Section 2.4, we found that the Hubble length at the time of last scattering subtends an angle of around 1 deg on the last-scattering surface itself. On scales significantly larger than this, we are directly studying the effects of perturbations on scales greater than the Hubble length at the time of decoupling, which therefore have retained their primordial form.

The crucial observations in this regime are those of the Cosmic Background Explorer (COBE) satellite, taken over four years, which rightly can be said to have revolutionized cosmology. These observations provided, for the first time, an estimate of the spectrum of inhomogeneities in the Universe on very large scales, of order thousands of megaparsecs.

The COBE satellite carried three separate experiments. The Far Infra-Red Absolute Spectrometer (FIRAS; Mather et al. 1990) provided what is by far the most accurate measurement of the frequency spectrum of the microwave background, confirming it as a blackbody within experimental limits. This strongly supports the Hot Big Bang model, as other models achieving thermalization at this level of accuracy is problematic. The temperature of the microwave background is measured as $T = 2.728 \pm 0.004$ K at 95 percent confidence (Fixsen et al. 1996), where the dominant error is the uncertainty in the temperature of the calibrating blackbodies carried on the satellite.

As well as showing that the maximum deviation from the blackbody spectrum across a wide range of frequencies from the Rayleigh–Jeans (low-frequency) end of the spectrum over to the Wien (high-frequency) region is an impressively small 50 parts per million (Fixsen et al. 1996), FIRAS also puts very strong constraints on two particular types of distortion, known as the μ and y distortions. The former arises when there is an effective chemical potential, caused by energy injection into the radiation fluid at a time when scattering processes can put the photons into kinetic equilibrium, but photon-number changing interactions are too slow to establish full thermodynamic equilibrium. Energy injected between redshifts of about 10^7 and 10^4 satisfies this condition (Sunyaev and Zel'dovich 1970; Danese and de Zotti 1977; Daly 1991), and so, the FIRAS limits on μ strongly constrain energy injection at this time. The y-distortion arises when microwave photons are Compton scattered by hot ionized gas after last scattering, boosting photons from the low-energy part of the spectrum to the high-energy part. The FIRAS limits are (Fixsen et al. 1996)

$$\mu < 9 \times 10^{-5}, \tag{9.1}$$
$$y < 15 \times 10^{-6}, \tag{9.2}$$

both at 95 percent confidence.

The Diffuse Infra-Red Background Experiment (DIRBE) sought to detect the cosmic infrared background, against domination by foreground sources. Although many interesting results have been obtained, and they also are used in the analysis of the results of both other instruments, they do not impinge directly on the topics of this book and we shall not discuss them.

The anisotropy experiment was the Differential Microwave Radiometer (DMR) experiment. This measured the anisotropies in three frequency bands at 31.5, 53, and 90 GHz, with two channels at each frequency. Having three frequencies allows discrimination against foreground sources, primarily galactic dust and synchrotron emission, while the two independent channels at each frequency enabled an estimate of the instrumental noise. The DMR experiment was the longest running on the satellite, with results released after one year (Smoot et al. 1992), two years (Bennett et al. 1994), and then finally the complete data set for four years of observations (Bennett et al. 1996 and refs therein). Intrinsic anisotropies were already detected in the first year's data, announced in April 1992, and confirmed by the subsequent years' data.

The radiometers did not measure the temperature of the sky absolutely, but instead compared the temperatures of two regions of the sky 60 deg apart. The beam size was 7 deg, with an accurately Gaussian profile in the centre but slightly wider side lobes. Each point on the sky was compared with many points on the ring 60 deg away. In principle, we can attempt to compare models with the full time-ordered COBE data set, but this is an immensely cumbersome task and the principal tool of analysis has been sky maps made by the COBE team from the full data set, bad data points having been removed and several systematic effects (such as the influence of the Earth's magnetic field on the beam switching) having been accounted for.

The sky maps are generated by projecting the sky onto a cube, giving a total of 6,144 pixels. In practice, however, the pixels lying too close to the galactic plane (within about 20 deg) are totally dominated by foreground emission and contain no useful information. The early COBE data simply featured a latitude cutoff, but the final four-year data set utilizes a custom galaxy cut to remove further undesirable foreground emission features. This cut leaves final maps consisting of 3,881 pixels.

The reduction of the full data set to sky maps has been done in two different coordinate systems – galactic and ecliptic. Results based on these have been found to be slightly different, suggesting a systematic error introduced by the reduction of the full data set. However, this uncertainty is somewhat smaller than the statistical error.

The most prominent feature in the microwave sky is a dipole pattern, which has been known since the 1970s. It is a Doppler shift, caused by the motion of the Earth relative to the rest frame of the cmb. The observed dipole is best measured by FIRAS, which gives a diurnal average velocity (corresponding to that of the Sun) of $371 \pm 1\,\mathrm{km\,s^{-1}}$ in the direction $(l, b) = (264$ deg, 48 deg) (Fixsen et al. 1996). DMR gives a result in excellent agreement, with a velocity of $368 \pm 3\,\mathrm{km\,s^{-1}}$ in the same direction (Bennett et al. 1996; Lineweaver et al. 1996). The preferred way of expressing this is to shift to the frame moving with the Local Group, giving a velocity of $627 \pm 27\,\mathrm{km\,s^{-1}}$ in the direction $l = 276$ deg, $b = 30$ deg (Lineweaver 1997). This turns out to be more or less what we predict on the basis of the matter distribution in the local region, a topic to which we return in Chapter 10.

9.1.1   Fitting to the COBE DMR observations

In the earliest incarnation, the first-year data, the COBE data could quite satisfactorily be condensed to a single number, the one usually quoted being the fluctuation on a scale of 10 deg. The COBE data contained very limited information on the slope of the spectrum because they covered only a small range of scales. At that time, it also was regarded as perfectly satisfactory, when considering critical-density models, to compute the theoretical spectrum in a given model using only the Sachs–Wolfe effect, which made the comparison completely analytical for a power-law initial spectrum.

With the release of the second year of data, and then ultimately the complete four-year data set, this approach became inadequate on both counts. It became appropriate instead to fit to the full COBE maps (Górski 1994; Górski et al. 1994), comprising 3,881 pixels of data once the region contaminated by galactic emission has been masked out. To obtain the strongest constraints, it is also necessary to fit full theoretical spectra to the COBE data, including all contributions to the anisotropies, as the full COBE data set retain good enough signal-to-noise in the region where the Sachs–Wolfe plateau is predicted to rise toward the first acoustic peak. The fitting of full model spectra is also particularly important if we generalize the discussion to include low-density models, either with a cosmological constant or with an open geometry; as we have seen, in such models there are also line-of-sight contributions to the anisotropies, particularly in the latter case, caused by the time-varying gravitational potentials, which can affect even the largest angular scales.

The required sophistication of analysis has made it hard for individual researchers to utilize the raw COBE data themselves. However, with the COBE satellite's mission now over, we do not expect new data on the largest scales for some time to come, and anyway, at the lowest multipoles the COBE observations are now cosmic-variance limited rather than instrumental-noise limited. That means that the observations are about as good as it will ever be possible to make them, and it is possible to quote results that should be fairly definitive.

The amplitude of anisotropies seen by COBE can always be matched by selecting the appropriate normalization of the spectra that generate them. In addition, the four-year COBE data cover a large enough range of scales to give a modest amount of additional information concerning the shape of the cmb spectrum, which then can be used to constrain cosmological parameters. For example, if we simply used the Sachs–Wolfe formula for power-law initial spectra in a critical-density Universe, then the index of those spectra is constrained by the four-year data to $n = 1.2 \pm 0.3$ at 1-sigma (Bennett et al. 1996; Górski et al. 1996). If we compute the full spectra while staying with a critical-density Universe, then the constraint (which to our knowledge has yet to be computed in detail) would change slightly and pick up a dependence on h and Ω_B from the rise to the acoustic peak, which modestly intrudes on the COBE scales. To a first approximation (Bond 1995; Bunn et al. 1995), the constraint on n changes to $n = 1.05 \pm 0.3$ when the acoustic peak correction is taken into account.

A more dramatic change is to go to a low-density Universe, which changes the predicted low multipoles, giving spectra further from scale invariance at the low ℓ-values COBE samples; see Figures 6.6 and 6.7 on pages 147 and 150. Again, the full constraints on parameters from COBE have not been derived, though it has been noted (Górski et al. 1998) that, even at quite low densities, COBE is unable to exclude open models, at least as long as the $n = 1$ assumption is preserved. No one has yet examined limits on n as a function of Ω_0 from COBE.

It is with good reason that these constraints have not been fully calculated – already one can obtain stronger constraints on the cosmological parameters simply by combining the COBE data with observations of galaxies and galaxy clusters. Such observations probe far smaller scales than COBE, and therefore provide a powerful "lever arm." In the remainder of this book, we examine the ways in which this sort of analysis can be carried out. Ultimately, the most accurate constraints, on n and several other parameters, will come from satellite measurements of the small-scale cmb anisotropy, though an understanding of galaxies and clusters will remain important in its own right.

9.1.2 The COBE normalization

For models in which the shape of the cmb spectrum provides a satisfactory fit to the COBE data (which is pretty much any model of interest, as we commented earlier), we wish to use the COBE observations to fix the normalization of the spectra of perturbations. In particular, we desire the normalization of the matter spectrum because its amplitude is crucial for carrying out comparison with a whole range of other observational data. Because COBE is the most accurate single piece of observational data, it has become standard to choose models to be COBE normalized. For many purposes, indeed, COBE is so accurate that we can ignore the uncertainty in normalization, which at 1-sigma is under 10 percent.

To fit an inflationary model to the COBE data, we need to know both the spectral index n [from Eq. (7.117)] and the parameter r specifying the relative magnitude of any gravitational wave contribution [from Eq. (7.150)]. In this subsection, we closely follow an analysis of the COBE four-year data, as applied to inflation-based models, given by Bunn et al. (1996). This work takes advantage of the detailed fitting techniques devised by Bunn and White (1995, 1997). Many others (Bennett et al. 1996; Górski et al. 1996, 1998; Hinshaw et al. 1996; Wright et al. 1996) also have investigated the normalization from the four-year COBE data, though without considering the gravitational wave contribution, and the results are in excellent agreement.

The normalizations discussed here only apply to models producing a spectrum of adiabatic perturbations that can be approximated by a power law; no matter of principle stops any other inflation model, such as one also leading to an isocurvature component, from being accurately normalized, but once we leave the simplest paradigm, the different inflation models would have to be examined on a case-by-case basis.

We specify the normalization of the matter spectrum using the quantity δ_H defined by Eq. (5.7). This proves extremely useful, because of its time independence in the critical-density case (we have to be a little more careful when generalizing to low density), which makes the COBE normalization independent of h to an excellent approximation.[1] Because COBE is mostly seeing primordial perturbations, unaffected by evolution, which depends on the choice of matter content, the COBE normalization specified through δ_H will be independent of Ω_B and any details of the dark matter content. This would not be true of a normalization specified on a shorter scale, say by σ_8, because the transfer function depends on all these parameters.

Critical-density Universes

We begin by considering a critical-density Universe with no gravitational waves. In the case of a scale-invariant spectrum, δ_H is independent of k as well as of time, and so is simply a number. Detailed fitting to COBE gives this number as (Bunn et al. 1996; Bunn and White 1997)

$$\delta_H = 1.94 \times 10^{-5}, \tag{9.3}$$

with a 1-sigma statistical error bar of 7 percent (though we suggest later that this should be increased to 9 percent). This result corresponds to an expected quadrupole $Q_{rms} = 17.1 \pm 1.5 \, \mu K$, as can be verified using the Sachs–Wolfe formula alone, that expression being an excellent approximation for the quadrupole. This is somewhat lower than the expected quadrupole obtained in a fit of the pure Sachs–Wolfe spectrum (e.g., Bennett et al. 1996) because of the rise to the acoustic peak.

Generalizing this to a non-scale-invariant spectrum and to the case in which gravitational waves are allowed raises several issues. First, we must choose a scale on which to specify the normalization, since δ_H is now k-dependent. Obviously, the result is independent of how we choose to express it; for convenience, we specify the normalization at the present Hubble scale, defining

$$\delta_H \equiv \delta_H(a_0 H_0). \tag{9.4}$$

Much more important is the choice of scale at which we take the inflationary prediction to apply. We assume, as discussed in Sections 7.5 and 7.7, that the two spectra are adequately described by power laws, with the spectral indices n and n_T, plus the relative amplitude r, given by Eqs. (7.117), (7.149), and (7.150). For a general inflation model, these parameters are only approximately constant, and to make this approximation as good as possible, it is desirable to consider them to be evaluated at the centre point of the data rather than at one end.

---

[1] To be precise, the Sachs–Wolfe plateau is independent of h, but the details of the rise to the acoustic peak, which is of secondary importance in determining the COBE normalization, is not. However, if we adopt some 'average' rate of rise given by central parameter values, the change as the parameters are varied is not very significant.

The centre (or *pivot*) point is defined by where COBE-normalized spectra tend to cross; this can be found only after fitting of course. Bunn et al. (1996) found that the pivot point (at fixed r) is at around $k_* = 7a_0 H_0$. Consequently, n, n_T, and r should be evaluated there.

Finally, we learned in Section 7.7 that the spectra are not independent, but rather are related by Eq. (7.151), the consistency equation. We must be careful to include this. The consistency equation allows us to eliminate n_T in favour of r; the normalization will depend on n and r, or, equivalently, on the slow-roll parameters ϵ and η.

Carrying out the fitting, Bunn et al. (1996) found that it can be condensed into a single fitting function

$$\delta_H(n, r) = 1.91 \times 10^{-5} \frac{\exp[1.01(1 - n)]}{\sqrt{1 + 0.75r}}. \tag{9.5}$$

The fit error is less than 1.5 percent for $0.7 < n < 1.3$ and $r < 2$. The different coefficient from Eq. (9.3) simply reflects the fitting error, which is nearly maximal at $n = 1$. The change from varying the other cosmological parameters (except the density parameter, which we discuss shortly) is within 4 percent. There is also an error of 3 percent from the way the COBE maps are made, with different results being obtained depending on whether galactic or ecliptic coordinates are used. Combining all these in quadrature with the 7 percent observational error suggests that a realistic estimate of the uncertainty is 9 percent at 1-sigma.

The terms in Eq. (9.5) have a simple interpretation. The n term simply reflects the pivot point mentioned earlier; at fixed r the spectra cross at $k_{\text{pivot}} = e^{2.02} a_0 H_0$. This is actually slightly higher than mentioned earlier, but the tensors give more weight to the lower multipoles, making $k_* = 7a_0 H_0$ a good choice for evaluating the spectra when they are present.

In the absence of gravitational waves, we have

$$\delta_H(k_{\text{pivot}}) = 1.91 \times 10^{-5}, \tag{9.6}$$

independent of n.

The denominator in Eq. (9.5) gives the reduction due to the gravitational waves. The factor 0.75 may come as a surprise because r was specifically defined in Section 7.7 to try and make that factor unity. However, the definition of r was based on use of the Sachs–Wolfe approximation for the density perturbations[2] and on the assumption that the Universe was perfectly matter dominated at last scattering. Neither of these approximations is good enough for a really accurate calculation, and in fact, the rise to the acoustic peak increases the density perturbation contribution to the C_ℓ at the pivot point by 5 percent, whereas correcting the latter assumption lowers the gravitational wave contribution by 20 percent. In combination, these give the 25 percent correction to the coefficient. The correction acts to make the impact of the gravitational waves less than the naïve estimate.

Low-density flat Universes

The generalization of these results to low-density flat Universes, for which the same assumptions about the inflationary spectra are valid, also can be made. It turns out that a good-quality

---

[2] And for the gravitational waves, but it is exact there.

Table 9.1. *Summary of COBE normalizations*

Model	Scaling	Eq.
Critical density	Scale-invariant	(9.3)
	Tilted	(9.5)
	Tilt with tensors	(9.5)
Low density, flat		(9.8)
Low density, open		(9.9)

fitting function can be obtained simply by attaching an Ω_0 dependence to the formula above, but first we need some generalization of the treatment of gravitational waves. In the critical-density case, their relative impact at the pivot point was given by $0.75r$; this changes with low density. By direct evaluation of the C_ℓ spectra, Bunn et al. (1996) found the fitting function

$$f(\Omega_0) \equiv \frac{C_{14}^{T}}{C_{14}^{S}} \simeq 0.75 - 0.13\Omega_\Lambda^2 \tag{9.7}$$

for the coefficient. In the scale-invariant case, a good fit is obtained by inserting an Ω_0 dependence; to allow for tilt and tensors, additional cross terms are required, yielding the fitting function (Bunn et al. 1996)

$$\delta_{\mathrm{H}}(n, r, \Omega_0) = 1.91 \times 10^{-5} \frac{\exp\left[1.01(1 - n)\right]}{\sqrt{1 + f(\Omega_0)r}} \Omega_0^{-0.80 - 0.05 \ln \Omega_0}$$

$$\times \left[1 - 0.18(1 - n)\Omega_\Lambda - 0.03r\Omega_\Lambda\right], \tag{9.8}$$

valid for the same range of n and r, for $0.2 \le \Omega_0 \le 1$, and with the same 9 percent uncertainty.

Recall when using this result that, in our notation, $\delta_{\mathrm{H}}(k)$ is already the present-day spectrum, incorporating the growth suppression factor, with the present density spectrum on large scales related to the initial curvature perturbation via Eq. (6.15). Equation (5.17) remains the correct one to obtain the dispersion σ.

Open Universes

There is no simple generalization of these results to open Universes, because there is no longer a standard form for the spectrum, and hence the C_ℓ; in addition to the standard spectrum of so-called sub-curvature modes, we also may have super-curvature modes and bubble-wall modes, each of which is highly model dependent. Despite this, they tend to contribute primarily only to the lowest multipoles, and not to those multipoles near the pivot point to which the normalization is primarily sensitive. Consequently, although they can seriously affect the goodness of fit to the COBE observations, typically they do not much affect the normalization of models that pass the fitting test.

We can simply postulate a power-law spectrum of density perturbations, extrapolated suitably to large scales as discussed in Section 8.6.2, and apply the fitting procedure. This was

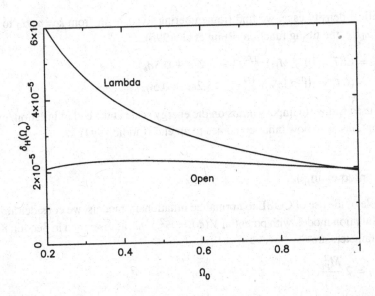

Fig. 9.1. COBE normalizations as a function of Ω_0, assuming $n = 1$ and $r = 0$. If the transfer functions have the same shape (e.g., CDM with the same Γ), then the relative normalization applies at all scales.

done by Bunn and White (1997), who obtained

$$\delta_H(n, \Omega_0) = 1.95 \times 10^{-5}\, \Omega_0^{-0.35-0.19\ln\Omega_0+0.17(1-n)}$$
$$\times \exp[(1-n) - 0.14(1-n)^2],\qquad(9.9)$$

with fit error 3 percent for $0.2 \leq \Omega_0 \leq 1$ and n as before. This can be taken as the result for either the "conformal" or "single-bubble" vacuum state; Yamamoto and Bunn (1996) demonstrated that, indeed, the effect of the presence or absence of super-curvature modes, plus the difference in long-wavelength behaviour of the sub-curvature modes, makes a negligible change to the COBE normalization. No open Universe result including gravitational waves is presently available.

Table 9.1 summarizes the relevant equations for the COBE normalization in different cosmological models, and Figure 9.1 shows the scaling with Ω_0.

9.1.3 Inflationary energy scale from COBE

By fixing the large-scale amplitude of density perturbations, COBE also fixes the energy scale of inflation at the time that those perturbations were generated, in terms of the slow-roll parameters. To evaluate this, we use the second-order expression for the density perturbation spectrum, Eq. (7.139).[3] We then can invert the expression for the COBE normalization, to obtain the potential V_* at the pivot point.

---

[3] For the COBE normalization, we needed only the first-order expression because the correction is almost independent of scale and so cancels out when we carry out the fitting to COBE.

In the critical-density case, we find (remembering to scale δ_H from $k = a_0 H_0$ to the pivot point $k = 7a_0 H_0$) the fitting function (Bunn et al. 1996)

$$V_*^{1/4} = (2.7 \times 10^{-2} \, M_{Pl}) \, \epsilon_*^{1/4} \, (1 - 3.2\epsilon_* + 0.5\eta_*)$$
$$= (6.6 \times 10^{16} \, \text{GeV}) \, \epsilon_*^{1/4} \, (1 - 3.2\epsilon_* + 0.5\eta_*). \tag{9.10}$$

This can be used further to impose limits on the energy scale at the end of inflation, by making some assumptions as to how inflation comes to an end (Liddle 1994).

9.1.4   A worked example

As an example of the use of COBE to normalize inflationary models, we consider the simplest single-field inflation model, with potential $V(\phi) = m^2\phi^2/2$, as discussed in Section 8.2.1. The slow-roll parameters are

$$\epsilon = \eta = 2\frac{M_{Pl}^2}{\phi^2}, \tag{9.11}$$

giving $\phi_{end} \simeq \sqrt{2}M_{Pl}$. Taking, for simplicity, the pivot point to correspond to $N_* = 50$, Eq. (3.17) gives $\phi_* = 14M_{Pl}$, from which Eqs. (7.117) and (7.150) give

$$n = 0.96, \qquad n_T = -0.02, \qquad r = 0.13. \tag{9.12}$$

The relevant fitting function, Eq. (9.5), yields

$$\delta_H = 1.90 \times 10^{-5}. \tag{9.13}$$

It seems that this model is very close to the scale-invariant limit. However, the short-scale observations that we discuss in the following chapters are at scales far enough removed for the tilt to make its presence felt. For example, if we evaluate the dispersion at $8h^{-1}$ Mpc, we find that, in comparison to a scale-invariant spectrum without gravitational waves, it is reduced by 13 percent, where we can interpret 8 percent as being due to the tilt and 5 percent as due to the gravitational waves. This difference is large enough to be significant, and many inflationary models give a larger effect.

Finally, the inflationary energy scale is

$$V_*^{1/4} = 8.6 \times 10^{-3} \, M_{Pl} = 2.1 \times 10^{16} \, \text{GeV}, \tag{9.14}$$

which fixes $m = 7.5 \times 10^{-6} \, M_{Pl} = 1.2 \times 10^{13} \, \text{GeV}$.

9.2   Degree-scale observations and acoustic oscillations

As a very young topic, with detections only having been made over the previous few years, the study of microwave anisotropies on degree scales is in rapid flux. Several different observational strategies are in play, the main goal being to minimize atmospheric noise. Ground-based experiments operate on sites of high altitude and extreme cold, such as the South Pole or northern Canada, where there is less water vapour in the atmosphere. An alternative strategy is to mount

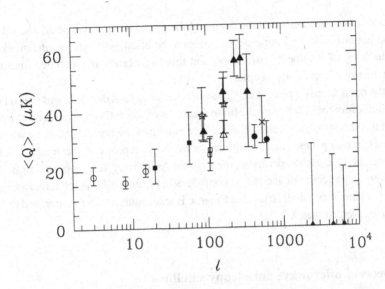

Fig. 9.2. A replot of Figure 5.9, showing a compilation of cmb data, as of early 1998 (data courtesy Martin White). The points correspond to various experiments as follows, where necessary reading left to right: The detections (1-sigma errors) are COBE (3 open circles); Tenerife and BAM (solid squares); Python (2 stars); Saskatoon (5 solid triangles); ARGO (2 open squares); MAX (open triangle), CAT (2 solid circles), and OVRO (cross). The 95 percent confidence upper limits are SuZIE, ACTA, and Ryle.

experiments on balloons, to attempt to get as far above the atmosphere as possible. Ultimately, it will be satellites that achieve the definitive results on these scales. We are in the fortunate position as we write this that two satellite proposals have been approved: NASA has funded the Microwave Anisotropy Probe (MAP) and the European Space Agency has funded Planck.

The aim of degree-scale observations is to study the first acoustic peak. A large number of experiments have detected anisotropies, largely concentrating around the degree scale but also probing smaller scales. A compilation, plotting $\langle Q \rangle^2 = \ell(\ell+1)C_\ell T_0^2/4\pi$, is shown in Figure 9.2 (a reproduction of Figure 5.9 on p. 125), though the field is moving so rapidly that it is doubtless out of date as you read this. There are already strong indications from completed experiments that such a peak does exist, though its height (or, more precisely, the rate of rise) is presently only very weakly determined and its location not at all. Unlike the Sachs–Wolfe plateau, the acoustic peak is sensitive to many cosmological parameters as well as the initial forms of the spectra. In particular, its location in ℓ-space is a powerful probe of the geometry of the Universe; in flat Universes, with or without a cosmological constant, the Hubble radius at last scattering subtends more or less the same angle, whereas in open Universes it has moved to smaller angular scales, approximately as $\Omega_0^{1/2}$ (see the discussion in Section 2.4). In terms of ℓ, in a flat Universe, one expects the peak to be located at $\ell \simeq 220$, whereas in an open $\Omega_0 = 0.3$ Universe it should be located at $\ell \simeq 500$.

Whereas the peak location depends only on geometry, its height depends on practically all the cosmological parameters. For a time, this was thought to lead to strong degeneracies, dubbed *cosmic confusion* (Bond et al. 1994), but it is now acknowledged that observations can be made with such astonishing accuracy that the details of the rise to the acoustic peak, and subsequent behaviour at larger ℓ, allow all these degeneracies to be lifted, with one significant exception. We study the issue of parameter estimation, and this degeneracy, in a detailed discussion on satellite experiments later in this chapter.

Beyond the main acoustic peak at degree scales lie further oscillations, and several of these may be visible before the cutoff from the finite thickness of the last-scattering surface intervenes to damp out the anisotropies. Ground-based observations of the types discussed earlier are not so well suited to probing these smaller angular scales; the most promising technology of that sort is interferometers. One such experiment, the Cosmic Anisotropy Telescope at Cambridge, has already proven very successful and can be considered a prototype for much larger experiments to come, such as the Very Small Array, the Cosmic Background Interferometer, and the Degree Angular-Scale Interferometer.

9.3   Aspects of microwave anisotropy satellites

The MAP satellite probes well beyond the first acoustic peak; its beam size is such that noise becomes dominant around the location of the fourth acoustic peak for a flat Universe, or the second one in an open Universe with $\Omega_0 = 0.3$ (see Figure 6.7 on p. 150). Planck will be extremely effective across several of the acoustic oscillations and far down the damping tail. Accurate measurements of these oscillations is crucial for obtaining the most accurate information about cosmological parameters, and even fairly crude measures such as their relative spacing can yield very strong information concerning the nature of the initial perturbations (Hu and White 1996b).

There are two main obstacles to obtaining detailed information. The first is whether emission from noncosmological (foreground) sources is strong enough to hide the cosmological information. The second is how to handle the extremely large correlated data sets that the satellites will generate. At one time, there was a third worry, which was that if the Universe reionized early enough, the liberated electrons might rescatter a large fraction of the microwave photons, erasing much of the information we want. This never seemed likely on the basis of extrapolation from the spectrum of the density perturbations downward in scale from what we observe (Tegmark et al. 1994; Liddle and Lyth 1995), and recent observations (see Figure. 9.2) showing evidence of a strong signal at the degree scales constrain this possibility directly. It looks unlikely to be a problem, but we discuss it briefly in Chapter 11.

The strategy then, illustrated schematically in Figure 9.3, is the following: Observations are made of the sky at a range of frequencies over time. MAP measures differential temperatures on the sky, whereas Planck measures absolute temperatures. The time-ordered data set is reduced to maps of the sky at each of the frequencies. These maps each contain a combination of genuine cmb signal and contamination from external foregrounds and instrument noise. From them, a map of the cmb anisotropy is to be extracted. This contains a lot of information, but if the perturbations are Gaussian, the information can be compressed, without loss, into the spectrum alone. Finally, from the spectrum, we can estimate the various cosmological parameters.

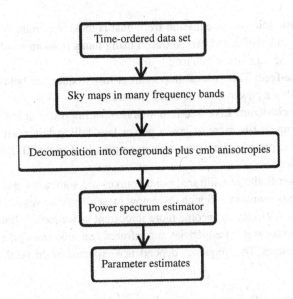

Fig. 9.3. Flowchart of the path from the original data set to estimates of the inflationary and cosmological parameters.

9.3.1 Foregrounds

There are several possible contaminants that might, in principle, degrade our ability to measure the cosmological signal. Fortunately, there exist a range of frequencies (from around 30 GHz to a little over 100 GHz) over which all the possible contaminants are subdominant to the expected cosmological signal. More pertinently, they typically have a very different spectral shape to the microwave background. The appropriate way to deal with them is therefore to have many frequency channels, allowing one to fit out several components of noise against the cosmological signal. Planck will have ten frequency channels extending all the way up to about 900 GHz, whereas MAP has five channels up to 90 GHz; Planck's extra range arises from the combination of bolometer and HEMT detectors, whereas MAP carries HEMT detectors alone. The lowest and highest frequencies are aimed at providing accurate estimates of the different foreground components, which then can be removed from the central frequency region where the cmb is dominant. In regions where the foregrounds are particularly strong (in particular, the galactic plane), it is likely to be preferable simply to mask out (i.e., ignore) those regions, as was done for COBE.

Although it is believed that foregrounds will not be an obstacle to extremely accurate determination of the cmb anisotropies, this statement does depend on some modelling, because available surveys are located at frequencies some distance away from the microwave region and extrapolation is required to estimate the significance in the frequency band of interest. A detailed analysis has been made by Tegmark and Efstathiou (1996), who studied not just the frequency dependence but also the angular dependence of several components:

- **Galactic dust:** Microwave emission from dust is most important at high frequencies of 200 GHz and above, and can be traced using maps from an all-sky survey carried out by the IRAS satellite at 100 μm.
- **Galactic free-free:** This is most important at low frequencies below about 30 GHz, and has roughly a power-law frequency spectrum.
- **Galactic synchrotron:** Like free-free, this is most important at low frequencies, and has a spectrum that is more or less a power law, falling faster with frequency than free-free and expected to be comparable in amplitude at about 30 GHz. Synchrotron is highly polarized.
- **Point sources:** Radio and infrared point sources are somewhat less well understood than the galactic emission, though there now exist all-sky surveys in both wavebands. Point sources are likely to become more important as we look to higher ℓ, with radio sources important at low frequencies and infrared emission from galaxies important at higher frequencies. The frequency dependence expected of the point sources is rather uncertain.

Tegmark and Efstathiou (1996) assume that all three galactic components have a coherence scale of order 10 deg. If this is correct, they give the strongest interference at $\ell \sim 10$. At the low-frequency end, the point sources become a more important contaminant from $\ell \sim 100$. However, for the entire ℓ range of Planck, there is a range of frequencies over which the cmb is expected to be the largest contributor.

With the wide frequency range available, subtraction of the various foregrounds should be very effective. Tegmark and Efstathiou (1996) conclude that the uncertainties on the estimate of C_ℓ caused by foregrounds should be at worst a few percent, and at the first acoustic peak, where the behaviour of the foregrounds is particularly favourable, they conclude that "even power accuracies of 1 percent would be an extremely pessimistic prediction."

There are some other effects that can amend or create anisotropies, which must be considered. The main two are gravitational lensing and the Sunyaev–Zel'dovich effect. Gravitational lensing acts to slightly smear out the acoustic oscillations (Kashlinsky 1988; Cole and Efstathiou 1989; Seljak 1996); it preserves the total power, but shifts it slightly from the peaks of the spectrum into the troughs. This effect is second order in the photon bending angle (Seljak 1996).[4] According to Zaldarriaga et al. (1997), it does not significantly reduce the ability to measure cosmological parameters. The Sunyaev–Zel'dovich effect (Sunyaev and Zel'dovich 1972, 1980; Rephaeli 1995) arises from the scattering of cmb photons off hot gas within clusters. It leads to a decrement in intensity on the low-frequency side and an increase on the high-frequency side of the blackbody distribution. In those specific directions, it can be quite a significant effect, and two-dimensional maps have been made (Jones et al. 1993; Myers et al. 1997), but because clusters subtend a small part of the total solid angle, the overall effect on the spectrum is small. The large frequency range of Planck is designed to allow very accurate measurements of the Sunyaev–Zel'dovich effect in many clusters, which

---

[4] Gravitational lensing may be observed more easily in cross correlations with the galaxy distribution (Suginohara et al. 1998), where the effect is first order in the photon bending angle for some observables.

promises many interesting scientific results. That the effect will be mapped in such detail ensures that it will not interfere with the determination of the cmb anisotropy.

9.3.2 Data compression

The full time-ordered data sets from the new cmb satellites will be far too vast to be analyzed directly, and significant compression of the data will be required. Fortunately, much progress has already been made for the case in which the perturbations are assumed to have Gaussian statistics. The first stage is the reduction of the time-ordered data set to a cmb sky map (Wright 1996a; Tegmark 1997a). For COBE, the bulk of the analysis has been done directly on the sky maps produced by the COBE team.

The standard methods for analyzing the COBE data involve likelihood analyses, perhaps using the Karhunen–Loève data compression technique (e.g., Bond 1995; Bunn and Sugiyama 1995; Bunn and White 1995, 1997; Tegmark et al. 1997b). For each model, this requires inversion of a large matrix, which is computationally time-consuming. The most simplistic "brute force" method (Tegmark and Bunn 1995) inverts a matrix of dimension $N_{\text{pixel}} \times N_{\text{pixel}}$, which is computationally just feasible for the 4,000 or so pixels in the COBE data set. Even so, with the large range of different parameter values that we might wish to explore, carrying out a separate inversion for each set of model parameters is an imposing task.

For Planck, the sky maps will have order 10^7 pixels, and thus standard matrix inversion is no longer feasible. Fortunately, if we assume that the underlying spectrum is Gaussian (something that can be tested for in a number of ways), then considerable compression is possible without throwing away any information, because we know that all the necessary information can be encoded in the spectrum. Tegmark (1997b) has shown that if the right spectrum estimator is chosen, the data compression can indeed be "lossless," in the sense that the error bars on determined parameters are as small as it is possible to obtain from the full data set. We can arrange for the error bars on neighbouring points of the spectrum to be uncorrelated (Hamilton 1997; Tegmark and Hamilton 1997). Devising computationally feasible data analysis strategies is an ongoing process (Bond et al. 1998; Borrill 1999; Oh et al. 1999).

9.3.3 Polarization

So far in this discussion, we have only mentioned the temperature anisotropy, but there is an additional source of information from the cmb, which is the polarization of the radiation, as discussed in Section 5.3. In addition to the temperature, we need two Stokes parameters to describe the polarization (in general, there would be one more to describe circular polarization, but it is not excited by Thomson scattering and so can be set to zero). The polarization turns out to be very sensitive to the reionization history, and if it can be measured it can be used to probe it. Both MAP and Planck will attempt to measure polarization.

In Chapter 5, we described the all-sky formalism for studying polarization (Hu and White 1997c; Kamionkowski et al. 1997a,b; Seljak and Zaldarriaga 1997; Zaldarriaga and Seljak 1997). In absolute generality, there are four spectra that can be determined: the temperature, the

E and B polarization, and the cross correlation between the temperature and the E polarization. Other cross correlations vanish (assuming parity invariance). The B polarization directly probes gravitational waves because it is not induced by density perturbations (Seljak 1997), but the anticipated signal is so small as to make it an extremely challenging observation (see, e.g., Hu and White 1997d). Nevertheless, polarization helps to test whether or not a component of the large-angle anisotropies originates from tensor perturbations, as the temperature–polarization cross correlation, which has a much higher amplitude, probes the reionization history and can lift the degeneracy between that and the tensor contribution that exists when we look at the temperature anisotropy alone (Hu and White 1997d; Zaldarriaga et al. 1997). Its observation significantly improves the determination of *all* the parameters (see Table 9.2) through this degeneracy breaking. A pedagogical discussion of polarization, including issues of foregrounds and detector sensitivities, has been given by Hu and White (1997d).

The HEMT detection technology of MAP and of the lower-frequency Planck channels automatically measures photons of one polarization, but, of course, two detectors are needed at each frequency to actually measure the polarization. The bolometers of Planck's high-frequency instrument measure the full intensity of incident light; to measure polarization in those channels, it is likely that one will have to throw away the "wrongly polarized" photons, so that the gain in polarization information is accompanied by a loss in temperature anisotropy information. The question of the (largely unknown) polarization of foregrounds is the crucial one; unlike the temperature anisotropy, the polarization anisotropy may be buried well below a contaminating foreground signal. The high polarization of synchrotron suggests that low-frequency detectors will find things difficult, but the dust foregrounds that intrude at the high frequencies of Planck may have only a small polarization (guesses are a few percent), which suggests that they will not be a problem. Theoretical calculations suggest the gain of measuring polarization with Planck will outweigh the loss (especially because, by then, other observations may have tied down the temperature spectrum rather well).

9.3.4   Parameter estimation

The crucial role of cmb experiments as far as inflationary cosmology is concerned is parameter estimation. Once we go to angular scales below the Sachs–Wolfe plateau, the predicted C_ℓ curve depends sensitively on a large number of parameters (Sunyaev and Zel'dovich 1970; Bond et al. 1994), and contains sufficient structure to allow many of them to be determined to unprecedented accuracy.

We have seen that structure formation in inflationary cosmologies typically leaves between about five and twelve parameters to be fixed by comparison to the observational data, namely,

> **Inflationary parameters:** $\delta_H, n, r, n_T, dn/d\ln k, \ldots,$
> **Cosmological parameters:** $h, \Omega_b, \Omega_0, \Omega_\Lambda, \Omega_{HDM}, g_*, \kappa, \ldots.$

Of course, one would be very unlucky if all of these were relevant. On the inflationary side, as an absolute minimum, we would consider a power-law spectrum of density perturbations, the first two parameters. The general result to lowest order in slow roll also would include gravitational waves, introducing r, though in most models this is expected to be negligible. At that order, n_T is not independent, being related to r by the consistency relation (7.151). However, we might wish to fit for r and n_T separately to try and test the consistency relation, though this seems unlikely to be feasible (see, e.g., Lidsey et al. 1997). Finally, we can examine departures from the power-law approximation by considering derivatives such as $dn/d \ln k$. In fact, it would be extremely interesting for inflation if these are significant, because adding another parameter should not significantly harm the overall fitting (because there are so many parameters anyway) and, if it can be determined, then it corresponds to additional information learned about inflation and hence about high-energy physics (Copeland et al. 1998).

On the cosmological side, as a minimum, h and Ω_b must be considered, the C_ℓ spectrum being quite sensitive to both. Especially if polarization is being measured, then the optical depth κ (equivalently, the reionization redshift z_{ion}) needs to be included. The remaining parameters could be fixed by assumption, for example, a critical-density, cold dark matter (CDM) Universe. Better though is first to fit for them, and then, if they prove consistent with a simpler hypothesis, the parameters can be refited under that assumption. The parameter g_* represents the number of relativistic species (which determines the epoch of matter–radiation equality), which can be assumed to have the Standard Model value ($g_* = 3.36$ at the photon temperature) or can be varied (Section 6.1.2). Other parameters that we might add are the helium mass fraction in the Universe and the cmb temperature. Finally, if reionization is significant, it may be necessary to model it in a more sophisticated way than just as a single redshift of instantaneous reionization.

In conclusion, the minimum set of parameters we would envisage is δ_H, n, h, Ω_b, and κ, and the maximum is all of the above and more.

The first crucial ingredient in parameter estimation is accurate theoretical predictions for the spectra as a function of the parameters. Fortunately, CMBFAST (Seljak and Zaldarriaga 1996) is publicly available and can predict the C_ℓ to within 1 percent. The vital information is how the C_ℓ curve varies as a function of parameters, which leads to a descriptor known as the Fisher information matrix, the inverse of which, the covariance matrix, leads to estimates of the errors on the different parameters and the correlations between them. The derivatives depend on the fiducial model chosen, and so, the estimates of the accuracy of parameter determination depend on the underlying model assumed. Estimates have been given for the Standard Cold Dark Matter (SCDM) model and also for some others – the variation actually is not huge.

The first attempts at estimating the accuracy of parameter estimation were made by Knox (1995), who considered only the inflationary parameters and also derived an important equation allowing one to deal with limited sky coverage. Interest was greatly stimulated by Jungman et al. (1996c), who looked at a large set of cosmological and inflationary parameters (having earlier [Jungman et al. 1996b] considered the density parameter alone). They considered two scenarios: one in which they assume that all but one parameter is known, and the more realistic situation in which they attempt to simultaneously fit for all the parameters. Their analysis was based on semianalytic approximation to the C_ℓ spectrum as a function of parameters, and a more accurate estimate of parameter errors requires the use of full numerical calculations to compute the derivatives with respect to parameters. This has been done by Bond et al. (1997) and Zaldarriaga et al. (1997), the latter including an estimate of the effect of gravitational lensing and also, more importantly, the inclusion of polarization.

Table 9.2. *Estimated parameter uncertainties from Planck, from Zaldarriaga et al. (1997), both with and without polarization.*

	Uncertainty	
Parameter	Without Polarization	With Polarization
Δh	0.006	0.004
$\Delta(\Omega_b h^2)$	0.0001	0.00008
$\Delta\Omega_\Lambda$	0.04	0.02
$\Delta\kappa$	0.09	0.006
$\Delta \ln \delta_H$	0.08	0.01
Δn	0.007	0.003
Δr	0.26	0.05
Δn_T	0.97	0.18

Note: These values assume a "correct" model, which is a tilted CDM model with $n = 0.98$ and $r = 0.1$, and an optical depth κ of 0.05. This optical depth is only "detected" if polarization is measured.

Many parameters are expected to be determined to exquisite accuracy. Bond et al. (1997) vary nine parameters about the SCDM model, and conclude that five of nine orthogonal combinations can be determined to a relative accuracy of 0.01 by Planck, with two more with 0.1. In practice, these are not completely reflected in the uncertainty of parameters, because each parameter is a linear sum of these "optimal" linear combinations, including some more poorly determined ones as well as the better determined ones. In Table 9.2, we quote from Zaldarriaga et al. (1997) the estimated errors from Planck on various parameters, each obtained by marginalizing over the other parameters, both with and without polarization. Because the size of the errors depends on the derivatives of the spectrum about a fiducial model, some assumption must be made – the results we show are for a tilted CDM model. Their paper shows error estimates for several other cosmologies.

Note the dramatic improvement in the determinations of κ, δ_H, and r when polarization is included, due to its vital role in breaking the degeneracy between them. If we were able to assume no gravitational waves, the uncertainty on κ and δ_H would drop further.

There is one significant degeneracy that cannot be lifted using the microwave background. Models with the same physical densities at last scattering, and with the same angular diameter distance to the last-scattering surface, give identical C_ℓ spectra (apart from the integrated Sachs–Wolfe terms on large scales and gravitational lensing effects on small scales, both of which are small corrections). This gives a degeneracy in the Ω_0–Ω_Λ plane, which cannot be broken by the cmb alone. Fortunately, including other data such as the magnitude–redshift relation for type Ia supernovae can break the degeneracy (Tegmark et al. 1998). Note also that the simplest models of inflation give a flat geometry, and we should restrict the parameters to the flat spatial geometry before trying to constrain inflationary parameters, which partly alleviates the degeneracy.

With the exception of this degeneracy, the accuracy attainable from the cmb is superior to that which can be achieved by any other means, though obviously the best results combine all the available data. The scientific results from MAP and Planck promise to be spectacular.

Examples

9.1 Equation (9.8) gives the COBE normalization for flat cosmologies. For $n = 1$ and $r = 0$, compare this result with what you would obtain from the approximate argument at the end of Section 6.2.2.

9.2 Using Eq. (9.9), compute σ_8 for some sample values of Ω_0 for COBE-normalized open CDM models with $\Gamma = 0.25$ and $n = 1$.

9.3 Consider Eq. (9.10), ignoring the higher-order corrections. Assuming that ϵ_H does not decrease after cosmologically interesting scales leave the horizon, limit $V_{end}^{1/4}$.

9.4 Compute the COBE normalization and the inflationary energy scale for the $\lambda \phi^4$ potential, assuming a critical-density model.

10 Galaxy motions and clustering

10.1 The clustering of galaxies

10.1.1 Overview

The developing understanding of the galaxy correlation function has played a pivotal role in guiding research in large-scale structure for decades. Early work involved the cataloguing of galaxy positions on the sky, which revealed a complex pattern of clustering, which motivated the early work on structure formation. Obtaining galaxy positions is a relatively easy observational task, and this strategy has continued to be popular; for example, the APM galaxy survey (Maddox et al. 1990, 1991) compiled using an automatic plate measuring device, played a crucial role in the downfall of the Standard Cold Dark Matter (SCDM) model. A positional catalogue is also an important part of the forthcoming Sloan Digital Sky Survey (SDSS; Loveday and Pier 1998; Margon 1998).

Such catalogues contain millions of galaxies, which accounts for their power in testing cosmological models. Also, despite their two-dimensional nature, they can be used to say something about the clustering of galaxies in three dimensions, provided that we make the normal assumption of statistical isotropy of the clustering pattern and the plausible assumption that the probability distribution for the brightness of a galaxy is independent of its position (to prevent the catalogue from becoming distorted near its magnitude limits). Then the angular correlation function can be deconvolved to give the full three-dimensional correlation function, via an equation known as Limber's equation. However, the angular correlation function cannot be used to say anything about the actual clustering pattern, only about its statistical properties, and even for them the numerical inversion problem is a difficult one.

To obtain full three-dimensional information, we require an estimate of the distance to a galaxy, which is most simply achieved by measuring its redshift. Observationally, this is much more time consuming than simply measuring positions, and, as we write, the number of galaxies with redshifts is of order 10^5 rather than in the millions. More important, these come from a wide variety of surveys with different survey strategies concerning depth and completeness, originating in addition from parent catalogues of different types (e.g., optical, infrared). This prevents them from being reliably combined to permit statistical analysis.

This will change with the appearance of large-scale systematic redshift surveys, well under way as we write. The 2dF galaxy redshift survey uses the 2dF (2-degree field) fiber spectrograph on the Anglo-Australian Telescope to obtain several hundred redshifts simultaneously, aiming at a total of 250,000 in the region of the sky of the APM survey. The SDSS uses a dedicated telescope aiming to redshift a million galaxies, covering π steradians to a depth of several hundred megaparsecs, selected from the parent positional survey mentioned earlier.

In the earlier literature, attention normally was focused on determining the two-point correlation function. We have seen already in Eq. (4.19) the definition of this for a continuous field such as the matter-density distribution. For a discrete field, such as a collection of galaxy locations, the definition must be phrased more carefully. The galaxy two-point correlation function ξ_{gal} can be defined in terms of the (joint) probability of finding two galaxies within small volumes dV_1 and dV_2 a distance r apart:

$$\text{Prob} = n^2[1 + \xi_{gal}(r)] \, dV_1 \, dV_2 \,, \tag{10.1}$$

where n is the mean galaxy number density. For a random (i.e., Poisson) distribution, $\xi_{gal}(r)$ is clearly 0, by definition of the mean density. In the limit of a continuous field, this definition coincides with that given earlier. Occasionally, quantities derived from the correlation function, such as the moment

$$J_3(R) = \int_0^R \xi_{gal}(r) r^2 \, dr \,, \tag{10.2}$$

are measured instead, the integral nature of such statistics, it is to be hoped, offering some control over statistical noise. Indeed, at one time the quantity $J_3(10h^{-1} \, \text{Mpc})$ was used as a possible normalization of theoretical models for the galaxy spectrum, as a rival to $\sigma(8h^{-1} \, \text{Mpc})$, which often was used to normalize the matter spectrum before the advent of the Cosmic Background Explorer (COBE).

Recently, however, it has been more fashionable to attempt to derive the spectrum of the galaxy distribution, rather than the correlation function and related quantities. Although $\xi(r)$ and $\mathcal{P}(k)$ are equally easy to compute for a theoretical model, being related by a Fourier transform, the latter has advantages for determination from a catalogue. Primarily, these are in the determination of the large-scale behaviour, given errors induced by the uncertainty in the underlying number density. This uncertainty affects the correlation function uniformly at large r, and the oscillations that we might hope to see can be completely subdominant. For the spectrum, this uncertainty only affects the modes very close to $k = 0$. A second advantage is that error estimation is easier for the spectrum [where analytical formulae can be applied (e.g., Feldman et al. 1994)]. Finally, the spectrum has the nice property of being everywhere positive; in correlation-function language, this constraint is manifested rather untidily as an infinite number of integral constraints.

10.1.2 Biased galaxy formation

Although galaxy catalogues in two and three dimensions form one of the most copious data sets in cosmology, they have not, as we write, led to extremely powerful constraints, though there is every reason to believe that the new large surveys mentioned earlier will change that. The principal reason for this is that there is no *a priori* reason why the galaxy distribution should be a good tracer of the mass distribution in the Universe. Indeed, observations show that it definitely cannot be; the correlation functions for, to give an example, galaxies selected optically and galaxies selected in the infrared are different and hence clearly cannot both trace the mass distribution accurately. This effect is known as *bias* in the galaxy distribution, and it seriously impairs our abilities to use it to constrain the matter spectrum.

The detailed problem of the formation of galaxies from initial density perturbations has proven a difficult one, and only now is beginning to be seriously addressed, through detailed analytical and semianalytical modelling [White (1994), and references therein; Kauffmann et al. (1997)] and through the enhancement of N-body simulations to include hydrodynamics (Cen and Ostriker 1992a,b; 1996; Katz et al. 1992). However, it is possible to develop some understanding of biasing without a detailed modelling of the galaxy formation process.

The key idea is that of peak biasing, introduced by Kaiser (1984). In fact, he introduced the idea in order to explain the clustering of galaxy clusters, but the same idea can be considered in the context of galaxy clustering. The assumption is that galaxies are fairly rare objects, forming from peaks in the matter distribution (smoothed on the appropriate scale). For a Gaussian random field, it is clear that the probability of a peak being near another peak is enhanced. The rarer the peaks, the stronger this effect. Extremely detailed analyses of this issue have appeared (Peacock and Heavens 1985; Bardeen et al. 1986). The conclusion is that, in this simple model, the correlation function of the galaxies will be enhanced relative to the mass by a scale-independent factor b, known as the *bias parameter*.

In fact, the literature contains at least three different usages of the bias parameter, which are not equivalent to one another, and we must always be careful as to which is being used. The simplest is

$$b \equiv \frac{\sigma_8^{\text{gal}}}{\sigma_8},$$

(10.3)

where $\sigma_8 \equiv \sigma(8h^{-1}\,\text{Mpc})$. Because the right-hand side is just a number, clearly this bias parameter is well defined. Historically, this definition often was used to normalize the theoretical matter spectrum because the variance of the galaxy distribution was observed to be about unity at $8h^{-1}$ Mpc. Therefore, it was demanded that $\sigma_8 = 1/b$, where b is more or less a free parameter. In this context, b invariably refers to the clustering of optically identified galaxies. This method of normalization has been almost totally eclipsed by the COBE normalization, which refers directly to the matter spectrum.

The second definition refers to the relative size of the galaxy and matter correlation functions

$$\xi_{\text{gal}}(r) = b^2 \xi(r).$$

(10.4)

Because the correlation functions need not have the same scale dependence, in general b need not be a constant. However, it may well be adequately represented by a constant across some range of scales, and indeed there is observational evidence supporting this (e.g. Peacock and Dodds 1994) as long as we look to large enough scales, which in practice more or less means scales in the linear regime.

The final definition relates the relative perturbations in the galaxy field and the matter field on a point-by-point basis:

$$\frac{\delta n}{n}(\mathbf{x}) = b\frac{\delta\rho}{\rho}(\mathbf{x}), \tag{10.5}$$

where n refers to the galaxy number density (obviously, some smoothing procedure is required to render this a continuous field).

The definition (10.5) implies Eq. (10.4), which in turn implies Eq. (10.3); however, the inverse implications do not apply. The first definition of b always exists and the second one may be a useful approximation over some range of scales. The final one, however, seems rather dubious even if considerable smoothing has been applied; nevertheless it has seen quite a lot of use, particularly in connection with relating the density field to the peculiar velocity field, because it is unclear what else one could do.

Another point that we must bear in mind concerning bias, which we alluded to earlier, is that the bias parameter may depend on the type of object under study. For example, the spectra of optically and infrared-identified galaxies satisfy the approximate relation (Peacock and Dodds 1994)

$$\mathcal{P}_{\text{opt}}(k) \simeq 1.7\,\mathcal{P}_{\text{I}}(k), \tag{10.6}$$

across a reasonable range of k, corresponding to, for example, $10h^{-1}$ Mpc up to $50h^{-1}$ Mpc. Here, the subscript I indicates either infrared or IRAS, the latter being the name of the satellite (Infra-Red Astronomical Satellite) that made the positional survey upon which existing infrared selected redshift catalogues (QDOT, 1.2 Jansky, and PSCz) were based. This implies that

$$b \simeq 1.3\,b_{\text{I}}, \tag{10.7}$$

where b without a subscript always refers to optical galaxies. The IRAS galaxies usually are interpreted as being younger galaxies, with the high infrared flux being due to more active star formation and with further circumstantial evidence coming from the fact that they are underrepresented in clusters as compared to optical galaxies. In this picture the IRAS galaxies would be forming later, and hence from lower peaks in the density distribution. The smaller bias parameter for those galaxies thus has a natural explanation in terms of the peak biasing model.

Later, when we come to consider clusters, we see that they have an exceptionally high bias parameter.

That the ratio of b to b_{I} is quite accurately independent of scale on measured scales from $10h^{-1}$ Mpc upward is encouraging because it provides circumstantial evidence that the assumption of a constant bias parameter relating the galaxy and matter spectra may be a reasonable one. More direct evidence for the constancy of the bias parameter has come from the analysis of the three-point function by Gaztañaga and Frieman (1994); although their analysis of the APM survey is open to several interpretations, the simplest is that the bias is quite accurately independent of scale on sufficiently large scales.

On the other hand, on short scales, there is quite considerable evidence that the bias is not scale independent. Hydrodynamical simulations (Cen and Ostriker 1992b) show a marked scale dependence below about $8h^{-1}$ Mpc and, more pertinently, Peacock (1996) claims to have detected the effect from a compilation of galaxy catalogues. Coles (1993) investigated local bias models in fairly general terms and concluded that the bias parameter must decrease with increasing scale. On the other hand, nonlocal bias mechanisms [e.g., inhibited galaxy formation in ionization spheres induced by quasars, as suggested by Babul and White (1991), or cooperative galaxy formation as discussed by Bower et al. (1993)] may be able to evade this trend.

As a final word on this topic, we mention the likely redshift dependence of biasing. There already exist some narrow-beam surveys measuring clustering to quite high redshift [e.g., the positional analysis of the Hubble Deep field (Villumsen et al. 1997), the angular spectrum of the Canada–France Redshift Survey in different redshift bins (Hudon and Lilly 1996), and the clustering of Lyman break galaxies (Adelberger et al. 1998)]. In interpreting these, we must ask whether the same biasing model still applies. Almost certainly it will not; at high redshifts, objects of a given characteristic (e.g., bright galaxies) will be rarer, corresponding to higher peaks in the density distribution, and hence are expected to exhibit stronger biasing. As time passes, lower peaks collapse to form objects of the same type, filling in the "gaps" in the distribution and hence lowering the bias toward its present value. It may well be that between moderate redshifts such as 3 and the present, the decreasing bias more or less balances the growth of the matter power spectrum, leading to a galaxy correlation function that is more or less independent of redshift. Matarrese et al. (1997) investigate a range of possible models for redshift-dependent biasing.

10.1.3 Redshift distortions

A crucial aspect of determining the spectrum of galaxies is that their locations are given not in real space but in redshift space. This is a consequence of their position being measured via the redshift; any peculiar velocity that the galaxy has leads to it being associated an incorrect distance s, related to its true distance by

$$s = r + H^{-1} v_r^{pec},$$

(10.8)

where v_r^{pec} is the component of the peculiar velocity in the radial direction away from the observer. Because the distortion acts only in the radial direction, it serves to make the observed redshift space spectrum anisotropic between the radial and angular coordinates. Because the observed clustering is due to the velocities, redshift distortions give a systematic correction to the spectrum.

Although sometimes redshift distortions are an irritant, preventing us from making a clean determination of the spectrum in real space, which is the most easily predicted quantity from a given model, in the long term they undoubtedly will be seen as enormously beneficial because the induced anisotropy in the spectrum gives access to the statistics of the peculiar velocities (though it says nothing about the peculiar velocities of individual galaxies), from the assumption that the real space spectrum should be isotropic. The advent of very large galaxy redshift surveys will provide both accurate spectral determinations and velocity information that can be used to probe biasing of the galaxy distribution and the growth rate of density perturbations.

The original analysis of redshift distortions was made by Kaiser (1987), who used a small-angle approximation that permitted a plane-wave approximation. He showed that, in the linear regime, the amplitude of a plane wave is enhanced, giving

$$\delta_{gal}^{obs}(k) = \left[1 + \frac{\Omega^{0.6}}{b}\mu^2\right]\delta_{gal}(k). \tag{10.9}$$

The bracketed term is the redshift distortion; here the Ω factor allows for the slower velocities in a low-density Universe [either open or flat, Eq. (6.17)], b allows for velocities being smaller due to biasing, and μ is the cosine of the angle between the wavevector of the perturbation and the line of sight. The effect is maximal for a wavevector aligned with the line of sight, and zero if the velocities are tangential to the line of sight.

To obtain an expression for the spectrum, we have to average over a uniform distribution for μ, which yields (Kaiser 1987)

$$\mathcal{P}_{gal}^{obs}(k) = b^2\left[1 + \frac{2}{3}\frac{\Omega^{0.6}}{b} + \frac{1}{5}\left(\frac{\Omega^{0.6}}{b}\right)^2\right]\mathcal{P}(k). \tag{10.10}$$

Most of the early work on redshift distortions was carried out using this approximation, even if the catalogues to which it was applied were so large that the small-angle approximation was highly dubious. More recently, a much more natural framework has been devised using a spherical expansion (along the lines of Section 4.3.2), which naturally captures the fact that the distortion operates radially (Fisher et al. 1994; Heavens and Taylor 1995). The redshift distortion couples modes of different k, and so, a relation such as Eq. (10.10), independent of the spectrum, cannot exist, though the Kaiser result can be obtained in the appropriate limit (Heavens and Taylor 1995).

So far, we have discussed linear theory, but there is another important effect that arises when objects collapse and virialize. They attain a Maxwellian distribution with some velocity dispersion, which smears out their redshift space position relative to real space. This effect is especially noticeable from clusters of galaxies in redshift surveys, where linear structures known as "fingers of God" are clearly visible. The effect on the observed density contrast can be modelled as an exponential decay on scales below the velocity dispersion (e.g., Peacock and Dodds 1994). The galaxy pairwise velocity at a separation of $1h^{-1}$ Mpc is a few hundred kilometers per second (as we see shortly), suggesting that the damping effect of virialized velocities is important for wavenumbers greater than a few tenths of inverse megaparsecs.

10.1.4 Topology of the galaxy distribution

A final use of the galaxy distribution, which we mention here but do not use further, is to explore the statistics of the perturbations, which we expect to be Gaussian in an inflationary cosmology for scales still in the linear regime. Nonlinear evolution, of course, necessarily generates non-Gaussian statistics even if the initial distribution is Gaussian, at the very least for no other reason than that the density contrast by definition exceeds -1, but is unbounded above.

One interesting technique for exploring the statistics is a topological analysis, sometimes called the genus test, developed originally by Gott and collaborators (Gott et al. 1986; Hamilton et al. 1986; Melott et al. 1989) and subsequently studied by many researchers in both a two- and a three-dimensional context [see Melott (1990) for a review]. The basic idea is to set a threshold level for the density field, and find the surfaces that are the contours of this density field at that threshold. The genus (or, more precisely, the genus density), which depends only on the topology and not the geometry of these surfaces, then is computed. The genus curve plots this quantity as a function of the threshold (in fact, normally it is the density of "genus minus one" that is plotted). The genus is a measure of the number of "handles" that the isosurface has; it is related to the Euler number.

The advantage of this technique is that the shape of the genus curve is always the same if the statistics are Gaussian; a different choice for the spectrum simply affects the overall amplitude. The Gaussian prediction is

$$g(\nu) \propto (1 - \nu^2) \exp\left(-\frac{\nu^2}{2}\right),$$ (10.11)

where ν is the threshold measured in terms of the number of standard deviations from the mean. This is, of course, symmetric between over- and underdensities.

A deviation from Gaussian statistics leads to a shift in the genus curve. A rightward shift is called a "swiss-cheese" shift, indicating that the low-density regions are predominantly simply connected within a high-density "sea," whereas a leftward shift is termed a "meatball shift."

At present, all estimates of the genus curve from observational data have proven consistent with Gaussian statistics (e.g., Colley 1997; Protogeros and Weinberg 1997). Unfortunately, this is not as strong as one would like, because the surveys that exist are not on so large a scale as to allow statistically strong measurements well within the linear regime. On short scales, there is an additional problem that substantial smoothing is needed to obtain a smooth distribution on which to carry out the analysis. On scales where the measurement carries reasonable weight, there is a tendency to show a slight meatball shift, consistent with expectations from nonlinear clustering. Because, generally, the amplitude of the underlying matter spectrum is unknown due to biasing, the precise size of the nonlinear effect is unknown and so it cannot be modelled effectively to permit a strong test of whether or not the initial distribution was Gaussian.

The genus statistic also has been applied to microwave background maps (Colley et al. 1996), again giving results consistent with Gaussianity.

A more general class of topological measures, the Minkowski functionals, has been introduced into cosmology (Schmalzing et al. 1995; Schmalzing and Buchert 1997), but they have yet to be applied widely.

10.1.5 Cluster correlation function

Studies of the correlation function of galaxy clusters have an extremely long history, dating back to the compilation of a catalogue of clusters in the Northern sky by Abell (1958). This catalogue included a **richness class** that estimated the number of galaxies lying within a certain radius, now known as the Abell radius, of the cluster centre. To be in the catalogue at all, a cluster requires sixty-five galaxies within an Abell radius, and such clusters have become known as Abell clusters. They are the most massive gravitationally bound objects in the Universe. Much more recently, a similar catalogue, the ACO catalogue, was compiled for the Southern sky (Abell et al. 1989), using more automated methods.

The original Abell catalogue yields a galaxy-cluster correlation function of a shape very similar to that of the galaxy correlation function. The correlation length was estimated by Bahcall and Soniera (1983) (see also Klypin and Kopylov 1983) as

$$\xi_{\text{Abell}}(r) = \left(\frac{r}{r_0}\right)^{-1.8}, \qquad r_0 \simeq 25h^{-1}\,\text{Mpc}, \tag{10.12}$$

which is a very strong correlation indeed, suggesting a cluster bias of 5 times the optical galaxy bias. Theoretical models do not anticipate such a high ratio. However, the usefulness of the Abell catalogue has been a source of some controversy for many years, the claim being that the correlations are enhanced artificially by systematic errors made during the compilation process, including projection effects. Sutherland (1988) revised the estimate of the correlation length down to $r_0 \simeq 14h^{-1}$ Mpc, though with potentially large uncertainties.

In attempts to eliminate the problems of the Abell catalogue, new surveys were used that identified the clusters automatically, thus giving a uniform sampling. Two such are the APM cluster catalogue (Dalton et al. 1992) and the Edinburgh/Durham cluster catalogue (Nichol et al. 1992). Both of these indicated substantially less clustering than the Abell catalogue; the former gives $r_0 \simeq (14 \pm 4)h^{-1}$ Mpc and the latter gives $r_0 \simeq (16 \pm 4)h^{-1}$ Mpc. They confirm the same shape as the galaxy correlation function. These new results suggest a cluster bias of around 3 times the optical galaxy bias, in much better agreement with expectation.

We also mention the European Southern Observatory (ESO) cluster survey (Katgert et al. 1996), which is the largest compilation of redshifts for galaxies in clusters. However, the aim there was primarily to redshift as many galaxies as possible within clusters (in several cases, well over 100) in order to explore cluster substructure, rather than to identify the distance to as many clusters as possible. This survey therefore is not so useful for determining cluster correlations.

For our purposes here, clusters are interesting for the same reason as galaxies are, as a (biased) tracer of the matter spectrum. Being much rarer objects than galaxies, they are to be associated with higher peaks in the density distribution, and so, it is natural to expect that they will be correlated more strongly than galaxies, as is borne out by the observations we have just discussed. In estimating the matter spectrum, we can treat clusters simply as another type of "galaxy."

10.1.6 Observational status

The spectrum now has been determined for a large number of different surveys, and the agreement in the regions of overlap is very good provided that we bear in mind that different types of object will feature different biasing. In our chapter on bringing observations together, we use the compilation of Peacock and Dodds (1994), as updated by Peacock (1996). A later paper by Peacock (1997) studied only APM and IRAS galaxies; in the linear regime, the agreement with the earlier work is good, and he also has interesting results in the nonlinear regime, which we do not discuss, however. Peacock and Dodds bring together several data sets from disparate sources, including Abell clusters, radio galaxies, optical galaxies, and IRAS galaxies. The observational data are corrected for redshift distortions and also nonlinear evolution of the spectrum on short scales, and are found to align well provided that the bias parameters follow the ratio (Peacock 1996)

$$b_{Abell} : b_{radio} : b_{opt} : b_I = 4.5 : 1.9 : 1.3 : 1. \tag{10.13}$$

Because the shorter scales show signs of nonlinear biasing (Peacock 1996), and because the nonlinear correction may vary depending on the cosmological model assumed (Smith et al. 1998), the results that we show in Chapter 12 include only those points on large enough scales for the linear approximation to apply.

One advantage of the Peacock–Dodds approach is that they draw from a wide range of sources that are subject to different systematic errors. However, there are advantages also in using a single source of input data to ensure the best possible control of selection criteria. Power spectra have been determined for several different surveys: QDOT (Feldman et al. 1994; Tadros and Efstathiou 1995), APM (deprojected) (Baugh 1996), Stromlo–APM (Tadros and Efstathiou 1996), and Las Campanas (Lin et al. 1996). The agreement with Peacock and Dodds is good in the regime where we use those data, with surveys, however, normally becoming controversial as the wavelength approaches the survey size. In the future, as we have intimated, the best results are likely to come from large single surveys, such as 2dF and SDSS, rather than from compilations, because they will comprise such large uniformly selected samples.

10.2   Galaxy velocities

The velocities of galaxies potentially provide more useful information than the galaxy locations concerning clustering in the Universe; because of the universal gravitational attraction of matter in any form, the galaxy velocity field directly probes the matter spectrum rather than the galaxy spectrum. Estimates therefore can be made that are independent of the bias parameter.

The drawback, however, is that galaxy velocities are far harder to measure than galaxy redshifts, because we require a reliable distance estimate to the galaxy. Combined with the redshift, this can be used to measure the radial component of the peculiar velocity. Distance estimates are available for only about 3,000 galaxies at present (the Mark III catalogue; Willick et al. 1995, 1996, 1997) and contain substantial uncertainties. Further, this is unlikely to be increased dramatically through large dedicated surveys, as galaxy redshifts will be. Nevertheless, it is interesting to see what can presently be said.

10.2.1   Bulk flows

For some years now, the leading method for analyzing the peculiar velocity field has been an ingenious technique known as POTENT, pioneered by Bertschinger, Dekel, and collaborators

(Bertschinger and Dekel 1989; Bertschinger et al. 1990; Dekel et al. 1990; Dekel 1994). This overcomes the problem that only the radial component of the peculiar velocity can be measured. The assumption made is that the peculiar velocity field takes the form of a potential flow, and hence can be written as the gradient of a scalar. We saw in Section 4.4.6 that in linear theory the growing mode is exactly of this form. The value of the scalar at a given point can be determined via a line integral of the peculiar velocity from our location to that point; in particular, if we choose the radial path, then only the radial component of the peculiar velocity is required. Having obtained the scalar, its gradient then gives the full three-dimensional velocity field.

The power and potential of this method has been demonstrated vividly via tests on simulated data. Unfortunately, so far the method has been thwarted to a large extent by the inadequacies of existing data, which suffer both from sparseness and noisiness. However, although the fine details of the reconstructed bulk flow probably are not to be believed, the POTENT reconstructions certainly capture the main characteristics of the local region as seen in galaxy redshift surveys. We return to that in Section 10.2.2. And, in particular, the reconstructions should provide good estimates of the bulk flow in spheres around our position.

Each component of the velocity field has a Gaussian distribution, inherited from the Gaussianity of the density field. The variance of the magnitude of the velocity field, V, smoothed on some scale R, is given by the usual expression

$$\sigma_V^2(R) = \int_0^\infty W^2(kR) \mathcal{P}_V(k) \frac{dk}{k}, \tag{10.14}$$

where $W(kR)$ is one of the usual window functions. This can be written in terms of the matter spectrum, using Eq. (4.121) and the velocity suppression factor (6.17) in the low-density case, as

$$\sigma_V^2(R) = H_0^2 f^2(\Omega_0) \int_0^\infty W^2(kR) \frac{\mathcal{P}_\delta}{k^2} \frac{dk}{k}, \tag{10.15}$$

where $f(\Omega_0) \simeq \Omega_0^{0.6}$, defined in Eq. (6.17), gives the growth rate of perturbations with $f(\Omega_0 = 1) = 1$. The dispersion of the velocity in one direction is given by the same expression multiplied by $1/\sqrt{3}$. As compared to the equivalent expression for the dispersion of the density contrast, Eq. (4.56), the integrand contains two extra powers of k that increase the relative influence of long wavelengths over short ones. So, the velocity field on a given scale samples larger wavelength modes than the density contrast on that scale. Consequently, when we compute the integrand, it is not as well dominated by scales of order of the smoothing scale as is the density contrast; rather, it still samples from a wide range of scales. That tends to correlate measurements of the velocity smoothed on different scales but measured at the same location.

To apply the POTENT method, the first step required is to generate a smooth velocity field over which the line-of-sight integral can be carried out. Because the data can be very sparse, especially in some regions of the sky, this smoothing distance must be quite large. The POTENT team use a Gaussian smoothing of $12h^{-1}$ Mpc, over and above the smoothing already given in Eq. (10.15). This is obtained by inserting an extra factor of $\exp[-(12kh^{-1} \text{ Mpc})^2]$ into the integrand.

Because the bulk flows over spheres of different radii are correlated, we might as well consider a single measurement, for example, that at $R = 40h^{-1}$ Mpc. The value from the Mark III POTENT analysis is (Dekel 1994)

$$V_{\text{POTENT}}(40h^{-1}\,\text{Mpc}) = 373 \pm 50\,\text{km} \cdot \text{s}^{-1}. \tag{10.16}$$

The quoted error arises from different ways of dealing with sampling-gradient bias, and estimates the systematic uncertainty in the analysis. In addition there is a 15-percent uncertainty from random distance errors. For comparison, the SCDM model gives

$$V_{\text{SCDM}}(40h^{-1}\,\text{Mpc}) = 409\,\text{km} \cdot \text{s}^{-1}, \tag{10.17}$$

in excellent agreement.

However, in making the comparison, the most important uncertainty is neither the systematic nor the random uncertainties mentioned earlier, but cosmic variance. Each component of the velocity is a Gaussian random field of the same variance, and so, the velocity squared obeys a χ^2 distribution with 3 degrees of freedom (i.e., a Maxwellian distribution), which possesses a large cosmic variance. Liddle et al. (1996b) estimate that cosmic variance, along with the other subdominant errors, yields an overall 95-percent uncertainty on the estimate of the magnitude of the density perturbation of -47 percent and $+295$ percent. The upper error bar is so large as to make it useless (i.e., the velocity field is consistent with density fields of normalizations much higher than permitted by other types of data), but the lower bar is able to constrain some models.

This simple analysis is obviously some way from making use of the entire data set, but, at this time, more complete analyses of the data have not been shown to offer significantly stronger constraints; see, for example, the later brief discussion of the velocity power spectrum.

We can use POTENT to illustrate the point that the velocity field samples longer scales than the density field. Liddle et al. (1996b) considered a set of CDM models with different spectral indices n, and normalized each so as to reproduce the POTENT result quoted above. They found that the dispersions of the density fields for the models thus normalized cross at the scale of $113h^{-1}$ Mpc, almost 3 times larger than we might guess from simply looking at the smoothing length.

One of the early motivations for studying the velocity field, as seen in Eq. (10.15) recalling $f(\Omega_0) \simeq \Omega_0^{0.6}$, is the possibility of probing Ω_0. What we see is that the velocities are expected to be substantially lower in a low-density Universe, for the same size of density perturbation. However, the shift to the use of the COBE normalization has rather changed the interpretation of this; COBE directly normalizes the matter spectrum and, what is more, the COBE normalization depends on Ω_0 in a way similar to that for the velocities.[1] COBE-normalized models therefore predict similar magnitudes of velocity flow whatever the value of Ω_0; the principal difference is actually the amplitude of the *matter* spectrum rather than the velocity spectrum. Unfortunately, the matter spectrum is not measured independently, so instead, we get an estimate of the

---

[1] To be more precise, from Eq. (9.8), the normalization in a low-density flat cosmology goes roughly as $\Omega_0^{-1.6}$, similar to the $\Omega_0^{-1.2}$ behaviour of the velocities, for any reasonable Ω_0. For open models, the scaling is weaker, and at lower densities, the COBE normalization behaves quite differently from the velocity flow.

magnitude of the bias. This is useful to know, but not an actual constraint unless the inferred bias proves unacceptably large or small. We investigate this further in Section 10.2.2.

More recent work (Kolatt and Dekel 1997; Zaroubi et al. 1997) has used POTENT to obtain the matter spectrum rather than the bulk flow. Although more or less in agreement with the galaxy spectrum and with the advantage of being independent of bias, the shape of the spectrum, codified by the shape parameter Γ, is much more weakly constrained than by galaxies. A model with free normalization is found to require $\Gamma = 0.4 \pm 0.20$ at 90-percent confidence (Zaroubi et al. 1997), with a slight change if models also are forced to satisfy the COBE normalization; this does not significantly constrain any favoured model. They also find $\sigma_8 = (0.88 \pm 0.15)\Omega_0^{-0.6}$, again at 90 percent, somewhat higher than other estimates (e.g., the cluster abundance discussed in Section 11.5) tend to give.

10.2.2 Velocity/Density comparisons

Because the peculiar velocities are supposed to be induced by the density perturbations, whose growth is of course induced by the velocity field they cause, it is interesting to ask whether or not the two fields are indeed consistent with one another. This offers one possible test of the gravitational instability picture, though it has been pointed out (Babul et al. 1994) that it probably is not as strong a test as we would like because, even if the structures were induced nongravitationally (e.g., by an explosion scenario), the galaxy motions still will be directed toward regions of high density – "cause and effect" has been reversed as compared to the gravitational instability picture. Nevertheless, a compelling agreement is seen to exist between the velocity and density fields (Kaiser et al. 1991; Dekel et al. 1993). This tempts us to go on and try to use the comparison to constrain cosmological parameters.

The main use of such comparisons has been to try and estimate the density parameter Ω_0 because, in a low-density Universe, peculiar velocities are smaller for a given size of density perturbation. However, because observations invariably sample the galaxy distribution rather than the matter distribution, the test becomes dependent on the biasing model. Normally, the point-by-point linear bias model of Eq. (10.5) is utilized, and in that case, the only free parameter in the velocity/density comparison, to be fixed by observations, is the combination $\beta \equiv \Omega^{0.6}/b$, where the bias parameter inserted must be that appropriate for the type of galaxies used, for example, $\beta_I \equiv \Omega^{0.6}/b_I$ for IRAS galaxies. Remember though that this entire bias model is rather dubious, which may be one way of accounting for the fact that different determinations of β are not always consistent, even if we bear in mind the different bias parameters for different classes of object. Extensive compilations of results from these methods can be found in the reviews of Dekel (1994) and Strauss and Willick (1995).

10.2.3 Cosmic Mach number

The cosmic Mach number is a statistic introduced by Ostriker and Suto (1990) to try to probe the shape of the spectrum using velocities, rather than its normalization. However, there has not been much success in obtaining useful observational constraints from this quantity, and it has not received much discussion in recent years. They defined a quantity

$$M_{OS}^2(R) = \frac{\langle |\mathbf{v}(R, \mathbf{x})|^2 \rangle}{\langle |\mathbf{v}(\mathbf{x}) - \mathbf{v}(R, \mathbf{x})|^2 \rangle},$$

(10.18)

where $\mathbf{v}(R, \mathbf{x})$ is the velocity field smoothed on a scale R and the average is over all observer positions $\mathbf{x}$. We can see that the normalization of the spectrum cancels out. The idea is that a large value of $M_{OS}(R)$, greater than 1, for example, indicates a fairly coherent velocity flow (where the velocity averaged about a point is not so different from the velocity at the point), whereas a small value indicates a fairly random, hot flow where the averaged velocity is much less than the velocity at a given point.

As always with such statistics, the problem arises from the need to smooth the existing data set very significantly to compute observational quantities, with a serious loss of discriminatory power (Ostriker and Suto 1990; Suto and Fujita 1990; Suto et al. 1990, 1992). Ostriker and Suto also found that the statistic quoted above was, in any case, hard to determine, and instead computed an alternative quantity

$$M(R) = \frac{\langle |\mathbf{v}(R)| \rangle}{\sigma_v(R)},$$

(10.19)

which, in any case, corresponds more closely to a conventional description of the Mach number. The observational estimates appear in agreement with those from the galaxy correlation function, but with considerably larger uncertainties.

10.2.4   Pairwise velocities

The pairwise velocity of two galaxies is the relative velocity projected along the line of sight joining them. It is estimated by studying redshift distortions in the two-point correlation function on short scales (Davis and Peebles 1983). The dispersion of this can be computed for any separation r; for example, in linear theory it is given, using Eqs. (4.121), (4.91), (4.16), and (4.20), by

$$\sigma_{\parallel}^2(r) \equiv \frac{1}{3} \langle |\mathbf{v}(\mathbf{x} + \mathbf{r}) - \mathbf{v}(\mathbf{x})|^2 \rangle$$

$$= \frac{2 H_0^2 f^2(\Omega_0)}{3} \int_0^\infty \frac{\mathcal{P}_\delta(k)}{k^2} \left[1 - \frac{\sin kr}{kr} \right] \frac{dk}{k},$$

(10.20)

where the factor of one-third accounts for the velocity being the line-of-sight component only, which is the observationally convenient property. However, no attempt has been made to estimate this quantity on a scale big enough to be in the linear regime; typically, it has been quoted on the scale $1h^{-1}$ Mpc. Therefore, this must be evaluated numerically, which has been attempted by various authors. In accordance with the preceding discussion of bulk flows, we expect that the dispersion of the pairwise velocity will be sensitive to longer scales than the dispersion of the density contrast itself, and indeed, this is borne out by simulations. For example, Gelb et al. (1993) found that, if they normalized models with different spectral shapes so as to produce the same pairwise velocity, then these models had more or less the same σ_8.

For a long time the standard measurement of the pairwise velocity was that of Davis and Peebles (1983), who obtained a value $340 \pm 40\,\text{km} \cdot \text{s}^{-1}$ at $1h^{-1}$ Mpc from the CfA1 survey. This turns out to be a very low value, requiring σ_8 in the range 0.3 to 0.5 for reasonable spectra in the critical-density case (Gelb et al. 1993), significantly lower than permitted by several other types of observation. Concerns about this low value were influential in model building for many years. Unfortunately, it has turned out that this original determination was somewhat incorrect because of a programming error in implementing a correction for the flow into the Virgo cluster (Somerville et al. 1997a). It now is accepted that the true value is significantly higher and has a larger uncertainty, the new estimate being $453 \pm 118\,\text{km} \cdot \text{s}^{-1}$. Further, it has been realized that there is substantial cosmic variance on this measurement (Somerville et al. 1997b), rendering a large uncertainty in the amplitude of the spectrum needed to match the data. This may partly explain why the literature now contains a very diverse set of estimates from different catalogues; see Somerville et al. (1997a) for a compilation.

Another intended use of the pairwise velocity was as a probe of Ω_0 because, in low-density Universes, we expect slower velocities for a given size of density perturbation. This too seems to have been thwarted, both by the large cosmic variance and because the pairwise velocity too strongly weights toward clusters, where we are far more likely to find two galaxies the requisite distance apart than in the field, and so, the pairwise velocity largely probes virialized systems rather than perturbation growth in a regime where the background cosmological evolution is important. A slightly different statistic, which is not pair weighted, has been devised (Davis et al. 1997) but has yet to be applied widely.

10.2.5 Peculiar velocities and the hubble flow

Peculiar velocities disturb the Hubble flow, raising the possibility that the Hubble flow measured in our local region may differ from the global value. However, provided that we look to a large enough scale, the dispersion of the measured Hubble constant is small. Pioneering work was done on this question using numerical simulations by Turner et al. (1992). On large enough scales, linear theory should be adequate, and Shi et al. (1996) showed that, in linear theory, the dispersion of the measured Hubble constant in a survey perfectly volume limited at radius R is

$$\left\langle \left(\frac{\Delta h}{h} \right)^2 \right\rangle = \int \mathcal{P}(k)\, W_{\text{hub}}^2(kR)\, \frac{dk}{k}, \tag{10.21}$$

where the window function is

$$W_{\text{hub}}(kR) = 3\, \frac{\sin(kR) - \int_0^{kR} dx\, \sin(x)/x}{(kR)^3}. \tag{10.22}$$

The conclusion from all these studies is that at $20h^{-1}$ Mpc, the approximate scale of the Virgo cluster, the dispersion can be tens of percent. This is well known to the observers, who use further distance indicators to jump to the Coma supercluster at nearly $100h^{-1}$ Mpc. On such a large scale, the theoretically expected dispersion is only a few percent (with some modest dependence on the cosmological model and initial spectrum assumed).

The idea that the local and global values might be different also has been dealt a blow by direct observations. Distance estimates of Type Ia supernovae, incorporating a correction due to light curve shape, have shown that the local and global values can differ by no more than 10 percent (Kim et al. 1997). The agreement of observation and theoretical expectation on this point seems good.

Example

10.1 An outdated normalization technique evaluated the J_3 integral of Eq. (10.2), setting

$$J_3^{\text{gal}}(10h^{-1}\,\text{Mpc}) = 270h^{-3}\,\text{Mpc}^{-3}.$$

Compare this with its historical rival $\sigma_8^{\text{gal}} = 1$, for the SCDM model.

10.2 For the SCDM model, compute the effective spectral index of the density perturbations and plot it as a function of scale.

11 The quasi-linear regime

11.1 Gravitational collapse

11.1.1 Overview

The bulk of this book has concerned the linear regime, and it is fair to say that, ultimately, our main understanding of the parameters describing the Universe as a whole will come via observations of that sort, especially those of microwave background anisotropies. However, we would not exist were it not for structures having entered the nonlinear regime, ultimately collapsing to form gravitationally bound objects, and so, it is interesting to ask about this regime. In fact, at present, one of the most powerful constraints on the spectrum comes from this regime, being the present abundance of galaxy clusters.

The nonlinear regime has been investigated using a variety of techniques. The most popular of all is numerical simulations, particularly simulations of the gravitational force alone, known as N-body simulations, though recently many codes have been developed that also include a range of hydrodynamical gas processes and sometimes even radiative processes such as photoionization. We discuss some of these in this chapter, but not in any depth because this is primarily a book about analytical and semianalytical techniques.

Among this latter class are theories designed to predict the number of collapsed objects of a given mass. We discuss two such theories: the Press–Schechter (1974) theory and the theory of peaks (Bardeen et al. 1986). These are semianalytic (meaning that there usually exist integrals that must be performed numerically to deal with realistic spectra, but that the basic structure of the theory is analytic), though it is fair to say that they are only used with confidence because they have been compared and calibrated using N-body simulations. Press–Schechter theory in particular proves very useful in imposing constraints on the matter spectrum from the observed abundances of objects, particularly if these observations can be made at a time when those objects are only just forming and hence are relatively rare, thus probing the exponential tail of the Gaussian density distribution. We study several constraints of this type, the most used being the present abundance of galaxy clusters.

In addition to these, there exist a number of approximation schemes intended to capture the nonlinear evolution of the entire density field at some level of approximation. These comprise a "halfway house" between linear theory and full N-body simulations; usually they require numerical computation, but not of the CPU-intensive type with which N-body simulations have become associated. Examples are the famous Zel'dovich approximation, along with various enhancements. For completeness, we give some brief discussion of these.

11.1.2 The spherical collapse model

A guideline to the behaviour of perturbations in the nonlinear regime is given by a simple analytical model, the spherical collapse model, which we implement in a pressureless critical-density Universe. In this model, we postulate a spherically symmetric perturbation with a top-hat profile of radius R and initial overdensity δ. The matter required to cause the overdensity is taken from a narrow shell of zero density, in order to preserve the mean density. According to the two famous theorems of Newton regarding spherical mass distributions, the Universe outside the perturbation continues to evolve as a critical-density Universe, and the perturbed region evolves independently of the behaviour of the exterior region, behaving as a Universe with initial density $\Omega = (1 + \delta)$. Such a closed Universe will recollapse in a finite time, remaining homogeneous in the absence of pressure. The most extensive analysis is that of Peebles (1980), following original calculations by Partridge and Peebles (1967) and Gunn and Gott (1972).

It would be nice to carry out the calculation using the analytic solution in terms of conformal time [the analytic continuation of Eq. (2.45) to the closed case], but unfortunately we will want to compare the evolution at fixed cosmological times rather than conformal times, and so, we need to use the parametric solution, which is the analytic continuation of Eqs. (2.46) and (2.47), namely

$$\frac{a(t)}{a(t_0)} = (1 - \cos\theta)\,\frac{\Omega_0}{2(\Omega_0 - 1)}, \tag{11.1}$$

$$H_0 t = (\theta - \sin\theta)\,\frac{\Omega_0}{2(\Omega_0 - 1)^{3/2}}, \tag{11.2}$$

where the development angle θ runs from 0 to 2π. Expressing this instead in terms of the scale factor and time of maximum expansion gives

$$\frac{a(t)}{a_{\max}} = \frac{1}{2}(1 - \cos\theta), \qquad \frac{t}{t_{\max}} = \frac{1}{\pi}(\theta - \sin\theta). \tag{11.3}$$

To study the linear regime, we need the small-parameter expansions of these to second order, giving

$$\frac{a(t)}{a_{\max}} \simeq \frac{\theta^2}{4} - \frac{\theta^4}{48}, \qquad \frac{t}{t_{\max}} \simeq \frac{1}{\pi}\left(\frac{\theta^3}{6} - \frac{\theta^5}{120}\right). \tag{11.4}$$

Combining these gives what we call the linearized scale factor,

$$\frac{a_{\lin}(t)}{a_{\max}} \simeq \frac{1}{4}\left(6\pi\,\frac{t}{t_{\max}}\right)^{2/3}\left[1 - \frac{1}{20}\left(6\pi\,\frac{t}{t_{\max}}\right)^{2/3}\right]. \tag{11.5}$$

This last expression is the important one. If we ignore the square-bracketed term, it gives the expansion of the background spatially flat Universe. If we include both terms, that gives us the *linear theory* expression for the growth of a perturbation. The full nonlinear evolution is given by the preceding parametric solution. Note that $a_{\max}$ is the maximum scale factor of the full solution, and that all three models are normalized to the same scale factor at early times when the perturbation is small. The three different scale factors are shown in Figure 11.1.

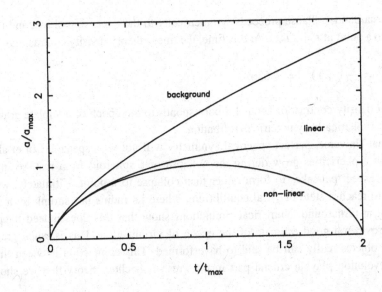

Fig. 11.1. The evolution of the three scale factors, namely the background evolution, the full nonlinear collapse, and the linearized evolution.

The first interesting event in the evolution of the full perturbation is **turnaround**, when it reaches maximum expansion at $\theta = \pi$. Up until that point the general expansion of the Universe has been dominating over the gravitational collapse and the physical size has continued to grow (though of course the comoving size is always decreasing). Suppose we had continued to use the linear-theory expression. Because we are assuming matter domination throughout, the energy density in our various model Universes always goes as a^{-3}. The linear density contrast (which, in this subsection, we subscript for extra clarity) therefore is given by

$$1 + \delta_{lin} = \frac{a_{back}^3}{a_{lin}^3},$$ (11.6)

where a_{back} is the background evolution given by the lowest-order truncation of Eq. (11.5). Substituting in the preceding expressions and linearizing again gives

$$\delta_{lin} = \frac{3}{20} \left(6\pi \frac{t}{t_{max}} \right)^{2/3}.$$ (11.7)

So, at turnaround, $t = t_{max}$, the linear density contrast is

$$\delta_{lin}^{turn} = \frac{3}{20} (6\pi)^{2/3} = 1.06.$$ (11.8)

Theoretically, it is always simplest to deal with linear theory; what this result tells us is that the end of the linear regime, where δ_{lin} reaches unity, corresponds roughly to the time that structures break away from the general expansion, but at that time, gravitationally bound structures have yet to form.

After turnaround, collapse proceeds symmetrically to the expansion phase, and the object collapses to a point at $t = 2\,t_{max}$. At this time, the linear-theory density contrast has become

$$\delta_{lin}^{coll} = \frac{3}{20}\,(12\pi)^{2/3} = 1.686. \tag{11.9}$$

So, a linear density contrast of about 1.7 corresponds to the epoch of complete gravitational collapse of a spherically symmetric perturbation.

In the real Universe, perfect spherical symmetry will not be respected. Lin et al. (1965) showed that asphericities grow during the collapse of a spheroid from rest, so in fact we generically expect pancakes to form rather than collapse to a point. Ultimately, we expect the object to reach a state of virial equilibrium, where its radius has shrunk by a factor of 2 from that at turnaround. Numerical simulations show that this does indeed happen, and that δ_{lin}^{coll} provides a good estimate of the epoch of virialization. This is when gravitationally bound objects really can be said to have formed. This connection between linear and nonlinear evolution plays a crucial part in the Press–Schechter theory that we shortly discuss.

We also can use the spherical collapse model to study the densities at various stages. The actual nonlinear density contrast at turnaround is just

$$1 + \delta_{nonlin}^{turn} = \frac{a_{back}^3}{a_{max}^3} = \frac{(6\pi)^2}{4^3} = 5.55. \tag{11.10}$$

In the spherical collapse model, the density goes infinite at the collapse time. However, if we assume that, instead, the collapsing object virializes at half the radius, its density will have gone up by a factor of 8. Meanwhile, the background density will have continued to fall, through expansion, by a further factor of 4. In combination then, we expect the overdensity at virialization to be

$$1 + \delta_{nonlin}^{vir} \simeq 178. \tag{11.11}$$

This number is remarkably well verified by numerical simulations as dividing virialized regions from those where matter is still infalling (Cole and Lacey 1996).

In a low-density Universe, the linear collapse threshold δ_{lin}^{coll} is almost unchanged, but the true density contrast at virialization is increased[1] to $1 + \delta_{nonlin}^{vir} \simeq 178\Omega_0^{-0.6}$ (White et al. 1993a; see also Eke et al. 1996).

Note that, although we used the expression for a homogeneous closed Universe, the result is actually somewhat more general than that because we can use the Newtonian theorems regarding spherical configurations to say that they apply to any spherically symmetric distribution, as long as the "shells" at different radii do not cross (in practice, this means that the density should decrease outward with radius). Then the collapse criterion can be applied separately shell by shell.

---

[1] Note that the actual density is decreased; the increase is caused by the lower background density used to form the density contrast.

The final information that we extract from the spherical collapse model is the virial velocity of the bound objects. The virial theorem states that

$$v^2 = \frac{GM}{R_g}, \tag{11.12}$$

where M is the mass of the system and R_g the gravitational radius defined by the requirement that the gravitational potential energy be $-GM^2/R_g$. The mass within an initial comoving radius R_{com} before collapse is

$$M = \frac{4\pi}{3}\rho_0 R_{com}^3. \tag{11.13}$$

The virial theorem tells us that the system will collapse by a factor of 2 from the radius at which it is at rest (assuming no dissipation), and so, $R_g = R_{turn}/2$. From Eq. (11.10), the turnaround radius is given in physical units by

$$R_{turn}^3 = \frac{1}{5.55}R_{phys}^3 = \frac{1}{5.55}\frac{1}{(1+z_{turn})^3}R_{com}^3. \tag{11.14}$$

Because the virialization time is twice the time to maximum and we assume matter domination, we have $1 + z_{turn} = 2^{2/3}(1 + z_{vir})$, leading to

$$R_g^3 = \frac{1}{178}\frac{1}{(1+z_{vir})^3}R_{com}^3. \tag{11.15}$$

Combining Eqs. (11.12), (11.13), and (11.15) and substituting in values gives the scaling relation

$$\left(\frac{v}{127\,\text{km}\cdot\text{s}^{-1}}\right)^2 = \left(\frac{M}{10^{12}h^{-1}M_\odot}\right)^{2/3}(1+z_{vir}). \tag{11.16}$$

The scalings in this expression are well verified by numerical simulations, though the precise normalization really needs to be fixed by them rather than by the approximate analysis above. Note that among objects of the same mass, those that form earlier have a higher virial velocity because they are more compact through forming when the Universe was smaller and denser. Galaxies have virial velocities of order $100\,\text{km}\cdot\text{s}^{-1}$, whereas those of clusters are of order $1,000\,\text{km}\cdot\text{s}^{-1}$.

There is also a scaling relation for bound gas in hydrostatic equilibrium. Because the temperature measures the kinetic energy of the gas, the appropriate scaling law comes from $T \propto v^2$. The constant of proportionality can be estimated from first principles (see, e.g., Padmanabhan 1993), or can be obtained via fitting to hydrodynamic simulations such as those of White et al. (1993b), which give

$$\frac{k_B T}{0.07\,\text{keV}} = \left(\frac{M}{10^{12}h^{-1}M_\odot}\right)^{2/3}(1+z_{vir}). \tag{11.17}$$

The most interesting application of this is to rich clusters, where the gas is sufficiently copious and hot enough to be observed in X rays, with energies of a few kilo-electron volts, by satellites such as Einstein, ROSAT, and ASCA, and, it is hoped that, by the time you read this, Chandra and XMM as well.

11.2   Press–Schechter theory

One of the most enduring theories of nonlinear evolution is Press–Schechter theory, introduced by Press and Schechter in 1974. Indeed, its popularity has increased in recent years as it has become apparent how well its results reproduce those of N-body simulations (Efstathiou et al. 1988; Lacey and Cole 1994), despite misgivings concerning its dynamical motivation. In fact, we might adopt the extreme view that the Press–Schechter formula, which we display shortly, is simply a fitting function for the results of numerical simulations, which happens to have some dynamical motivation.

Press–Schechter theory simply asserts that if we smooth the *linear theory* density field on some mass scale M, then the fraction of space in which the smoothed density field exceeds some threshold δ_c is in collapsed objects of mass greater than M. This is illustrated in Figure 11.2. Under the assumption of Gaussian statistics for the linear-theory density field, this is an extremely easy quantity to calculate. The only problem is that, for a Gaussian field, half the volume of the Universe is necessarily underdense, and will never exceed the threshold regardless of how much the density field evolves. Consequently, only half the mass in the Universe is available to form bound objects, clearly not the case for the real Universe. To remedy this difficulty, Press and Schechter simply multiplied the volume above the threshold by a factor of 2. Although this generates the desired answer, it is obviously rather dubious on dynamical grounds. Nevertheless, as stated earlier, this "corrected" version of Press–Schechter theory gives a very good fit to the results of N-body simulations. In fact, justification of the factor 2 also has come from analytic arguments using a variety of techniques (Peacock and Heavens 1990; Bond et al. 1991; Bower 1991; Jedamzik 1995; Yano et al. 1996). A heuristic explanation is as follows: A given point may turn out to be below the threshold when the smoothing is carried out on a scale M_1, yet turn out to be above the threshold at a larger scale M_2 – clearly it should be identified as being in an object of mass above M_1, but is not. This is sometimes known as the "cloud-in-cloud" problem.

To complete our description of Press–Schechter theory, we need to discuss two further issues: What choice of filter should be used for the smoothing and what choice of threshold should be used? The answer to the first is that we are free to choose different filters. In the technical analysis of Press–Schechter theory, a popular choice is the sharp k-space filter (i.e., a filter chosen to give a top hat in Fourier space), which allows the proof of various results (Peacock and Heavens 1990; Bond et al. 1991) because of the ease of evaluating the dispersion and because the independence of the different Fourier modes means that as the filter width increases the variance undergoes a random walk. Unfortunately, in real space, this filter has an oscillatory nature (just as the k-space version of the real-space top hat) that makes the physical interpretation difficult. Principally, then, the choice has been between the real-space top-hat filter and the Gaussian filter, Eqs. (4.52) and (4.53) respectively.

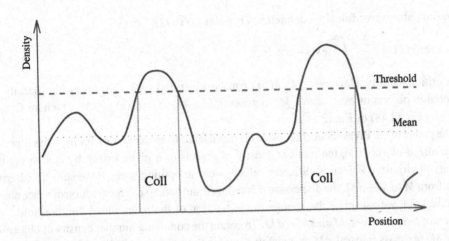

Fig. 11.2. Schematic of Press–Schechter theory applied to the density field smoothed on some scale. The volume in regions above the threshold, indicated by 'Coll' for collapsed, is identified with objects of the smoothing mass and above. If we smoothed further, more regions would drop below the threshold, giving a smaller total mass in objects above the higher smoothing scale.

As regards the choice of threshold, the first point is that it will depend on the choice of filter used. A range of thresholds is advertised in the literature, with the differences between them primarily being the choice of filter. The spherical top-hat collapse model discussed in Section 11.1 suggests that virialized objects form when the linear-theory density contrast reaches $\delta_c = 1.69$, from Eq. (11.9), a result most naturally associated with the top-hat filter. This value was used in the original Press and Schechter (1974) paper for that reason, and in fact stands up well to comparison with numerical simulations for any reasonable value of the spectral index n (Lacey and Cole 1994). The Gaussian filter requires a value that is quite a bit smaller, around $\delta_c \simeq 1.4$, and the required δ_c also has a nonnegligible dependence on the spectral index.[2] In each case, the uncertainty at 1-sigma is about ± 0.1, and the dependence on Ω_0 is negligible (White et al. 1993a; Eke et al. 1996). We use the top-hat filter throughout. An additional complication exists if we are comparing with objects that may not be fully virialized, such as high-redshift absorption systems, in which case the threshold must be reduced.

According to the prescription described above, the energy density in collapsed objects, as a fraction of the total density, is given by integrating over the tail of the Gaussian probability configuration (4.22), with the additional factor 2 multiplier, giving

$$f(>M(R), z) = \text{erfc}\left(\frac{\delta_c}{\sqrt{2}\,\sigma(R, z)}\right), \tag{11.18}$$

where $\sigma(R, z)$ is the dispersion of the smoothed density field given by Eq. (4.56). For generality, we include the possibility of evaluating this at different redshifts. Here, erfc is the

---

[2] Alternatively, Lacey and Cole (1994) suggest a high δ_c value, but with the mass associated with the Gaussian filter increased. This improves the fit to simulations while further undermining the physical motivation.

complementary error function, defined by (Press et al. 1992):

$$\text{erfc}(x) = \frac{2}{\sqrt{\pi}} \int_x^\infty \exp(-u^2)\, du. \tag{11.19}$$

In a critical-density Universe, $f(>M, z)$ can be identified with $\Omega(>M, z)$, the density of material in objects of mass above M; in general, $\Omega(>M, z) = \Omega(z) f(>M, z)$ where $\Omega(z)$ is given by Eq. (2.48) or Eq. (2.51).

The power of the Press–Schechter theory is that it leads to an expression for the mass function of virialized objects (i.e, the number density of objects of a given mass), by comparing the density in virialized objects at different values of the smoothing scale. If we shift the filtering scale from M to $M + dM$, the dispersion σ is reduced and hence so is the fraction of space above the threshold; we associate the change in the fraction of the mass above the threshold with objects of mass between M and $M + dM$. To obtain the comoving number density of objects of mass M, per mass interval dM, at redshift z, we do the following. Differentiating Eq. (11.18) with respect to the filtering mass gives the change in the volume fraction of the Universe, brought about by increasing the filter mass by dM. Multiplying this by ρ_0 gives the change in comoving mass density[3] above the threshold, and because that mass density is to be assigned to objects of mass M, we divide by M to get the comoving number density. Carrying this out through seemingly endless application of the chain rule yields (Press and Schechter 1974)

$$\frac{dn(M, z)}{dM} dM = -\sqrt{\frac{2}{\pi}} \frac{\rho_0}{M} \frac{\delta_c}{\sigma^2(M, z)} \frac{d\sigma(M, z)}{dM} \exp\left(-\frac{\delta_c^2}{2\sigma^2(M, z)}\right) dM, \tag{11.20}$$

from which the number density of objects of mass above M is

$$n(>M, z) = \int_M^\infty \frac{dn(M, z)}{dM} dM. \tag{11.21}$$

For typical spectra, this has to be carried out numerically, though it can be done analytically if we assume that the dispersion is given by a power law.

In the regions where the smoothed density field is still well in the linear regime, the number density produced is extremely sensitive to the precise value of the dispersion because we are operating well out on the tail of the Gaussian. For the applications that we consider, this is of great advantage because it means that even if there are large uncertainties in the observed densities (even by orders of magnitude in some cases), this typically translates into only a small uncertainty in the dispersion and hence into a useful constraint. The theoretical uncertainty in the appropriate value of δ_c, in contrast, appears within the exponential, and the fractional uncertainty in δ_c translates to the same fractional uncertainty in σ.

11.3   Theory of peaks

An alternative to the Press–Schechter theory is the theory of peaks in the density field. In this approach, it is the number of peaks in the smoothed density field above a given threshold

---

[3] It is the *present* density that appears here, because we are using comoving units and, during matter domination, the comoving mass density is a constant. We could equally well have used the physical mass density $\rho(z)$ and included a factor $(1 + z)^{-3}$ to convert the density to comoving.

that is identified with the number of objects of that mass and above. Early discussion of this possibility was made by Peacock and Heavens (1985), and the theory was taken more or less to its logical conclusion in a classic paper by Bardeen et al. (1986), one of the few papers instantly identified by the authors' initials (BBKS).

BBKS computed a number of statistics of three-dimensional Gaussian random fields. In particular, they quoted the number densities for three quantities as a function of threshold. The first is the Euler number of surfaces bounding regions at a certain threshold. The second is the number density of upcrossing points on those surfaces, defined as one where $\nabla\delta$ points along some arbitrarily chosen reference direction. Finally, they quoted the number density of peaks at least ν standard deviations high. They claimed that the number of upcrossing points is more or less equivalent to the number of peaks, and easier to compute. In the limit of very high peaks, $n_{peak} = n_{up} = n_{Euler}/2$; in this limit the bounding surfaces are all spheres (hence with Euler number $+2$) surrounding single peaks. At lower densities the isodensity surfaces can take on a more complicated geometrical or topological form, and this simple set of relationships is broken.

Typically, the number density of objects predicted by the peaks theory is similar to that predicted by Press–Schechter theory; see, for example, Liddle and Lyth (1993a) for a comparison for some sample spectra. In the limit of very high peaks, the peaks theory begins to predict a much larger number, and we can make arguments that, in this limit, the Press–Schechter theory is underestimating the number of objects (Thomas and Couchman 1992; Liddle and Lyth 1993a). Even there though, the difference probably is not significant in terms of obtaining constraints because other uncertainties dominate.

In the rest of this book, we use Press–Schechter rather than peaks theory in computing the number densities of objects. The advantage is that Press–Schechter theory is easier to apply and is known to be in good agreement with numerical simulations. Thus far, peaks theory has not been subjected to numerical testing of the same level of rigour.

However, something that is accessible via peaks theory and not Press–Schechter is estimates of the bias parameter; if a given class of objects forms from significantly high peaks, then those peaks will exhibit excess clustering. In addition to this bias from the peak locations in the initial density field, there is also a contribution from additional nonlinear clustering after the objects form. If both effects are included, the bias parameter takes on quite a complex form, depending both on the shape of the underlying spectrum on the relevant scale and on an asphericity parameter measuring the failure of the peaks to be perfectly spherically symmetric. Full definitions can be found in BBKS and Liddle and Lyth (1993a). Generally speaking, the bias parameters predicted by this method tend to be reasonable, but with too large an uncertainty to say anything very precise, especially because further astrophysical effects are expected to contribute to biasing (see, e.g., Kauffmann et al. 1997).

11.4   Numerical simulations

Although this is primarily a book about linear perturbation theory, for completeness we say something here about numerical techniques that can be used to explore the nonlinear regime. Numerical simulation of cosmological models has become a vast industry, and allows a model to be subjected to the most stringent possible tests. The drawback is that numerical simulations remain time consuming, and individual simulations are only able to test a particular combination of parameters. As we have seen, structure formation models have many free parameters available for variation, concerning both the cosmological model and the form of the initial perturbations, and it has not proven possible so far to use numerical simulation to properly sample this parameter space.

We begin with a brief description, which we keep qualitative, of some approximate methods for dealing with nonlinear evolution. Although these approximate methods can have an analytic form, in most cases in practice, it is necessary to implement them numerically, the chief advantage being computational speed as compared to that of full numerical simulations. An extensive review is that of Sahni and Coles (1995); see also Coles and Lucchin (1995).

11.4.1   Approximate methods

Second-order perturbation theory

Linear theory applies from the time the perturbation was very small up until it reaches order unity. It has become quite popular to extend linear theory by continuing the perturbation expansion to second order (Peebles 1980; Fry 1984; Bouchet et al. 1992; Scoccimarro and Frieman 1996), even though this clearly has validity only in a very limited regime before all the higher-order terms become important. Second-order perturbation theory comes in two versions, Eulerian or Lagrangian, depending on whether the coordinates are fixed or allowed to follow the fluid. The most important use of these techniques at the moment appears to be the analytic computation of the development of skewness and higher moments, as the density field begins to become non-Gaussian on entering the nonlinear regime.

Zel'dovich approximation

The Zel'dovich approximation (Zel'dovich 1970; Shandarin and Zel'dovich 1989) is a remarkably successful approximation to the entry to the nonlinear regime. It is used routinely to set up the initial conditions for full numerical simulations. It takes the form of a mapping from initial conditions to the final situation, of the form

$$\mathbf{x} = \mathbf{q} + D(t)\mathbf{u}. \tag{11.22}$$

In this expression, $\mathbf{x}$ and $\mathbf{q}$ are the final and initial (comoving) coordinates of the particle, $\mathbf{u}$ is the initial velocity field, and $D(t)$ is the growth factor for density perturbations, given by $t^{2/3}$ if the Universe has critical density [Eq. (4.99)], and including the growth suppression [Eq. (6.12) or Eq. (6.24)] in the low-density case. Assuming domination by the growing mode, the velocity term is the gradient of the potential, Eq. (4.120). In the Zel'dovich approximation, the particles move on straight-line trajectories.

The Zel'dovich formula reproduces linear theory when the displacements are small; the crucial observation is that it might continue to be a reasonable approximation as the nonlinear regime is entered, even though the particle motions are governed entirely by the initial conditions and have no awareness of the developing structures. This indeed turns out to be the case, as verified by detailed comparison to N-body simulations; the approximation holds well up until "shell crossing," the time at which particle motions first intersect. At that point the density becomes formally infinite. The shell crossing occurs on two-dimensional surfaces known as Zel'dovich pancakes or caustics, which are the first structures to form. In the Zel'dovich approximation, the particles are unaware that this has happened, and continue to drift out the far side of the pancake, broadening it in a way that is not seen in full simulations.

If we are interested in large scales, for example, to study galaxy clusters, we can prevent the caustic formation by first smoothing out the short-scale power in the spectrum. This is known as the truncated Zel'dovich approximation, and the improvement is seen in comparison to N-body simulations.

Numerically, the power of the Zel'dovich approximation is that a simulation involves a single time step, taking us directly from the initial configuration to the final one.

Finally, we remark that although the standard treatment is Newtonian, the Zel'dovich approximation can be recovered from general relativity in the appropriate limit (Salopek et al. 1994).

Adhesion approximation

The adhesion approximation deals with the problem of caustics in another way, which is to assume a viscosity that causes particles to "stick" when they reach caustics, mimicking the enhanced gravitational forces that will be present there. Mathematically, this leads to Burger's equation, and the geometric interpretation of the approximation is also very elegant.

Frozen-potential approximation

The frozen-potential approximation (Brainerd et al. 1993; Bagla and Padmanabhan 1994) takes advantage of the fact that, in a critical-density Universe, the gravitational potential is constant in the linear regime, from Eqs. (4.91) and (4.99), to make the assumption that the potential might remain more or less constant even in the nonlinear regime. Here, the particle motion at a given point during the simulation is governed by the gravitational potential at that point *at the beginning of the simulation*. Unlike the Zel'dovich approximation, a simulation therefore does involve a series of time steps, but, compared to what is required for a full simulation, there is a huge saving in not having to recompute the gravitational potential at each time step. The physical basis of the approximation is also much stronger.

Because the caustics tend to develop in the minima of the potential, the particles that reach them oscillate about the caustic, with their velocities damped by the expansion, giving a behaviour quite similar to that seen in N-body simulations.

There is also an approximation scheme known as frozen flow (Matarrese et al. 1992), where it is assumed that the initial velocity field is fixed, but this approximation seems to fare the worst in comparison with simulations.

11.4.2 N-body simulations

Simulations that attempt to model only the gravitational force are known as N-body simulations. Very early programs operated by direct calculation of the forces between all the particles (von Hoerner 1960; Aarseth 1963), but this proves computationally costly. Modern codes tend to be one of two types: **P^{3}M** (particle–particle, particle–mesh) codes compute the gravitational potential on a grid that is used for long-distance interactions, with only close interactions computed using direct force summation. Tree codes, instead, group collections of particles into a hierarchical structure, and compute long-distance forces by replacing the force from groups of particles with a single force from the group. Both types of codes can be made "adaptive," meaning that they automatically increase the grid resolution in those high-density regions that most need it.

Simulations now can be run with very large numbers of particles indeed. Simulations of 64^3 and 128^3 particles are commonplace, and a group calling themselves the Virgo Consortium has run the first billion-particle simulation (Colberg et al. 1998). Typically, the simulation box is order 100 Mpc on a side for a full cosmological simulation, the size being necessary to ensure that all of the necessary large-scale power from a cold dark matter (CDM)-like spectrum is included. Smaller-scale simulations are used to probe situations such as cluster formation and virialization, or in an attempt to model galaxy formation.

One very fast adaptive N-body code with gas hydrodynamics, which is publically available, is Hydra (Couchman et al. 1995, 1996). Simulation data obtained from large runs of this code also are available (Thomas et al. 1996), which can be used, for example, to calibrate data analysis techniques.

The use of N-body simulations, especially by Davis, Efstathiou, Frenk, and White in a series of papers (White et al. 1983a,b, 1987; Davis et al. 1985; Frenk et al. 1988, 1990), was the principal driving force in the development of the standard CDM (SCDM) model, because observations available in the mid-eighties primarily referred to scales in the nonlinear regime. In particular, these simulations brought an understanding of the galaxy correlation function, and led to the introduction of biasing to explain the observations of the time. Since the demise of SCDM, simulations have been employed to explore the nonlinear evolution in the many variants that have surfaced in its wake. Although, undoubtedly, useful information has become available this way, the present inability to probe the full parameter space available for structure formation models means that simulations can be of use only when allied to the more economical techniques to which the bulk of this book is devoted.

11.4.3 Hydrodynamical simulations

Recently, much of the work on numerical simulations has been devoted to extending N-body simulations to include hydrodynamical processes influencing the behaviour of the baryonic component of the matter in the Universe. Again, two techniques are presently available. Smoothed particle hydrodynamics (SPH) models the baryons as particles that carry certain properties, such as temperature, at each time step generating a fluid by smoothing over neighbouring gas particles (Evrard 1988; Hernquist and Katz 1989; Katz et al. 1992, 1996a; Pearce and Couchman 1997). The alternative technique, called Eulerian hydrodynamics, is to track the gas behaviour on a grid (Cen 1992; Cen and Ostriker 1992a,b; Cen and Ostriker 1996).

The introduction of hydrodynamics introduces the possibility of analyzing a variety of effects that are inaccessible to purely gravitational codes. For example, in hydrodynamical codes, galaxy formation can be modeled more properly in terms of cooling conditions, whereas, in

N-body simulations, galaxies are identified by some ad hoc linking method (e.g., Lacey and Cole 1994). This enables an investigation of biasing, including its scale dependence (Cen and Ostriker 1992b). Other properties that can be studied include the gas distribution in clusters (e.g., Navarro et al. 1995) and quasar absorption systems (Hernquist et al. 1996; Katz et al. 1996b; Zhang et al. 1998).

One worry with hydrodynamical simulations is that once you begin to put in hydrodynamical processes you have to wonder where to stop, because so many processes are potentially of importance. For example, only in some codes is photoionization of the intergalactic gas by ultraviolet photons included, even though it is known that this process must completely reionize[4] the Universe by a redshift of 5, which surely influences the subsequent development of structure. Energy feedback into the intergalactic gas from supernovae is another process that almost certainly has important consequences.

11.5 Applications of Press–Schechter theory

Press–Schechter theory provides the shortest-scale constraints on the linear spectrum that we have. Its most reliable application is on the scale that is just going nonlinear, so that the abundance of the objects studied is small enough to be in the regime where it is very sensitive to the matter spectrum normalization, yet large enough to be measured to reasonable accuracy. At the present epoch, this means scales around $8h^{-1}$ Mpc, corresponding to a mass of around $10^{14} M_\odot$ (depending somewhat on Ω_0 and h), the mass of a cluster of galaxies. The galaxy cluster abundance is one of the most important cosmological constraints, in particular, providing extremely strong evidence against the SCDM model. However, the technical details of applying Press–Schechter theory to the cluster abundance are in fact the hardest of all, and so, we leave them until last.

At the present epoch, galaxy clusters are the only type of object that can be studied by Press–Schechter theory. However, we can go to shorter scales, provided that the observations are made at an earlier epoch, when the matter spectrum was less evolved. Any high-redshift object that can be identified as rare can be studied using Press–Schechter theory because the rarity immediately implies that the scale is in the quasi-linear regime. Three types of object have been studied, damped Lyman-alpha systems (DLASs), quasars, and, more recently, high-redshift galaxies identified using the Lyman break. All of them sample the matter spectrum in the redshift range 3 to 4. Of these, the first is the easiest to deal with theoretically and also the most powerful, and so, we begin there.

11.5.1 Damped Lyman alpha systems

DLASs are prominent absorption features seen in quasar spectra. They correspond to rather high column densities of neutral hydrogen of upward of 10^{20} atoms per square centimeter, and are the largest features seen in quasar spectra. They are believed to correspond to gravitationally

---

[4] We discuss reionization in a semianalytic context later in this chapter.

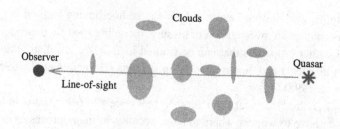

Fig. 11.3. Schematic of the origin of features in quasar absorption spectra. Light from the quasar redshifts on route to us, and so, the absorption feature from resonant Lyman-alpha scattering moves toward the blue end of the spectrum, creating a complex series of absorption lines.

bound clouds of gas, with masses at least of small galaxies. Figure 11.3 illustrates this situation. Two other types of absorption features are classified as Lyman limit systems and Lyman-alpha forest lines; these correspond to lower column densities. DLASs contain most of the neutral hydrogen in the absorption systems, and it is believed that the neutral hydrogen dominates over the ionized hydrogen (which is not directly observed), because the column density is so high that the outer regions shield the bulk of the object from ionizing radiation. The total mass of neutral hydrogen in these systems has been estimated as a function of redshift up to redshift 3 by Lanzetta et al. (1995), and to redshift 4 by Storrie-Lombardi et al. (1996), who revised the earlier results downward somewhat.

The DLAS constraint is simple to apply because the absorption systems directly estimate the density of neutral gas in bound objects, allowing us to use Eq. (11.18) rather than the more complicated Eq. (11.21). We give a simplified outline here, loosely following Liddle et al. (1996b).

For the remainder of this discussion, we assume a critical-density Universe; the generalization to low-density Universes can be found in Liddle et al. (1996c). In a critical-density Universe, the redshift 4 data from Storrie-Lombardi et al. (1996) gives a density of neutral hydrogen of

$$\Omega_{HI}(z = 4) = (9.4 \pm 3.8) \times 10^{-4} h^{-1}. \tag{11.23}$$

However, the Press–Schechter formula predicts the total mass in bound objects above a certain mass, and so, we need further assumptions. To be conservative, we can assume that all of the hydrogen is in its neutral state, and also lower the threshold δ_c to 1.5 to allow for the possibility that the absorption is caused by objects that are not fully virialized [as supported by some numerical simulations, e.g. Katz et al. (1996b)]. To obtain the total density in bound objects, something must be assumed about the way the baryonic matter traces the dark matter; the simplest assumption is that they trace each other accurately, in which case Eq. (11.23) is simply multiplied by Ω_b^{-1}. Finally, we need an estimate of the mass to which these objects correspond. This is actually not so important because the strongly constrained models are those with little short-scale power, which implies an abundance that does not depend too sensitively on the minimum mass considered. The normal assumption is a mass of around $10^{10} h^{-1} M_\odot$, based on the requirement that the collapsing protospheroids are massive enough to give rise to rotationally supported gaseous discs. The corresponding linear scale is $0.2h^{-1}$ Mpc, from Eq. (4.54).

If we put all of this together and take the central nucleosynthesis value for Ω_b, the observations give

$$f(>10^{10}h^{-1}M_\odot, z = 4) = (0.059 \pm 0.026)\,h. \tag{11.24}$$

The error bar is pretty big, and so, assuming Gaussian errors is not a good thing; the observational procedure suggests taking the error as Gaussian in the logarithm giving a 95-percent lower limit $f > 0.018\,h$. At some level, it does not really matter what we do, because the result is so insensitive. Inserting this limit into Eq. (11.18), for $h = 0.5$, gives

$$\sigma(0.2h^{-1}\,\text{Mpc}, z = 4) > 0.65. \tag{11.25}$$

In Chapter 12, we use the constraint given by Liddle et al. (1996b), which fits the h-dependent 95-percent confidence lower limit as

$$\sigma(0.2h^{-1}\,\text{Mpc}, z = 4) > 0.50 + 0.2h. \tag{11.26}$$

This can be compared directly to the theoretical prediction. In a CDM Universe, this can be scaled to the present simply by multiplying by $1 + z = 5$; in a Cold plus Hot Dark Matter (CHDM) Universe, the slower growth rate of perturbations on these short scales also could be taken into account (as in, e.g., Liddle et al. 1996b). We therefore obtain a linear-theory constraint on scales that, at the present epoch, are certainly far from linear.

The importance of the DLAS constraint is that it applies on very short scales, and excludes models whose spectra fall very sharply in this limit. The CHDM model has this property, and the constraint was applied first by Mo and Miralde-Escudé (1994) and by Kauffmann and Charlot (1994), who claimed that the then-standard CHDM model with 30-percent hot dark matter was excluded. This was investigated further using numerical simulations by Ma and Bertschinger (1994) and Klypin et al. (1995), and extended semianalytically to a wide class of models by Liddle et al. (1996a,b,c). We look at the constraints in detail in Chapter 12.

11.5.2 Quasars

Historically, quasars were the first objects to be used to constrain the spectrum in this way, by Efstathiou and Rees (1988). At that time the favoured structure formation model was a highly biased CDM model at well below the Cosmic Background Explorer (COBE) normalization, and these two features in combination made the situation quite marginal. The models now under consideration are less biased and at higher normalization, and they pass the quasar constraint comfortably. It turns out to be quite a bit weaker than the DLAS constraint.

Quasars are believed to be supermassive black holes powered by accretion near the Eddington limit, with the bright ones hosted in galaxies of masses around $10^{12}M_\odot$ and above. A detailed study has been given by Haehnelt (1993). At redshift 4, the comoving number density of bright quasars is about $5 \times 10^{-8}(h^{-1}\,\text{Mpc})^{-3}$. To compare with theory, we therefore have to evaluate

Eq. (11.21). This gives (Liddle et al. 1996b)

$$\sigma(0.95h^{-1}\,\mathrm{Mpc}, z = 4) \geq 0.26, \tag{11.27}$$

which is quite a bit weaker than the DLAS constraint quoted in Section 11.5.1. We should not be surprised; this calculation makes the conservative assumption that every suitably massive galaxy will host a quasar, and that only a single generation of quasars is required to supply the observed number density. In reality, both assumptions are likely to fail, and indeed, the weakness of the constraint suggests that they do.

11.5.3 High-redshift galaxies

An exciting development is the identification of high-redshift galaxies between redshifts 3 and 3.5 via the location of the redshifted Lyman continuum break (Giavalisco et al. 1996; Steidel et al. 1996a,b). Follow-up spectroscopy confirms that almost all candidates identified as dropouts through Hubble Space Telescope filters are indeed in the redshift range given above. They have been used by Mo and Fukugita (1996) and White et al. (1996) to constrain cosmologies.

The comoving number density, taken to be at redshift 3.25, is observed to be 3×10^{-3} $(h^{-1}\,\mathrm{Mpc})^{-3}$, significantly higher than the quasar density, though at a slightly lower redshift. If the observed velocity dispersions of around 200 to 300 km $\cdot$ s^{-1} are entirely due to gravitational motions, the masses have to be about $10^{12} M_\odot$ [see Eq. (11.16)]. Additional line broadening, such as shocks, would lower this mass estimate. White et al. (1996) found for a certain class of cosmological models that the constraint could be quite stringent if the masses are as high as $10^{12} M_\odot$, but is comfortably passed if they are an order of magnitude lighter.

11.5.4 Galaxy clusters

The abundance of galaxy clusters is presently the most important application of Press–Schechter theory, giving a matter spectrum constraint with uncertainty not much greater than COBE and of course applying at a much shorter length scale. The technique was introduced by Evrard (1989) and Henry and Arnaud (1991) and was popularized through the paper of White et al. (1993a). Since then, attention has been focused on studying as wide a range of cosmologies as possible, recent papers including Bond and Myers (1996), Eke et al. (1996), Viana and Liddle (1996), and Pen (1998). The calculations have reached a level of sophistication that is impossible to reproduce here; we provide a short summary and refer the reader to those papers for a full description.

The number density of clusters selected by different means is fairly accurately established; the principal uncertainty is in determining the masses of the clusters. Three methods are currently in use:

(1) **Virial mass.** Redshifts of cluster galaxies are used to determine the virial velocity, which is converted into a mass using more sophisticated versions of Eq. (11.16). The

main drawback of this approach is that the virial velocity can be systematically over-estimated through projection effects (see, e.g., van Haarlem et al. 1997), for example, a small separate cluster in the foreground or background.

(2) **X-ray mass.** X-ray images of the hot gas in clusters allow modelling of the gravitational potential confining the gas. This technique is insensitive to projection, but may underestimate the mass if the modelling of the gas distribution is too simplistic, with extra mass needed to counteract additional sources of pressure support. The earliest papers used this method, with data available from Edge et al. (1990) and Henry and Arnaud (1991).[5]

(3) **Lensing mass.** Weak gravitational lensing distorts the shapes of background galaxies, stretching the images tangentially to the cluster centre (Kaiser and Squires 1993; Fahlman et al. 1994). Potentially, this is the most unambiguous mass determination, the main problem at present being that existing charge-coupled device (CCD) cameras are too small to allow a large enough image to be made with sufficient detail to measure the mass at a large radius.

As we indicated, X-ray masses tend to be lower than the other two, which more or less agree, on average (see, e.g., Wu and Fang 1997), though with some dispersion on a cluster-by-cluster basis. The detailed results obtained therefore depend on precisely the type of observation used.

The simplest application is in a critical-density Universe, because there, density perturbations have been growing significantly right up to the present and we can assume that rare objects are just forming. In low-density Universes, structure ceased growing at moderate redshift, and the clusters we see will have formed over a range of redshifts. From the usual scaling law (11.16), clusters of the same mass but forming at different redshifts have different temperatures/virial velocities, and a sophisticated treatment can attempt to allow for this.

For the X-ray temperature, calibration of the scaling laws via hydrodynamical simulations is required, and this has been carried out by White et al. (1993b). The data can be treated in different ways. Viana and Liddle (1996) concentrated on the most X-ray bright clusters (where the assumption of spherical collapse is expected to be best). Eke et al. (1996) used numerical simulations to fit to the entire X-ray temperature function, and noted that the bright clusters have a systematically higher number density than permitted by a fit to the full data.

The main application of the cluster abundance is to fix the power spectrum normalization, usually expressed via the dispersion σ_8 given by a top-hat smoothing on a scale $8h^{-1}$ Mpc. We quote the results of Viana and Liddle (1996), updated in Viana and Liddle (1999), which we use in the next chapter to constrain models. Because they make less use of numerical simulations than Eke et al. (1996), the uncertainties are somewhat greater due to the extra use of scaling laws.

To illustrate the simplest application of the method, we discuss only a simplified application in the critical-density case. The number density of clusters at temperature 7 keV at the present epoch is (Henry and Arnaud 1991)

$$n(7\,\text{keV}, z = 0) = 2.0^{+2.0}_{-0.1} \times 10^{-7} h^3 \,\text{Mpc}^{-3} \,\text{keV}^{-1}, \tag{11.28}$$

---

[5] Eke et al. (1996) note that there are two errors in the data analysis by Henry and Arnaud (1991); although each is quite significant, they fortunately nearly cancel.

where the uncertainty acts logarithmically. The simulations of White et al. (1993b) indicate that, in a critical-density Universe, a cluster with X-ray temperature $T = 7.5$ keV has a mass within an Abell radius (i.e., $1.5h^{-1}$ Mpc) of $M_{\text{Abell}} = (1.1 \pm 0.2) \times 10^{15} h^{-1} M_\odot$. Two things have to be done to this result to get what we want. First, we want the mass within the virial radius, not an Abell radius, which comes from assuming that this encloses an overdensity of 178 [from Eq. (11.11)] and a standard mass profile $\rho \propto r^{-2.4}$ from simulations (White et al. 1993b; Navarro et al. 1995). Second, we need to scale from 7.5 keV to 7 keV, the temperatures being close enough that the scaling in Eq. (11.17) is easily good enough. These give an equivalent virial mass (Viana and Liddle 1996) of

$$M_{\text{vir}} = (1.2 \pm 0.3) \times 10^{15} h^{-1} M_\odot.$$

$$(11.29)$$

If we assume top-hat smoothing so that $M = 4\pi\rho_0 R^3/3$, this mass corresponds to a scale of $10h^{-1}$ Mpc. Equation (11.17) gives the transformation of the interval dT to dM, namely,

$$d(k_B T) = \frac{2}{3} \frac{k_B T}{M} dM.$$

$$(11.30)$$

To evaluate the Press–Schechter formula, we need to compute the derivative of $\sigma(M)$; this could be done using the full transfer function, but an adequate power-law fit (White et al. 1993a; Viana and Liddle 1996) is

$$\sigma(R) = \sigma_8 \left(\frac{R}{8h^{-1}\,\text{Mpc}}\right)^{-\gamma},$$

$$(11.31)$$

where, for a CDM spectrum,

$$\gamma \simeq 2.9(0.3\Gamma + 0.2).$$

$$(11.32)$$

We take the shape parameter to be $\Gamma = 0.25$, giving $\gamma = 0.80$.

The Press–Schechter formula (11.20) becomes

$$\frac{dn(M)}{dM} dM = \sqrt{\frac{2}{\pi}} \frac{\rho_0}{M^2} \frac{0.8}{3} \frac{\delta_c}{\sigma(M)} \exp\left(-\frac{\delta_c^2}{2\sigma^2(M)}\right) dM.$$

$$(11.33)$$

Converting to temperatures, approximating the differentials as finite differences, and substituting in the values from Eqs. (2.14), (11.28), and (11.29) gives

$$\frac{\delta_c}{\sigma(10h^{-1}\,\text{Mpc})} \exp\left(-\frac{\delta_c^2}{2\sigma^2(10h^{-1}\,\text{Mpc})}\right) = 0.018.$$

$$(11.34)$$

The solution is $\delta_c/\sigma(10h^{-1}\,\text{Mpc}) = 3.2$, which, for the usual assumption $\delta_c = 1.7$, gives

$$\sigma(10h^{-1}\,\text{Mpc}) = 0.53.$$

$$(11.35)$$

Conventionally, the cluster abundance is quoted at the scale of $8h^{-1}$ Mpc; if we continue to assume a CDM spectrum with $\Gamma = 0.25$, the ratio of powers at $8h^{-1}$ Mpc and $10h^{-1}$ Mpc is 1.20, giving the final result,

$$\sigma_8 = 0.64.$$

$$(11.36)$$

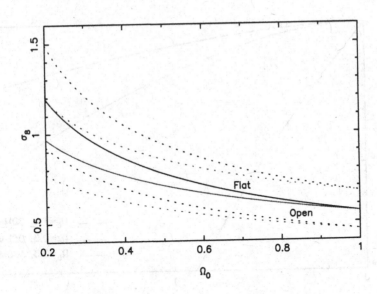

Fig. 11.4. Abundance constraint on σ_8 for flat and open cosmologies, with the central value shown as solid lines and the 95-percent confidence range delimited by the dashed lines.

Despite the crudeness of the approach just described, it is in pretty good agreement with the results of more sophisticated analyses.

For this approach to be useful, we also need to keep track of the uncertainties, as well as to improve the sophistication of the method and to generalize to a range of cosmologies. The results found by Viana and Liddle (1996, 1999) apply to both flat and open cosmologies. They quote the result in terms of a fitting function

$$\sigma_8 = 0.56\Omega_0^{-C} \quad \text{with} \quad C = 0.34 \quad \text{(open)}, \qquad C = 0.47 \quad \text{(flat)}. \tag{11.37}$$

The uncertainty quoted is $+20\Omega_0^{0.1\log_{10}\Omega_0}$ and $-18\Omega_0^{0.1\log_{10}\Omega_0}$ in the open case, whereas it is $+20\Omega_0^{0.2\log_{10}\Omega_0}$ and $-18\Omega_0^{0.2\log_{10}\Omega_0}$ in the flat case, both at 95-percent confidence. This is shown in Figure 11.4. The semianalytic nature of their calculation makes the uncertainty quite large; other authors (Eke et al. 1996; Pen 1998) quote rather smaller uncertainties.

All of the preceding refers to clusters in the present Universe. A potentially very strong diagnostic of the density parameter Ω_0 is the dependence of the cluster abundance on redshift (Frenk et al. 1990; Oukbir and Blanchard 1992; Richstone et al. 1992), about which little is known, the best observations being those of Luppino and Gioia (1995), Carlberg et al. (1997), and Henry (1997). Tentative indications of a population of high-redshift clusters come from Deltorn et al. (1997), Donahue et al. (1998), and Jones et al. (1997). In a critical-density Universe, the density perturbation has been growing significantly right up to the present, and so, we expect far fewer clusters at high redshift (for clusters, "high" meaning around $z \simeq 1$), whereas, in a low-density Universe, the perturbations have frozen out and a similar number of clusters are expected at redshift 1 to redshift 0. Viana and Liddle (1996) concluded by

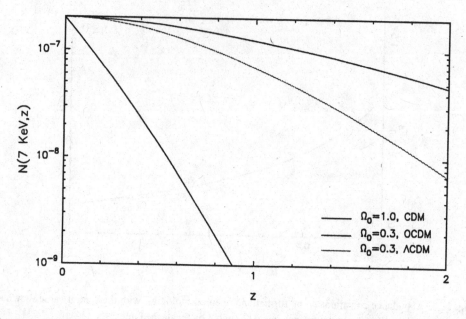

Fig. 11.5. Abundance of rich clusters as a function of redshift, with the leftmost line showing critical density and the others showing open and flat models with $\Omega_0 = 0.3$. The curves are normalized to reproduce the present observed abundance. In low-density Universes, the anticipated number of high-redshift clusters is vastly higher than in critical-density models, with the open model giving the highest abundance.

modelling mergers that we expect around 100 times as many rich clusters at redshift 1 in a low-density model as in a critical-density one, assuming both are normalized to the same number at redshift; see Figure 11.5. An extremely detailed analysis of a wide range of cluster properties that may be useful at high redshift is that of Eke et al. (1996).

11.6   Reionization of the universe

The final topic of nonlinear evolution that we study is reionization of the Universe, a phenomenon of nonlinear collapse that can feed back into the linear-theory tests by affecting the microwave background anisotropies as well as early structure formation. The effect of reionization on the microwave background is particularly important for this book and we study its effect in Section 5.4. The aim of this section is to complete the circle, by using quasi-linear theory to estimate the likely redshift of reionization.

The crucial piece of observational input is the Gunn–Peterson test [Gunn and Peterson (1965); see Peebles (1993) for a recent discussion]. When we look at a quasar spectrum shortward of the Lyman-alpha emission, there is a complicated collection of absorption lines (the Lyman-alpha forest, Lyman limit, and DLASs), but, between these absorption lines, the intensity returns more or less to a continuum level (normally obtained by extrapolation from longward of the Lyman-alpha emission from the quasar). If we posit a smooth intergalactic medium (IGM), any neutral hydrogen component would absorb (and quickly isotropically reemit) this continuum. Conservatively, observations allow an optical depth in the continuum of no more than 10 percent at any redshift (Steidel and Sargent 1987; Webb et al. 1992); this translates into the extraordinarily strong bound (Peebles 1993)

$$\Omega_{HI}(z) \lesssim 2 \times 10^{-7} \Omega^{1/2}(z) h^{-1} (1+z)^{-3/2}, \tag{11.38}$$

where $\Omega_{HI}(z)$ is the (redshift-dependent) density of neutral hydrogen in the form of a smooth IGM (i.e., contributions from identified lines, such as the DLAS abundance quoted earlier, are not included). Because the total baryon density is considerably higher than this, the conclusion is that the Universe is already in a very high state of ionization by the redshifts of the most distant known quasars, at $z \simeq 5$.

In practice, the situation is complicated in that there is no particularly sharp distinction between the IGM and the absorption lines in a hierarchical clustering model [see, e.g., the simulations by Hernquist et al. (1996) and Miralda-Escudé et al. (1996)], but the limit is so strong that the conclusion remains the same.

Ionizing the entire Universe requires a considerable amount of energy, in the form of ionizing ultraviolet photons. Despite the large energy requirement, there are actually three commonly discussed sources able to do the job:

(1) **Massive stars.** Very massive stars have temperatures sufficient to emit primarily in the ultraviolet, and have lifetimes so short that they can be taken as negligible. Because the energy release per nucleon during hydrogen burning is of order 1 MeV, a single fusion event within a star can provide energy sufficient to ionize many thousands of atoms remaining in the IGM.

(2) **Gravitational energy.** Black-hole accretion, for example onto quasars, is potentially an even more efficient way of converting the energy of a nucleon into radiation to provide reionization.

(3) **Decaying particles.** A completely different scheme is where the ionizing photons are generated by late-decaying particles. The best-known example is the decaying-neutrino scenario (Sciama 1990, 1993).

The decaying-neutrino scenario leads to several important changes in the structure formation picture that we have been building. Because it has been described extensively by Sciama (1993), we do not consider it further here. There is also a fourth possibility that has received less attention recently, which is collisional heating from explosive galaxy formation (Ikeuchi and Ostriker 1986), and both cosmic rays and black hole evaporation products have also been mentioned. Here we assume that one of the first two options must be responsible for reionization. It generally is thought [though see Meiksin and Madau (1993)] that, although quasars may well provide sufficient radiation to keep the Universe ionized at low redshift, it is likely to be massive stars that bring about the original reionization. We make that assumption here.

Studies of reionization date back to pioneering work by Couchman and Rees (1986), who were interested primarily in the effect of the energy feedback on further structure formation.

More recent treatments are by Fukugita and Kawasaki (1994), Shapiro et al. (1994), Tegmark et al. (1994), and Haiman and Loeb (1997). Our simplified description mostly follows Tegmark et al. (1994), who made a semianalytic treatment of the SCDM model and some other cosmologies [see also Tegmark and Silk (1995)]. However, these cosmologies were not COBE normalized, something to which the reionization redshift is sensitive; Liddle and Lyth (1995) used the techniques of Tegmark et al. (1994) to study a range of models that were all COBE normalized.

In a hierarchical model, where structure forms exponentially quickly at first, reionization is expected to proceed rapidly. Reionization therefore can be considered to have happened at a single redshift, where the ionization fraction jumps from almost 0 to almost unity. The time at which this happens is tightly tied to the masses of the first objects that form, which is governed by a complex interaction of dynamical and cooling rates. Early work typically assumed that the lightest objects to form would be about $10^6 M_\odot$ (the Jeans mass at recombination), though Couchman and Rees (1986) argued for an order of magnitude lower. More recently, detailed treatments by Tegmark et al. (1997a) and Haiman and Loeb (1997) have appeared, the former more or less confirming the mass range given in the preceding sentence, the latter arguing for smaller values. For illustration, we take the minimum mass to be $M_{min} = 10^6 M_\odot$. Lowering it to $10^5 M_\odot$ gives a smaller uncertainty than others in the calculation (Liddle and Lyth 1995).

The computation of the fraction of baryons that have to be in collapsed objects involves a number of rather uncertain "efficiency factors." We need to estimate what fraction of the baryons end up in stars massive enough to emit ultraviolet photons, and how effective these photons are at escaping from the galaxy in which they are born into the IGM. Tegmark et al. (1994) put all of this together to come up with a best estimate,

$$f_{ion} \simeq 0.008, \tag{11.39}$$

with an uncertainty of potentially 2 orders of magnitude in either direction. Fortunately, these large modelling uncertainties are not reflected in the accuracy of the final answer for the reionization redshift.

The Press–Schechter formula (11.18) immediately gives us the collapse fraction as a function of redshift; we want the redshift z_{ion} such that

$$\sigma(M_{min}, z_{ion}) = \frac{\sqrt{2}}{\delta_c} \, \text{erfc}^{-1}(f_{ion}). \tag{11.40}$$

It is because $\text{erfc}^{-1}(f)$ is such a weak function of f for small arguments that the huge modelling uncertainties are not disastrous.

If we simplify further to a critical-density matter-dominated Universe, we have $\sigma \propto 1/(1+z)$ giving

$$1 + z_{ion} = \frac{\sqrt{2}\,\sigma(M_{min}, 0)}{\delta_c} \, \text{erfc}^{-1}(f_{ion}) \simeq 1.6\,\sigma(M_{min}, 0). \tag{11.41}$$

From Eq. (2.14), our choice of minimum mass corresponds to a radius $R_{min} = 7 \times 10^{-3} h^{-1}$ Mpc (for $h = 0.5$), by some way the shortest scale that we have yet considered. For a CDM-like

Table 11.1. *Estimated reionization redshifts in a range
of cosmologies, with a rough guide to the uncertainties*

Model	Estimated Reionization Redshift		
	Low	Central	High
SCDM	23	37	56
CDM, $\Gamma = 0.25$	7	13	19
CDM, $n = 0.7$	5	8	13
CHDM, $\Omega_\nu = 0.2$	8	13	20
ΛCDM, $\Omega_0 = 0.3$	24	40	61
OCDM, $\Omega_0 = 0.3$	19	33	49

Note: Based on Tegmark et al. (1994) and Liddle and Lyth
(1995). SCDM is included for illustration though it provides
a poor fit to observational data.

spectrum with $\Gamma = 0.25$ normalized to COBE, $\sigma(0.007h^{-1}\,\mathrm{Mpc}) = 8.5$, which gives an
estimated reionization redshift $z_{ion} = 13$. This value is typically what is expected in any critical-
density CDM model because the matter spectrum shape is well determined by observations,
at least from a scale of 1 Mpc upward.

In other cosmologies, the appropriate redshift dependence of the dispersion must be taken
into account (Liddle and Lyth 1995). For CHDM, this is not much different from CDM,
and the estimated reionization redshift is very similar. In low-density models, the effect is
much more dramatic because of the growth suppression at low redshifts from Eqs. (6.12)
and (6.24), which is particularly important in the open case, plus the different COBE normalizations (9.8) and (9.9), which are much higher in the flat case. There is also a slight
reduction because we have to go to a higher length scale to enclose the minimum mass. If we
choose $\Omega_0 = 0.3$ and keep $\Gamma = 0.25$, then the redshift of reionization is predicted to increase
to

$$z_{ion} = 33 \quad \text{(open)}, \tag{11.42}$$

$$z_{ion} = 40 \quad \text{(flat)}. \tag{11.43}$$

Estimated reionization redshifts, with ranges, are shown in Table 11.1.

The preceding analysis is rather simplistic, though arguably the uncertainties are too great
to merit a high level of sophistication. The most detailed treatment to date, that by Haiman and
Loeb (1997), suggests somewhat earlier reionization than the preceding estimate, primarily
because they assume a lower minimum galaxy mass and hence form structures earlier.

We saw in Section 9.3 that cosmic microwave background (cmb) satellite observations with
polarization may be able to measure the optical depth to within 1 percent. This is good enough
to measure the reionization redshift whatever it turns out to be, given the Gunn–Peterson
constraint.

Examples

11.1 Evaluate the integral Press–Schechter mass function $n(>M)$ for the case in which the dispersion $\sigma(M)$ is a power law in M. What range of power laws might be relevant for observations? Plot the differential mass function dn/dM for such a power law and describe its properties.

11.2 What is the fraction of the Universe in bound objects of mass above M if $\sigma(M) = 0.1$, 0.5, 1, 2, assuming top-hat smoothing. Which of these results is most reliable?

11.3 How massive would the DLAS have to be to rule out a $\Gamma = 0.25$ COBE-normalized critical-density CDM model, assuming Eq. (11.25) scaled to the appropriate mass.

11.4 Estimate the reionization redshift obtained in critical-density CDM models for some Γ values in the range 0.2 to 0.5. In each case, compute the percentage suppression of the cmb acoustic peak, using Eq. (5.93), assuming $h = \Gamma$ and Ω_b given by nucleosynthesis.

12 Putting observations together

In this chapter, we put together a set of observations presently available to us (i.e., in 1999), which can be interpreted using the linear and quasi-linear approaches that we have described. At present, no single type of observation is dominant in providing constraints on models of structure formation; instead, the best results come from compiling as wide a set of data as possible, covering a range of scales from our present observable Universe down to the scales of galaxies.

No doubt, the observational details will be superseded quickly, but the general approaches to using them are well established. We also give this discussion here as a post hoc motivation for the models that we considered earlier, both the inflationary aspects and the structure formation scenarios.

A detailed comparison of models with observations requires numerical investigation, to probe the nonlinear regime where many of the observations of galaxy correlations and velocities are made. However, we have seen that there remain a very significant number of undetermined parameters on which the formation of structure depends; we might consider three inflationary parameters, δ_H, n, and r, and several cosmological parameters such as h, Ω_0, Λ, Ω_b, a possible admixture of hot dark matter Ω_{HDM}, and the redshift of reionization z_{ion}. In all the models that we discuss, we need cold dark matter (CDM); inflation-based models do not appear to work without it. It always has whatever density is required to make the total add up correctly. Unless otherwise stated, the baryon density always has the standard nucleosynthesis value $\Omega_b h^2 = 0.016$, from Eq. (2.67).

Present computer technology does not permit a full numerical investigation of such a large parameter space, and only a very small set of models, corresponding to specific choices of these parameters, has been studied. To probe the parameter space more widely, it is necessary to use more economical techniques, accessible to linear and quasi-linear theory (the latter assisted by calibration to numerical simulations for specific parameter values). We have used such techniques to investigate a range of cosmologies (Liddle et al. 1996a,b,c; White et al. 1996), and this chapter is based largely on those works, which can be consulted if the reader desires more detail.

As we see, present observations are not powerfully constraining, and all of the types of model discussed in Chapter 6 are presently at least marginally viable, though it remains to be seen how far into the future that will continue to be the case. Because many possibilities remain, detailed numerical investigation has not yet come into its own as a way of further analyzing promising models, and hence we do not mention this technique further, though it undoubtedly will become more important as observations further improve.

12.1 Observations

In the preceding three chapters, we outlined the most crucial observations accessible to semi-analytical techniques. We use the following:

- **Large-angle cosmic microwave background (cmb) anisotropies.** The crucial Cosmic Background Explorer (COBE) observations are discussed in some depth in Section 9.1. They give an accurate normalization for the matter spectrum applying on very large scales, summarized for the different cosmologies in Table 9.1. The shape of the spectrum n and the actual amount of gravitational waves r are not significantly constrained by COBE.

- **Degree-scale cmb anisotropies.** Many detections of intermediate-angle cmb anisotropies have been reported, mostly around the 1-deg scale ($\ell \simeq 200$). They indicate, in a broad sense, that the rise to an acoustic peak from the Sachs–Wolfe plateau predicted by theory does occur (see Figure 5.9), though the height of the peak is only very weakly constrained. White (1996; see also Liddle et al. 1996c) used a compilation of the data to produce a constraint on the ratio of power at $\ell = 220$ and $\ell = 10$. Because the height of the acoustic peak can be lowered by early reionization (Section 11.6), which is not otherwise constrained, only the upper limit is of direct use.

- **Galaxy and cluster correlations.** We use the data compilation of Peacock and Dodds (1994), updated by Peacock (1996), described in Section 10.1. We drop the four data points corresponding to the shortest scales (in practice, those within the nonlinear regime), which may be contaminated by nonlinear biasing (Peacock 1996). It also has been commented (Smith et al. 1998) that the inversion of nonlinear evolution required for these points may not be as independent of the assumed cosmological model as one would desire. The galaxy data are useful primarily in constraining the shape of the spectrum rather than its amplitude, the latter being uncertain due to the bias parameter. The shape is well determined by the data, assuming that the linear biasing model is adequate on those scales [as suggested by higher-order correlations (Gaztañaga and Frieman 1994) and numerical simulations (Cen and Ostriker 1992b)]. For scale-invariant CDM models, this can be expressed as a constraint on the shape parameter $\Gamma = 0.23 \pm 0.04$ at 2-sigma (Liddle et al. 1996b), but we exhibit constraints from direct fitting to the data points.

- **Galaxy bulk flows.** The best measures of large-scale motions come from the POTENT method described in Section 10.2. Attempts have been made to determine the matter spectrum from the velocity field (Kolatt and Dekel 1997) but present constraints are very weak. POTENT provides estimates of the bulk flow in spheres of various radii about our position. Because velocities sample a wider range of scales than the density field, the points are correlated quite strongly and we use only one, the bulk flow in a sphere of radius $40h^{-1}$ Mpc. The interpretation of these data is given in Section 10.2.

- **Galaxy cluster abundance.** After COBE, the galaxy cluster abundance is the most crucial piece of existing data. Luckily, clusters are rare enough to be in the regime where their number density is very sensitive to the matter spectrum normalization, but common enough for their number density to be determined accurately. In fact, the biggest uncertainty comes from determining the masses of the clusters. Many authors have looked at this constraint, beginning with Evrard (1989) and Henry and Arnaud (1991), with the technique being popularized by a paper by White et al. (1993a). We use the constraints obtained by Viana and Liddle (1996, 1999). Their results have the advantage of being available for a wide range of cosmologies, but the disadvantage that the use of numerical calibration was minimal, which leads to larger modelling uncertainties than those of other authors (e.g., Eke et al. 1996; Cole et al. 1997) and hence rather conservative constraints. Conventionally, the cluster abundance is quoted as a limit on σ_8, and their results are quoted in Section 11.5, Eqs. (11.37), (11.37) and (11.37).

- **Quasar, damped Lyman-alpha system (DLAS) and Lyman-break galaxy abundance.** Galaxy clusters are the smallest objects in the present Universe accessible to linear theory. However, we can reach smaller scales by looking to high-redshift observations, which can probe scales around $1h^{-1}$ Mpc at a time when they were linear (by definition, if we can find any rare object, it must be probing the quasi-linear regime). Three types of object have been probed so far. Quasars were the first to be studied (Efstathiou and Rees 1988), but have so many modelling uncertainties (host galaxy mass, lifetime, formation probability) that only very weak constraints are found. Their use has been superseded by DLAS abundance, which has a very easy interpretation in Press–Schechter theory, and by the abundance of high-redshift galaxies identified by the Lyman break.

Figure 12.1 shows the observational data, in terms of the dispersion $\sigma(R)$. This figure is valid only for critical density, and is only for illustration and not for use in any statistical comparison. Degree-scale cmb observations cannot be plotted in this way because they depend too strongly on the cosmological parameters. This figure shows that observations exist across 4 orders of magnitude in scale, covering 6 orders of magnitude in the dispersion. It is impressive to see that the observational data occupy a smooth curve, and indeed one that is similar to some theoretical curves plotted earlier (Figures 5.4 and 6.3).

To see more clearly what is going on, it is best to normalize this figure to a fiducial model, which we take to be the Standard Cold Dark Matter (SCDM) model. We know that this model does not fit the data in detail, but it is good to a factor of a few. In Figure 12.2, we see that the data do indeed fall well below the prediction of SCDM on short scales, and that the galaxy correlation function has the wrong shape. However, the data still track a fairly simple curve, and in the figure, we illustrate a couple of ways in which a better fit can be obtained.

The data we have discussed are too disparate to be combined statistically, for example, in some form of chi-square test. Instead, we constrain models by contouring the 95-percent confidence levels in the parameter space for each type of data, and take the permitted models to be those not excluded at more than 95 percent on any single piece of data. This gives rather more conservative constraints than a statistical combination of the data, but is more secure against the results being biased by badly estimated error bars on one of the sets of observations.

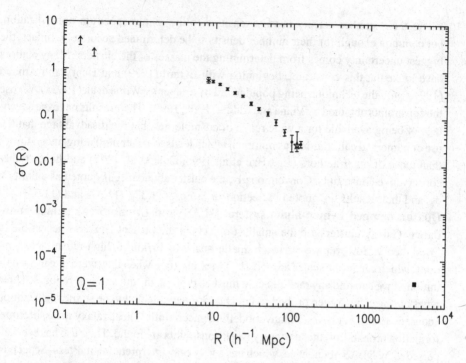

Fig. 12.1. Observational data, with 1-sigma uncertainties and 95-percent confidence lower limits. We represent the COBE data schematically at $4,000\,h^{-1}$ Mpc; they are indicated by a solid square whose size represents the uncertainty. The galaxy correlation function data are shown by circles; an uncertainty in overall normalization is not illustrated. The bulk flow constraint is represented by a star, and the cluster abundance constraint by a cross. The lower limits are DLAS (left) and quasars (right).

12.2   Critical-density models

12.2.1   CDM models

The simplest paradigm, which we used as the basis for our discussion in Chapters 4 and 5, is a Universe of critical density in which all the dark matter is cold. The most famous model of this type is SCDM, which, in addition to those assumptions, takes $n = 1$, $r = 0$, and $h = 0.5$. It is now well known to be excluded; for example, the galaxy correlation function has an unacceptable shape ($\Gamma = 0.44$, assuming Ω_b from nucleosynthesis), and it wildly overproduces clusters when normalized to COBE. This is immediately apparent from Figure 12.2; if SCDM is to be right, then all of the data should line up horizontally.

It would be highly desirable if this simplest scenario could be retained, by capitalizing on the freedom to vary the other parameters listed above. Introducing gravitational waves proves no help (Liddle et al. 1996b), though their presence is not strongly constrained, and so, the main strategy is to lower n, which removes the excess short-scale power in the spectrum that leads to the cluster overproduction. This strategy has become known as tilting the spectrum,

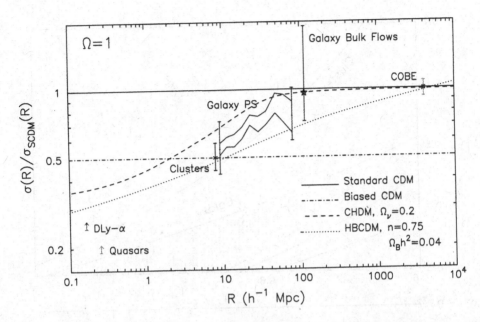

Fig. 12.2. Data normalized to the prediction of the SCDM model. The solid line is SCDM; the others are some sample modifications, namely biased CDM, a Cold plus Hot Dark Matter (CHDM) model, and a titled model with high baryon content. All theoretical models are precisely COBE normalized except the biased CDM model. The galaxy correlation function data are now represented by a 1-sigma band about the (unplotted) central values. We can shift the entire galaxy correlation function data set vertically, corresponding to changing the bias, under control of the large error bars at the end of the band.

and has a long history (Vittorio et al. 1988; Bond 1992; Cen et al. 1992; Liddle et al. 1992). Unfortunately, unless we lower h as well, it is necessary to reduce n very significantly, to below 0.6 or so (Liddle et al. 1996b). Such a strong tilt has a disastrous effect on the predicted height of the acoustic peak in the cmb spectrum, eliminating it completely, which is in conflict with observations (see Figure 5.10 on page 126). The interaction of the constraints is shown in Figure 12.3, and the data can only be fitted if h is extremely low, below about 0.4, which few would regard as tenable (though see Bartlett et al. 1995). Note the crucial role of the acoustic peak height in imposing this constraint.

However, there is one remaining way out, which appears to be the last chance for the simplest model. That is, if the baryon density turns out to be substantially higher than the standard nucleosynthesis value used in Figure 12.4 (White et al. 1995b, 1996). A higher baryon density has two beneficial effects: The extra damping provides an additional reduction in the short-scale matter spectrum [recall that Γ is roughly proportional to $\exp(-2\,\Omega_b)$; see Eq. (5.14)], and the higher pressure at recombination leads to an enhancement of the acoustic peak in the cmb spectrum. In combination, these allow a fit to the data at higher h and n. White et al. (1996) suggested that $h = 0.5$ can be achieved, with $n \simeq 0.75$ to 0.8, provided the baryon density lies somewhere around 12 percent rather than the conventional 6 percent; this is shown in Figure 12.4. In fact,

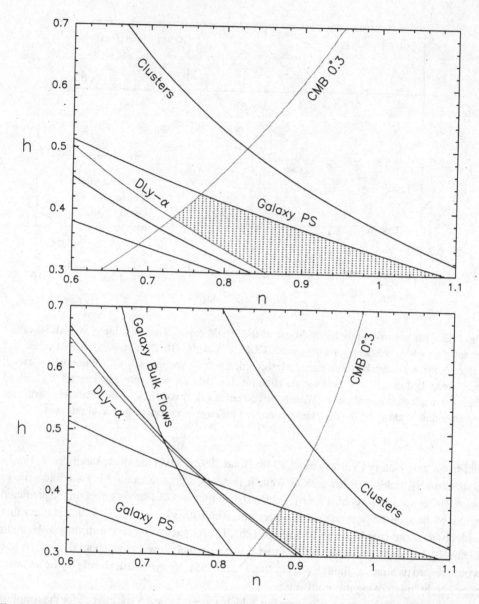

Fig. 12.3. Critical-density CDM models. The top panel is without gravitational waves, the lower one includes those as generated by a specific model – power-law inflation. All constraints are plotted at 95-percent confidence.

this does not seem too unlikely; deuterium observations have been forcing the nucleosynthesis prediction upward, and such a high fraction is probably necessary to explain the observed cluster baryon fraction under the critical-density assumption. On the other hand, even with such a high baryon density, only very restrictive values of h and n are viable. As we write, the critical-density CDM model is struggling to survive even when we just consider large-scale

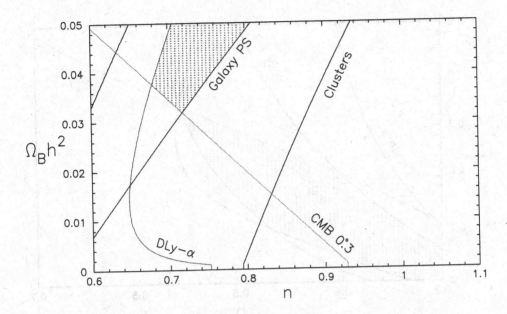

Fig. 12.4. Retaining $h = 0.5$, and salvaging CDM by increasing the baryon content.

structure data, as Figure 12.4 shows: This is without bringing in evidence such as the supernovae magnitude–redshift relation which argue against the entire critical-density paradigm.

12.2.2 Cold plus hot dark matter (CHDM) models

An alternative way to try to salvage critical density is to abandon the CDM hypothesis and introduce an admixture of hot dark matter (HDM), Ω_{HDM}, to give the CHDM model (Bonometto and Valdarnini 1984; Fang et al. 1984; Shafi and Stecker 1984; Holtzman 1989; Schaefer et al. 1989). The freestreaming property of the HDM that we discussed in Chapter 6 naturally produces the desired reduction in short-scale power. This model became very popular during the post-COBE era, partly because it is tenable even retaining the $n = 1, r = 0$ assumptions. The original version had $\Omega_{\mathrm{HDM}} = 0.3$ (Davis et al. 1992b; Schaefer and Shafi 1992, 1994; Taylor and Rowan-Robinson 1992; Klypin et al. 1994), but this fell foul of the DLAS limit because the freestreaming was too effective (it probably also oversteepened the galaxy correlation function).[1] Preferred values now tend to lie around $\Omega_{\mathrm{HDM}} \simeq 0.2$ (Liddle and Lyth 1993b; Klypin et al. 1995; Pogosyan and Starobinsky 1995; Liddle et al. 1996b). Figure 12.5 shows the allowed region. We see that $h \simeq 0.5$ is already allowed, and we have not even used our freedom to vary n. If we do so, then values of h up to about 0.6 are allowed (Liddle et al. 1996b); because this reduces the age of the Universe to below 11 Gyr, which presumably is as far as we would want to go. We also could shift Ω_b away from the nucleosynthesis value.

[1] This is only marginally excluded in Figure 12.5, which shows quite a conservative constraint. Numerical simulations (e.g., Klypin et al. 1995; Ma et al. 1997) are needed to accurately delineate the allowed region.

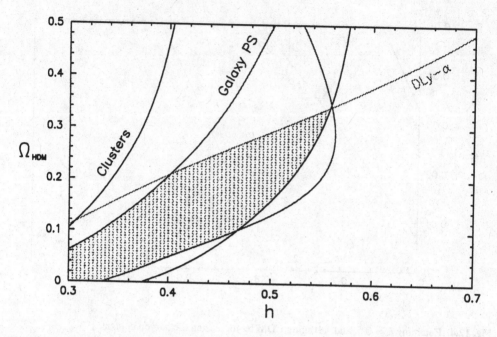

Fig. 12.5. Scale-invariant CHDM models. Note that the DLAS constraint imposes tight upper limits on Ω_{HDM}. All parameters in this figure pass the acoustic peak height constraint, because their prediction accurately matches CDM with the same h.

12.3 Low-density models

Retaining the CDM hypothesis, another simple strategy for altering the shape of the matter spectrum is to lower the matter density. This shifts the epoch of matter–radiation equality, lowering the shape parameter Γ. All observations need to be re-evaluated to allow for the lower density, and we must be careful to distinguish the case of a genuinely open Universe from a flat Universe with a cosmological constant.

The COBE normalization changes quite dramatically in low-density Universes. With a cosmological constant the present normalization becomes higher as Ω_0 is reduced (relative to a critical-density model with the same Γ). In an open Universe, it initially becomes higher, but at $\Omega_0 \lesssim 0.4$ it begins to decline (see Figure 9.1 on page 251) as the line-of-sight term becomes dominant in the Sachs–Wolfe effect and the growth suppression factor (6.12) becomes large. Equations (9.8) and (9.9) give the normalizations.

The very different normalization suggests that low-density models with Ω_0 around 0.3 ought to be quite different from, and hence easily distinguished from, critical-density models. Unfortunately, however, most other observations tend also to require a higher normalization if the Universe has a low density. For example, the cluster abundance normalization goes up, as we saw in Section 11.5, because with a lower density the material required to assemble the cluster has to come from a larger volume. Note that Figure 12.2 would have to be redrawn if the value of Ω_0 were changed.

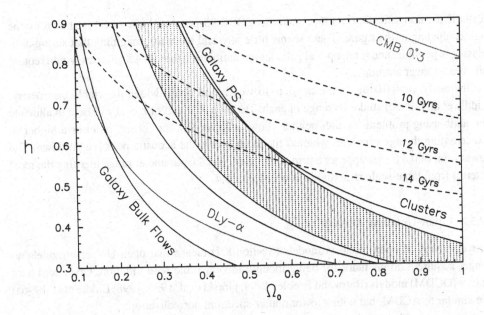

Fig. 12.6. Scale-invariant ΛCDM models. Contours of constant age are shown. The more lightly shaded area toward the top contains models that must be antibiased, even if the COBE normalization is taken to its 95-percent lower limit.

12.3.1 Models with a cosmological constant

The natural way to keep low-density models compatible with the simplest models of inflation, which generate a flat Universe, is to introduce a cosmological constant, giving the ΛCDM model (Peebles 1984; Turner et al. 1984; Efstathiou et al. 1990; Kofman et al. 1993; Krauss and Turner 1995; Ostriker and Steinhardt 1995; Liddle et al. 1996c). The presence of such a cosmological constant is favoured by measurements of the magnitude–redshift relationship in high-redshift type Ia supernovae (Perlmutter et al. 1998; Schmidt et al. 1998). Inflation says nothing about whether there might be a residual vacuum energy at the present epoch. A cosmological constant also maximizes the age of the Universe for a given value of the Hubble constant, Eq. (2.63), though, post-Hipparcos, such a strategy does not really seem necessary. In fact, as we note in Chapter 2, low-density models that fit the large-scale structure observations are actually younger than critical-density models that do so, the smaller Ω_0 being more than compensated by the larger h that is required. This is clear from Figure 12.6.

Figure 12.6 shows the allowed region for cosmological-constant models based on a scale-invariant initial spectrum (Liddle et al. 1996c). Contours of constant age also are shown. We see a substantial allowed region, going down to $\Omega_0 \simeq 0.3$ with no particular preference for any value of Ω_0 within that range. Below that value, the situation begins to become problematic for several reasons. The required h value becomes larger than given by direct measurement, which leads also to a suspiciously young age. Also, the matter-spectrum normalization begins to require a strong antibiasing of optical galaxies; the region where this definitely is required (even if the normalization is lowered to the COBE 95-percent lower limit) is shown by the lighter shading in the figure. There seems little physical basis for imagining that strong antibiasing is possible, and if the optical galaxies are antibiased, then IRAS galaxies would require an even stronger antibias.

Obviously, capitalizing on the freedom to vary n and r will widen the viable parameters; Liddle et al. (1996c) studied a range of each. Taking $n < 1$ and/or $r > 0$ will help to alleviate the antibiasing problem, though neither allows a lower Ω_0; the former requires a higher h, exacerbating the age problem, whereas the latter brings the acoustic peak height down. The lower limit of 0.3 on Ω_0 appears a firm conclusion, and the parameter range meriting the most interest from large-scale structure is now at $\Omega_0 \simeq 0.4$.

12.3.2 Open-Universe models

The discovery of open inflation models (Section 8.7) means that open-Universe models no longer acquire a minus mark for being thought incompatible with inflation. Open cold dark matter (OCDM) models (Ratra and Peebles 1994; Górski et al. 1995, 1998; Liddle et al. 1996a) are similar to ΛCDM, but with a lower matter-spectrum normalization.

As we saw in Chapters 6 and 8, open inflation models potentially lead to a more complicated phenomenology, with extra types of mode function (super-curvature and bubble wall) that also should be included. However, these influence only the very largest angular scales, and the only danger would be a conflict between the predicted spectral shape of cmb anisotropies for $\ell \lesssim 10$ and the COBE data. However, Górski et al. (1998) find that the COBE 4-year data are not able to exclude any of the interesting range. Fortunately, having passed this hurdle, all other observations, including the COBE normalization scale ($\ell \simeq 14$), are on short enough scales that the extra modes can be ignored. Indeed, even spatial curvature can be ignored, allowing the use of the normal flat-space mode functions, as long as we include its effect on the expansion rate via the Friedmann equation. The different expansion rate leads to the suppression of density perturbation growth at late times, Eq. (6.24), and changes the COBE normalization as described in Section 9.1.

Figure 12.7 shows results for the case of a scale-invariant initial spectrum. Overall, it has a look similar to that of the same figure for the cosmological-constant model; in particular, it does not single out a preferred Ω_0, but does impose a lower limit around 0.3 (this time though, in addition to the age, there is a lack of clusters, caused by the falling matter-spectrum normalization at very low densities in this model). Again, the additional freedom to vary n and r extends the range of viable parameters (Liddle et al. 1996a); White and Silk (1996) argue that $\Omega_0 = 0.3$ is possible if $n > 1$.

12.4 Other options

The models that we have discussed are the standard ones, and they provide a range of alternatives capable of fitting present data. However, they do not comprise a complete set of the possibilities that have been investigated, and for completeness we list some more options here.

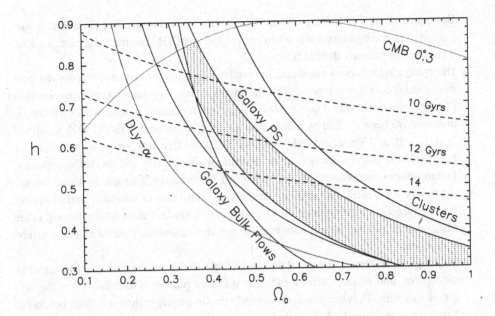

Fig. 12.7. Scale-invariant OCDM models. Constant-age contours are shown.

- **Extra massless species.** Low-density models fit the data well because they shift matter–radiation equality to a later epoch, lengthening the time during which short-scale perturbations have their growth suppressed. Recall that the present density in relativistic species, Eq. (2.31), is assumed, not measured, because the neutrino background is not detectable. So, another way of shifting matter–radiation equality is to add extra massless species (Bardeen et al. 1987; Bond and Efstathiou 1991) to delay the onset of matter domination. However, these are strongly constrained by nucleosynthesis, and so, some trickery involving particle decays is necessary (Dodelson et al. 1994; McNally and Peacock 1995; White et al. 1995a) to generate the extra relativistic energy density late on. Normally, the tau neutrino does the decaying, and in the popular parameter region it has a mass in the kilo-electron volt range and a lifetime of order of years, injecting extra energy density into nonstandard neutrino species. An alternative (Dodelson et al. 1994) is a much shorter lifetime, which can alter the nucleosynthesis dynamics sufficiently to allow many extra massless species with a thermal energy density without overproducing helium.

  To fit the data, an approximate doubling of the number of massless species is required because $\Gamma_{eff} \propto g_*^{-1/2}$. This generates a spectrum that looks like that of a low Hubble constant, while keeping the actual Hubble constant that would be directly measured higher.

- **Nonthermal hot component.** The standard CHDM scenario fixes the relation between the neutrino energy density and its mass (and hence freestreaming length) using the standard relativistic decoupling abundance calculation. However, more elaborate particle physics model building may violate this, allowing extra freedom (Kaiser et al. 1993;

Bonometto et al. 1994; Pierpaoli and Bonometto 1995; Pierpaoli et al. 1996). A particularly interesting situation is when both HDM and CDM are the same particle, with a nonthermal velocity distribution.

- **Decaying cosmological constant.** In "realistic" particle physics models for the cosmological constant, it is possible that the energy density is actually a decaying function of time, sometimes called quintessence. Structure formation models with such contributions have been studied by Coble et al. (1997), Turner and White (1997), Caldwell et al. (1998), and Viana and Liddle (1998). The extra freedom compared to ΛCDM of being able to choose a decay rate for this extra matter clearly widens the possibilities.

- **Isocurvature component.** Models based on isocurvature fluctuations alone do not seem very successful, but nothing prevents an admixture of adiabatic and isocurvature perturbations. Stompor et al. (1996) recently studied this, noting among other things that the COBE 4-year data are not very discriminating against an isocurvature component.

- **Topological defects.** Topological defects (Kibble 1976) such as cosmic strings, global monopoles, and textures offer a radically different paradigm to inflation for the origin of structure. They are much more predictive in principle than inflation, but so far have proven technically difficult and only recently has theoretical development begun to catch up with observations. Their observational situation looks unpromising (Allen et al. 1997; Pen et al. 1997). For reviews of defects, see Vilenkin and Shellard (1994) and Hindmarsh and Kibble (1995).

12.5   **Summary**

12.5.1   Implications for cosmological parameters

At present, compiled large-scale structure observations do not strongly constrain cosmological parameters, though they do argue against certain combinations.

The density parameter is subject to a fairly firm lower bound of $\Omega_0 \simeq 0.30$, regardless of whether there is a cosmological constant, which is competitive with bounds from other sources (see, e.g., Dekel 1994).

The Hubble parameter is not constrained, but a measured value would have important implications, which independently reinforce conclusions drawn from the age–expansion rate conflict. For example, if h is much above about 0.5, then critical-density CDM models will be excluded. For h up to about 0.60, we will be able to retain critical density, but the CDM hypothesis must be given up, CHDM being the natural alternative. To get h as high as 0.60 also will require that the initial spectrum be tilted, to n significantly less than 1. If h is higher yet, both the age and structure formation evidence will independently require a low-density Universe.

The baryon density is, at best, very weakly constrained, but especially if the Universe is to have a critical density it had better be high rather than low, for a variety of reasons, the most compelling of which is probably the cluster baryon fraction (White et al. 1993b) rather than large-scale structure and the cmb spectrum.

12.5.2 Implications for inflation

For inflation the situation is more interesting – there are two strong conclusions. First, there are models of inflation that fit the observational data, and second, there are models that do not! Without the first, of course, this book would not exist; so, perhaps we can take it as read, though we should remember that that could quite easily not have been true. However, it is the second that reminds us that, nowadays, inflation is very much a proper, falsifiable science, and that models that appear perfectly well founded in principle simply may not be relevant to our Universe. The best example of this situation at present is the extended inflation model (La and Steinhardt 1989) and its variants (Kolb 1991), which were very popular when first introduced. However, as we saw in Chapter 8, this model gives power-law inflation, a model that can produce copious gravitational waves, and what is more, the dynamics of the phase transition ending inflation require it to be far from the scale-invariant limit. This renders it incapable of producing perturbations that may seed structure formation (Liddle and Lyth 1992; Green and Liddle 1996).

The main division of inflationary models is between those that predict $n < 1$ and those that predict $n > 1$. The former is required if a critical-density CDM Universe is to be viable, and indeed quite a large tilt is required without significant gravitational wave production.

Several hybrid inflation models predict $n > 1$. For them, we already have the strong conclusion that there cannot be a critical density of CDM; it appears that a hot component, an open Universe, or a cosmological constant is necessary. The last offers the highest n of all; Liddle et al. (1996c) found models allowing n as high as 1.4, though those models may be ruled out through excessive black-hole production if the power law perseveres to very short scales.

One of the most distinctive signatures of inflation would be if a gravitational-wave component could be identified (Copeland et al. 1993, 1994a). Unfortunately, the implication of modern inflation model building, in which the scalar field is not allowed to travel more than order of the Planck mass, is that the gravitational waves are likely to be negligible (Lyth 1997). Present large-scale structure observations do not show any evidence in favour of gravitational waves, but typically they have to be included at a very high amplitude before their effect on the COBE normalization becomes disastrous, and so, they are weakly constrained, roughly to $r \lesssim 0.5$ depending on the cosmology chosen. Future cmb experiments, especially with polarization, may be able to make quite a strong statement in this regard.

If the Universe turns out to be open, that immediately forces us to look at the much narrower class of open inflation models. Many of these models give spectra with $n < 1$, which at $\Omega_0 \simeq 0.3$ is quite strongly disfavoured because the matter spectrum normalization is already low (White and Silk 1996). At $\Omega_0 \simeq 0.4$ though, things are fine unless n is very low, as we can guess from Figure 12.7. If indeed Ω_0 proves to be very low, it may be necessary to have $n > 1$, for which models do exist (García-Bellido and Linde 1997).

Ultimately, the matter spectrum probably will not prove to be the most useful way of constraining models, partly because it is difficult to measure and partly simply because the theoretical models under discussion do not predict it to have much structure; they simply give a smooth curved line. By contrast, the microwave anisotropy spectrum contains significant structure in the acoustic oscillations, promising much greater sensitivity to all of the inflationary and cosmological parameters.

13 Outlook for the future

The present situation looks very good for the inflationary cosmology. As we hope that we have demonstrated, the picture of structure formation based on the inflationary cosmology is very complete and elegant, bringing together quantum theory for the origin of the perturbations, general relativity for their evolution, and a variety of astrophysical processes to make the final link to observations.

Inflation-based structure formation models appear an excellent paradigm within which to understand the inhomogeneous Universe in which we live, even though, as we write, it is rather unclear precisely which type of model might best fit the data. For instance, should the Universe have critical density, and if so, is a component of hot dark matter necessary? Should it instead have lower density, with or without a cosmological constant? What *is* clear is the success of the basic paradigm, that everything originates from a Gaussian, adiabatic, density perturbation with spectral index close to 1.

For the future, we await improved observations to see if this picture can be sustained. Many types of observation will probe the cosmological modelling and, for inflationary cosmologists, none is more eagerly awaited than improved measurements of the microwave background anisotropies, which directly probe linear perturbation theory. The Microwave Anisotropy Probe and Planck satellites promise data of unprecedented quality when it comes to constraining inflationary and cosmological parameters, and before then, many ground- and balloon-based experiments should accurately probe at least the first acoustic peak; for them, any significant spatial curvature is easy prey, and other parameters may be pinned down too.

Concerning the density perturbation spectrum, let us first stress that it is very easily possible that the satellites might see something that is incompatible with any reasonable inflation model. This is true even if the cosmological parameters are fairly undetermined, because the C_ℓ spectrum contains far more information than the various parameters. One adverse feature would be a radical departure from the $n \simeq 1$ paradigm, unless it were of a rather special kind, and another would be a significant departure from Gaussianity. It will be a striking success for inflation if it is still healthy after the satellite results are in.

If inflation does survive, then the prize should be a highly accurate (± 0.01) measurement of the spectral index of the density perturbations. As we saw in Chapter 8, such an accurate measurement is highly discriminatory between inflationary models, and at a stroke will rule most of them out. Inflationary model building will become a highly constrained business.

As we write, there is a strong renaissance in inflationary model building, incorporating ideas from supersymmetry and its generalizations. At the moment, such model building is carried out with considerable freedom, reproduction of the correct amplitude of density perturbations being the only serious constraint. The implication of such models is that we are unlikely to see the effect of gravitational waves in the microwave background, even if polarization is used in addition to the temperature anisotropies. It is crucial, however, to test this expectation; if gravitational waves are present, then they provide vital extra information, including a direct measurement of the energy scale of inflation, which is independent of the form of the inflation model.

Within the next few years, we ought to understand the details of the Universe in which we live in a truly quantitative way, rather than the qualitative description we currently possess. Obtaining definitive answers to such questions as the age of the Universe, and its ultimate fate, will be dramatic indeed. Further, the direct link that inflation has forged between physics at high energies and observations of the cosmic microwave background may provide the most striking results that high-energy physics has seen for some considerable time. As cosmologists, we live in interesting times!

14 Advanced topic: Cosmological perturbation theory

In these final two chapters, we derive from first principles the equations describing the evolution of the cosmological perturbations. In the present chapter, we focus on the energy density, pressure, and anisotropic stress, deriving and generalizing some equations that were quoted without proof in Chapters 4, 5, and 6. Then, in the next chapter, we describe diffusion and freestreaming.

In both chapters, we use the same symbol to denote a quantity and its Fourier components. For example, δ (without any argument) can denote either the density contrast $\delta(\mathbf{x}, t)$ or its Fourier component $\delta_\mathbf{k}(t)$ or $\delta(\mathbf{k}, t)$.

We begin with an overview of special and general relativity. Particular attention is paid to the concept of a locally orthonormal frame, because it will be crucial when we come to describe diffusion and freestreaming.

14.1 Special relativity

Special relativity can be viewed as a local description of physics, which leaves to general relativity the task of matching up the different local descriptions. For simplicity, we make the idealization that it holds in an infinite region of space-time, though its regime of validity is always finite. In the context of cosmology, it holds in a space-time region that is small compared to the Hubble scale.

14.1.1 Inertial frames

A starting point for special relativity is provided by the space-time interval ds^2 between nearby points. Taking it for the moment as a given concept, an **inertial frame** is defined as a coordinate system (t, x, y, z), in which

$$ds^2 = -dt^2 + dx^2 + dy^2 + dz^2. \tag{14.1}$$

This also is called a Minkowski coordinate system. It is convenient to use the index notation

$(x^0, x^1, x^2, x^3) = (t, x, y, z)$, and to denote a generic coordinate by x^μ.[1] Then we can define a **metric tensor** $\eta_{\mu\nu}$ as the diagonal matrix with elements $(-1, 1, 1, 1)$. Using it, Eq. (14.1) can be written

$$ds^2 = \eta_{\mu\nu}\, dx^\mu dx^\nu. \tag{14.2}$$

We adopt the summation convention, that there is a sum over every pair of identical space-time indices. In this case there is a sum over μ and another sum over ν. The formalism requires that one member of the pair be in the lower position while the other is in the upper position.

Note that here the x^i are Cartesian space coordinates, corresponding to physical distance. In most parts of this book, x^i are comoving coordinates in the expanding Universe, in terms of which the Cartesian coordinates (assuming spatial flatness) are $r^i = a(t)x^i$.

It is clear that ds^2 will not have the form of Eq. (14.1) in a generic coordinate system. For instance, replacing (x, y, z) by spherical polar coordinates, it becomes

$$ds^2 = -dt^2 + dr^2 + r^2(\sin^2\theta\, d\theta^2 + d\phi^2). \tag{14.3}$$

The fact that we can choose coordinates that make Eq. (14.1) true is telling us something about the nature of space-time, as specified by the interval ds^2. This is reminiscent of the situation for a two-dimensional surface, where the distance between nearby points is of the form $dl^2 = dx^2 + dy^2$ in Cartesian coordinates. Such coordinates exist if the surface is flat, but in general fail to exist if it is curved. By analogy, space-time is said to be flat if we can find an inertial frame. Just like the surface, space-time can be regarded as flat in a "small" region, and special relativity holds in such a region.

By definition, a coordinate transformation from one inertial frame to another preserves the form of Eq. (14.1). The most general such transformation is linear, so that if we keep the origin fixed, it is of the form

$$x'^\mu = \frac{\partial x'^\mu}{\partial x^\nu} x^\nu. \tag{14.4}$$

This is a **Lorentz transformation**, characterized by the requirement

$$\eta_{\mu\nu} = \frac{\partial x'^\alpha}{\partial x^\mu} \frac{\partial x'^\beta}{\partial x^\nu} \eta_{\alpha\beta}. \tag{14.5}$$

Excluding the discrete transformations $t' = -t$ (time reversal) and $x'^i = -x^i$ (parity), the most general Lorentz transformation is a rotation and/or a boost.

A boost along, for example, the x axis corresponds to

$$x' = \gamma(x - vt), \qquad t' = \gamma(t - vx), \tag{14.6}$$

where

$$\gamma = (1 - v^2)^{-1/2}. \tag{14.7}$$

---

[1] We take Greek letters (μ, ν, ...) to run over the values 0, 1, 2, 3, and italic letters (i, j, ...) to run over the values 1, 2, 3.

Recall that we set $c = 1$. In elementary discussions the term Lorentz transformation denotes a boost only, but the standard usage is that it denotes a boost and/or a rotation. Allowing only boosts with $v \ll 1$ gives the Galilean transformation $t' = t$ and $x' = x - vt$. This is the Newtonian description of space-time, in which there is a universal time coordinate.

An interval is timelike if $ds^2 < 0$, spacelike if $ds^2 > 0$, and lightlike (also sometimes called "null") if $ds^2 = 0$. A line is timelike, lightlike, or spacelike if ds^2 has that property for all nearby points. The set of all lightlike lines passing through a given space-time point is the lightcone, which we illustrated in Figure 4.3 on page 88. It divides space-time into a future ($ds^2 < 0$ with $t > 0$), past ($ds^2 < 0$ with $t < 0$), and elsewhere ($ds^2 > 0$). (We are taking the arrow of time as a given concept, and choosing t to increase with time.) Timelike lines represent the possible motions of massive objects, such as observers, and often are called worldlines.

14.1.2   4-Vectors and tensors

An old-fashioned but perfectly serviceable definition of a 4-vector A^μ is that it is a 4-component object transforming like x^μ:

$$A'^\mu = \frac{\partial x'^\mu}{\partial x^\nu} A^\nu. \tag{14.8}$$

In a given inertial frame, each 4-vector is of the form $A^\mu = (A^0, \mathbf{A}) = (A^0, A^i)$. The components A^i form a 3-vector, defined as a quantity transforming under rotations in the same way as x^i. We denote the magnitude of $\mathbf{A}$ by A, so that $A^2 \equiv \sum_i (A^i)^2$. The inner product of two vectors is defined as $\eta_{\mu\nu} A^\mu B^\nu$, the vectors being orthogonal if it vanishes. It is the same in all inertial frames, known as **Lorentz invariant**; such a quantity is called a **scalar**.

If $x^\mu(\lambda)$ is a parameterized line in space-time, there is a tangent 4-vector $e^\mu = dx^\mu/d\lambda$ at each point on the line. If the line is timelike (a worldline), it is a possible particle trajectory. Then e^μ can be normalized to become the 4-velocity $u^\mu = dx^\mu/dt_{\mathrm{pr}}$, where $dt_{\mathrm{pr}} = \sqrt{-ds^2}$ is the **proper time interval**, so that $\eta_{\mu\nu} u^\mu u^\nu = -1$. The 4-acceleration is

$$a^\mu \equiv \frac{du^\mu}{dt_{\mathrm{pr}}} \equiv \frac{d^2 x^\mu}{dt_{\mathrm{pr}}^2}. \tag{14.9}$$

This also can be written

$$a^\mu \equiv u^\nu \partial_\nu u^\mu. \tag{14.10}$$

The worldline is a **geodesic** if $a^\mu = 0$.

In a generic inertial frame, we can write

$$u^\mu = u^0(1, v^i), \tag{14.11}$$

where $v^i \equiv dx^i/dt$ is the 3-velocity. The time component is

$$u^0 = dt/dt_{\mathrm{pr}} = (1 - v^2)^{-1/2}, \tag{14.12}$$

where the final equality comes from $\eta_{\mu\nu}u^{\mu}u^{\nu} = -1$. This is the time dilation formula for a moving clock. At a given point on the worldline, we can choose a **local rest frame**, defined as an inertial frame in which **v** instantaneously vanishes. Then $dt = dt_{\mathrm{pr}}$, $u^{\mu} = (1, 0)$, and $a^{\mu} = (0, a^{i})$, where $a^{i} = dv^{i}/dt$ is the 3-acceleration.

The unit tangent vector of a spacelike line is dx^{μ}/dl, where $dl = \sqrt{ds^{2}}$ is the proper distance, and the line is a geodesic if $d^{2}x^{\mu}/dl^{2} = 0$. (A spacelike geodesic represents the shortest proper distance $l \equiv \int dl$ between nearby points, and a timelike one the longest proper time.) The 4-momentum of a particle with mass m is $p^{\mu} = mu^{\mu}$, and we can say that a timelike geodesic corresponds to constant 4-momentum. This definition works also for lightlike geodesics, corresponding to the limit of zero mass.

From all this, we learn that the coordinate lines of an inertial frame are orthogonal geodesics.[2]

For any 4-vector, it is useful to define the lower-component object $A_{\mu} = \eta_{\mu\nu}A^{\nu}$; more explicitly, $A_{0} = -A^{0}$, $A_{i} = A^{i}$. Then the scalar product is $A^{\mu}B_{\mu}$. Because $A_{i} = A^{i}$, there is no need to distinguish between upper and lower spatial indices for 3-vectors. We generally use lower indices, with a sum over repeated spatial indices understood even if they are both in the lower position. The transformation law for A_{μ} is

$$A'_{\mu} = \frac{\partial x^{\nu}}{\partial x'^{\mu}} A_{\nu}. \tag{14.13}$$

This is a special case of Eq. (14.44), which is proved later.

A product $A^{\mu}B^{\nu}$ provides the simplest example of a second-rank tensor. More generally, a second-rank tensor is any quantity $C^{\mu\nu}$ that transforms like a product $A^{\mu}B^{\nu}$ of two vectors. We can lower any component with $\eta_{\mu\nu}$, for instance, $C^{\mu}{}_{\nu} = \eta_{\nu\lambda}C^{\mu\lambda}$. Third- and higher-rank tensors are defined in the same way, and it is useful to regard vectors and scalars as, respectively, first- and zeroth-rank tensors.

From Eq. (14.5), the metric tensor $\eta_{\mu\nu}$ is indeed a tensor, but a very special one. Its components are the same in every inertial frame, and $\eta^{\mu}{}_{\nu} = \eta_{\nu}{}^{\mu} = \delta^{\mu}_{\nu}$, where δ^{μ}_{ν} is the Kronecker delta, equal to 1 for equal indices and to 0 for unequal ones. Also, $\eta^{\mu\nu}$ has the same components as $\eta_{\mu\nu}$.

Given some vectors and/or tensors, more complicated ones can be built by addition and multiplication. Also, summing over a pair of identical indices ("contraction") changes a second-rank tensor into a scalar, a third-rank tensor into a vector, a fourth-rank tensor into a second-rank tensor, and so on, and we can repeat the process. A simple example is the expression $\eta_{\mu\nu}A^{\mu}B^{\nu}$, which is a scalar.

If the components of a vector or tensor vanish in one coordinate system, they vanish in all coordinates systems (we see soon that this is true also for noninertial coordinates). This means that a vector or tensor is defined uniquely by giving its components in any coordinate system.

The 4-gradient $\partial_{\mu} \equiv \partial/\partial x^{\mu}$ is a 4-vector with lower-index components because it transforms according to Eq. (14.13) (the chain rule of partial differentiation). It turns a scalar (defined in a region of space-time) into a vector, a vector into a second-rank tensor, and so on.

---

[2] For any coordinate system, a coordinate line is a line along which only one of the coordinates is varying.

14.1.3 The laws of physics

The special relativity formalism, which we have just summarized, is supposed to be adequate in the approximation that gravity is negligible. The equations describing fundamental laws of physics are assumed to have the same form in any inertial frame. This is called relativistic invariance, and to achieve it, each equation has to be of the form "tensor $= 0$."

Let us see which laws we need for the purpose of this book. We need the statement that for a free particle moves along a geodesic,

$$a^\mu \equiv \frac{du^\mu}{dt_{\text{pr}}} = 0, \tag{14.14}$$

where $u^\mu = dx^\mu/dt_{\text{pr}}$ is the 4-velocity and a^μ is the 4-acceleration. In this case the left-hand side is a vector (first-rank tensor).

We need some field theory, as described in Sections 7.1 and 7.2. Let us mention Eq. (7.7) for a scalar field. It can be written

$$\Box\phi + V'(\phi) = 0, \tag{14.15}$$

where $\Box$ is the d'Alembertian

$$\Box \equiv \eta^{\mu\nu}\partial_\mu\partial_\nu = 0. \tag{14.16}$$

In this case, the left-hand side is a scalar (zero-rank tensor).

We also need the fluid-flow formalism developed in the next section, and the gas dynamics described in Section 15.2. Finally, we need the description of Thomson scattering in Section 15.5.3.

At the deepest level, all of this, as well as the rest of nongravitational physics, is supposed to follow from the Lagrangian density $\mathcal{L}_{\text{SM}}$ of the Standard Model of the fundamental interactions. It specifies the action $S = \int d^4x\mathcal{L}$, which gives the classical equations of motion through the action principle $\delta S = 0$, and quantum field theory through, for example, the path integral formalism. Through Eq. (14.84), it also gives the stress–energy tensor, to which we now turn.

14.2 Fluid flow in special relativity

Relativistic fluid flow is described using the stress–energy tensor $T^{\mu\nu}$. This is ultimately defined by asserting that $8\pi G T^{\mu\nu}$ is the source of gravity, standing on the right-hand side of the Einstein field equation (14.74). At the moment, we are considering the limit of flat space-time, where Newton's gravitational constant G tends to 0, but this definition of $T^{\mu\nu}$ still makes sense.

From the Einstein equation, we learn that the stress–energy tensor is symmetric:

$$T^{\mu\nu} = T^{\nu\mu}. \tag{14.17}$$

At any point in space-time, the energy density is defined as T^{00}, the momentum density is defined as $T^{0i} = T^{i0}$, and the stress tensor is defined as $T_{ij} = T_{ji}$. If the stress is isotropic, the pressure P is defined by $T_{ij} = \delta_{ij} P$. Remember that there is no distinction between upper and lower components for spatial indices, in the inertial coordinate systems that we are considering at the moment.

The Einstein equation also implies that

$$\partial_\mu T^{\mu\nu} = 0. \tag{14.18}$$

As we see shortly, this is the stress–energy conservation law, encoding both energy and momentum conservation.

For a particular system, there are usually expressions for the stress–energy tensor that make no explicit reference to gravity. In particular, we see in Section 15.2 how to calculate it for a relativistic gas in terms of the masses and velocities of the constituent particles. Also, we see how Eq. (14.84) gives the stress–energy tensor of a field. However, for the moment, we keep the discussion general, using only the two properties (14.17) and (14.18).

To see that Eq. (14.18) is the energy–momentum conservation law, let us warm up with the simpler case of charge conservation. The electromagnetic 4-current is $j^\mu = (\rho, j^i)$, where ρ is the charge density and j^i is the current density. It is defined as the source of electromagnetism appearing in Maxwell's equations, which imply that

$$\partial_\mu j^\mu \equiv \partial_t \rho + \partial_i j^i = 0. \tag{14.19}$$

This is the charge conservation equation because the charge in a fixed volume V is $Q = \int_V \rho \, d^3 r$, and the preceding equation plus the divergence theorem assert that its rate of decrease is the current $\int_{\partial V} j^i dS_i$ flowing out of V. (Here, ∂V is the boundary of V and dS_i points outward.) Taking the boundary to enclose an isolated system, so that $j^i = 0$, we learn that Q is time independent.

Equation (14.18) has four components, corresponding to energy and the three momentum components

$$\partial_t T^{00} + \partial_j T^{0j} = 0 \qquad \text{(energy conservation)}, \tag{14.20}$$

$$\partial_t T^{i0} + \partial_j T^{ij} = 0 \qquad \text{(momentum conservation)}. \tag{14.21}$$

The 4-momentum in a fixed volume V is

$$p^\mu = \int_V T^{\mu 0} \, d^3 r, \tag{14.22}$$

and its rate of decrease is

$$-\frac{dp^\mu}{dt} = \int_{\partial V} T^{\mu j} \, dS_j. \tag{14.23}$$

We see that the momentum density T^{0j} corresponds to the flow of *energy*, and the stress T^{ij} to the flow of the ith component of *momentum*. Defining force **F** as the rate of increase of momentum, we see that the contribution to F^i from area dS_j is $-T^{ij}dS_j$, which is the usual definition of stress. For an isolated system, $T^{\mu\nu}$ vanishes on the boundary, making energy and momentum time independent.

We are taking the stress–energy tensor to be a smoothly varying function of position, which is equivalent to saying that we are dealing with a fluid. At any point in space-time, the **local rest frame** of the fluid is defined as the frame in which the momentum density T^{i0} vanishes:[3]

$$T^{0i} = 0 \quad \text{(local rest frame)}. \tag{14.24}$$

At each space-time point, the energy density in the local rest frame is denoted by ρ:

$$\rho \equiv T^{00}. \tag{14.25}$$

The fluid 4-velocity in this frame is $u^{\mu} = (1, 0, 0, 0)$. A worldline with this 4-velocity is said to be **comoving** (moving with the fluid).

In a local rest frame, with coordinates (t, r_i), Eqs. (4.108), (4.109), and (4.110) define the shear and vorticity of the fluid flow, as well as a locally defined expansion parameter H, in terms of the velocity gradient.

A **perfect fluid** is defined as one that is isotropic in the local rest frame, $T_{ij} = P\delta_{ij}$, where P is the pressure. (The force on area dS_i is $-PdS_i$, the usual definition of pressure.) More generally,

$$T_{ij} = P\delta_{ij} + \Sigma_{ij}, \tag{14.26}$$

where the second term is traceless ($\Sigma_{ii} = 0$) and is called the anisotropic stress.

For a perfect fluid, in a generic inertial frame, we have

$$T^{\mu\nu} = (\rho + P)u^{\mu}u^{\nu} + P\eta^{\mu\nu}. \tag{14.27}$$

This expression must be correct because it has the required form in the local rest frame. In any inertial frame where the fluid 3-velocity satisfies $v \ll 1$, Eqs. (14.25) and (14.26) remain valid to first order in v, and, to the same order, the momentum density is

$$T^{i0} = (\rho + P)v^i. \tag{14.28}$$

We see that both energy density and pressure contribute to the "inertial mass density," defined as the momentum density divided by the 3-velocity.

Using these results, we find for a perfect fluid that energy conservation $\partial_{\mu}T^{\mu 0} = 0$ corresponds, in the local rest frame, to the continuity equation

$$\partial_t\rho = -3H(\rho + P) \quad \text{(continuity)}. \tag{14.29}$$

We saw after Eq. (2.8) that this is the familiar relation $dE = -PdV$. Momentum conservation $\partial_{\mu}T^{\mu i} = 0$ corresponds to the Euler equation

$$a_i = -\frac{\partial_i P}{\rho + P} \quad \text{(Euler)}. \tag{14.30}$$

---

[3] Such a frame always exists in practice, though we can devise idealized situations in which it does not, such as the case of a plane electromagnetic wave in the vacuum. At the moment we are working with the total stress–energy tensor, so that there is a uniquely defined local rest frame. Later we split the stress–energy tensor into pieces corresponding to photons, baryons, etc., and then each component has its own local rest frame. Finally, as we noted earlier, each timelike line passing through the point has its own local rest frame.

This is just the Newtonian $\mathbf{F} = m\mathbf{a}$, except that the mass density is replaced by $\rho + P$.

If there is anisotropic stress, Eq. (14.27) becomes

$$T^{\mu\nu} = (\rho + P)u^\mu u^\nu + P\eta^{\mu\nu} + \Sigma^{\mu\nu}, \tag{14.31}$$

where, in the local rest frame, the only nonzero components of $\Sigma^{\mu\nu}$ are the space–space components. When this expression is used, the Euler equation becomes

$$a_i = -\frac{\partial_i P + \partial_j \Sigma_{ij}}{\rho + P} \qquad \text{(Euler)}. \tag{14.32}$$

In any inertial frame where the fluid 3-velocity satisfies $v \ll 1$, Eqs. (14.29) and (14.32) receive corrections of only second order in v. As we see, this means that they can be used in cosmological perturbation theory.

14.3 Special relativity using generic coordinates

In general relativity, we have to learn how to use generic coordinate systems because there are no inertial ones. It is helpful to do this first in the familiar context of special relativity, where inertial coordinates do exist. Consider therefore a generic coordinate system x^μ, related to an inertial one x'^μ by

$$dx'^\mu = \frac{\partial x'^\mu}{\partial x^\nu} dx^\nu, \tag{14.33}$$

where now the transformation is arbitrary. In this coordinate system,

$$ds^2 = g_{\mu\nu} dx^\mu dx^\nu, \tag{14.34}$$

where the metric tensor is now

$$g_{\mu\nu} = \frac{\partial x'^\alpha}{\partial x^\mu} \frac{\partial x'^\beta}{\partial x^\nu} \eta_{\alpha\beta}. \tag{14.35}$$

Note that $g_{\mu\nu} = g_{\nu\mu}$. Expression (14.34) is called the **line element**, and it is a convenient way of defining the metric tensor in a given coordinate system.

14.3.1 Vectors, tensors, and the space-time volume

The transformation (14.33) remains valid when both x^μ and x'^ν are generic. In the generic case, it leads to

$$g_{\mu\nu} = \frac{\partial x'^\alpha}{\partial x^\mu} \frac{\partial x'^\beta}{\partial x^\nu} g'_{\alpha\beta}. \tag{14.36}$$

In generic coordinates, the components of a 4-vector, defined at a given point, are taken to transform like dx^μ:

$$A^\mu = \frac{\partial x^\mu}{\partial x'^\nu} A'^\nu. \tag{14.37}$$

The archetype is the tangent vector $t^\mu = dx^\mu/d\lambda$ of a line $x^\mu(\lambda)$, a particular example being the 4-velocity $u^\mu = dx^\mu/dt_{\rm pr}$ of a worldline.

The lower-component object is defined as

$$A_\mu = g_{\mu\nu}A^\nu. \tag{14.38}$$

The inverse transformation is

$$A^\mu = g^{\mu\nu}A_\nu, \tag{14.39}$$

where $g^{\mu\nu}$ is the matrix inverse of $g_{\mu\nu}$,

$$g^{\mu\lambda}g_{\lambda\nu} = \delta^\mu_\nu. \tag{14.40}$$

Let us find the transformation law for A_μ. In a new coordinate system, $A'_\mu = g'_{\mu\nu}A'^\nu$. Using Eqs. (14.36), (14.37), and (14.38), this becomes

$$A'_\mu = \frac{\partial x^\alpha}{\partial x'^\mu}\frac{\partial x^\beta}{\partial x'^\nu}\frac{\partial x'^\nu}{\partial x^\lambda}g_{\alpha\beta}g^{\lambda\sigma}A_\sigma. \tag{14.41}$$

Using

$$\frac{\partial x^\beta}{\partial x'^\nu}\frac{\partial x'^\nu}{\partial x^\lambda} = \frac{\partial x^\beta}{\partial x^\lambda} = \delta^\beta_\lambda, \tag{14.42}$$

this becomes

$$A'_\mu = \frac{\partial x^\alpha}{\partial x'^\mu}g_{\alpha\lambda}g^{\lambda\sigma}A_\sigma. \tag{14.43}$$

Finally, Eq. (14.40) gives

$$A'_\mu = \frac{\partial x^\nu}{\partial x'^\mu}A_\nu. \tag{14.44}$$

Tensors are defined to transform like products of vectors, so that for a second-rank tensor

$$C'_{\mu\nu} = \frac{\partial x^\alpha}{\partial x'^\mu}\frac{\partial x^\beta}{\partial x'^\nu}C_{\alpha\beta}. \tag{14.45}$$

A tensor that vanishes in one coordinate system vanishes in all. Therefore, tensors having equal components in one system have equal components in all.

In inertial coordinates, the space-time volume element is defined by the product $d^4x = dx^0dx^1dx^2dx^3$. In generic coordinates it becomes Jd^4x, where d^4x is still given by the same formula and the Jacobian $J \equiv |\partial x'^\alpha/\partial x^\mu|$. From Eq. (14.35), regarded as the product of three matrices, it follows that $J = \sqrt{-g}$, where g is the determinant of $g_{\mu\nu}$. In generic coordinates the space-time volume element is therefore $\sqrt{-g}\, d^4x$.

14.3.2 Covariant derivative

If f is a scalar, $\partial_\mu f$ defines a 4-vector in any coordinate system because it has the correct transformation law (14.44). But if A^μ is a 4-vector, $\partial_\nu A^\mu$ does *not* define a tensor because it has the wrong transformation law. The **covariant derivative** $D_\nu A^\mu$ is defined as the tensor that reduces to $\partial_\nu A^\mu$ in an inertial frame. This means that, in a generic coordinate system x^μ,

$$D_\nu A^\mu \equiv \frac{\partial x'^\alpha}{\partial x^\nu} \frac{\partial x^\mu}{\partial x'^\beta} \frac{\partial A'^\beta}{\partial x'^\alpha}, \tag{14.46}$$

where the prime denotes an arbitrarily chosen inertial frame. $A'^\beta = (\partial x'^\beta / \partial x^\lambda) A^\lambda$ and Eq. (14.42) gives

$$D_\nu A^\mu = \partial_\nu A^\mu + \Gamma^\mu_{\nu\alpha} A^\alpha, \tag{14.47}$$

where the **connection** (Christoffel symbol) is defined by

$$\Gamma^\mu_{\nu\alpha} = \frac{\partial^2 x'^\mu}{\partial x^\nu \partial x^\alpha}. \tag{14.48}$$

It is symmetric in the lower indices, $\Gamma^\mu_{\nu\alpha} = \Gamma^\mu_{\alpha\nu}$.

The covariant derivative of a lower-index vector or a tensor is defined in the same way, and it always can be expressed in terms of the connection. For example, we find

$$D_\nu A_\mu = \partial_\nu A_\mu - \Gamma^\alpha_{\mu\nu} A_\alpha, \tag{14.49}$$

$$D_\mu(A_\nu B_\lambda) = \partial_\mu(A_\nu B_\lambda) - \Gamma^\alpha_{\mu\nu} A_\alpha B_\lambda - \Gamma^\alpha_{\mu\lambda} A_\nu B_\alpha, \tag{14.50}$$

$$D_\mu(A_\nu B^\lambda) = \partial_\mu(A_\nu B^\lambda) - \Gamma^\alpha_{\mu\nu} A_\alpha B^\lambda + \Gamma^\lambda_{\mu\alpha} A_\nu B^\alpha. \tag{14.51}$$

Acting on a scalar, D_μ is equivalent to ∂_μ.

The last two expressions give the covariant derivative of a tensor and, in particular,

$$D_\mu g_{\nu\lambda} = \partial_\mu g_{\nu\lambda} - \Gamma^\alpha_{\mu\nu} g_{\alpha\lambda} - \Gamma^\alpha_{\mu\lambda} g_{\nu\alpha}. \tag{14.52}$$

However, in an inertial frame, $g_{\mu\nu}$ has the constant value $\eta_{\mu\nu}$, and $D_\mu g_{\nu\lambda}$ vanishes in that and hence all other frames, implying that

$$\partial_\mu g_{\nu\lambda} = \Gamma^\alpha_{\mu\nu} g_{\alpha\lambda} + \Gamma^\alpha_{\mu\lambda} g_{\nu\alpha}. \tag{14.53}$$

This relation makes no mention of the inertial coordinates x'_μ, and it can be inverted to give the connection in terms of the metric[4]

$$\Gamma^\gamma_{\beta\mu} = \frac{1}{2} g^{\alpha\gamma} (\partial_\mu g_{\alpha\beta} + \partial_\beta g_{\alpha\mu} - \partial_\alpha g_{\beta\mu}). \tag{14.54}$$

Note that the connection is *not* a tensor; its complicated transformation law, which may be derived from Eq. (14.54), is not the same as the one for a product of three vector components.

14.3.3 Orthogonal coordinate systems

Typically, though not always, the coordinate system $x^\mu = (t, x^i)$ is chosen so that the coordinate lines are orthogonal. If the time-coordinate lines (the lines along which only t varies) are

---

[4] To obtain this, cyclically permute $(\mu\nu\lambda)$ in Eq. (14.53) to give two more equations. Adding two of them and subtracting the third, remembering that $\Gamma^\mu_{\nu\alpha} = \Gamma^\mu_{\alpha\nu}$, gives $2g_{\alpha\nu}\Gamma^\nu_{\beta\mu} = \partial_\mu g_{\alpha\beta} + \partial_\beta g_{\alpha\mu} - \partial_\alpha g_{\beta\mu}$. Then, contracting with $g^{\gamma\alpha}$ gives the desired result.

geodesics, we can choose t to coincide with the proper time t_{pr} along each line. If they are not geodesics, this is impossible because t_{pr} and t increase at different rates (Figure 4.6).

To see this, consider a point in space-time, and use an inertial coordinate system (local rest frame) $x'^{\mu} = (t_{pr}, x'^i)$, in which the time-coordinate line of the coordinate system $x^{\mu} = (t, x^i)$ has zero velocity. Nearby time-coordinate lines have velocity $\mathbf{v}$, and 4-velocity $u'^{\mu} = (1, \mathbf{v})$ (to first order in v). Because this 4-velocity is orthogonal to the surfaces of constant t, it is proportional to the 4-gradient of $t(x'^{\mu})$:

$$f u'_{\mu} = \frac{\partial t}{\partial x'^{\mu}} \tag{14.55}$$

for some f. Contracting both sides with u'^{μ}, and remembering that $u'^{\mu} = dx'^{\mu}/dt_{pr}$, we learn that $f = dt/dt_{pr}$. Differentiating with respect to x'^{ν} and antisymmetrizing, we find that

$$\frac{\partial(f u'_{\nu})}{\partial x'^{\mu}} - \frac{\partial(f u'_{\mu})}{\partial x'^{\nu}} = 0. \tag{14.56}$$

Near the space-time point under consideration, $u'_{\mu} = (-1, \mathbf{v})$. Evaluating the space–space components of Eq. (14.56) at the origin, we find that the time-coordinate lines define a flow with zero vorticity:

$$\frac{\partial v_j}{\partial x'^i} - \frac{\partial v_i}{\partial x'^j} = 0. \tag{14.57}$$

It is not difficult to show that, conversely, any threading that satisfies this equation is orthogonal to some slicing.

The space-time components give

$$\frac{\partial f}{\partial x'^i} + a_i f = 0, \tag{14.58}$$

where $a_i = \partial v_i / dt_{pr}$ is the 3-acceleration of the time-coordinate line. Because $f = -(dt/dt_{pr})$, this is Eq. (4.152) that we wrote down earlier. It shows that, indeed, we can choose $t = t_{pr}$ if and only if the acceleration vanishes.

As we note later, the preceding argument works also in the context of general relativity if x'^{μ} is taken to be a locally inertial frame.

In this book we have in mind the case that the time-coordinate lines are comoving, so that their 4-velocity is that of the cosmic fluid.

14.3.4 Laws of nongravitational physics

Starting with equations that are valid in an inertial frame, we obtain equations that are valid in a generic coordinate system by making the following substitutions:

$$\eta_{\mu\nu} \to g_{\mu\nu}, \tag{14.59}$$
$$\partial_{\mu} \to D_{\mu}. \tag{14.60}$$

Let us see how this works in some simple cases.

Geodesics

The most basic law of special relativity is that the worldline of a free particle (one without any interaction) is a geodesic. Let us see how to write this law in a generic coordinate system.

In such a system, the 4-velocity of an object with worldline $x^\mu(t_{pr})$ continues to be $u^\mu \equiv dx^\mu/dt_{pr}$, but Eq. (14.10) becomes

$$a^\mu \equiv u^\nu D_\nu u^\mu. \tag{14.61}$$

A timelike geodesic may be defined as a worldline with $a^\mu = 0$ and, if we use Eqs. (14.47) and (14.54), this becomes

$$\frac{du^\mu}{dt_{pr}} = g^{\mu\nu}\left(\frac{1}{2}\partial_\nu g_{\alpha\beta} - \partial_\beta g_{\nu\alpha}\right)u^\alpha u^\beta. \tag{14.62}$$

In terms of the 4-momentum $p^\mu = mu^\mu$, we can write

$$\frac{dp^\mu}{dx^0} = g^{\mu\nu}\left(\frac{1}{2}\partial_\nu g_{\alpha\beta} - \partial_\beta g_{\nu\alpha}\right)\frac{p^\alpha p^\beta}{p^0}. \tag{14.63}$$

This equation remains valid for zero-mass particles.

Fluid flow and the stress–energy tensor

In a generic coordinate system, Eq. (14.31) becomes

$$T^{\mu\nu} = (\rho + P)u^\mu u^\nu + Pg^{\mu\nu} + \Sigma^{\mu\nu}. \tag{14.64}$$

The energy–momentum conservation law $\partial_\mu T^{\mu\nu} = 0$ becomes

$$D_\mu T^{\mu\nu} = 0. \tag{14.65}$$

Scalar field equation

The scalar field equation (14.15) remains the same, with the d'Alembertian now $\Box \equiv g^{\mu\nu}D_\mu\partial_\nu$. After some manipulation, the latter can be written in the more convenient form (Misner et al. 1973)

$$\Box\phi = -\frac{1}{\sqrt{-g}}\partial_\mu[\sqrt{-g}\, g^{\mu\nu}\partial_\nu\phi]. \tag{14.66}$$

14.4 General relativity

14.4.1 Locally orthonormal and locally inertial frames

In general relativity, the laws of physics do not pick out a globally preferred class of coordinate systems, and in particular, there are no inertial coordinate systems. However, near any point in space-time, it is postulated that there are **locally orthonormal** coordinate systems (also called locally orthonormal frames), in which $ds^2 = \eta_{\mu\nu}dx^\mu dx^\nu$. Transforming to a generic coordinate system, this becomes $ds^2 = g_{\mu\nu}dx^\mu dx^\nu$, where $g_{\mu\nu}$ is given by Eq. (14.34). At a given point in space-time, a locally orthonormal frame is distinguished from other coordinate systems by having $g_{\mu\nu} = \eta_{\mu\nu}$ at the point.[5] For convenience, this point is chosen as the origin of coordinates, $x^\mu = 0$.

Starting with a locally orthonormal frame, we can define the components of vectors and tensors in a generic coordinate system as in Section 14.3.1. To define the *derivative* of a vector or tensor, as in Section 14.3.2, we need to start with a **locally inertial** frame. At any space-time point, a locally inertial frame is a coordinate system in which $g_{\mu\nu} = \eta_{\mu\nu}$ (the requirement for a locally orthonormal frame), and in addition, $\partial_\lambda g_{\mu\nu} = 0$ (equivalently, the connection $\Gamma^\mu_{\alpha\beta}$ vanishes).

The existence of a locally inertial frame, at every point, may be proved by construction as follows: Let x^μ be locally orthonormal at a given point, taken to be the origin, and define new coordinates by

$$x'^\mu = x^\mu - \Gamma^\mu_{\alpha\beta} x^\alpha x^\beta, \tag{14.67}$$

where $\Gamma^\mu_{\alpha\beta}$ is the connection defined by Eq. (14.54). The new metric components $g'_{\mu\nu}$ are given by Eq. (14.34) and, differentiating this expression, we find that, at the origin,

$$\frac{\partial g'_{\mu\nu}}{\partial x'^\lambda} = \frac{\partial g_{\mu\nu}}{\partial x^\lambda} - \Gamma^\alpha_{\lambda\mu} g_{\alpha\nu} - \Gamma^\alpha_{\lambda\nu} g_{\mu\alpha}. \tag{14.68}$$

This vanishes according to Eq. (14.53), so that the new coordinates are indeed locally inertial. The existence of the locally inertial frame is called the **local flatness theorem**.

According to general relativity, this is as far as we can go; it is impossible to choose coordinates so that all second derivatives of $g_{\mu\nu}$ vanish. As a result, we cannot choose globally inertial coordinates, in which $g_{\mu\nu} = \eta_{\mu\nu}$ throughout space-time. This is analogous to the impossibility of choosing Cartesian coordinates on a spherical surface, and we say that space-time is curved.

Although we cannot find a globally inertial coordinate system, it *is* possible to find a globally orthogonal coordinate system, in which $g_{\mu\nu}$ is diagonal. Pathological space-times exist in which this is impossible, but it is always possible in situations of physical interest. Near a given point X^μ, locally orthogonal coordinates then are given by

$$x'^\mu = g_{\mu\mu}^{-1/2}(x^\mu - X^\mu) \tag{14.69}$$

(with no summation over μ). This explicit construction of a locally orthonormal frame is frequently useful. By contrast, there is little need of an explicit construction [e.g., using Eq. (14.67)] of a locally inertial frame.

---

[5] We are taking the interval ds^2 initially as a given concept. An alternative route into general relativity is to take the tensor $g_{\mu\nu} = g_{\nu\mu}$ as a given object, which then would define $ds^2 = g_{\mu\nu}\,dx^\mu\,dx^\nu$. We can choose $g_{\mu\nu}$ to be symmetric and, because any real, symmetric matrix can be diagonalized, we can find at any space-time point, coordinates in which $g_{\mu\nu}$ is diagonal with each element ± 1. However, we still have to postulate that (with a suitable choice for the sign of ds^2 and for the ordering of coordinates) the elements are actually $(-1, 1, 1, 1)$.

14.4.2 Special relativity as the local theory

In the context of general relativity, the laws of nongravitational physics are supposed to be valid at each point in space-time, in a locally inertial frame. Given that this is true, the laws can be written in a global coordinate system if we know the metric $g_{\mu\nu}$, using the substitutions $\eta_{\mu\nu} \to g_{\mu\nu}$ and $\partial_\mu \to D_\mu$.

We already gave a brief survey of these laws in Section 14.3.4, in the framework of special relativity. The only significant change, in the context of general relativity, is that the geodesic law applies to a free-falling particle (one with only the gravitational interaction). The "free particle" of special relativity, defined as one with no interaction whatsoever, does not exist according to general relativity.[6]

In the case of equations that do not involve derivatives of 4-vectors or tensors, the substitution $\partial_\mu \to D_\mu$ is unnecessary. In that case, it is enough to use a locally orthonormal frame. We make essential use of the latter, in deriving the equations of cosmological perturbation theory.

Going the other way, it is crucial that the laws of nongravitational physics be written in such a way that each vector and tensor is differentiated once at most. If repeated derivatives were present, the replacement $\partial_\mu \to D_\mu$ generally would be ambiguous (Misner et al. 1973) because, as we are about to see, repeated covariant derivatives do not commute.

14.4.3 Space-time curvature

Because space-time is curved, repeated covariant differentiation does not commute. Rather, we find from Eq. (14.51) that

$$D_\alpha(D_\beta A^\mu) - D_\beta(D_\alpha A^\mu) = R^\mu{}_{\nu\alpha\beta} A^\nu, \tag{14.70}$$

where the **curvature tensor** is defined by[7]

$$R^\mu{}_{\nu\alpha\beta} \equiv \partial_\alpha \Gamma^\mu_{\nu\beta} - \partial_\beta \Gamma^\mu_{\nu\alpha} + \Gamma^\mu_{\sigma\alpha}\Gamma^\sigma_{\nu\beta} - \Gamma^\mu_{\sigma\beta}\Gamma^\sigma_{\nu\alpha}. \tag{14.71}$$

From the curvature tensor, we can form the Ricci tensor

$$R_{\mu\nu} \equiv R^\lambda{}_{\mu\lambda\nu}, \tag{14.72}$$

and the curvature scalar

$$R \equiv R^\mu{}_\mu. \tag{14.73}$$

---

[6] The geodesic law follows from momentum conservation for a free-falling particle. This, in turn, follows from the Einstein field equation (14.74), insofar as the particle motion can be described adequately using a locally inertial frame. (Roughly speaking, that is the case when the spin of the particle has negligible gravitational effect.) As has often been remarked, the above situation may be summarized by saying that the field equation gives both the effect of matter on the space-time curvature *and* the effect of space-time curvature on matter. In other words, it gives the gravitational field due to matter *and* the motion of matter in a gravitational field. In Newtonian gravity, the Poisson equation that does the former job says nothing about the motion of matter.

[7] We follow the conventions of Misner, Thorne & Wheeler (1973).

To see that $R^\alpha{}_{\beta\mu\nu}$ is actually a tensor, note first that Eq. (14.70) defines it uniquely in every coordinate system because it is valid for all values of the indices α, β, and μ, and for all 4-vector fields A^ν. On the other hand, it is clear that Eq. (14.70) remains valid after transforming to a new coordinate system, if we assume that $R^\alpha{}_{\beta\mu\nu}$ transforms as a tensor. Because Eq. (14.70) uniquely defines $R^\alpha{}_{\beta\mu\nu}$, it follows that this assumption is correct.

It is reasonable to regard the curvature tensor as the relativistic analogue of the Newtonian gravitational field. Special relativity makes the idealization that it vanishes, and applies only to nongravitational physics. Any physics involving the curvature tensor can be regarded as being, by definition, gravitational.

A given region of space-time is practically flat if we can choose almost-inertial coordinates ($g_{\mu\nu} \simeq \eta_{\mu\nu}$), such that all components of the curvature tensor are much less than 1. The maximum size of such a region is called the *curvature scale*, and is typically of order $|R|^{-1/2}$.

It does not make sense to take linear combinations of vectors defined in different regions of space-time (separated by more than the curvature scale) because the transformation coefficients $\partial x'^\mu / \partial x^\nu$ are different. The strictly local nature of vectors in curved space-time has striking consequences. In cosmology, it prevents us from defining a useful "relative velocity" between ourselves and a galaxy at a distance $\gtrsim H_0^{-1}$. It also prevents us from adding up the 4-momenta of everything in the observable Universe, to obtain a total 4-momentum (in particular, a total energy).

In a given coordinate system, all of this can be repeated on a slice of fixed t, replacing Greek indices with Roman ones. In this way, we arrive at the curvature scalar $R^{(3)}$ of the slice.

14.4.4   Einstein field equation

Einstein's field equation is

$$R_{\mu\nu} - \frac{1}{2}g_{\mu\nu}R = 8\pi G\, T_{\mu\nu}. \tag{14.74}$$

The quantity $T_{\mu\nu}$ is the stress–energy tensor, the source of gravity, and G is Newton's gravitational constant, which we are using instead of the equivalent $M_{\rm Pl}^2 = (8\pi G)^{-1}$.

The Einstein equation implies the energy–momentum conservation equation

$$D_\mu T^{\mu\nu} = 0 \tag{14.75}$$

because there is an identity

$$D_\mu\left(R^{\mu\nu} - \frac{1}{2}g^{\mu\nu}R\right) \equiv 0. \tag{14.76}$$

To prove it, we focus on one point in space-time, and work in a locally inertial frame, so that we need only prove

$$\partial_\nu\left(R^{\mu\nu} - \frac{1}{2}\eta^{\mu\nu}R\right) = 0. \tag{14.77}$$

(The more general expression is of the form "tensor $=0$," and it reduces to Eq. (14.77) in a locally inertial frame, and so, it is enough to prove the latter.) In the locally inertial frame,

$D_\mu = \partial_\mu$ and $\Gamma^\mu_{\alpha\beta} = 0$. Differentiating Eq. (14.71) gives

$$\partial_\lambda R^\mu_{\ \nu\alpha\beta} = \partial_\lambda \partial_\alpha \Gamma^\mu_{\nu\beta} - \partial_\lambda \partial_\beta \Gamma^\mu_{\nu\alpha}, \tag{14.78}$$

and therefore

$$\partial_\lambda R^\mu_{\ \nu\alpha\beta} + \partial_\beta R^\mu_{\ \nu\lambda\alpha} + \partial_\alpha R^\mu_{\ \nu\beta\lambda} = 0. \tag{14.79}$$

This is called the Bianchi identity. Contracting μ with α, and λ with ν, gives

$$2\partial_\nu R^\nu_{\ \mu} - \partial_\mu R = 0. \tag{14.80}$$

Contracting with $\eta^{\alpha\mu}$ now gives the desired Eq. (14.77).

14.4.5 General relativity and the action principle

During inflation the Universe contains fields as opposed to particles, and indeed, at the quantum level, we can say that this is always the case because a particle can be regarded as a quantized field oscillation.

With fields as the source of gravity, general relativity can be derived from the action principle that we discussed on page 166. Because the volume element in generic coordinates is $d^4x\sqrt{-g}$, the action of a system will be of the form

$$S = \int d^4x \sqrt{-g}\ \mathcal{L}, \tag{14.81}$$

where the Lagrangian density $\mathcal{L}$ is a scalar quantity.

The simplest possibility is

$$\mathcal{L} = \frac{1}{2} M_{\text{Pl}}^2 R, \tag{14.82}$$

where the space-time curvature R is to be regarded as a function of $g_{\mu\nu}$, and its first and second space-time derivatives. The reduced Planck mass appears as a constant of proportionality, normalized so that $\mathcal{L}_{\text{mat}}$ below has the canonical form. Making a small variation $\delta g_{\mu\nu}$, whose first derivatives vanish at infinity, the action principle $\delta S = 0$ gives the Einstein equation (14.74) with $T^{\mu\nu} = 0$.

The most general possibility leading to the Einstein equation is the **Einstein–Hilbert** action,

$$\mathcal{L} = \frac{1}{2} M_{\text{Pl}}^2 R + \mathcal{L}_{\text{mat}}, \tag{14.83}$$

where the second term is a function only of $g_{\mu\nu}$ and its *first* space-time derivatives. Varying $g_{\mu\nu}$ in the action principle gives the Einstein equation with (Misner et al. 1973)

$$T_{\mu\nu} = -2\frac{\partial \mathcal{L}_{\text{mat}}}{\partial g^{\mu\nu}} + g_{\mu\nu}\mathcal{L}_{\text{mat}}. \tag{14.84}$$

Why should we expect $\mathcal{L}_{\text{mat}}$ to involve only first derivatives of $g_{\mu\nu}$? The idea is to identify $\mathcal{L}_{\text{mat}}$ with the special-relativistic Lagrangian density, describing fields in the absence of gravity. The only change that we are supposed to make to the special-relativistic expression is to replace $\eta_{\mu\nu}$ by $g_{\mu\nu}$, and ∂_μ by D_μ. The latter contains only $g_{\mu\nu}$ and (for fields other than scalar fields) its first derivatives. Acting on a scalar, D_μ is just ∂_μ, and not even first derivatives of $g_{\mu\nu}$ appear.[8]

Varying the fields in the action principle gives equations of motion, which reduce to the special relativity (flat space-time) equations when $g_{\mu\nu}$ is replaced by $\eta_{\mu\nu}$. In this way, we can arrive at Maxwell's equations for the electromagnetic field, as well as its contribution to the stress–energy tensor.

We need the case of a scalar field. Its Lagrangian density in a locally orthonormal frame is given by Eq. (7.3), and in a generic coordinate system it is

$$\mathcal{L}_{\text{mat}} = -\frac{1}{2}g^{\mu\nu}\partial_\mu\phi\,\partial_\nu\phi - V(\phi).$$
(14.85)

The stress–energy tensor of the scalar field is therefore

$$T_{\mu\nu} = \partial_\mu\phi\,\partial_\nu\phi - g_{\mu\nu}\left[\frac{1}{2}g^{\alpha\beta}\partial_\alpha\phi\,\partial_\beta\phi + V(\phi)\right].$$
(14.86)

In a locally orthonormal frame ($g_{\mu\nu} = \eta_{\mu\nu}$), the momentum density is $T^{0i} = -\dot\phi\partial_i\phi$, as quoted in Eq. (7.10). Taking it to vanish, Eqs. (14.25) and (14.26) define the energy density ρ and pressure P:

$$\rho_\phi = \frac{1}{2}\dot\phi^2 + V(\phi),$$
(14.87)

$$P_\phi = \frac{1}{2}\dot\phi^2 - V(\phi).$$
(14.88)

These are the equations that we used in Chapter 3, as a starting point for studying inflaton field dynamics.

Varying ϕ in the action principle gives Eq. (7.6), valid now for generic coordinates. Using Eq. (14.85), we find the scalar field equation (14.15), with the d'Alembertian given by Eq. (14.66).

14.4.6 Beyond Einstein gravity?

The modern view is that we do not expect the Einstein–Hilbert action to be exactly correct, or for that matter the action of a renormalizable field theory, such as the Standard Model. Rather, the present paradigm is that the action will have the most general form consistent with some set of symmetries, which allows an infinite number of terms.

For terms involving only the curvature, the only requirement is that $\mathcal{L}$ be a scalar. The simplest possibility is to add terms of the form $\lambda_n M_{\text{Pl}}^{2(1-n)} R^{n+1}$, with n a positive integer and

---

[8] We will not need to define the covariant derivative acting on fermionic fields (spinors), but all the above statements remain true in that case.

λ_n a dimensionless constant.[9] When considering gravity on a distance scale $l \gg M_{\text{Pl}}^{-1}$, the effect of such a term will be suppressed by a factor of order $(l\, M_{\text{Pl}})^{-2n}$, relative to the Einstein–Hilbert term $M_{\text{Pl}}^2 R/2$. A similar argument applies to other terms involving only the curvature tensor. We conclude that, in the absence of fields, Einstein gravity is likely to be a good approximation at distances bigger than the Planck scale.

This is the same type of argument that we use in field theory. Considering, for instance, the potential equation (8.2), we expect the nonrenormalizable terms to be negligible, for energies and field values that are small compared with M_{Pl}. There, we are, of course, ignoring gravity.

If we consider both the fields and gravity, the situation becomes more complicated. Using a scalar field, we can write down reasonable-looking modifications of Einstein gravity, known generically as scalar–tensor theories of gravity. The classic tests of general relativity provide strong constraints on such theories at the present epoch, and we noted in Section 8.6.4 that the cosmic microwave background (cmb) anisotropy constrains them also during inflation.

14.5 Cosmological perturbations

14.5.1 Coordinate versus "covariant" approaches

We are now ready to tackle cosmological perturbation theory. In the rest of this chapter the focus is on the stress–energy tensor, and in the next on particle diffusion and freestreaming.

Roughly speaking, there are two approaches. The first one is based on globally defined coordinates in the perturbed Universe. For the stress–energy tensor, this approach was given in more or less complete form in the remarkable paper of Lifshitz (1946). For diffusion and freestreaming, it has evolved gradually, an influential early paper being that of Peebles and Yu (1970). As we see, it is essential in the latter case to consider also a locally orthonormal frame at each space-time point, whose time direction is lined up with that of a suitable coordinate system.

The other approach, generally known as the covariant approach, does not invoke any globally defined coordinates. All that is used is a slicing and a threading (Section 4.6.3), plus at each point in space-time a locally orthonormal frame. The basis for this approach was developed by Ehlers (1961), Hawking (1966), and Ellis (1971). Explicit equations for the density perturbation and related quantities were first given for a perfect fluid by Olson (1976), who used a slicing that turned out (Lyth and Stewart 1990b) to be a particular synchronous slicing that we mention in Section 14.6.7. The corresponding equations for the comoving slicing were first given by Lyth and Mukherjee (1988). Anisotropic stress was included by Ellis and Bruni (1989), and later works extended the analysis to noncritical density, uncoupled perfect fluids, peculiar velocity, gravitational waves, diffusion, and freestreaming. An account focusing on the last two topics is that of Durrer (1996).

---

[9] We cannot tolerate $n < -1$ because then $\mathcal{L}$ would blow up in the flat space-time limit. The case $n = -1$ corresponds to a cosmological constant, which nowadays is regarded as a contribution to $\mathcal{L}_{\text{mat}}$ rather than as a modification of Einstein gravity.

In both approaches, the evolution of the perturbation in the stress–energy tensor, which is the subject of the present chapter, is given in principle by the Einstein field equation. However, it is also useful to consider the energy–momentum conservation equation $D_\mu T^{\mu\nu} = 0$. Although this is a consequence of the Einstein field equation, it is simpler because it contains only first derivatives of $g_{\mu\nu}$, whereas the field equation contains second derivatives. Yet, at the same time, it encapsulates much of the physical information contained in the field equations. The time component ($\nu = 0$) corresponds to the continuity equation (energy conservation) and the space components to the Euler equation (momentum conservation, equivalent to the relativistic acceleration law).

The covariant approach is essentially the one that we used to describe the density perturbation in Chapter 4. In that context, the covariant approach is simpler than the usual metric perturbation approach, but as we go through the list of topics mentioned at the end of the second to last paragraph, it becomes progressively more complicated, and by the time we get to diffusion and freestreaming, the usual approach is far simpler. This is the one that we use from now on.

14.5.2 The unperturbed Universe

In the unperturbed Universe, the slices of fixed t can be chosen to be homogeneous. Its metric is called the Robertson–Walker metric, and the unperturbed Universe is called the Robertson–Walker Universe or, alternatively, the Friedmann–Lemaitre–Robertson–Walker (FLRW) Universe (Friedmann 1922; Lemaitre 1927; Robertson 1936; Walker 1936). The last two authors emphasized that the metric is a consequence of the spatial homogeneity, whether or not the Einstein equation is valid.

Only one parameter is needed to specify the geometry of a homogeneous three-dimensional space, which we can take to be the parameter K appearing in the Friedmann equation. If $K = 0$, the spatial slices are flat and the line element can be written

$$ds^2 = -dt^2 + a^2(t)[dx^2 + dy^2 + dz^2]. \tag{14.89}$$

If K is positive, the spatial slices are closed, and a plane bisecting space has the geometry of a sphere of radius $aK^{1/2}$. This leads almost immediately to the expression

$$ds^2 = -dt^2 + a^2(t)\left[dx^2 + \frac{1}{K}\sin^2(K^{1/2}x)(d\theta^2 + \sin^2\theta\, d\phi^2)\right]. \tag{14.90}$$

If K is negative, the spatial slices are open and the line element can be written

$$ds^2 = -dt^2 + a^2(t)\left[dx^2 + \frac{1}{|K|}\sinh^2(|K|^{1/2}x)(d\theta^2 + \sin^2\theta\, d\phi^2)\right]. \tag{14.91}$$

In all cases, x is the distance of a given point from the origin, and $a|K|^{-1/2}$ is the curvature scale, as defined in Section 2.2.3. The curvature *scalar*, calculated using the formulae of Section 14.4.3, is $R^{(3)} = a^2/6K^2$, but the curvature scale is the thing with the simple significance.

By changing the definition of the radial coordinate, we can bring these line elements into the form

$$ds^2 = -dt^2 + a^2(t)\frac{(dx^2 + dy^2 + dz^2)}{1 + \frac{1}{4}K^2(x^2 + y^2 + z^2)}. \tag{14.92}$$

Despite the sum of squares in the numerator, x, y, and z are not Cartesian coordinates because they also appear in the denominator. Another simple form often quoted is

$$ds^2 = -dt^2 + a^2(t)\left[\frac{d\tilde{x}^2}{1 - K\tilde{x}^2} + \tilde{x}^2(d\theta^2 + \sin^2\theta \, d\phi^2)\right]. \tag{14.93}$$

Using any of these forms for the metric, we can evaluate the connection and curvature tensor, as well as the components of the stress–energy tensor given by Eq. (14.64). Then we can derive the results about the unperturbed Universe that we quoted earlier. Equation (7.64) for a scalar field is just Eq. (14.15). The time component of $D_\mu T^{\mu\nu} = 0$ gives the continuity equation (2.8). Its space components vanish identically, corresponding to the fact that momentum is conserved automatically in an isotropic Universe. The Friedmann equation (2.12) is the time–time component of the Einstein field equation, and the other components yield no further information.

It is often convenient to use conformal time $d\tau = dt/a$, so that a^2 multiplies the entire metric. In the spatially flat case,

$$ds^2 = a^2(\tau)(-d\tau^2 + dx^2 + dy^2 + dz^2). \tag{14.94}$$

For the rest of this chapter, we consider only this case, and generally use conformal time τ. *For the rest of this chapter, an overdot will denote $\partial/\partial\tau$, not $\partial/\partial t$.*

14.5.3 Perturbed metric

As we discussed in Section 4.6.3, there is no uniquely preferred coordinate system in the presence of perturbations. The only essential requirement is that the coordinates reduce to those of Eq. (14.94) in the limit of zero perturbation. A choice meeting this requirement is called a gauge, and the transformation from one gauge to another is called a gauge transformation.[10]

For the moment we leave open the gauge choice, and consider the most general first-order perturbation:

$$ds^2 = a^2(\tau)\{-(1 + 2A)d\tau^2 - B_i \, d\tau \, dx^i + [(1 + 2D)\,\delta_{ij} + 2E_{ij}]dx^i dx^j\}. \tag{14.95}$$

The term in square brackets specifies the curvature perturbation $2(D\delta_{ij} + E_{ij})$, and E_{ij} is taken to be traceless so that the separation of the two terms is unique. The term B_i is the **shift function**; as we show later, it specifies the relative velocity between the threading and the

---

[10] In practice, a family of gauges may be regarded as equivalent. An example is the synchronous gauge discussed later.

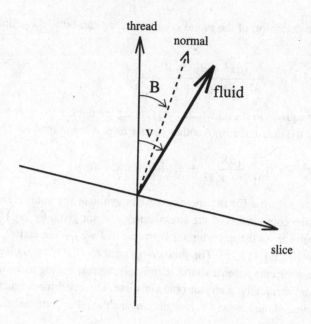

Fig. 14.1. The coordinates (τ, x^i) define a threading and a slicing (corresponding, respectively, to fixed x^i and fixed τ). As indicated, the slicing generally is not orthogonal to the threading. The time direction of the locally orthonormal frame considered in the text is lined up with the time coordinate line (thread). The space directions of the locally orthonormal frame, not shown, are orthogonal to the time direction, and they coincide with the space coordinate lines to zero order in the perturbation. In the locally orthonormal frame, the velocity of a comoving observer (fluid velocity) is **v**, and the velocity of the worldline that is normal (orthogonal) to the slices is **B**, where B_i is the shift function defined by Eq. (14.95).

worldlines orthogonal to the slicing. The term A is the **lapse function**, which specifies the relation between τ and the proper time t_{pr} along the threading. To first order,

$$\frac{dt_{\mathrm{pr}}}{d\tau} = \sqrt{1+2A} \simeq 1 + A. \tag{14.96}$$

14.5.4 Perturbed stress–energy tensor

The chosen coordinate system $x^\mu = (\tau, x^i)$ applies throughout space-time; it is a global coordinate system. In addition, at each space-time point we need a locally orthonormal coordinate system (t, r^i) (locally orthonormal frame) with the following properties: The time directions of the global and local coordinate systems are lined up to first order, and the spatial directions are lined up to zero order. From the perturbed metric (14.95), this is equivalent to

$$dt = a(1 + A)\, d\tau \qquad \text{(first order)}, \tag{14.97}$$

$$dr^i = a\, dx^i \qquad \text{(zero order)}. \tag{14.98}$$

As illustrated in Figure 14.1, the fluid velocity in the locally orthonormal frame is $v^i = dr^i/dt$. Reverting to the global coordinates, the components $u^\mu = dx^\mu/dt$ of the fluid

4-velocity are

$$u^0 = -1 \qquad \text{(zero order)}, \qquad (14.99)$$

$$u^i = v^i \equiv v_i \qquad \text{(first order)}. \qquad (14.100)$$

There is no distinction between upper and lower indices for the 3-velocity v^i, but there is for the 4-velocity. Indeed, for $u_\mu = g_{\mu\nu} u^\nu$, we find

$$u_0 = 1 \qquad \text{(zero order)}, \qquad (14.101)$$

$$u_i = v_i - B_i \qquad \text{(first order)}. \qquad (14.102)$$

Using these expressions, we can work out Eq. (14.64) to first order. To simplify the expressions, it is best to work with $T^\mu{}_\nu \equiv T_\nu{}^\mu$, and we then find

$$T^0{}_0 = -(\rho + \delta\rho), \qquad (14.103)$$

$$T^0{}_i = (\rho + P)(v_i - B_i), \qquad (14.104)$$

$$T^i{}_0 = -(\rho + P)v^i, \qquad (14.105)$$

$$T^i{}_j = (P + \delta P)\delta^i_j + \Sigma^i{}_j. \qquad (14.106)$$

(Raising and lowering indices has no effect on either v_i or Σ_{ij} because they are spatial components defined in a locally orthonormal frame. Note that $T^i{}_j = T^j{}_i$.) It is usual to define a dimensionless version of the anisotropic stress by

$$\Pi_{ij} \equiv \frac{\Sigma_{ij}}{P}. \qquad (14.107)$$

We note for future reference that, in the locally orthonormal frame, the velocity of the worldline orthogonal to the fixed-τ slice is B^i. This is illustrated in Figure 14.1, and the proof is very simple. Working with the coordinates (τ, x^i), the 4-velocity of the worldline with velocity B_i is to first order of the form $B^\mu = B^0(1, B^i)$. On the other hand, any 4-vector lying in the fixed-τ slice is of the form $e^\mu = (0, e^i)$. One can check easily that indeed $g_{\mu\nu} B^\mu e^\nu$ vanishes to first order, for all choices of e^i.

14.5.5 Scalar–vector–tensor decomposition

Using the metric perturbations A, B_i, D, and E_{ij}, which were defined by Eq. (14.95), we can work out the perturbation in the left-hand side of the Einstein equation, and equate it with $8\pi G \, \delta T^{\mu\nu}$. In this way, we obtain differential equations involving the metric perturbation and the perturbations in the stress–energy tensor. They consist of evolution equations (those that involve time derivatives) and constraint equations (those that do not involve time derivatives).

It turns out that the equations break down into three independent sets, involving what are called **scalar**, **vector**, and **tensor** perturbations, as listed in Table 14.1. The perturbations A, D,

Table 14.1. *Scalar, vector, and tensor perturbations*

	Metric	Stress–Energy Tensor
Scalars	A, B, D, E	$\delta\rho, \delta P, V, \Pi$
Vectors	B_i^{V}, E_i	v_i^{V}, Π_i
Tensors	$E_{ij}^{\mathrm{T}} \equiv a^{-2}h_{ij}^{\mathrm{T}}$	Π_{ij}^{T}

$\delta\rho$, and δP were defined already. To define the others, let us begin with v_i. As in the Newtonian case (Section 4.4.6), it can be decomposed as

$$v_i = v_i^{\mathrm{S}} + v_i^{\mathrm{V}},\tag{14.108}$$

where v_i^{S} is of the form

$$v_i^{\mathrm{S}} = -\frac{ik_i}{k}\,V,\tag{14.109}$$

and $k_i v_i^{\mathrm{V}} = 0$. The superscripts S and V stand for scalar and vector, the idea being that v_i^{S} can be obtained by differentiating a scalar whereas v_i^{V} cannot.

An analogous decomposition can be made for the traceless, symmetric anisotropic stress Π_{ij}. It can be written

$$\Pi_{ij} = \Pi_{ij}^{\mathrm{S}} + \Pi_{ij}^{\mathrm{V}} + \Pi_{ij}^{\mathrm{T}},\tag{14.110}$$

where Π_{ij}^{S} is of the form

$$\Pi_{ij}^{\mathrm{S}} = \left(-\frac{k_i k_j}{k^2} + \frac{1}{3}\delta_{ij}\right)\Pi,\tag{14.111}$$

Π_{ij}^{V} is of the form

$$\Pi_{ij}^{\mathrm{V}} = -\frac{i}{2k}(k_i \Pi_j + k_j \Pi_i),\tag{14.112}$$

with $k_i \Pi_i = 0$, and finally $k_i \Pi_{ij}^{\mathrm{T}} = 0$. Again, S and V stand for scalar and vector, and now we have, in addition, T standing for tensor. These words are appropriate because Π_{ij}^{S} can be obtained by differentiating a scalar, and Π_{ij}^{V} can be obtained by differentiating a vector, but Π_{ij}^{T} cannot be obtained in either way.

This decomposition can always be made; choose the z axis to be along $\mathbf{k}$ to give $\mathbf{k} = (0, 0, k)$, leading to

$$\Pi_{ij}^{\mathrm{S}} = \frac{1}{3}\begin{pmatrix} \Pi & 0 & 0 \\ 0 & \Pi & 0 \\ 0 & 0 & -2\Pi \end{pmatrix},\tag{14.113}$$

$$\Pi_{ij}^V = -\frac{i}{2} \begin{pmatrix} 0 & 0 & \Pi_1 \\ 0 & 0 & \Pi_2 \\ \Pi_1 & \Pi_2 & 0 \end{pmatrix}, \tag{14.114}$$

$$\Pi_{ij}^T = \begin{pmatrix} \Pi^\times & \Pi^+ & 0 \\ \Pi^+ & -\Pi^\times & 0 \\ 0 & 0 & 0 \end{pmatrix}. \tag{14.115}$$

It is clear that any traceless, symmetric matrix can be uniquely decomposed in this form. In Eq. (14.115), we indicated the independent components by Π^+ and $\Pi^\times$, which is the standard notation for the similar case of gravitational waves.

Analogous decompositions can be made for the metric perturbations. We write $B_i = B_i^S + B_i^V$, where B_i^S is of the form

$$B_i^S = -\frac{ik_i}{k} B, \tag{14.116}$$

and $k_i B_i^V = 0$. Also, $E_{ij} = E_{ij}^S + E_{ij}^V + E_{ij}^T$, where E_{ij}^S is of the form

$$E_{ij}^S = \left(-\frac{k_i k_j}{k^2} + \frac{1}{3} \delta_{ij} \right) E, \tag{14.117}$$

E_{ij}^V is of the form

$$E_{ij}^V = -\frac{i}{2k} (k_i E_j + k_j E_i), \tag{14.118}$$

with $k_i E_i = 0$, and finally, $k_i E_{ij}^T = 0$.

The three types of perturbation have a simple physical significance. The scalar perturbations are the ones generated by the vacuum fluctuation of the inflaton field and are our main concern. The tensor metric perturbations are gravitational waves, and the anisotropic stress that can interact with them. They are generated at some level during inflation. Finally, the vector perturbations consist of the relativistic generalization of purely rotational fluid flow, together with the anisotropic stress that can interact with them; neither of these is generated by inflation and we assume that they are absent.

Because the equations for the three types of perturbation are decoupled, we can assume that the other two vanish when considering one type, even if that is not true in Nature. The equations are written down in Section 14.6, but first we want to consider the effect of gauge transformations.

14.5.6 Gauge transformations

Recall that a gauge transformation is a first-order change in the coordinates,

$$\tilde{x}^0 = x^0 + \delta\tau(x^\mu), \qquad \tilde{x}^i = x^i + \delta x^i(x^\mu). \tag{14.119}$$

As in Eq. (4.141), the gauge transformation shifts the space-time position, but because we now deal with tensors, it also transforms the components. Using the transformation law (14.36), the result to first order for any tensor $B_{\mu\nu}$ is

$$\tilde{B}_{\mu\nu}(\tau, x^i) = \frac{\partial x^\alpha}{\partial \tilde{x}^\mu} \frac{\partial x^\beta}{\partial \tilde{x}^\nu} B_{\alpha\beta}(\tau - \delta\tau, x^i - \delta x^i)$$

$$= B_{\mu\nu}(\tau, x^i) + B_{\alpha\nu}\partial_\mu \delta x^\alpha + B_{\mu\alpha}\partial_\nu \delta x^\alpha - \delta x^\lambda \partial_\lambda B_{\mu\nu}. \tag{14.120}$$

In this formula, we can decompose δx_i into vector and scalar parts, $\delta x_i = \delta x_i^S + \delta x_i^V$, with $\delta x_i^S = -i(k^i/k)\delta x$ and $k_i \delta x_i^V = 0$. These give, respectively, perturbations in the scalar and vector parts of the metric tensor and the stress–energy tensor, but not in the tensor parts. *The tensor perturbations are gauge invariant.*

We are not interested in vector perturbations, which leaves only the scalar perturbations. Applying Eq. (14.120) to the metric tensor (14.95) gives

$$\tilde{A} = A - (\delta\tau)\dot{} - aH\delta\tau, \tag{14.121}$$

$$\tilde{B} = B + (\delta x)\dot{} + k\delta\tau, \tag{14.122}$$

$$\tilde{D} = D - \frac{k}{3}\delta x - aH\delta\tau, \tag{14.123}$$

$$\tilde{E} = E + k\delta x. \tag{14.124}$$

Applying it to the stress–energy tensor, given by Eqs. (14.103), (14.104), (14.105), and (14.106), gives

$$\tilde{V} = V + (\delta x)\dot{}, \tag{14.125}$$

$$\tilde{\delta} = \delta + 3(1 + w)aH\delta\tau, \tag{14.126}$$

$$\tilde{\delta P} = \delta P - \dot{P}\delta\tau, \tag{14.127}$$

$$\tilde{\Pi} = \Pi. \tag{14.128}$$

In the second equation, we used $\dot{\rho} = -3aH(\rho + P)$ and the definition $w \equiv P/\rho$.

14.5.7 Spatial curvature for scalar perturbations

Let us consider the spatial curvature of the slices of fixed τ, keeping only the scalar perturbations. The spatial metric tensor is the coefficient of $dx^i dx^j$ in Eq. (14.95). Using the formulae of Section 14.4.3, now applied to the spatial slices, we can show that the corresponding curvature scalar is

$$R^{(3)} = 4\frac{k^2}{a^2}\left(D + \frac{1}{3}E\right). \tag{14.129}$$

From Eqs. (14.123) and (14.124),

$$\left(\tilde{D} + \frac{1}{3}\tilde{E}\right) = \left(D + \frac{1}{3}E\right) - aH\delta\tau. \tag{14.130}$$

This is independent of the change in the spatial coordinates, defined by δx, in accordance with the fact that the spatial curvature scalar $R^{(3)}$ appearing in Eq. (14.129) is independent of these coordinates.

By choosing δx appropriately, we can make $\tilde{E}$ vanish, and then Eq. (14.95) gives the spatial curvature tensor in terms of $\tilde{D}$, and hence in terms of the spatial curvature scalar. (This is not generic for 3-space; it holds only for scalar cosmological perturbations to first order.)

The quantity $\mathcal{R}$, which has played a key role in this book, is simply $D + \frac{1}{3}E$, evaluated with any coordinate choice that makes the slicing comoving:

$$\mathcal{R} = D + \frac{1}{3}E \quad \text{(comoving)}. \tag{14.131}$$

From Eq. (14.129), this definition of $\mathcal{R}$ is equivalent to saying that the spatial curvature scalar of the comoving slicing is

$$R^{(3)} = 4\frac{k^2}{a^2}\mathcal{R}. \tag{14.132}$$

We see soon that this is equivalent to the definition in terms of the locally defined Friedmann equation (4.167), which we used in the earlier chapters.

Equation (14.130), with the left-hand side equal to 0, shows that $\mathcal{R} = H \delta t$, where $\delta t = a\delta\tau$ is the time displacement going from the flat to the comoving slicing. This is Eq. (4.143), which we used in earlier chapters.

14.6 Evolution of the perturbations

Using all this, it is straightforward to work out the components of the Einstein equations to first order. We are interested only in tensor and scalar perturbations, and the former turn out to be very simple.

14.6.1 Evolution of the tensor perturbations

For tensor perturbations, the metric is simply

$$ds^2 = a^2(\tau)\left[-d\tau^2 + \left(\delta_{ij} + 2E_{ij}^{\mathrm{T}}\right)dx^i dx^j\right]. \tag{14.133}$$

The tensor perturbations appear only in the space–space components of the Einstein equation, which give

$$\left(E_{ij}^{\mathrm{T}}\right)'' + 2aH\left(E_{ij}^{\mathrm{T}}\right)' + k^2 E_{ij}^{\mathrm{T}} = 8\pi G a^2 P \Pi_{ij}^{\mathrm{T}}. \tag{14.134}$$

After horizon entry, E_{ij}^{T} oscillates, and represents a gravitational wave, with amplitude

$$h_{ij} \equiv a E_{ij}^{\mathrm{T}}. \tag{14.135}$$

If Π_{ij}^T is negligible, as it is after matter domination, the evolution of each Fourier component in terms of cosmological time t is given by

$$\frac{\partial^2 h_{ij}}{\partial t^2} + 3H\frac{\partial h_{ij}}{\partial t} + \left(\frac{k}{a}\right)^2 h_{ij} = 0. \tag{14.136}$$

This is the same equation as for a massless scalar field.

14.6.2 Scalar perturbations: General considerations

To define the scalar perturbations, we need to choose a gauge. Several are used in the literature (all coinciding well after horizon entry) and the choice between them is a matter of convenience and personal taste. To treat diffusion and freestreaming in Chapter 10, we use the conformal Newtonian gauge, defined in Section 14.6.3.

In a given gauge, the first-order perturbation in the Einstein field equation provides the evolution/constraint equations for the perturbations. However, much of the information is encoded more usefully in the continuity and Euler equations. We saw how this works in the unperturbed Universe, at the end of Section 14.5.2. With the continuity equation in place, the field equation provides only one piece of additional information, which is the Friedmann equation. In that case, the Euler equation is trivial, because momentum has to be conserved in an isotropic Universe.

A similar thing happens in the perturbed Universe, except that both the conservation condition and the Euler equation are nontrivial. With them in place, the field equations provide *no* further evolution equations. Rather, they provide a set of **constraint equations**, giving the metric perturbation (specified by A, B, D, and E) in terms of the quantities $\delta\rho$, δP, Π, and V that specify the stress–energy tensor.[11]

A slicing is needed to define the perturbations $\delta\rho$ and δP, a threading is needed to define the velocity perturbation V, and the anisotropic stress Π is gauge invariant.

14.6.3 Conformal Newtonian gauge

The mathematically simplest choice is the conformal Newtonian (also called longitudinal) gauge, defined by the line element[12]

$$ds^2 = a^2(\tau)[-(1 + 2\Psi)d\tau^2 + (1 - 2\Phi)\,\delta_{ij}\,dx_i\,dx_j]. \tag{14.137}$$

In this gauge the slicing and threading are orthogonal, and the worldlines defining the latter have zero shear because the spatial part of the metric perturbation is isotropic.

---

[11] We emphasize that this need apply only in a completely defined gauge. The synchronous gauge (Section 14.6.7) is not completely defined.

[12] Starting with an arbitrary gauge, we can indeed uniquely choose δx so that E vanishes, Eq. (14.124), and then uniquely choose $\delta\tau$ so that B vanishes, Eq. (14.122).

A subscript N is used to denote the conformal Newtonian gauge, and this line element is equivalent to

$$A_N \equiv \Psi, \tag{14.138}$$

$$D_N \equiv -\Phi, \tag{14.139}$$

$$B_N = E_N = 0. \tag{14.140}$$

This perturbation is used in the metric tensor, making it is easy to work out the components of the connection to first order. Then we can use Eqs. (14.25), (14.26), and (14.28) for the stress–energy tensor, to find the first-order perturbations in the continuity equation $D_\mu T^\mu{}_0 = 0$ and in the Euler equation $D_\mu T^\mu{}_i = 0$. These perturbations give, respectively, the two equations

$$\dot{\delta}_N = -(1+w)(kV_N - 3\dot{\Phi}) + 3aHw\delta_N - 3aH\frac{\delta P_N}{\rho}, \tag{14.141}$$

$$\dot{V}_N = -aH(1-3w)V_N - \frac{\dot{w}}{1+w}V_N + k\frac{\delta P_N}{\rho + P} - \frac{2}{3}k\frac{w}{1+w}\Pi + k\Psi. \tag{14.142}$$

The subscript N again reminds us that we are in the conformal Newtonian gauge.

The only additional information coming from the field equation consists of the constraint equations

$$k^2\Phi = -4\pi Ga^2\rho\left[\delta_N + 3\frac{aH}{k}(1+w)V_N\right], \tag{14.143}$$

$$k^2(\Psi - \Phi) = -8\pi Ga^2 P\,\Pi. \tag{14.144}$$

14.6.4 Total-matter gauge

In the earlier chapters, we used the comoving slicing. This is not orthogonal to the zero-shear threading, yet remains useful. Let us therefore displace the slicing of the conformal Newtonian gauge so that it becomes the comoving slicing, while leaving unaltered both the threading and the spatial coordinates that label it. (The reader may find it useful to refer back to Figure 14.1 on page 336 at this point.) The resulting gauge has been called the **total-matter gauge**. (The term "comoving gauge" is used occasionally to denote any gauge with the comoving slicing.)

The gauge transformation from the conformal Newtonian to the total-matter gauge is defined by

$$\delta x = 0, \qquad \delta\tau = \frac{V}{k}. \tag{14.145}$$

We see from Eq. (14.125) that the requirement of keeping V the same makes δx a function of the position coordinates only. Such a δx would relabel the threads unless it vanished. Also, putting $\delta\tau = V/k$ into Eq. (14.122) gives $V - B = 0$, which is the condition that the slicing be comoving.

Perturbation in the stress–energy tensor

In the total-matter gauge,

$$V = V_{\mathrm{N}}, \tag{14.146}$$

$$\delta = \delta_{\mathrm{N}} + 3\frac{aH}{k}(1+w)V_{\mathrm{N}}, \tag{14.147}$$

$$\delta P = \delta P_{\mathrm{N}} - \frac{\dot{P}V_{\mathrm{N}}}{k}, \tag{14.148}$$

where the subscript N again denotes the conformal Newtonian gauge. We use these quantities, but *retain the original metric perturbations* Φ and Ψ, so that the perturbations in the continuity and Euler equations give

$$\dot{\delta} - 3waH\delta = -(1+w)kV - 2aHw\Pi, \tag{14.149}$$

$$\dot{V} + aHV = k\frac{\delta P}{\rho + P} - \frac{2}{3}\frac{w}{1+w}k\Pi + k\Psi, \tag{14.150}$$

and the constraint equations become

$$k^2\Phi \equiv -4\pi Ga^2\rho\delta, \tag{14.151}$$

$$k^2(\Psi - \Phi) \equiv -8\pi Ga^2 P\Pi. \tag{14.152}$$

These are the four equations that we use in the next chapter, as part of a complete system of equations.

Equations (14.151) and (14.152) are identical to Eqs. (4.181) and (4.182). Further, Eqs. (14.149) are identical to Eqs. (4.179) and (4.180), provided that the Newtonian relation $3\delta H = kV/a$ is satisfied [Eq. (4.118)]. Because the two sets of equations were derived independently, this proves the last result. As we discussed in Section 4.9.2, that result in turn implies that $\mathbf{v} \equiv -i(\mathbf{k}/k)V$ is the general-relativistic version of peculiar velocity. We used it in Chapter 5 to derive the Sachs–Wolfe effect. Note though that the usual approach, on which we are now embarking, is self-contained and, in particular, also will give the Sachs–Wolfe effect. From this viewpoint the relation $3\delta H = kV/a$ is superfluous, and it is not even mentioned in most accounts of cosmological perturbation theory.

Metric perturbation

Equations (14.124) and (14.145) show that the metric perturbation E continues to vanish in the total-matter gauge. Because the slicing in this gauge is comoving, Eq. (14.131) becomes $D = \mathcal{R}$. Therefore, Eqs. (14.123), (14.138), and (14.145) give

$$\mathcal{R} = -\Phi - V\frac{aH}{k}. \tag{14.153}$$

Use of $3\delta H = kV/a$ makes this the alternative definition of $\mathcal{R}$ [Eq. (4.167) along with the Poisson equation], which we used in earlier chapters. For the spatially flat case, we now have derived all of the formulae assumed earlier in the book.

Finally, let us note that Eqs. (14.145) and (14.121) give the lapse function A as

$$A = \Psi - k^{-1}\dot{V} - \frac{aH}{k}V. \tag{14.154}$$

We use Eq. (14.150) so that this becomes

$$A = -\frac{\delta P - \frac{2}{3}P\Pi}{\rho + P}. \tag{14.155}$$

This provides an alternative derivation of Eq. (4.177), through use of Eq. (14.96).

14.6.5 "Gauge-invariant" equations

Not all of the quantities appearing in a given equation need to be defined in the same gauge. We already encountered an example in Section 14.6.4, which contains quantities defined in the total-matter gauge, but also the potentials Φ and Ψ that originally were defined in the conformal Newtonian gauge. In that particular case the potentials can equally well be regarded as defined within the total-matter gauge by Eqs. (14.151) and (14.152). We encountered a much less trivial example in Eq. (4.6), which relates the inflaton field perturbation on the spatially flat slicing to the spatial curvature perturbation in the total-matter gauge (comoving slicing).

From these examples, it is clear that mixing gauges in the same equation can be a very useful device, and we employ it again in the next chapter. It was first used systematically by Bardeen (1980) for the stress–energy tensor perturbation, by Kodama and Sasaki (1984, 1987) for diffusion and freestreaming, and by Sasaki (1986) for the inflaton field perturbation. These authors actually used a gauge-invariant formalism, in which all quantities are regarded as being defined in a single gauge, with the form of the equations the same in all gauges. For instance, instead of defining δ as the density contrast on the comoving slices, we can define it in a generic gauge as the following combination of the density perturbation, the peculiar velocity, and the shift function:

$$\delta \equiv \tilde{\delta} + 3(1+w)\frac{aH}{k}(\tilde{V} - \tilde{B}), \tag{14.156}$$

where the tilde denotes the generic gauge. From Eqs. (14.122), (14.125), and (14.126), the right-hand side has the same form in every gauge (is gauge-invariant), and it reduces to the density perturbation in a gauge in which the slicing is comoving. It is a matter of taste whether this viewpoint is a useful one. On the other hand, the term gauge-invariant to describe the actual equations is firmly established. So, in the usual language, we are using a gauge-invariant formalism in this book.

14.6.6 Separate fluids

In addition to the total stress–energy tensor, in the next chapter we need to consider the contribution of separate fluids. In this context, a fluid is a component of the gaseous Universe

Table 14.2. *The four fluids*

Type of fluid	Matter	Radiation
Noninteracting	CDM	Neutrinos
Interacting	Baryons	Photons

with a well-defined equation of state $P_i(\rho_i)$, where i labels the fluid. Each fluid satisfies the continuity and Euler equations, except that there may be collision terms describing the exchange of energy and/or momentum between the fluids. The total stress–energy tensor is the sum of the components

$$\delta\rho = \sum_i \delta\rho_i, \tag{14.157}$$

$$(\rho + P)V = \sum_i (\rho_i + P_i)V_i, \tag{14.158}$$

$$\delta P = \sum_i \delta P_i, \tag{14.159}$$

$$P\Pi = \sum_i P_i\Pi_i. \tag{14.160}$$

As summarized in Table 14.2, we are dealing with four fluids if the nonbaryonic dark matter is cold. Two of these are "matter" with $P_i = 0$, and two are "radiation" with $P_i = \rho_i/3$. For matter, $\delta P_i = 0$ and $P_i\Pi_i = 0$. The term "baryons" designates ordinary matter, including both nuclei and electrons, and it constitutes a single fluid because electrical neutrality requires the density contrast of the electrons to be the same as that of the protons.

For the individual fluids, the transformation from the conformal Newtonian to the total-matter gauge is

$$V_i = V_{\mathrm{N}i}, \tag{14.161}$$

$$\delta_i = \delta_{\mathrm{N}i} + 3\frac{aH}{k}(1 + w_i)V_{\mathrm{N}}, \tag{14.162}$$

$$\delta P_i = \delta P_{\mathrm{N}i} - \frac{\dot{P}_i V_{\mathrm{N}}}{k}. \tag{14.163}$$

Note that the total $V_{\mathrm{N}} (=V)$, defining the shift in the slicing, appears on the right-hand side of the last two equations.

14.6.7 The synchronous gauge

The synchronous gauge was used by almost everyone before Bardeen's 1980 paper, and is still quite popular. It corresponds to $A = B = 0$, so that the only nonzero perturbations are D and E. The condition $A = 0$ says that the threading consists of geodesics, and $B = 0$ says that the slicing is orthogonal to them. The threading is not unique, and we really should talk about *a* synchronous gauge. There is a whole class of gauges with the property $A = B = 0$.

To impose an adiabatic or isocurvature initial condition in a simple way, the threading is chosen to coincide with the comoving one far outside the horizon (Lyth and Stewart 1990b). This corresponds to choosing particular solutions of the perturbed Einstein equations, a procedure known as "dropping the gauge modes" because it picks out a particular gauge.

Formulating the initial conditions requires considerable care; in addition, the equations are more complicated than if there were no gauge modes. (The relation between the metric perturbation and the density perturbation is not algebraic, but takes the form of a differential equation.) Although this caused great confusion historically, the synchronous gauge is now well understood, and is perfectly serviceable as a way of understanding the evolution of the perturbations after the "initial" epoch a few *e*-folds before horizon entry. For numerical computation, it is at least as good as the gauge-invariant formalism, and better than the conformal Newtonian gauge. The CMBFAST code uses it.

Examples

14.1 Verify that the Lorentz boost (14.6) satisfies Eq. (14.5). Do the same for a rotation around the z axis.

14.2 Sections 14.1–14.3 remain valid in the context of general relativity, provided that the inertial frame is replaced by a locally inertial frame. In many cases, we can use instead the more primitive concept of a locally orthonormal frame. Say whether that is possible, for each of the following cases:
(a) Eq. (14.2),
(b) Eq. (14.9),
(c) Eq. (14.10),
(d) Eq. (14.14),
(e) Eqs. (14.15) and (14.16),
(f) Eqs. (14.18)–(14.21),
(g) Eqs. (14.26)–(14.29),
(h) Eq. (14.32),
(i) Eq. (14.35),
(j) Eq. (14.46),
(k) Eqs. (14.56) and (14.57).

14.3 Show that the gauge transformation of a generic 4-tensor $B_{\mu\nu}$ is given by Eq. (14.120). For the metric tensor, show that this leads to Eqs. (14.121) and (14.124). For the stress-energy tensor, verify that Eq. (14.125) is just the gauge transformation (4.187) of a 3-vector, whereas each of Eqs. (14.126) and (14.127) is just the gauge transformation (14.128) of a scalar.

15 Advanced topic: Diffusion and freestreaming

In the preceding chapter, we found equations for the perturbation in the stress–energy tensor. They generally cannot be solved in isolation, because there are too many unknowns. This can be traced to the fact that the stress–energy tensor of a gas comes from the total effect of all particles in a given volume element. To specify the state of the gas completely, we need to specify the momentum distribution function of each particle species. We also need this function (for the photons) to calculate the cosmic microwave background (cmb) anisotropy. In this chapter we develop a complete system of equations for the scalar perturbations, which can be solved given the primordial curvature perturbation. The similar case of tensor perturbations is mentioned briefly.

We develop the equations in the conformal Newtonian gauge, though it is necessary to refer back to the comoving slicing, at least for the initial conditions and preferably also for numerical computation. We focus on the case in which the nonbaryonic dark matter is purely cold, though the formalism is readily extended to handle a hot component or decaying dark matter.

15.1 Matter

15.1.1 Cold Dark Matter (CDM)

CDM is easily dealt with. Its random motion is, by definition, negligible, so that its pressure is practically zero and the only quantities of interest are its density contrast, δ_c, and its peculiar velocity, V_c.[1] It satisfies by itself the energy and momentum conservation equations (14.141) and (14.142):

$$\frac{\partial \delta_c^N}{\partial \tau} = -kV_c + 3\frac{\partial \Phi}{\partial \tau}, \tag{15.1}$$

$$\frac{\partial V_c}{\partial \tau} = -aHV_c + k\Psi. \tag{15.2}$$

Once we have the potentials Φ and Ψ, this tells us everything we need to know about the CDM.

---

[1] To be precise, the peculiar velocity of a species i is given by $\mathbf{v}_i = -i(\mathbf{k}/k)V_i$. For brevity, we also call V_i the peculiar velocity.

15.1.2 Baryons

The nuclei and electrons have a common density contrast δ_b because the Coulomb interaction forces the charge density to be practically zero. As a result they also have a common peculiar velocity V_b. Their energy conservation equation is the same as that of the CDM, but their momentum conservation equation is different because of the pressure gradient, and also because Thomson scattering transfers momentum between the baryons and photons.

Let us deal with the pressure first. It comes from the electrons because they are far lighter than the nuclei. Baryons and photons are in thermal equilibrium, and there is a definite adiabatic equation of state $dP_b/d\rho_b = c_s^2$, so that $\delta P_b = c_s^2\delta\rho_b$, with c_s the sound speed. We are dealing with a monatomic gas whose density is $\rho_b = nm$, where n is the number density of atoms, and $m \simeq 1\,\text{GeV}$ is the mean atomic mass number, whose time dependence can be ignored in this context. Its pressure is $P_b = nT_b$, and we can take n to refer to the unperturbed Universe corresponding to $n \propto a^{-3}$. (Remember that the unperturbed Universe expands adiabatically because there can be no heat flow.) It follows that

$$c_s^2 = \frac{dP_b}{d\rho_b} = \frac{T_b}{m}\left(1 - \frac{1}{3}\frac{d\ln T_b}{d\ln a}\right). \tag{15.3}$$

The baryon temperature evolves according to

$$\frac{dT_b}{d\tau} = -2aHT_b + \frac{8}{3}\frac{\mu}{m_e}\frac{\rho_\gamma}{\rho_b}an_e\sigma_T(T_\gamma - T_b). \tag{15.4}$$

This follows (Peebles 1993, Chap. 6) from the first law of thermodynamics,

$$dQ = \frac{3}{2}d\left(\frac{P_b}{\rho_b}\right) + P_bd\left(\frac{1}{\rho_b}\right), \tag{15.5}$$

with specific heating rate

$$\frac{dQ}{d\tau} = 4\frac{\rho_\gamma}{\rho_b}an_e\sigma_T(T_\gamma - T_b). \tag{15.6}$$

Because the baryons have $P_b \ll \rho_b$, the pressure can be dropped from the continuity and Euler equations, except for the gradient term in the latter equation, which gives an additional contribution $c_s^2k\delta_b^N$.

Including Thomson scattering, the final result is

$$\frac{\partial\delta_b^N}{\partial\tau} = -kV_b + 3\frac{\partial\Phi}{\partial\tau}, \tag{15.7}$$

$$\frac{\partial V_b}{\partial\tau} = -aHV_b + c_s^2k\delta_b^N + k\Psi + \frac{4\rho_\gamma}{3\rho_b}an_e\sigma_T(V_\gamma - V_b). \tag{15.8}$$

We see later that the final term, coming from Thomson scattering, amounts to the statement that, in the electron/baryon rest frame, an average colliding photon loses all of its momentum.

15.1.3 Recombination and reionization

The preceding equations contain the free electron density n_e, and so do the ones that we develop later to describe the photons. At early times, there is full ionization, so that n_e is equal to the total electron number density. Then first ^{4}He and later H nuclei combine with electrons to form atoms; in the usual parlance they "recombine," though that term is not really appropriate because atoms never existed before.

At recombination, the function $n_e(t)$ falls abruptly to a small value, but we need to calculate it accurately. Its profile during the abrupt fall is crucial for an accurate description of the cmb anisotropy. The tail is important for determining the baryon temperature, and hence the important result that there are enough free electrons to hold the baryons and photons at the same temperature until $z \sim 100$. It is this result that determines the mass of the first baryonic objects, estimated roughly as the Jeans mass (4.130).

To describe the recombination process for helium, it is a good approximation to use the thermal equilibrium distribution (15.23), which for atoms and nuclei leads to an equation called the **Saha equation**. For hydrogen, we need to solve the coupled Boltzmann equations describing the transitions between different levels (Peebles 1968; Jones and Wyse 1985; Pequignot et al. 1991).

According to the recombination calculation, n_e becomes completely negligible after $z \sim 100$. However, this calculation considers only the unperturbed Universe, which means that it ignores structure formation. When enough baryonic structures have formed, practically complete reionization takes place, and normally we assume that this happens suddenly at some redshift z_{ion}. The various models described in Chapters 4 to 6, with parameters in the observationally allowed regime, predict a value $z = z_{ion} \sim 10$ to 20 for critical density, rising to roughly around 30 in low-density Universes, as we saw in Section 11.6. Future satellite measurements of the cmb polarization should be able to determine z_{ion} even if it is at the lower end of this range.

15.2   Gas dynamics in flat space-time

Our main task is to consider the radiation, and to do that, we need to treat diffusion and freestreaming. As in the preceding chapter, the strategy is first to treat them in the context of special relativity. The treatment remains valid in general relativity, at each space-time point, in a locally orthonormal or locally inertial frame. The latter frame is needed if vectors or tensors are being differentiated; otherwise the former is sufficient.

15.2.1   Distribution function

If we are uninterested in particle spin, the properties of an ideal gas are specified completely by the *distribution function* $f(t, \mathbf{r}, \mathbf{p})$, which gives the number dN of particles with position $\mathbf{r}$ and momentum $\mathbf{p}$ in specified ranges. We normalize it so that

$$dN = \frac{1}{(2\pi)^3} g_i f(t, \mathbf{r}, \mathbf{p}) \, d^3r \, d^3p, \tag{15.9}$$

where g_i is the number of spin states. Because the density of momentum states is $(2\pi)^{-3}$, this makes f the number of particles per quantum state, often called the occupation number.

For a massive particle species with spin s, there are $2s + 1$ spin states, and the same for the antiparticle. For a photon, there are two spin states, which can be taken to correspond to plane polarization in orthogonal directions. Neutrinos have left-handed spin and antineutrinos right-handed, and counting them both gives $g_i = 2$ for each neutrino species.

If $E = \sqrt{m^2 + p^2}$ is the particle energy, then the quantities Ed^3r and d^3p/E are Lorentz invariant. (Starting in the rest frame of a given particle and boosting along the z axis, for example, we see that, indeed, $dz \propto 1/E$, corresponding to the Lorentz contraction, and that $dp_z \propto E$.) As a result, $d^3r d^3p$ is Lorentz invariant, and so is f. The number density, which is not Lorentz invariant, is given by

$$n(t, \mathbf{r}) = \frac{g_i}{(2\pi)^3} \int f(t, \mathbf{r}, \mathbf{p}) \, d^3p. \tag{15.10}$$

15.2.2 Stress–energy tensor

All of the properties of the gas are determined by specifying the distribution function for each particle species. We are interested in the stress–energy tensor, but let us warm up with the simpler example of the electromagnetic 4-current.

Suppose that the particles all have the same charge q, and focus on a small region of space. Consider first only those particles with momentum in some range d^3p centred on a particular value $\mathbf{p}$. In the local rest frame, where $\mathbf{p} = 0$, they contribute $\rho(\mathbf{r}) = qf d^3p$ and $\mathbf{j}(\mathbf{r}) = 0$. In an arbitrary frame, their contribution is therefore

$$j^\mu = \frac{g_i}{(2\pi)^3} q f p^\mu \left(\frac{d^3p}{E}\right). \tag{15.11}$$

(The last factor is a scalar, making j^μ a 4-vector that reduces to the required form in the local rest frame.) Integrating over momentum gives

$$j^\mu = \frac{g_i}{(2\pi)^3} q \int f p^\mu \frac{d^3p}{E}. \tag{15.12}$$

To get the stress–energy tensor, all we have to do is replace the charge q of each particle by its 4-momentum p^ν. This gives

$$T^{\mu\nu} = \frac{g_i}{(2\pi)^3} \int f p^\mu p^\nu \frac{d^3p}{E}. \tag{15.13}$$

The 4-momentum density, as it should be, is

$$T^{\mu 0} = \frac{g_i}{(2\pi)^3} \int f p^\mu d^3p. \tag{15.14}$$

Using Eqs. (14.25), (14.26), and (14.28), we find that for each species the energy density ρ, momentum density $(\rho + P)\mathbf{v}$, pressure P, and anisotropic stress Σ_{ij} are given in terms of the occupation number by

$$\rho = \frac{g_i}{(2\pi)^3} \int f E \, d^3 p, \tag{15.15}$$

$$(\rho + P)\mathbf{v} = \frac{g_i}{(2\pi)^3} \int f \mathbf{p} \, d^3 p, \tag{15.16}$$

$$P = \frac{g_i}{3(2\pi)^3} \int f p^2 \frac{d^3 p}{E}, \tag{15.17}$$

$$\Sigma_{ij} = \frac{g_i}{(2\pi)^3} \int f \left(p_i p_j - \frac{1}{3} p^2 \delta_{ij} \right) \frac{d^3 p}{E}. \tag{15.18}$$

These expressions are valid only if $\mathbf{v}$ is nonrelativistic, but that is the only case that we need. Note that for radiation, $p = E$, and we have the familiar result $P = \rho/3$. If f is independent of direction, then Σ_{ij} vanishes and we have a perfect fluid. This is true in particular in thermal equilibrium.

15.2.3 Collisionless Boltzmann equation

We now develop equations giving the evolution of the distribution functions. If there are no collisions, each particle has constant momentum $\mathbf{p}$. In that case the distribution function is constant along a given particle trajectory $\mathbf{r}(t)$:

$$\frac{df}{dt} = 0. \tag{15.19}$$

This is the **Liouville equation**, also sometimes called the collisionless Boltzmann equation.

To see why f is constant, consider an initially rectangular region of phase space. As time passes, the region moves and its shape is distorted, but the number of particles in it remains constant; thus f will be constant if the volume of the region is constant. To see that this is indeed the case, assume without loss of generality that the particle motion is in the r^1 direction, as illustrated in Figure 15.1, and focus on the initially rectangular pair of faces with sides dr^1 and dp^1. The coordinate p^1 of each corner is constant, but the other coordinate r^1 moves with speed proportional to p^1 ($p^i = m \, dr^i / dt$). The rectangle therefore becomes a parallelogram with the passage of time, but its area remains constant and so does the volume that we are considering. (A more formal way of seeing the constancy of the volume is to calculate the Jacobian of the transformation of variables that corresponds to the elapse of a time interval dt.)

The constancy of f is a special case of the Liouville theorem, which says that f is constant along each trajectory in phase space, if the interactions are described by a Hamiltonian, and f is defined so that $dN = d^3 q \, d^3 p$, where q and p are canonically conjugate variables. This more general statement is not needed for cosmology because, in the absence of particle collisions, the only interaction is gravity, which can be eliminated by going to a locally inertial frame.

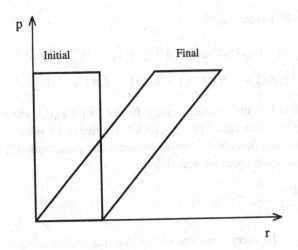

Fig. 15.1. The volume of a phase-space element does not change if the particles are moving with constant velocity.

15.2.4 Collision term

If there are collisions, particles will enter and leave the chosen volume of phase space. However, in the ideal gas limit, which is always good in cosmology, a given particle in the volume travels a long way before it collides. (To be precise, it travels much farther than the mean particle spacing.) As a result, it still makes sense to follow f along the trajectory of a particle that does collide, but now its rate of change no longer vanishes. We call the rate of change the *collision term*, and if we denote it by $C[f]$, the Liouville equation becomes

$$\frac{df}{dt} = C[f]. \tag{15.20}$$

This is the **Boltzmann equation** for f.

The outcome of a collision depends on the particle spins, but in many situations we can ignore this fact. Then, the collision term for a given particle species can be calculated if we know the distribution functions of all species whose collisions create or destroy the given species. This leads to a set of coupled Boltzmann equations, which can be solved to give the distribution functions.

Let us write down the collision term appearing in the coupled Boltzmann equations, in fully relativistic form but ignoring the spin dependence of the collision rate. Consider a particle species a, and suppose for the moment that its only significant interaction is a process of the form ab↔cd, with all four species different. Corresponding to this process is a transition rate $|M|^2$, which, by virtue of time-reversal invariance, is the same in both directions. Depending on the circumstances, it may be calculable from the appropriate theory (quantum field theory for elementary particles, atomic physics for atoms, nuclear physics for nuclei) and/or it may be measurable in the laboratory. Apart from the collision rate, we need to know whether the particles are bosons or fermions; for definiteness, let a and c be fermions and b and d be bosons. Then the collision term is

$$C[f_a] = \frac{1}{2E_a} \int Dp_b \, Dp_c \, Dp_d (2\pi)^4 \delta^4 \big(p_a^\mu + p_b^\mu - p_c^\mu - p_d^\mu\big)|M|^2$$
$$\times [f_c f_d(1 - f_a)(1 + f_b) - f_a f_b(1 - f_c)(1 + f_d)]. \tag{15.21}$$

The right-hand side is Lorentz invariant, except for the factor $1/E_a$, which accounts for the time dilation of the time interval dt [in Eq. (15.20)] relative to its value in the rest frame of particle a. The Dirac delta function δ^4 ensures 4-momentum conservation. The quantity Dp is the Lorentz-invariant phase-space element defined by

$$Dp = \frac{d^3 p}{(2\pi)^3 2E(p)}, \tag{15.22}$$

with $E = \sqrt{m^2 + p^2}$ the energy (we saw earlier that it is indeed Lorentz invariant). Inside the square brackets in Eq. (15.21), the first term represents the process cd $\rightarrow$ ab, causing a flow of particles *into* the momentum-space element $d^3 p_a$, and the second term represents the process ab $\rightarrow$ cd, causing an outward flow of particles. Focusing on the second term, there are factors f_a and f_b because the rate is proportional to the occupation number of each initial state. There is also a factor $1 - f_c$ because transition to an occupied fermion state is forbidden, and a factor $1 + f_d$ because transition to an occupied boson state is amplified (this is the "stimulated emission" that makes a laser work).

In general, the collision term is a sum of terms, each describing a separate collision process that may have an arbitrary number of particles in the initial and final states. Provided that neither the final nor the initial state contains identical particles, the collision term is the direct generalization of the one we wrote down, with M the appropriate amplitude and the delta function modified to include all of the 4-momenta. For each initial particle i, there is a factor $Dp_i f_i$, and for each final particle, there is a factor $Dp_i(1 \pm f_i)$. If there are identical particles, the expression has a well-known modification corresponding to the symmetry or antisymmetry of the corresponding wavefunction.

15.2.5 Thermal equilibrium

If appropriate processes occur with a sufficiently fast transition rate, the coupled Boltzmann equations will lead to time-independent distribution functions, corresponding to thermal equilibrium in which the square brackets of Eq. (15.21) vanish. (For exact thermal equilibrium the temperature is position independent, but in practice, there is approximate local thermal equilibrium, with T a function of position.) We assume that the angular momentum vanishes, and consider an inertial frame in which the momentum vanishes as well. Then the distribution functions in equilibrium are of the form

$$f = \frac{1}{\exp\left(\dfrac{E - \mu}{T}\right) \pm 1}, \tag{15.23}$$

with minus for a boson and plus for a fermion. At a given temperature T, the chemical potential μ specifies the number density.

The chemical potentials are "conserved" in each process; for instance, if a process ab↔cd is in equilibrium, then $\mu_a + \mu_b = \mu_c + \mu_d$.[2] Also, total energy is conserved. As a result, Eq. (15.23) indeed gives a vanishing square bracket in Eq. (15.21) and its generalizations. Moreover, Eq. (15.23) is the most general form for f that makes the square bracket vanish (e.g., Bowler 1982), because energy and momentum are the only mechanical conserved quantities.

If a particle species carries no conserved charge, it can be freely created and destroyed. If the relevant interactions are in equilibrium, its chemical potential then vanishes. A familiar example is the photon, which is freely created and destroyed whenever a charged particle accelerates.

Most particles carry a conserved charge. Electric charge is conserved, and so are baryon number and the three lepton numbers (at least in the regime of the Standard Model of particle interactions). Assuming that relevant interactions are in equilibrium, the particle and antiparticle then have an equal and opposite chemical potential, which vanishes if and only if the relevant charge density *carried by that species* vanishes. With enough interactions in equilibrium, there will be so many equations that all of the chemical potentials are determined once the total charge densities are specified, and in particular they will all vanish if the densities do. It is widely assumed that the chemical potentials indeed vanish in the very early Universe, with perhaps only electric charge being actually conserved. Then we have to create baryon number at the appropriate level, probably also creating lepton number at the same level.

In any case, we know that just before nucleosynthesis, the radiation (relativistic species) in the Universe consists of the three neutrinos, photons, electrons, and positrons. The Standard Model tells us which interactions take place, and we find that, indeed, there are enough reactions to determine the chemical potentials in terms of the electric charge density and the three lepton number densities. To have successful nucleosynthesis, we need $|\mu| \ll T$ for each species, as long as it is relativistic. (This is in conformity with the scenario outlined in the preceding paragraph for the very early Universe.) As a result, their occupation numbers are given by

$$f(p) = \frac{1}{\exp(E/T) \pm 1},\tag{15.24}$$

with $E = p$.

15.2.6 Evolving the density matrix

The collision term cannot fully handle particle spin, or the mixing of different particle species. Focusing on the spin, we can try to take account of it by letting the particle label i include the spin state. However, we then are faced with the problem that different spin states interfere, so that we should be adding the *amplitudes M* and not the rates $|M|^2$. In this situation, we need

---

[2] We put "conserved" in quotation marks because the chemical potentials are generally time dependent. For a genuinely conserved quantity, they would be time independent.

to replace the occupation number f by the density matrix $\mathcal{D}$, with dimensionality equal to the number of spin states. The expectation value of an observable relating to spin, with matrix A, is equal to the trace of $\mathcal{D}A$. In particular, the occupation number is simply $f = \text{Tr}\,\mathcal{D}$. The density matrix satisfies a Boltzmann equation, with the collision term of each matrix element involving all of the others in general.

This complication can be ignored in many circumstances. It can be ignored in thermal equilibrium because the interference terms will cancel, which shows that each spin state has the same thermal distribution. Out of thermal equilibrium, it is often a good approximation to use the Boltzmann equation, taking $|M|^2$ as the unpolarized rate (the rate averaged over initial spins and summed over final spins).

In our case, the collision term comes from Compton scattering, $\gamma e \to \gamma e$. To good accuracy, the electrons can be taken to be nonrelativistic, so that we are dealing with Thomson scattering (the classical limit of Compton scattering, which conserves photon energy in the electron rest frame). However, Thomson scattering *does* depend strongly on polarization, and we need the density matrix of the photons for an accurate calculation. Solving the coupled Boltzmann equations obtained from the unpolarized cross-section incurs errors in microwave anisotropy predictions of order up to roughly 10 percent. To specify the photon density matrix, we need the distribution function and the three Stokes parameters defined in Section 5.3. The collision term for the photon then depends on f and the Stokes parameters.

15.3   Gas dynamics in the perturbed Universe

We now consider gas dynamics in the perturbed, expanding Universe. We work in the conformal Newtonian gauge, whose coordinates are denoted by (τ, x^i). We also consider, at each space-time point, the locally orthonormal frame considered in Section 14.5.4.

In the absence of perturbations, the particle momentum $\mathbf{p}$ in this frame "redshifts" like $1/a$. We are assuming that collisions do not alter the energy of a species, and do not create or destroy particles. This is the case for Thomson scattering, which is the only collision relevant in our application. It is therefore convenient, even in the presence of perturbations, to make the following definitions:

$$\mathbf{q} \equiv a\mathbf{p}, \tag{15.25}$$

$$\mathbf{n} \equiv \frac{\mathbf{q}}{q} \quad \text{(unit vector)}, \tag{15.26}$$

$$\epsilon \equiv aE = (q^2 + m^2a^2)^{1/2}. \tag{15.27}$$

15.3.1   The geodesic equation and the Sachs–Wolfe effect

Our first task is to evaluate $dq/d\tau$, giving the deviation from the unperturbed "redshift" $p \propto 1/a$. To handle it, we need to relate (E, p^i), the 4-momentum components in the locally orthonormal frame, to the 4-momentum components in the conformal Newtonian coordinate system (τ, x^i). We denote the latter by P^μ. The relation is[3]

$$P^0 = g_{00}^{1/2} E = a(1 + \Psi)E, \tag{15.28}$$

$$P^i = g_{ii}^{1/2} p^i = a(1 - \Phi)p^i, \tag{15.29}$$

with no summation over i. The geodesic equation (14.63) is

$$\frac{dP^\mu}{d\tau} = g^{\mu\nu}\left(\frac{1}{2}\frac{\partial g_{\alpha\beta}}{\partial x^\nu} - \frac{\partial g_{\nu\alpha}}{\partial x^\beta}\right)\frac{P^\alpha P^\beta}{P^0}. \tag{15.30}$$

This is evaluated along the particle trajectory, $dx^i/d\tau = n^i$. Evaluating, for example, the time component, we find to first order

$$\frac{1}{q}\frac{dq}{d\tau} = \frac{\partial\Phi}{\partial\tau} - \frac{\epsilon}{q}n_i\frac{\partial\Psi}{\partial x^i}. \tag{15.31}$$

This result leads to a simple derivation of the Sachs–Wolfe effect, Eq. (6.18). For a photon, $\epsilon = q$, and after matter domination, $\Psi = \Phi$. Since

$$\frac{d\Phi}{d\tau} \equiv \frac{\partial\Phi}{\partial\tau} + \frac{dx^i}{d\tau}\frac{\partial\Phi}{\partial x^i}, \tag{15.32}$$

this leads to

$$\frac{1}{q}\frac{dq}{d\tau} = 2\frac{\partial\Phi}{\partial\tau} - \frac{d\Phi}{d\tau}. \tag{15.33}$$

To evaluate the Sachs–Wolfe effect, we start from Eq. (5.48):

$$\frac{\delta T}{T} = \Theta(t_{ls}, \mathbf{x}_{ls}, \mathbf{n}) + \left(\frac{\delta T}{T}\right)_{jour}. \tag{15.34}$$

In Section 5.2.6, we use this equation with the comoving slicing. Now we use it instead with the conformal Newtonian slicing. With this slicing,

$$\left[\frac{\delta T(\mathbf{e})}{T}\right]_{jour} = \int_{\tau_{ls}}^{\tau_0}\frac{1}{q}\frac{dq}{d\tau}d\tau$$

$$= \Phi(\mathbf{x}_{ls}, \tau_{ls}) - \Phi(0, \tau_0) + 2\int_{\tau_{ls}}^{\tau_0}\frac{\partial\Phi(\mathbf{e}x, \tau)}{\partial\tau}d\tau. \tag{15.35}$$

As before, the second term of this expression does not contribute to the anisotropy.

To evaluate the first term of Eq. (15.34), we again can keep only the monopole, which is $\Theta_0^N = \delta_\gamma^N/4$, the suffix N reminding us to use the conformal Newtonian slicing. If we use the adiabatic condition again, and remember that decoupling occurs during matter domination, we obtain $\delta_\gamma^N/4 = \delta^N/3$. Finally, we go to the comoving slicing using Eqs. (14.146) and (14.147), and

$$V = \frac{2}{3}\frac{k}{aH}\Phi, \tag{15.36}$$

---

[3] This relation assumes that alignment of the two coordinate systems is exact. That cannot be the case for the spatial directions because the spatial metric is perturbed, but the difference can be ignored when considering first-order perturbations.

which is the growing solution of Eq. (14.150).[4] This gives

$$\Theta_0^N(\mathbf{x}_{ls}, \tau_{ls}) = \frac{1}{3}\delta(\mathbf{x}_{ls}, \tau_{ls}) - \frac{2}{3}\Phi(\mathbf{x}_{ls}, \tau_{ls}).$$

(15.37)

The first term is negligible, as we saw in Section 5.2.6, and the second one combines with Eq. (15.35) to give Eq. (6.18), namely,

$$\left[\frac{\delta T(\mathbf{e})}{T}\right]_{SW} = \frac{1}{3}\Phi(\mathbf{e}\mathbf{x}_{ls}) + 2\int_{\tau_{ls}}^{\tau_0}\frac{\partial\Phi(\mathbf{e}\mathbf{x}, \tau)}{\partial\tau}\, d\tau,$$

(15.38)

simplifying to Eq. (4.10) if Φ is time independent.

15.3.2 Distribution function

Recall that, in addition to the global coordinates $x^\mu = (\tau, x^i)$, we are invoking at each spacetime point the locally orthonormal frame defined in Section 14.5.4. The latter frame defines the distribution function $f(\tau, \mathbf{x}, \mathbf{n}, q)$. For the argument leading to the collisionless Boltzmann equation, we need the corresponding locally inertial frame. We also need that frame to work out the collision term appearing in the Boltzmann equation (15.20). Knowing the collision term, the chain rule gives

$$\frac{\partial f}{\partial\tau} + \frac{\partial f}{\partial x^i}\frac{dx^i}{d\tau} + \frac{\partial f}{\partial q}\frac{dq}{d\tau} + \frac{\partial f}{\partial n^i}\frac{dn^i}{d\tau} = \frac{dt_{pr}}{d\tau}\frac{df}{dt_{pr}}.$$

(15.39)

On the right-hand side, df/dt_{pr} is the collision term that we previously denoted as $C[f]$. This is the Boltzmann equation, now valid in the context of general relativity. Having justified it, we have no further need of a locally inertial frame, and we can revert to the locally orthonormal frame that defines $f(\tau, \mathbf{x}, \mathbf{n}, q)$.

Consider first the unperturbed Universe, where f is independent of $\mathbf{x}$ and $\mathbf{n}$. If each collision conserves the energy of the species under consideration, and neither creates nor destroys it, then $q = ap$ is time independent and so is f. This is the case for us because the only significant process is Thomson scattering. Expressed as functions of q, the photon and neutrino distributions corresponding to Eq. (15.24) are

$$f_\gamma(q) = \frac{2}{\exp(q/T_0) - 1},$$

(15.40)

$$f_\nu(q) = \frac{2}{\exp(q/T_{\nu 0}) + 1},$$

(15.41)

where $T_{\nu 0} = (4/11)^{1/3}T_0$ and T_0 is the present photon temperature. The second expression is also valid for a massive neutrino species, at early times when the mass is negligible, provided that $T_{\nu 0}$ is *defined* to be $(4/11)^{1/3}T_0$.

---

[4] In Eqs. (14.147) and (14.150), we set $\Pi = \delta P = w = 0$ and $\Psi = \Phi$, appropriate to matter domination. Note also that the present flat-space formalism is adequate even in the case of an open Universe, because we are evaluating quantities at the epoch τ_{eq}.

Now consider the perturbation. Taking q and $\mathbf{n}$ as the variables, we split f into an unperturbed part plus a perturbation,

$$f(\tau, \mathbf{x}, \mathbf{n}, q) = f(\tau, q) + \delta f(\tau, \mathbf{x}, \mathbf{n}, q). \tag{15.42}$$

For photons and massless neutrinos, the unperturbed quantity $f(\tau, q)$ is given by Eq. (15.40) or Eq. (15.41), and is independent of τ.

Because the collision term in Eq. (15.39) affects only perturbations, we can use the zero-order result $dt_{pr}/d\tau = a$ for its coefficient. The last term on the left-hand side is of second order, and so, we drop it. In the second term, $\partial f/\partial x^i$ is of first order. We therefore can identify $dx^i/d\tau$ with the particle velocity $dr^i/dt = q^i/\epsilon = n^i q/\epsilon$. In the third term, $dq/d\tau$ is given by Eq. (15.31). It is of first order, and so, we can replace $\partial f/\partial q$ by $df(q)/dq$, where $f(q)$ is now the unperturbed quantity given by Eq. (15.40) or Eq. (15.41).

Putting all this into Eq. (15.39) gives

$$\frac{\partial \delta f}{\partial \tau} + \frac{q}{\epsilon} n^i \partial_i(\delta f) + q \frac{df}{dq} \left(\frac{\partial \Phi}{\partial \tau} - \frac{\epsilon}{q} n^i \partial_i \Psi \right) = a \frac{d}{dt} \delta f. \tag{15.43}$$

As usual, we make the Fourier expansion for δf by writing

$$\delta f(\tau, \mathbf{x}, \mathbf{n}, q) = \frac{1}{(2\pi)^{3/2}} \int d^3 k \, \delta f(\tau, \mathbf{k}, \mathbf{n}, q) e^{i \mathbf{k} \cdot \mathbf{x}}. \tag{15.44}$$

The Boltzmann equation becomes

$$\frac{\partial \delta f}{\partial \tau} + i k \mu \frac{q}{\epsilon} \delta f + q \frac{df}{dq} \left(\frac{\partial \Phi}{\partial \tau} - i k \frac{\epsilon}{q} \mu \Psi \right) = a \frac{d}{dt} \delta f, \tag{15.45}$$

where μ is the cosine of the angle between $\mathbf{n}$ and $\mathbf{k}$,

$$\mu = \hat{\mathbf{k}} \cdot \mathbf{n} = \frac{\mathbf{k} \cdot \mathbf{n}}{k} = \frac{\mathbf{k} \cdot \mathbf{q}}{kq}. \tag{15.46}$$

15.3.3 Brightness function

In the case of photons, it is useful to work with the brightness function $\Theta(\tau, \mathbf{x}, \mathbf{n})$, defined as the fractional perturbation in the effective temperature of the distribution function (Section 5.2). As we now see, the evolution equation of the brightness function does not involve q, and because the initial conditions do not either, *the brightness function is independent of q*.

To work out the evolution equation, note first that

$$\delta f(\tau, \mathbf{x}, \mathbf{n}, q) = -q \frac{df}{dq} \Theta(\tau, \mathbf{x}, \mathbf{n}, q). \tag{15.47}$$

This is the result of substituting $T_0 \to T_0(1 + \Theta)$ into the unperturbed expression (15.40). Now substitute this into Eq. (15.45). The collision term on the right-hand side is a linear functional of δf with no explicit q dependence.[5] We therefore have

$$\frac{\partial \Theta}{\partial \tau} + ik\mu\Theta - \left(\frac{\partial \Phi}{\partial \tau} - ik\mu\Psi\right)\Theta = a\frac{d\Theta}{dt}. \tag{15.48}$$

As advertised, this equation has no q dependence.

There is an alternative expression for the brightness function, coming from the relation

$$\int_0^\infty f(\tau, \mathbf{x}, \mathbf{n}, q)q^3 dq = \text{const} \times \frac{T^4(\tau, \mathbf{x}, \mathbf{n})}{a^4}. \tag{15.49}$$

[We don't need the numerical constant, which in fact is given by Eqs. (15.15) and (2.20).] It gives

$$4\Theta = \frac{\int_0^\infty \delta f \, q^3 dq}{\int_0^\infty f q^3 dq}. \tag{15.50}$$

In the literature, this usually is taken as the definition of the brightness function (sometimes without the factor 4), but to do that is to hide the important prediction that the fractional temperature perturbation is independent of q. Nevertheless, Eq. (15.50) is of crucial importance because, as we see in Eqs. (15.67)–(15.69), it directly relates the brightness function to the perturbation in the stress–energy tensor. For massless neutrinos, there is an analogous brightness function Θ_ν satisfying Eq. (15.48) with no collision term.

15.4 Multipoles and the Boltzmann hierarchy

15.4.1 Transfer function

We are going to end up with a closed system of equations, which determines everything in terms of the initial curvature perturbation $\mathcal{R}$. Alternatively, we could have considered an isocurvature initial condition, with $\mathcal{R} = 0$ replaced by some nonzero entropy perturbation S. If both were present, they would correspond to independent Gaussian perturbations with the following applying to each one separately.

The form of the equations is invariant under rotations, and the result for the presently observed cmb anisotropy will be of the form

$$\Theta(\tau_0, \mathbf{k}, \mathbf{n}) = T_\Theta(k, \mu)\mathcal{R}(\mathbf{k}). \tag{15.51}$$

It is convenient to make the Legendre expansion

$$T_\Theta(k, \mu) = \sum_\ell (-i)^\ell (2\ell + 1) T_\Theta(k, \ell) P_\ell(\mu). \tag{15.52}$$

The brightness function is then of the form

$$\Theta(\tau_0, \mathbf{x}, \mathbf{n}) = \frac{1}{(2\pi)^{3/2}} \sum_\ell (-i)^\ell (2\ell + 1) \int T_\Theta(k, \ell) P_\ell(\mu)\mathcal{R}(\mathbf{k}) e^{i\mathbf{k}\cdot\mathbf{x}} d^3k. \tag{15.53}$$

---

[5] To be more precise, it is a linear functional of δf and the corresponding quantity specifying the photon polarization. The specific form of the collision term is given later.

We can take our position as the origin $\mathbf{x} = 0$. Then, inserting Eq. (4.38) and using the addition theorem (4.45), we find Eq. (5.25) for the multipoles $a_{\ell m}$.

15.4.2 Boltzmann hierarchy

We also make the Legendre expansion for the brightness function

$$\Theta(\tau, \mathbf{k}, \mathbf{n}) = \sum_{\ell=0}^{\infty} (-i)^\ell (2\ell + 1) \Theta_\ell(\mathbf{k}, \tau) P_\ell(\mu). \tag{15.54}$$

Then the transfer function $T_\Theta(k, \ell)$ is just $\Theta_\ell(\mathbf{k}, \tau_0)$, evaluated with the initial condition $\mathcal{R}_\mathbf{k} = 1$. If we use

$$(\ell + 1) P_{\ell+1}(\mu) = (2\ell + 1)\mu P_\ell(\mu) - l P_{\ell-1}(\mu), \tag{15.55}$$

the Boltzmann equation becomes

$$\frac{\partial \Theta_0}{\partial \tau} = -k\Theta_1 + 4\frac{\partial \Phi}{\partial \tau} + a\frac{d\Theta_0}{dt}, \tag{15.56}$$

$$\frac{\partial \Theta_1}{\partial \tau} = \frac{k}{3}(\Theta_0 - 2\Theta_2) + \frac{k}{3}\Psi + a\frac{d\Theta_1}{dt}, \tag{15.57}$$

$$\frac{\partial \Theta_\ell}{\partial \tau} = \frac{k}{2\ell + 1}[\ell\Theta_{\ell-1} - (\ell + 1)\Theta_{\ell+1}] + a\frac{d\Theta_\ell}{dt} \qquad (\ell \geq 2). \tag{15.58}$$

The last term in each equation is the collision term. This set of equations is called the Boltzmann hierarchy.

Identical equations are satisfied by the neutrino brightness function, with no collision term. That is all we need for the pure CDM model, but otherwise we need to consider particles that start out as radiation (relativistic motion corresponding to negligible mass) and end up as matter (nonrelativistic motion). An example that we study in some detail in Chapter 6 is that of Hot Dark Matter, in the form of massive neutrinos.

In such a case, there is no brightness function and we have to work with the original distribution function. We can define multipoles F_ℓ through the expansion

$$\frac{\delta f}{f} = \sum_{\ell=0}^{\infty} (-i)^\ell (2\ell + 1) F_\ell(\tau, \mathbf{k}, q) P_\ell(\mu). \tag{15.59}$$

The Boltzmann hierarchy for them is

$$\frac{\partial F_0}{\partial \tau} = -\frac{qk}{\epsilon} F_1 - \frac{\partial \Phi}{\partial \tau}\frac{d\ln f}{d\ln q} + af^{-1}\frac{dF_0}{dt}, \tag{15.60}$$

$$\frac{\partial F_1}{\partial \tau} = \frac{qk}{3\epsilon}(F_0 - 2F_2) - \frac{\epsilon k}{3q}\Psi\frac{d\ln f}{d\ln q} + af^{-1}\frac{dF_1}{dt}, \tag{15.61}$$

$$\frac{\partial \delta f_\ell}{\partial \tau} = \frac{qk}{(2\ell + 1)\epsilon}[\ell F_{\ell-1} - (\ell + 1)F_{\ell+1}] + af^{-1}\frac{dF_\ell}{dt} \qquad (\ell \geq 2). \tag{15.62}$$

15.4.3 Stress–energy tensor

In the locally orthonormal frame, we can calculate the perturbation in the stress–energy tensor through Eqs. (15.15) to (15.18). Using Eqs. (14.109) and (14.111), we find that the vacuum fluctuation generates only the scalar part of v_i and Π_{ij}, so that the perturbation is defined by δ^N, V, δP^N, and Π. The contribution of a given particle species is

$$(\rho + \delta \rho^N) = \frac{4\pi g_i}{(2\pi)^3 a^4} \int \epsilon f(1 + F_0) q^2 dq, \tag{15.63}$$

$$V(\rho + P) = \frac{4\pi g_i}{(2\pi)^3 a^4} \int q f F_1 q^2 \, dq, \tag{15.64}$$

$$(P + \delta P^N) = \frac{4\pi g_i}{3(2\pi)^3 a^4} \int \frac{q^2}{\epsilon} f(1 + F_0) q^2 dq, \tag{15.65}$$

$$P\Pi = \frac{4\pi g_i}{(2\pi)^3 a^4} \int \frac{q^2}{\epsilon} f F_2 q^2 \, dq. \tag{15.66}$$

The superscripts N remind us that we are, at this point, defining the density and pressure perturbations on the conformal Newtonian slicing.

For massless particles, we can use Eq. (15.50) to write these equations in terms of the brightness function. For photons one has, remembering $g_i = 2$,

$$\delta^N \equiv \frac{\delta\rho^N}{\rho} = \frac{\delta P^N}{P} = 4\,\Theta_0, \tag{15.67}$$

$$V = 3\,\Theta_1, \tag{15.68}$$

$$\Pi = 12\,\Theta_2. \tag{15.69}$$

Identical expressions hold for massless neutrinos.

This is all the information we need to write down the Boltzmann hierarchy for the neutrinos in final form, because there is no collision term. The hierarchy reads

$$\frac{\partial \delta_\nu^N}{\partial \tau} = -\frac{4}{3} k V_\nu, \tag{15.70}$$

$$\frac{\partial V_\nu}{\partial \tau} = -a H V_\nu k \left(\frac{1}{4} \delta_\nu^N - \frac{1}{6} \Pi_\nu \right) + k\Psi, \tag{15.71}$$

$$V_\nu \equiv 3\Theta_{\nu 1}, \tag{15.72}$$

$$\Pi_\nu \equiv 12\Theta_{\nu 2}, \tag{15.73}$$

$$\frac{\partial \Theta_{\nu \ell}}{\partial \tau} = \frac{k}{2\ell + 1} \big[\ell \Theta_{\nu(\ell-1)} - (\ell + 1)\Theta_{\nu(\ell+1)} \big] \qquad (\ell \geq 2). \tag{15.74}$$

The first two are just the energy and momentum conservation equations (14.141) and (14.142). Similar equations hold for the photons, with the addition of the collision term which we now go on to evaluate.

15.4.4 Collision term: Simplified treatment ignoring polarization

The only significant collision is the Compton scattering process $\gamma e \to \gamma e$. As discussed in Section 15.2.6, there is no real need to go beyond the classical limit of Thomson scattering, but it *is* necessary to take polarization into account. We do that in the next section, but first let us give a simplified treatment that ignores both polarization and the angular dependence of Thomson scattering.

If there are n_e free electrons per unit volume, with zero velocity, the probability per unit time (rate) for an unpolarized photon to undergo Thomson scattering is just $n_e\sigma_T$. Here σ_T is the cross-section for Thomson scattering, given by

$$\sigma_T \equiv \frac{8\pi}{3}\alpha^2, \tag{15.75}$$

where $\alpha = e^2/4\pi = 1/137$ is the fine-structure constant. With c restored, the rate is $n_e c\sigma_T$, and $\sigma_T = 6.652 \times 10^{-25}\ \mathrm{cm}^2$. In the rest frame of the electron, a Thomson-scattered photon loses, on average, all of its momentum, but none of its energy.

We are ignoring polarization, and pretending that in the electron rest frame the scattered wave is isotropic. The photon distribution function f is a scalar. Because df/dt is a perturbation, we can take t to be frame independent to first order, making the collision term a scalar as well. It therefore can be evaluated in any frame. Working in the electron rest frame, Eq. (15.21) is simply

$$\frac{df(\mathbf{n})}{dt} = \frac{n_e\sigma_T}{4\pi}\int d^2n'\{f(\mathbf{n'})[1+f(\mathbf{n})] - f(\mathbf{n})[1+f(\mathbf{n'})]\}$$

$$= \frac{n_e\sigma_T}{4\pi}\int d^2n'\,[\widetilde{\delta f}(\mathbf{n'}) - \widetilde{\delta f}(\mathbf{n})] = \frac{d}{dt}\widetilde{\delta f}(\mathbf{n}). \tag{15.76}$$

The random motion of the electrons, ignored at this point, will cancel because the final result is linear in the electron velocity. To obtain the second line, we cancelled a pair of factors $f(\mathbf{n'})f(\mathbf{n})$, remembering that the unperturbed quantity is independent of $\mathbf{n}$. The tilde denotes the electron rest frame, and is necessary because the separation of f into an average plus a perturbation is frame dependent.

The collision term becomes

$$\frac{d\widetilde{\Theta}}{dt} = \frac{n_e\sigma_T}{4\pi}\int d^2n'\,(\widetilde{\Theta}(\mathbf{n'}) - \widetilde{\Theta}(\mathbf{n})) \tag{15.77}$$

$$= n_e\sigma_T\left(-\widetilde{\Theta}(\mathbf{n}) + \frac{1}{4\pi}\int d^2n'\,\widetilde{\Theta}(\mathbf{n'})\right). \tag{15.78}$$

This gives $d\widetilde{\Theta}_0/dt = 0$, and $d\widetilde{\Theta}_\ell/dt = -n_e\sigma_T\widetilde{\Theta}_\ell$ for $\ell \geq 1$.

We are defining Θ in the locally orthonormal frame lined up with the coordinates, in which the electron velocity is V_b. This affects only the dipole, Θ_1, or equivalently, V_γ, and in terms of the latter we have $V_\gamma = \widetilde{V}_\gamma + V_b$.

To summarize, the simplified treatment gives

$$\frac{\partial \delta_\gamma}{\partial \tau} = -\frac{4}{3} k V_\gamma, \tag{15.79}$$

$$\frac{\partial V_\gamma}{\partial \tau} = -aHV_\gamma + k\left(\delta_\gamma + \frac{1}{6}\Pi_\gamma\right) + k\Psi + an_e\sigma_T(V_b - V_\gamma), \tag{15.80}$$

$$\Pi_\gamma = 12\Theta_2, \tag{15.81}$$

$$\frac{\partial \Theta_\ell}{\partial \tau} = \frac{k}{2\ell + 1}\left[\ell\Theta_{(\ell-1)} - (\ell + 1)\Theta_{(\ell+1)}\right] - an_e\sigma_T\Theta_\ell \quad (\ell \geq 2). \tag{15.82}$$

15.5   Polarization

15.5.1   Spin-weighted spherical harmonics

As described in Section 5.3, the plane polarization of electromagnetic radiation travelling from the direction **e** is described by two Stokes parameters, Q and U. To define them, we need an orthogonal pair of axes x and y, perpendicular to **e**. We take them to be right-handed about the direction **e** of observation, and so, the combinations $Q_\pm \equiv (Q \pm iU)$ transform as $e^{\mp 2i\psi(e)}$ under a clockwise rotation of the axes through ψ radians.

The neatest way of working with these quantities is to use the spin-weighted harmonics $_sY_{\ell m}$, with $s = \pm 2$. These harmonics are defined in terms of the rotation matrices $\mathcal{D}^\ell_{-s,m}$ as

$$_sY_{\ell m}(\theta, \phi) = \left(\frac{2\ell + 1}{4\pi}\right)^{1/2} \mathcal{D}^\ell_{-s,m}(\phi, \theta, 0). \tag{15.83}$$

The explicit expression is

$$_sY_{\ell m}(\theta, \phi) = e^{im\phi}\left[\frac{(\ell+m)!(\ell-m)!}{(\ell+s)!(\ell-s)!}\frac{2\ell+1}{4\pi}\right]^{1/2} \sin^{2\ell}(\theta/2)$$

$$\times \sum_r \binom{\ell-s}{r}\binom{\ell+s}{r+s-m}(-1)^{\ell-r-s+m}\cot^{2r+s-m}(\theta/2). \tag{15.84}$$

They share the orthonormality and completeness properties of the ordinary harmonics $Y_{\ell m} \equiv {}_0Y^\pm_{\ell m}$. They also transform in the same way except that they pick up the phase factor $e^{-is\psi}$, where $\psi(\mathbf{e})$ is the rotation of the coordinate lines. The transformation is unitary, and so, Eq. (4.44) generalizes to

$$\sum_m {}_rY^*_{\ell m}(\theta_1, \phi_1)\,{}_sY_{\ell m}(\theta_2, \phi_2) = e^{i(r\psi_1 - s\psi_2)}\sum_m {}_rY^*_{\ell m}(\theta'_1, \phi'_1)\,{}_sY_{\ell m}(\theta'_2, \phi'_2). \tag{15.85}$$

The conjugation condition is $_sY^*_{\ell m} = (-1)^{m+s}\,{}_{-s}Y_{\ell m}$, and under a reversal of direction $_sY_{\ell m}$ becomes $(-1)^\ell\,{}_{-s}Y_{\ell m}$.

We use the notation $_{\pm 2}Y_{\ell m} \equiv Y^\pm_{\ell m}$. The generalization of Eq. (4.33) is

$$Y^\pm_{\ell 0}(\theta) = \sqrt{\frac{2\ell+1}{4\pi}\frac{(\ell-2)!}{(\ell+2)!}}\,P^2_\ell(\cos\theta), \tag{15.86}$$

Table 15.1. *Quadrupole* ($\ell = 2$) *harmonics for spin-0 and spin-2*

m	Y_{2m}	$_2Y_{2m}$
2	$\frac{1}{4}\sqrt{\frac{15}{2\pi}}\sin^2\theta\, e^{2i\phi}$	$\frac{1}{8}\sqrt{\frac{5}{\pi}}(1-\cos\theta)^2\, e^{2i\phi}$
1	$\sqrt{\frac{15}{8\pi}}\sin\theta\,\cos\theta\, e^{i\phi}$	$\frac{1}{4}\sqrt{\frac{5}{\pi}}\sin\theta\,(1-\cos\theta)\, e^{i\phi}$
0	$\frac{1}{2}\sqrt{\frac{5}{4\pi}}(3\cos^2\theta - 1)$	$\frac{3}{4}\sqrt{\frac{5}{6\pi}}\sin^2\theta$
-1	$-\sqrt{\frac{15}{8\pi}}\sin\theta\,\cos\theta\, e^{-i\phi}$	$\frac{1}{4}\sqrt{\frac{5}{\pi}}\sin\theta\,(1+\cos\theta)\, e^{-i\phi}$
-2	$\frac{1}{4}\sqrt{\frac{15}{2\pi}}\sin^2\theta\, e^{-2i\phi}$	$\frac{1}{8}\sqrt{\frac{5}{\pi}}(1+\cos\theta)^2\, e^{-2i\phi}$

where P_ℓ^2 is the associated Legendre function (*not* the Legendre function squared).
The $\ell = 2$ harmonics are shown in Table 15.1.

15.5.2 Transfer function

The presently observed Stokes parameters are given by a transfer function

$$Q_\pm(\tau_0, \mathbf{k}, \mathbf{n}) = T_\pm(k, \mu)\,\mathcal{R}(\mathbf{k}). \tag{15.87}$$

The appropriate generalization of Eq. (15.52) is

$$T_\pm(k, \mu) = 4\pi \sum_\ell (-i)^\ell\, T_\pm(k, \ell) Y_{\ell 0}(0) Y_{\ell 0}^\pm(\mu). \tag{15.88}$$

We are going to find that $T_\pm(k, \ell)$ are equal, and we denote them as $T_E(k, \ell)$. Then, using Eq. (15.85), we find that $T_E(k, \ell)$ is indeed the transfer function appearing in Eq. (5.79).

15.5.3 Polarized Thomson scattering

Given the Stokes parameters of an incident wave, we then can calculate those of a Thomson-scattered wave. The parameter V decouples from the others and, in particular, Thomson scattering does not generate circular polarization if there is none initially. Henceforth we drop V.

Consider first the Thomson scattering of a wave with definite frequency, with components given by Eq. (5.54). A free electron, oscillating about the origin under the influence of the electric field, emits dipole radiation. At large distances, using spherical polar coordinates (r, θ, ϕ) and setting $\phi = 0$, the scattered wave has the form

$$E_x^s(t) = \frac{\alpha}{r} E_x \cos(\omega t - \delta_1)\cos\theta, \tag{15.89}$$

$$E_y^s(t) = \frac{\alpha}{r} E_y \sin(\omega t - \delta_2), \tag{15.90}$$

where α is the fine-structure constant.

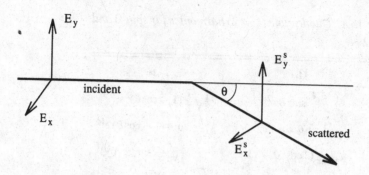

Fig. 15.2. Schematic of Thomson scattering. The scattering angle is aligned with the x axis.

We temporarily adopt the standard definition (5.55) of the quantities (I, Q, U). Using the above formula, we can work out the effect of Thomson scattering on these quantities. We choose the same y axis for the scattered wave (as in Figure 15.2). Then, after averaging over time and frequency, we find that the parameters of the scattered wave are related to those of the incoming wave by

$$I^s + Q^s = \frac{3\sigma_T}{8\pi r^2}(I + Q)\cos^2\theta, \tag{15.91}$$

$$I^s - Q^s = \frac{3\sigma_T}{8\pi r^2}(I - Q), \tag{15.92}$$

$$U^s = \frac{3\sigma_T}{8\pi r^2}U\cos\theta. \tag{15.93}$$

The superscript "s" refers to the scattered wave, and we used Eq. (15.75) to replace α by σ_T. For unpolarized incoming radiation, this gives the differential cross-section specifying the fraction of power scattered into solid angle $d\Omega$,

$$\frac{d\sigma_T}{d\Omega} = r^2\frac{I_s}{I} = \frac{3\sigma_T}{16\pi}(1 + \cos^2\theta). \tag{15.94}$$

Integrating over solid angle reproduces the total cross-section σ_T.

For the cosmological application, Eqs. (15.91)–(15.93) can be taken to apply to the perturbation in the cmb, with the unperturbed part playing no role. We revert to the cosmological redefinition (5.63) of the Stokes parameters, so that we are dealing with (Θ, Q, U). Also, it is better to work with the combinations $Q_\pm \equiv (Q \pm iU)$, which simply acquire a phase factor $e^{\mp 2i\psi}$ under a rotation of the axes. As we are working in the electron rest frame, we should write $\widetilde{\Theta}$ and $\widetilde{Q}_\pm$, but as before we drop the tildes, since the only effect of the electron velocity will be to change V_γ to $V_\gamma - V_b$ in the final equations. The effect of Thomson scattering is then

$$\frac{d\Theta}{d\Omega} = \frac{\sigma_T}{4\pi}\left[\frac{3}{4}(1 + \cos^2\theta)\Theta + \frac{3}{8}(\cos^2\theta - 1)Q_+ + \frac{3}{8}(\cos^2\theta - 1)Q_-\right], \tag{15.95}$$

$$\frac{dQ_\pm}{d\Omega} = \frac{\sigma_T}{4\pi}\left[\frac{3}{4}(\cos^2\theta - 1)\Theta + \frac{3}{8}(\cos\theta \pm 1)^2 Q_+ + \frac{3}{8}(\cos\theta \mp 1)^2 Q_-\right]. \tag{15.96}$$

We need to consider a generic initial photon direction $\mathbf{n}'$ and final photon direction $\mathbf{n}$, referred to some spherical polar coordinate system (θ, ϕ), with the Stokes parameters for a given direction defined using an x axis pointing away from the pole. The preceding expression applies if the scattering plane corresponds to $\phi = \phi' = 0$, with $\theta' = 0$ (strictly speaking, with θ' infinitesimally small so that the x axis is well defined). Using the spin-weighted harmonics, it is easy to write down the generalization of Eqs. (15.95) and (15.96). All we need are expressions with the right transformation properties, which reduce to those equations when $\theta' = \phi = 0$. It follows from Table 15.1 that the appropriate expressions are

$$\frac{d\Theta}{d\Omega} = \sigma_T \sum_m \left[\left(\frac{\delta_{m0}}{4\pi} + \frac{1}{10} Y_{2m} \overline{Y}'_{2m} \right) \Theta' - \frac{1}{10} \sqrt{\frac{3}{2}} \left(Y_{2m} \overline{Y}^{+'}_{2m} Q'_+ + Y_{2m} \overline{Y}^{-'}_{2m} Q'_- \right) \right],$$
(15.97)

$$\frac{dQ_\pm}{d\Omega} = \frac{1}{10} \sigma_T \sum_m \left(-\sqrt{6} Y^\pm_{2m} \overline{Y}'_{2m} \Theta' + 3 Y^\pm_{2m} \overline{Y}^{+'}_{2m} Q'_+ + 3 Y^\pm_{2m} \overline{Y}^{-'}_{2m} Q'_- \right).$$
(15.98)

Here, unprimed quantities are functions of $\mathbf{n}$, primed quantities are functions of $\mathbf{n}'$, and an overline denotes complex conjugation.

15.5.4 Collision term

The generalization of Eq. (15.78) to include polarization and the angular dependence of the scattering is

$$\frac{d\Theta(\mathbf{n})}{dt} = n_e \left[-\sigma_T \Theta(\mathbf{n}) + \int d^2 n' \frac{d\Theta(\mathbf{n}, \mathbf{n}')}{d\Omega'} \right],$$
(15.99)

and similarly for $(Q \pm iU)$.

For a given Fourier component, there is azimuthal symmetry about the $\mathbf{k}$ axis, so that we only need $Y^\pm_{\ell 0}$ given by Eq. (15.86). Also, the plus and minus multipoles are the same. We denote them as E_ℓ, and define them analogously to Eq. (15.88):

$$Q_\pm(\tau, \mathbf{k}, \mathbf{n}) = 4\pi \sum_\ell (-i)^\ell E_\ell Y_{\ell 0}(0) Y^\pm_{\ell 0}(\mu).$$
(15.100)

Then the transfer function $T_E(k, \ell)$ is simply $E_\ell(k)$, evaluated with initial condition $\mathcal{R}_\mathbf{k} = 1$. However, the collision terms and the Boltzmann hierarchy become simpler if we use

$$\hat{E}_\ell \equiv \sqrt{\frac{(\ell - 2)!}{(\ell + 2)!}} E_\ell.$$
(15.101)

Projecting out the multipoles, we find the collision terms

$$\frac{d\Theta_\ell}{dt} = \sigma_T n_e \left(-\Theta_\ell + \delta_{\ell 0} \Theta_0 + \frac{1}{10} \delta_{\ell 2} A \right),$$
(15.102)

$$\frac{d\hat{E}_\ell}{dt} = \sigma_T n_e \left(\hat{E}_\ell - \frac{1}{20} \delta_{\ell 2} A \right),$$
(15.103)

where

$$A \equiv \Theta_2 - 12\hat{E}_2. \tag{15.104}$$

This classical treatment needs extending to the quantum regime if we wish to keep higher-order terms in the electron random motion (Compton as opposed to Thomson scattering). Such a treatment starts by taking the Stokes parameters to define the 2×2 photon density matrix, whose collision term then is computed using quantum field theory.

15.5.5 Boltzmann hierarchy

Using the identities

$$(\ell + 1)P_{\ell+1}(\mu) = (2\ell + 1)\mu P_\ell(\mu) - \ell P_{\ell-1}(\mu), \tag{15.105}$$

$$\mu P_\ell^2(\mu) = \frac{1}{2\ell + 1} \left[(\ell + 2)P_{\ell-1}^2 + (\ell - 2)P_{\ell+1}^2 \right], \tag{15.106}$$

we now can obtain a Boltzmann hierarchy. It reads

$$\frac{\partial \delta_\gamma}{\partial \tau} = -\frac{4}{3}kV_\gamma, \tag{15.107}$$

$$\frac{\partial V_\gamma}{\partial \tau} = -aHV_\gamma + k\left(\frac{1}{4}\delta_\gamma - \frac{1}{6}\Pi_\gamma\right) + k\Psi + an_e\sigma_T(V_b - V_\gamma), \tag{15.108}$$

$$\Pi_\gamma = 12\Theta_2, \tag{15.109}$$

$$\frac{\partial \Theta_2}{\partial \tau} = \frac{2}{15}kV_\gamma - \frac{3}{5}k\Theta_3 + an_e\sigma_T\left(-\Theta_2 + \frac{1}{10}A\right), \tag{15.110}$$

$$\frac{\partial \Theta_\ell}{\partial \tau} = \frac{k}{2\ell + 1}[\ell\Theta_{(\ell-1)} - (\ell + 1)\Theta_{(\ell+1)}] - an_e\sigma_T\Theta_\ell \quad (\ell \geq 3), \tag{15.111}$$

$$\frac{\partial \hat{E}_\ell}{\partial \tau} = \frac{k}{2\ell + 1}[(\ell - 2)\hat{E}_{(\ell-1)} - (\ell + 3)\hat{E}_{(\ell+1)}] - an_e\sigma_T\left[\hat{E}_\ell + \frac{1}{20}\delta_{\ell 2}A\right]. \tag{15.112}$$

Had we retained separate multipoles for $Q_\pm$, they would have appeared symmetrically in the Boltzmann hierarchy, leading to identical solutions.

15.6   Initial conditions and the transfer functions

We finally have a closed system of equations. These are

(1) the energy and momentum conservation equations for the CDM [Eqs. (15.1) and (15.2)] and for the baryons [Eqs. (15.7) and (15.8)];
(2) the Boltzmann hierarchy for the neutrino brightness function [Eqs. (15.70)–(15.74)], and for the photon brightness function and polarization [Eqs. (15.107)–(15.113)];
(3) Eqs. (14.157)–(14.160), which give δ^N, V, δP^N, and Π, along with Eq. (14.147), which then gives δ (defined on comoving slices);

(4) the algebraic expressions (14.151) and (14.152), giving Φ in terms of δ and $\Psi - \Phi$ in terms of Π;

(5) the various equations describing the evolution of the homogeneous background Universe, including the reionization redshift z_{ion}, which we can take to be given either by the quasi-linear calculation of Section 11.6 or as a free parameter.

15.6.1 Initial conditions

For each scale $\mathbf{k}$, this system of equations can be solved uniquely, given suitable initial conditions during radiation domination and well before horizon entry. As discussed in Section 4.8, we suppose that each comoving region of the Universe looks practically the same as some unperturbed Universe. In the absence of an isocurvature contribution, this means that $\mathcal{R}_\mathbf{k}$ has some constant value, with the density contrasts of the various species satisfying the adiabatic condition. In this subsection, we restore the subscripts $\mathbf{k}$.

In the presence of anisotropic stress, Eq. (4.169) becomes

$$\frac{2}{3} H^{-1} \dot{\Phi}_\mathbf{k} + \frac{5 + 3w}{3} \Phi_\mathbf{k} = -(1 + w)\mathcal{R}_\mathbf{k} + 2w \left(\frac{aH}{k} \right)^2 \Pi_\mathbf{k}. \tag{15.113}$$

Well outside the horizon $\Pi_\mathbf{k} \propto (k/aH)^2$, and during an era of constant w we have

$$\mathcal{R}_\mathbf{k} = -\frac{5 + 3w}{3 + 3w} \Phi_\mathbf{k} + \frac{2w}{1 + w} \left(\frac{aH}{k} \right)^2 \Pi_\mathbf{k}. \tag{15.114}$$

From the definition (4.167) of $\mathcal{R}_\mathbf{k}$ and Eqs. (14.149), (14.151), and (14.152) (with $w = 1/3$), we find the following results well before horizon entry:

$$\Psi_\mathbf{k} = -\frac{2}{3} \frac{1}{1 + \frac{4}{15} X_\nu} \mathcal{R}_\mathbf{k}, \tag{15.115}$$

$$\Phi_\mathbf{k} = -\frac{2}{3} \frac{1 + \frac{2}{5} X_\nu}{1 + \frac{4}{15} X_\nu} \mathcal{R}_\mathbf{k}, \tag{15.116}$$

$$V_\mathbf{k} = \frac{1}{2} \frac{k}{aH} \Psi_\mathbf{k}, \tag{15.117}$$

$$\Pi_\mathbf{k} = \Pi_{\nu\mathbf{k}} = \frac{2}{5} \left(\frac{k}{aH} \right)^2 \Psi_\mathbf{k}, \tag{15.118}$$

$$\delta_\mathbf{k} = -\frac{2}{3} \left(\frac{k}{aH} \right)^2 \Phi_\mathbf{k}. \tag{15.119}$$

Here, $X_\nu = \rho_\nu/(\rho_\gamma + \rho_\nu) = (7/8)(4/11)^{4/3} N_\nu \simeq 0.68$, where $N_\nu = 3$ is the number of neutrino species. The peculiar velocity of each species is $V_i = V$, and its density contrast is given by the adiabatic condition ($\delta_i = \delta$ for radiation, $\delta_i = 3\delta/4$ for matter). The photon contribution to $\Pi_\mathbf{k}$ is negligible because of the frequent collisions (perfect fluid), and all higher moments of the brightness function are negligible for photons and neutrinos.

To use these conditions as the starting point for numerical integration, it is necessary to modify the system of equations somewhat. Staying within the conformal Newtonian gauge, we can replace one (Ma and Bertschinger 1995) or both (Seljak 1994) of Eqs. (14.151) and (14.152) by components of the Einstein field equation involving time derivatives of the potentials. Alternatively (Kodama and Sasaki 1984; Sugiyama and Gouda 1992), we can use Eqs. (14.162) and (14.163) for each species individually, before implementing the equations on a computer. This latter procedure constitutes what generally is called the gauge-invariant formalism, and is generalized to the case of massive neutrinos by Schaefer and de Laix (1996).

Alternatively, we might set $\mathcal{R}_\mathbf{k}$ equal to zero and consider a constant initial entropy perturbation, such as the quantity $S_\mathbf{k}$ of Section 6.6. The isocurvature initial condition is very simple; all perturbations are negligible except for $S_\mathbf{k}$, which is practically equal to δ_c (the same in all gauges). With this condition, the gauge-invariant equations can be integrated numerically.

15.6.2   Transfer functions

The object of all this is to calculate the transfer functions. To obtain them, we should solve the system of equations with the initial condition $\mathcal{R}_\mathbf{k} = 1$. Then, as we noted earlier, the cmb transfer functions T_Θ and T_E are, respectively, Θ_ℓ and E_ℓ evaluated at the present epoch. The matter-density transfer function $T(k)$ defined by Eq. (5.4) is just $\mathcal{R}_\mathbf{k}^{(m)}$.

15.6.3   Line-of-sight integral

The matter transfer function is easy to evaluate because it involves only the lowest multipole of the matter density contrast, which means that we can truncate the Boltzmann hierarchy at a low order in ℓ.

The cmb transfer functions are more difficult because we are interested in high multipoles. Codes have been written that solve the Boltzmann hierarchy to sufficiently high order (Bond and Efstathiou 1987; Hu et al. 1995; Ma and Bertschinger 1995), but they are too slow to permit an exploration of the parameter space. In the approximation of ignoring reionization, an alternative is to decompose the anisotropy as in Eq. (5.48), choosing a surface just after last scattering so that only a few multipoles need to be calculated from the Boltzmann hierarchy (Bond and Efstathiou 1987; Schaefer and de Laix 1996). This is still slow and, more seriously, reionization is likely to be important.

The most efficient approach, which is also of interest in its own right, is to express the transfer functions $T_\Theta(k, \mu)$ and $T_E(k, \mu)$ as integrals along the line of sight (Seljak and Zaldarriaga 1996; Zaldarriaga and Seljak 1997). After making the Legendre expansion, we find expressions of the following form:

$$T_{\Theta,E}(k, \ell) = \int_0^{\tau_0} S_{\Theta,E}(k, \tau) j_\ell[k(\tau_0 - \tau)] \, d\tau, \qquad (15.120)$$

where

$$S_\Theta(k, \tau) = S_0 g + S_1 \dot{g} + S_2 \ddot{g} + e^{-\kappa}(\dot{\Phi} + \dot{\Psi}), \tag{15.121}$$

$$S_E(k, \tau) = -\frac{3A_2}{4k^2} g. \tag{15.122}$$

Here $A_2 \equiv \Theta_2 - 12\hat{E}_2$ is gauge invariant, and so are the three terms S_i. Evaluated in the conformal Newtonian gauge, they are

$$S_0(k, \tau) = \Theta_0^N - \frac{\dot{V}_b^N}{k} + \Psi - \frac{A_2}{4} - \frac{3\ddot{A}_2}{4k^2}, \tag{15.123}$$

$$S_1(k, \tau) = -\left(\frac{V_b^N}{k} + \frac{3\dot{A}_2}{4k^2}\right), \tag{15.124}$$

$$S_2(k, \tau) = -\frac{3A_2}{4k^2}. \tag{15.125}$$

The quantities S_Θ and S_E involve only low-order multipoles, which can be obtained by solving the Boltzmann hierarchy.

In these expressions, κ is the optical depth and $g \equiv (d\kappa/d\tau)e^{-\kappa}$ is the visibility function. We encountered these quantities in Section 5.4.1, except that there, we defined $g \equiv (d\kappa/dt)e^{-\kappa}$ (i.e., coordinate rather than conformal time), which is a^{-1} times the present g. In the absence of reionization, κ is practically zero after decoupling, whereas the scattering probability $g(\tau) \simeq \delta(\tau - \tau_{ls})$ is peaked there. Reionization moves some of the scattering probability to the era $z < z_{ion}$.

Because S_Θ is practically zero before decoupling, it is quite a good approximation to set $\Psi = \Phi$, with both constant. Then, because $\int g \, d\tau = 1$, the term $g(\Theta_0^N + \Psi)$ gives the ordinary Sachs–Wolfe effect [Eq. (15.37) plus Eq. (15.35)]. The term $e^{-\kappa}(\dot{\Phi} + \dot{\Psi})$ is the integrated Sachs–Wolfe effect, corrected for reionization.

The expressions can be derived from the formalism we have presented, or by a more direct, intuitive approach (Seljak and Zaldarriaga 1996). The latter does not make the full formalism redundant, because we still need it to actually evaluate the expressions.

The output of these equations, with most of the relevant parameters as freely chosen input, is publicly available as the CMBFAST code [Seljak and Zaldarriaga (1996), partly built on an earlier code described by Bertschinger (1995)], and it has been used for all the numerical calculations in this book.

Examples

15.1 Verify that, with the distribution function of the form Eq. (15.23), the collision term given by Eq. (15.21) vanishes if and only if $\mu_a + \mu_b = \mu_c + \mu_d$.

15.2 Verify the geodesic equation (15.31), for a free particle in the perturbed Universe.

15.3 In the perturbed Universe, the angle-dependent photon temperature seen by any observer is related to the photon occupation number by Eq. (15.49). Calculate the constant appearing in this expression.

Appendix: Constants and parameters

As experienced readers of Kolb and Turner (1990), we know which part of their book is used the most! Here is our much less extensive version. See also Scott et al. (1999) for even more constants and formulae. Where necessary, our numerical values assume a current microwave background temperature of $T_0 = 2.728$ K (Fixsen et al. 1996) and three massless neutrino species. Many of the parameters are derived in Chapter 2.

Constants

Reduced Planck constant	$\hbar = 1.055 \times 10^{-27} \, \text{cm}^2 \cdot \text{g} \cdot \text{s}^{-1}$
Speed of light	$c = 2.998 \times 10^{10} \, \text{cm} \cdot \text{s}^{-1}$
Newton's constant	$G = 6.672 \times 10^{-8} \, \text{cm}^3 \, \text{g}^{-1} \, \text{s}^{-2}$
Reduced Planck mass	$M_{\text{Pl}} = 4.342 \times 10^{-6} \, \text{g}$
	$= 2.436 \times 10^{18} \, \text{GeV}/c^2$
(Planck mass	$m_{\text{Pl}} = \sqrt{8\pi} \, M_{\text{Pl}} = 2.177 \times 10^{-5} \, \text{g})$
Reduced Planck length	$L_{\text{Pl}} = 8.101 \times 10^{-33} \, \text{cm}$
Reduced Planck time	$T_{\text{Pl}} = 2.702 \times 10^{-43} \, \text{s}$
Boltzmann constant	$k_{\text{B}} = 1.381 \times 10^{-16} \, \text{erg} \, \text{K}^{-1}$
	$= 8.618 \times 10^{-5} \, \text{eV} \, \text{K}^{-1}$
Thomson cross section	$\sigma_{\text{T}} = 6.652 \times 10^{-25} \, \text{cm}^2$
Electron mass	$m_e = 0.511 \, \text{MeV}/c^2$
Neutron mass	$m_n = 939.6 \, \text{MeV}/c^2$
Proton mass	$m_p = 938.3 \, \text{MeV}/c^2$
Solar mass	$M_\odot = 1.99 \times 10^{33} \, \text{g}$
Megaparsec	$1 \, \text{Mpc} = 3.086 \times 10^{24} \, \text{cm}$

Parameters

Hubble constant	$H_0 = 100 \, h \, \text{km} \cdot \text{s}^{-1} \, \text{Mpc}^{-1}$
Present Hubble distance	$cH_0^{-1} = 2998 \, h^{-1} \, \text{Mpc}$
Present Hubble time	$H_0^{-1} = 9.78 \, h^{-1} \, \text{Gyr}$
Present critical density	$\rho_{c,0} = 1.88 \, h^2 \times 10^{-29} \, \text{g} \cdot \text{cm}^{-3}$
	$= 2.775 \, h^{-1} \times 10^{11} \, M_\odot/(h^{-1} \, \text{Mpc})^3$
	$= (3.000 \times 10^{-3} \, \text{eV}/c^2)^4 \, h^2$
Present photon density	$\Omega_{\gamma,0} \, h^2 = 2.48 \times 10^{-5}$

Present relativistic density	$\Omega_{R,0}\, h^2 = 4.17 \times 10^{-5}$
Baryon-to-photon ratio	$\eta = 2.68 \times 10^{-8}\, \Omega_b h^2$
Matter–radiation equality	$1 + z_{eq} = 24\,000\, \Omega_0 h^2$
Hubble length at equality	$(a_{eq} H_{eq})^{-1} = 14\, \Omega_0^{-1} h^{-2}\, \mathrm{Mpc}$
Top-hat filter/$10^{12}\, M_\odot$	$M(R) = 1.16\, h^{-1}(R/1\, h^{-1}\, \mathrm{Mpc})^3$
Gaussian filter/$10^{12}\, M_\odot$	$M(R) = 4.37\, h^{-1}(R/1\, h^{-1}\, \mathrm{Mpc})^3$

Conversion from natural units

$$1\,\mathrm{cm} = 5.068 \times 10^{13}\, \mathrm{GeV}^{-1} \hbar$$
$$1\,\mathrm{s} = 1.519 \times 10^{24}\, \mathrm{GeV}^{-1} \hbar/c$$
$$1\,\mathrm{g} = 5.608 \times 10^{25}\, \mathrm{GeV}/c^2$$
$$1\,\mathrm{erg} = 6.242 \times 10^{2}\, \mathrm{GeV}$$
$$1\,\mathrm{K} = 8.618 \times 10^{-14}\, \mathrm{GeV}/k_B$$

Numerical solutions and hints
for selected examples

2.1 Thirty-six families of neutrinos would be required.

2.4 Figure 2.1 indicates $g_*(1\,\mathrm{MeV}) \simeq 10$. Use this in the Friedmann equation.

  Horizon size at nucleosynthesis: $10^{-12}\,\mathrm{Mpc} \simeq 10^8\,\mathrm{km}$.

  Corresponding scale today: $10^{-3}\,\mathrm{Mpc}$.

  Horizon mass at nucleosynthesis: $10^7 M_\odot$ (see also page 114).

  Mass within that scale now: $100 M_\odot$ (the radiation has redshifted away, leaving only the matter).

3.1 First find t_{eq}, the time when matter and curvature contribute equally to the Friedmann equation. You should find $t = 9.7 \times 10^{-35}\,\mathrm{s}$ when $\Omega = 0.01$.

3.2 At decoupling, the physical separation is $0.17 h^{-1}\,\mathrm{Mpc}$

  Stretched to $180 h^{-1}\,\mathrm{Mpc}$ by the present.

  Corresponding to about 2 deg.

  Giving about 10 000 separate regions.

3.3 $\Lambda = \rho / M_{\mathrm{Pl}}$.

3.4 $\phi - \phi_0 = f_0 t_0 \ln(t/t_0)$, where $f_0 \equiv \dot\phi(t_0)$; $a(t) = (t/t_0)^{1/3}$.

3.5 Begin by noting $\Omega_{\mathrm{mon}}(t_0) = \Omega_{\mathrm{mon}}(t_{\mathrm{eq}}) = \Omega_{\mathrm{mon}}(T) \times T/T_{\mathrm{eq}}$. You then need to use the radiation density formula (2.20), to work out both Ω_{mon} and the horizon distance. The final answer is that the number of monopoles per horizon volume should be less than about 10^{-20}. This suggests that inflation by a factor of 10^7 or so (16 e-foldings) will solve the monopole problem.

3.6 Use the slow-roll equations to substitute and check for consistency.

3.7 The slow-roll breakdown is not at the same ϕ. $N = \phi_i^2/8M_{\mathrm{Pl}}^2 - 1$.

3.8 At sixty e-foldings, $P_\phi/\rho_\phi = -0.995$.

4.2 Use the value of ρ_c in the Appendix.

4.3 Equation (4.57) is exact if $2\alpha^m = \Gamma(m/2)$, provided $m > 0$.

4.4 Choose the z axis to be in the $\mathbf{k}$ direction.

4.6 Check that the expression has the correct dependence on a, and then work out the numerical coefficient at the present epoch using values quoted in the Appendix.

4.7 At matter–radiation equality, $c_s = \sqrt{4/21}$.

5.1 Although you are interested in radiation domination, you should not set the matter density equal to zero. Use Eq. (2.42) for the Hubble parameter.

5.4 The present electron density is about the present baryon density, $2 \times 10^{-7}\,\text{cm}^{-3}$. At decoupling, the density is 1100^3 larger. H_{dec} is given by Eq. (2.64). Ignoring the Ω_0 and h dependencies, you should get about 10 Mpc.

5.7 To get $\kappa \simeq 1$, we need

$$z_{\text{ion}} \sim 100 \left(\frac{h\,\Omega_b}{0.03} \right)^{-2/3} \Omega_0^{1/3}. \tag{5.1}$$

This is the same for flat or open, because during the relevant epochs $\Omega \simeq 1$.

6.1 Check that Eq. (6.3) has the correct dependence on redshift, and then evaluate it at $z = z_{\text{nr}}$ using the Appendix, assuming an rms velocity of c and matter domination from z_{nr} to the present.

7.1 In the particle interpretation, each particle has energy ω and momentum k in the $+z$ direction. Remember the relation between mass, energy, and momentum.

7.3 Simply integrate Eq. (7.105) with a constant $\delta_H^2(k)$.

8.4 Assuming that sixty e-foldings are required to give $\Omega_0 \sim 1$, it needs to tunnel at $\phi \simeq 15 M_{\text{Pl}}$.

9.1 Agreement is good to a few percent down to $\Omega_0 = 0.3$, then deteriorates.

9.2 For $\Omega_0 = 1$, we have $\sigma_8 = 0.60$. For the rest, the power spectrum is the same, and so, the relative normalization is just given via Eq. (9.9).

9.3 See (Liddle 1994) for a full discussion of this question.

9.4 You should find $n \simeq 0.94$ and $r \simeq 0.24$, for which the relevant COBE normalization is $\delta_H = 1.87 \times 10^{-5}$. This gives $V_*^{1/4} = 9.6 \times 10^{-3} M_{\text{Pl}}$, and so, $\lambda \simeq 5 \times 10^{-14}$.

10.1 The J_3 normalization is about 10 percent lower.

10.2 The required definition is Eq. (7.109). The effect of the transfer function rolls it from $+1$ on large scales to around -3 on short ones, the transition happening around k_{eq}.

11.1 The easy case is if the power law exceeds -1, for which the answer is an incomplete gamma function.

11.2 The fraction would be 10^{-63}, 7×10^{-4}, 0.09, and 0.4. The second is most likely to be reliable because it is in both the linear regime and the regime tested by numerical simulations.

11.3 For this model, $\sigma(r) = 3.25$ at $r = 0.31 h^{-1}$ Mpc. This length scale corresponds to a mass of $3.5 \times 10^{10} M_\odot$, from Eq. (4.54). However, allowing for the uncertainty in the COBE normalization, and other theoretical uncertainties, will increase this.

11.4 Using Eq. (11.41), $z_{\text{ion}} = 8$, 13, 18, 24, 31, 38, and 47, for $\Gamma = 0.2$, 0.25, ..., 0.5. The corresponding peak suppressions from Eq. (5.93) are 15, 23, 28, 34, 42, 47, and 54 percent.

14.2 We can always use the locally orthonormal frame for expressions that do not involve the space-time gradient of a vector or tensor. Cases (a), (b), (d), (g), and (i) are of this form, and require only the locally orthonormal frame. Case (k) is a tricky one; space-time derivatives of vectors do appear, but they are always antisymmetric (4-curls) so that the Christoffel symbols cancel and we can still use a locally orthonormal frame. A locally inertial frame is essential in the remaining cases. Note that in cases (i) and (j), the frame is invoked only on the right-hand side of the equation.

14.3 Writing $\tilde{x}^\mu = x^\mu + \delta x^\mu$, and taking the right-hand side to be a function of $\tilde{x}^\mu$, work out $(\partial x^\alpha / \partial \tilde{x}^\mu)(\partial x^\beta / \partial \tilde{x}^\nu)$ to first order.

15.3 In Eq. (15.49), f and T are related by Eq. (5.22), with $p = aq$. The integral over q can be read off from the blackbody formula $\rho = \pi^2 T^4 / 15$.

References

As well as journal references, wherever possible we also have given a reference to the **e-print** archive at http://xxx.lanl.gov/ which has mirror sites in several countries.

Aarseth, S. J., 1963, Mon. Not. R. Astron. Soc. **126**, 223.

Abbott, L. F., and Wise, M. B., 1984, Nucl. Phys. B **244**, 541.

Abbott, L. F., Fahri, E., and Wise, M. B., 1982, Phys. Lett. B **117**, 29.

Abell, G. O., 1958, Astrophys. J. Supp. **3**, 211.

Abell, G. O., Corwin, H. G., and Olowin, R. P., 1989, Astrophys. J. Supp. **70**, 1.

Adelberger, K., Steidel, C., Giavalisco, M., Dickinson, M., Pettini, M., and Kellogg, M., 1998, Astrophys. J. **505**, 18, astro-ph/9804236.

Adler, R. J., 1981, *The Geometry of Random Fields*, Wiley, Chichester, UK.

Aghanim, N., Desert, F. X., Puget, J. L., and Gispert, R., 1996, Astron. Astrophys. **311**, 1, astro-ph/9604083.

Agrawal, V., Barr, S. M., Donoghue, J. F., and Seckel, D., 1998, Phys. Rev. Lett. **80**, 1822, hep-ph/9801253.

Albrecht, A., 1997, in *Critical Dialogues in Cosmology*, ed. N. Turok, World Scientific, Singapore, astro-ph/9612017.

Albrecht, A., and Steinhardt, P. J., 1982, Phys. Rev. Lett. **48**, 1220.

Albrecht, A., Brandenberger, R. H., and Matzner, R. A., 1985, Phys. Rev. D **32**, 1280.

Albrecht, A., Ferriera, P., Joyce, M., and Prokopec, T., 1994, Phys. Rev. D **50**, 4807, astro-ph/9303001.

Allen, B., Caldwell, R. R., Dodelson, S., Knox, L., Shellard, E. P. S., and Stebbins,

A., 1997, Phys. Rev. Lett. **79**, 2624, astro-ph/9704160.

Alpher, R. A., Bethe, H. A., and Gamov, G., 1948, Phys. Rev. **73**, 803.

Amendola, L., Baccigalupi, C., and Occhionero, F., 1996, Phys. Rev. D **54**, 4760, gr-qc/9609032.

Babul, A., and White, S. D. M., 1991, Mon. Not. R. Astron. Soc. **253**, 31p.

Babul, A., Weinberg, D. H., Dekel, A., and Ostriker, J. P., 1994, Astrophys. J. **427**, 1, astro-ph/9311052.

Bagla, J. S., and Padmanabhan, T., 1994, Mon. Not. R. Astron. Soc. **266**, 227.

Bahcall, N., and Soniera, R., 1983, Astrophys. J. **270**, 20.

Bailin, D., and Love, A., 1994, *Supersymmetric Gauge Field Theory and String Theory*, Institute of Physics Publishing, Bristol.

Banks, T., Kaplan, D. B., and Nelson, A., 1994, Phys. Rev. D **49**, 779, hep-ph/9308292.

Bardeen, J. M., 1980, Phys. Rev. D **22**, 1882.

Bardeen, J. M., Steinhardt, P. J., and Turner, M. S., 1983, Phys. Rev. D **28**, 679.

Bardeen, J. M., Bond, J. R., Kaiser, N., and Szalay, A. S., 1986, Astrophys. J. **304**, 15 [BBKS].

Bardeen, J. M., Bond, J. R., and Efstathiou, G., 1987, Astrophys. J. **321**, 28.

Barrow, J. D., 1988, Nucl. Phys. B **296**, 697.

Barrow, J. D., 1990, Phys. Lett. B **235**, 40.

Barrow, J. D., and Cotsakis, S., 1988, Phys. Lett. B **214**, 515.

Barrow, J. D., and Maeda, K., 1990, Nucl. Phys. B **341**, 294.

Barrow, J. D., and Liddle, A. R., 1997, Gen. Rel. Grav. **29**, 1503, gr-qc/9705048.

Bartlett, J. G., Blanchard, A., Silk, J., and Turner, M. S., 1995, Science **267**, 980, astro-ph/9407061.

Bastero-Gil, M., and King, S. F., 1998, Phys. Lett. B **423**, 27, hep-ph/9709502.

Baugh, C. M., 1996, Mon. Not. R. Astron. Soc. **280**, 267, astro-ph/9512011.

Bennett, C. L., et al., 1994, Astrophys. J. **436**, 423, astro-ph/9401012.

Bennett, C. L., et al., 1996, Astrophys. J. **464**, L1, astro-ph/9601067.

Bertschinger, E., 1995, "COSMICS," astro-ph/9506070.

Bertschinger, E., and Dekel, A., 1989, Astrophys. J. **336**, L5.

Bertschinger, E., Dekel, A., Faber, S. M., Dressler, A., and Burstein, D., 1990, Astrophys. J. **364**, 370.

Binetruy, P., and Dvali, G., 1996, Phys. Lett. B **388**, 241, hep-ph/9606342.

Binetruy, P., and Gaillard, M. K., 1986, Phys. Rev. D **34**, 3069.

Blumenthal, G. R., Faber, S. M., Primack, J. R., and Rees, M. J., 1984, Nature **311**, 517.

Bond, J. R., 1992, in *Highlights in Astronomy*, Vol. 9, *Proceedings of the IAU Joint Discussion*, ed. J. Berjeron, Kluwer, Dordrecht.

Bond, J. R., 1995, Phys. Rev. Lett. **74**, 4369, astro-ph/9407044.

Bond, J. R., and Efstathiou, G., 1987, Mon. Not. R. Astron. Soc. **226**, 655.

Bond, J. R., and Efstathiou, G., 1991, Phys. Lett. B **265**, 245.

Bond, J. R., and Myers, S. T., 1996, Astrophys. J. Supp. **103**, 63.

Bond, J. R., Cole, S., Efstathiou, G., and Kaiser, N., 1991, Astrophys. J. **379**, 440.

Bond, J. R., Crittenden, R., Davis, R. L., Efstathiou, G., and Steinhardt, P. J., 1994, Phys. Rev. Lett. **72**, 13, astro-ph/9309041.

Bond, J. R., Efstathiou, G., and Tegmark, M., 1997, Mon. Not. R. Astron. Soc. **291**, L33, astro-ph/9702100.

Bond, J. R., Jaffe, A. J., and Knox, L., 1998, Phys. Rev. D **57**, 2117, astro-ph/9708203.

Bonometto, S. A., and Valdarnini, R., 1984, Phys. Lett. A **103**, 369.

Bonometto, S. A., Gabbiani, F., and Masiero, A., 1994, Phys. Rev. D **49**, 3918, astro-ph/9305010.

Borrill, J., 1999, Phys. Rev. D **59**, 027302, astro-ph/9712121.

Bouchet, F. R., Juszkiewicz, R., Columbi, S., and Pellat, R., 1992, Astrophys. J. **394**, L5.

Bower, R. G., 1991, Mon. Not. R. Astron. Soc. **248**, 332.

Bower, R. G., Coles, P., Frenk, C. S., and White, S. D. M., 1993, Astrophys. J. **405**, 403.

Bowler, M. G., 1982, *Lectures on Statistical Mechanics*, Pergamon Press, Oxford.

Boyanovsky, D., de Vega, H., Holman, R., and Salgado, J. F. J., 1996, "Preheating and reheating in inflationary cosmology," astro-ph/9609007.

Brainerd, T. G., Scherrer, R. J., and Villumsen, J. V., 1993, Astrophys. J. **418**, 570.

Bucher, M., and Cohn, J. D., 1997, Phys. Rev. D **55**, 7461, astro-ph/9701117.

Bucher, M., and Turok, N., 1995, Phys. Rev. D **52**, 5538, hep-ph/9503393.

Bucher, M., Goldhaber, A. S., and Turok, N., 1995, Phys. Rev. D **52**, 3314, hep-ph/9411206.

Bunn, E. F., and Sugiyama, N., 1995, Astrophys. J. **446**, 49, astro-ph/9407069.

Bunn, E. F., and White, M., 1995, Astrophys. J. **450**, 477, [E] **460**, 1071, astro-ph/9503054.

Bunn, E. F., and White, M., 1997, Astrophys. J. **480**, 6, astro-ph/9607060.

Bunn, E. F., Scott, D., and White, M., 1995, Astrophys. J. **441**, L9, astro-ph/9409003.

Bunn, E. F., Liddle, A. R., and White, M., 1996, Phys. Rev. D **54**, 5917R, astro-ph/9607038.

Caldwell, R. R., Dave, R., and Steinhardt, P. J., 1998, Phys. Rev. Lett. **80**, 1586, astro-ph/9708069.

Carlberg, R. G., Morris, S. L., Yee, H. K. C., and Ellingson, E., 1997, Astrophys. J. **479**, L19, astro-ph/9612169.

Carr, B. J., 1975, Astrophys. J. **201**, 1.

Carr, B. J., 1985, in *Observational and Theoretical Aspects of Relativistic Astrophysics and Cosmology*, ed. J. L. Sanz and L. J. Goicoechea, World Scientific, Singapore.

Carr, B. J., 1994, Ann. Rev. Astron. Astrophys. **32**, 531.

Carr, B. J., Gilbert, J. H., and Lidsey, J. E., 1994, Phys. Rev. D **50**, 4853, astro-ph/9405027.

Carroll, S. M., Press, W. H., and Turner, E. L., 1992, Ann. Rev. Astron. Astrophys. **30**, 499.

Cen, R., 1992, Astrophys. J. Supp. **78**, 341.

Cen, R., and Ostriker, J. P., 1992a, Astrophys. J. **393**, 22.

Cen, R., and Ostriker, J. P., 1992b, Astrophys. J. **399**, L113.

Cen, R., and Ostriker, J. P., 1996, Astrophys. J. **464**, 27, astro-ph/9601021.

Cen, R., Gnedin, N. Y., Kofman, L. A., and Ostriker, J. P., 1992, Astrophys. J. **399**, L11.

Chaboyer, B., Demarque, P., Kernan, P., and Krauss, L. M., 1998, Astrophys. J., **494**, 96, astro-ph/9706128.

Chandrasekhar, S., 1960, *Radiative Transport*, Dover, New York.

Coble, K., Dodelson, S., and Frieman, J. A., 1997, Phys. Rev. D **55**, 1851, astro-ph/9608122.

Cohn, J. D., 1996, Phys. Rev. D **54**, 7215, astro-ph/9605132.

Colberg, J. M., et al., 1998, "Galaxy clusters in the Hubble volume simulations," astro-ph/9808257.

Cole, S., and Efstathiou, G., 1989, Mon. Not. R. Astron. Soc. **239**, 195.

Cole, S., and Lacey, C., 1996, Mon. Not. R. Astron. Soc. **281**, 716, astro-ph/9510147.

Cole, S., Weinberg, D. H., Frenk, C. S., and Ratra, B., 1997, Mon. Not. R. Astron. Soc. **289**, 37, astro-ph/9702082.

Coleman, S., and de Luccia, F., 1980, Phys. Rev. D **21**, 3305.

Coles, P., 1993, Mon. Not. R. Astron. Soc. **262**, 1065.

Coles, P., and Lucchin, F., 1995, *Cosmology: The Origin and Evolution of Cosmic Structure*, Wiley, Chichester, UK.

Colley, W. N., 1997, Astrophys. J. **489**, 471, astro-ph/9612106.

Colley, W. N., Gott, J. R., and Park, C., 1996, Mon. Not. R. Astron. Soc. **281**, L82, astro-ph/9601084.

Copeland, E. J., Kolb, E. W., Liddle, A. R., and Lidsey, J. E., 1993, Phys. Rev. D **48**, 2529, hep-ph/9303288.

Copeland, E. J., Kolb, E. W., Liddle, A. R., and Lidsey, J. E., 1994a, Phys. Rev. D **49**, 1840, astro-ph/9308044.

Copeland, E. J., Liddle, A. R., Lyth, D. H., Stewart, E. D., and Wands, D., 1994b, Phys. Rev. D, **49**, 6410, astro-ph/9401011.

Copeland, E. J., Easther, R., and Wands, D., 1997, Phys. Rev. D **56**, 874, hep-th/9701082.

Copeland, E. J., Grivell, I. J., and Liddle, A. R., 1998, Mon. Not. R. Astron. Soc. **298**, 1233, astro-ph/9712028.

Copi, C. J., Schramm, D. N., and Turner, M. S., 1995a, Science **267**, 192, astro-ph/9407006.

Copi, C. J., Schramm, D. N., and Turner, M. S., 1995b, Phys. Rev. Lett. **75**, 3981, astro-ph/9508029.

Cornish, N. J., Spergel, D. N., and Starkman, G., 1996, Phys. Rev. Lett. **77**, 215, astro-ph/9601034.

Couchman, H. M. P., and Rees, M., 1986, Mon. Not. R. Astron. Soc. **221**, 53.

Couchman, H. M. P., Pearce, F. R., and Thomas, P. A., 1995, Astrophys. J. **452**, 797, astro-ph/9409058.

Couchman, H. M. P., Pearce, F. R., and Thomas, P. A., 1996, astro-ph/9603116.

Covi, L., and Lyth, D. H., 1999, Phys. Rev. D **59**, 063515, hep-ph/9809562.

Dalton, G. B., Efstathiou, G., Maddox, S. J., and Sutherland, W. J., 1992, Astrophys. J. **390**, L1.

Daly, R. A., 1991, Astrophys. J. **371**, 14.

Danese, L., and de Zotti, G., 1977, Riv. Nuovo Cimento **7**, 277.

Davis, M., and Peebles, P. J. E., 1983, Astrophys. J. **267**, 465.

Davis, M., Efstathiou, G., Frenk, C. S., and White, S. D. M., 1985, Astrophys. J. **292**, 371.

Davis, M., Efstathiou, G., Frenk, C. S., and White, S. D. M., 1992a, Nature **356**, 489.

Davis, M., Summers, F., and Schlegel, D., 1992b, Nature **359**, 393.

Davis, M., Miller, A., and White, S. D. M., 1997, Astrophys. J. **490**, 63, astro-ph/9705224.

de Oliveira-Costa, A., Smoot, G. F., and Starobinsky, A. A., 1996 Astrophys. J. **468**, 457, astro-ph/9510109.

Dekel, A., 1994, Ann. Rev. Astron. Astrophys. **32**, 371, astro-ph/9401022.

Dekel, A., Bertschinger, E., and Faber, S. M., 1990, Astrophys. J. **364**, 349.

Dekel, A., Bertschinger, E., Yahil, A., Strauss, M. A., Davis, M., and Huchra, J. P., 1993, Astrophys. J. **412**, 1.

de Carlos, B., Casas, J. A., Quevedo, F., and Roulet, E., 1993, Phys. Lett. B **318**, 447, hep-ph/9308325.

Deltorn, J.-M., Le Fevre, O., Crampton, D., and Dickinson, M., 1997, Astrophys. J. **483**, L21, astro-ph/9704086.

Dine, M., and Riotto, A., 1997, Phys. Rev. Lett. **79**, 2632, hep-ph/9705386.

Dodelson, S., and Jubas, J., 1995, Astrophys. J. **439**, 503, astro-ph/9308019.

Dodelson, S., Gyuk, G., and Turner, M. S., 1994, Phys. Rev. Lett. **72**, 3754, astro-ph/9402028.

Dolgov, A. D., 1992, Phys. Rep. **222**, 309.

Dolgov, A. D., and Linde, A. D., 1982, Phys. Lett. B **116**, 329.

Donahue, M., Voit, G. M., Gioia, I. M., Luppino, G., Hughes, J. P., and Stocke, J. T., 1998, Astrophys. J. **502**, 550, astro-ph/9707010.

Durrer, R., 1996, Helv. Phys. Acta **69**, 417, astro-ph/9610234.

Dvali, G., and Riotto, A., 1998, Phys. Lett. B **417** 20, hep-ph/9706408.

Dvali, G., Shafi, Q., and Schaefer, R., 1994, Phys. Rev. Lett. **73**, 1886, hep-ph/9406319.

Easther, R., 1995, Class. Quant. Grav. **13**, 1775, astro-ph/9511143.

Edge, A. C., Stewart, G. C., Fabian, A. C., and Arnaud, K. A., 1990, Mon. Not. R. Astron. Soc. **245**, 559.

Efstathiou, G., 1995, Mon. Not. R. Astron. Soc. **274**, L73.

Efstathiou, G., and Bond, J. R., 1986, Mon. Not. R. Astron. Soc. **218**, 103.

Efstathiou, G., and Rees, M. J., 1988, Mon. Not. R. Astron. Soc. **230**, 5p.

Efstathiou, G., Frenk, C. S., White, S. D. M., and Davis, M., 1988, Mon. Not. R. Astron. Soc. **235**, 715.

Efstathiou, G., Sutherland, W. J., and Maddox, S. J., 1990, Nature **348**, 705.

Ehlers, J., 1961, Abh. Mainz. Akad. Wiss. Lit. (Math. Nat. Kl.) **11**, 792.

Eisenstein, D. J., 1997, "An analytic expression for the growth function…," astro-ph/9709054.

Eisenstein, D. J., and Hu, W., 1999, Astrophys. J. **511**, 5, astro-ph/9710252.

Eke, V., Cole, S., and Frenk, C. S., 1996, Mon. Not. R. Astron. Soc. **282**, 263, astro-ph/9601088.

Ellis, G. F. R., 1971, in *General Relativity and Cosmology, Proceedings of the 47th Enrico Fermi Summer School*, ed. R. K. Sachs, Academic Press, New York.

Ellis, G. F. R., and Bruni, M., 1989, Phys. Rev. D **140**, 1804.

Ellis, J., Nanopoulos, D., and Quirós, M., 1986, Phys. Lett. B **174**, 176.

Ellis, G. F. R., Lyth, D. H., and Mijić, M. B., 1991, Phys. Lett. B **271**, 52.

Evrard, A. E., 1988, Mon. Not. R. Astron. Soc. **235**, 911.

Evrard, A. E., 1989, Astrophys. J. **341**, L71.

Fahlman, G., Kaiser, N., Squires, G., and Woods, D., 1994, Astrophys. J. **437**, 56, astro-ph/9402017.

Falco, E. E., Kochanek, C. S., and Munoz, J. M., 1998, Astrophys. J. **492**, L9, astro-ph/9707032.

Fang, L. Z., Li, S. X., and Xiang, S. P., 1984, Astron. Astrophys. **140**, 77.

Feast, M. W., and Catchpole, R. M., 1997, Mon. Not. R. Astron. Soc. **286**, L1.

Feldman, H. A., Kaiser, N., and Peacock, J. A., 1994, Astrophys. J. **426**, 23, astro-ph/9304022.

Fisher, K. B., Scharf, C. A., and Lahav, O., 1994, Mon. Not. R. Astron. Soc. **266**, 219, astro-ph/9309027.

Fixsen, D. J., Cheng, E. S., Gales, J. M., Mather, J. C., Shafer, R. A., and Wright, E. L., 1996, Astrophys. J. **473**, 576, astro-ph/9605054.

Freedman, W. L., 1997, in *Critical Dialogues in Cosmology*, ed. N. Turok, World Scientific, Singapore, astro-ph/9612024.

Freese, K., Frieman, J. A., and Olinto, A. V., 1990, Phys. Rev. Lett. **65**, 3233.

Frenk, C. S., White, S. D. M., Davis, M., and Efstathiou, G., 1988, Astrophys. J. **327**, 507.

Frenk, C. S., White, S. D. M., Efstathiou, G., and Davis, M., 1990, Astrophys. J. **351**, 10.

Friedmann, A., 1922, Z. Phys. **10**, 377.

Fry, J. N., 1984, Astrophys. J. **279**, 499.

Fukugita, M., and Kawasaki, M., 1994, Mon. Not. R. Astron. Soc. **269**, 563, astro-ph/9309036.

Futamase, T., and Tanaka, M., 1999, Phys. Rev. D **60**, 063511, hep-ph/9704303.

García-Bellido, J., 1996, Phys. Rev. D **54**, 2473, astro-ph/9510029.

García-Bellido, J., and Liddle, A. R., 1997, Phys. Rev. D **55**, 4603, astro-ph/9610183.

García-Bellido, J., and Linde, A. D., 1997, Phys. Rev. D **55**, 7480, astro-ph/9701173.

García-Bellido, J., and Wands, D., 1995, Phys. Rev. D **52**, 6739, gr-qc/9506050.

García-Bellido, J., Liddle, A. R., Lyth, D. H., and Wands, D., 1995, Phys. Rev. D **52**, 6750, astro-ph/9508003.

García-Bellido, J., Linde, A. D., and Wands, D., 1996, Phys. Rev. D **54**, 6040, astro-ph/9605094.

García-Bellido, J., Liddle, A. R., Lyth, D. H., and Wands, D., 1997, Phys. Rev. D **55**, 4596, astro-ph/9608106.

Garriga, J., 1996, Phys. Rev. D **54**, 4764, gr-qc/9602025.

Garriga, J., 1998, "Open inflation and the singular boundary," hep-th/9803210.

Garriga, J., and Mukhanov, V. F., 1997, Phys. Rev. D **56**, 2439, astro-ph/9702201.

Garriga, J., Montes, X., Sasaki, M., and Tanaka, T., 1998, Nucl. Phys. B **513**, 343, astro-ph/9706229.

Gasperini, M., 1997, in *Fourth Paris Cosmology Colloquium*, eds. H. J. de Vega and N. Sanchez, World Scientific, Singapore, gr-qc/9706037.

Gasperini, M., and Veneziano, G., 1993a, Astropart. Phys. **1**, 317, hep-th/9211021.

Gasperini, M., and Veneziano, G., 1993b, Mod. Phys. Lett. **A8**, 3701, hep-th/9309023.

Gaztañaga, E., and Frieman, J. A., 1994, Astrophys. J. **437**, L13, astro-ph/9407079.

Gelb, J. M., Gradwöhl, B.-A., and Frieman, J. A., 1993, Astrophys. J. **403**, L5, hep-ph/9208239.

Giavalisco, M., Steidel, C. C., and Macchetto, F. D., 1996, Astrophys. J. **470**, 189, astro-ph/9603062.

Górski, K. M., 1994, Astrophys. J. **430**, L85, astro-ph/9403066.

Górski, K. M., et al., 1994, Astrophys. J. **430**, L89, astro-ph/9403067.

Górski, K. M., Ratra, B., Sugiyama, N., and Banday, A. J., 1995, Astrophys. J. **444**, L65, astro-ph/9502034.

Górski, K. M., et al., 1996, Astrophys. J. **464**, L11, astro-ph/9601063.

Górski, K. M., Ratra, B., Stompor, R., Sugiyama, N., and Banday, A. J., 1998, Astrophys. J. Supp. **114**, 1, astro-ph/9608054.

Gott, J. R., 1982, Nature **295**, 304.

Gott, J. R., Melott, A. L., and Dickinson, M., 1986, Astrophys. J. **306**, 341.

Gott, J. R., and Statler, T. S., 1984, Phys. Lett. B **136**, 157.

Gradshteyn, I. S., and Rhyzik, I. M., 1994, *Tables of Integrals, Series and Products*, Academic Press, London.

Green, A. M., and Liddle, A. R., 1996, Phys. Rev. D **54**, 2557, astro-ph/9604001.

Green, A. M., and Liddle, A. R., 1997a, Phys. Rev. D **55**, 609, astro-ph/9607166.

Green, A. M., and Liddle, A. R., 1997b, Phys. Rev. D **56**, 6166, astro-ph/9704251.

Grishchuk, L. P., 1974, Zh. Eksp. Teor. Fiz. **67**, 825 [Sov. Phys. JETP **40**, 409].

Grishchuk, L. P., 1993, Phys. Rev. Lett. **70**, 2371, gr-qc/9304001.

Grishchuk, L. P., and Zel'dovich, Ya. B., 1978, Astron. Zh. **55**, 209 [Sov. Astron. **22**, 125 (1978)].

Grivell, I. J., and Liddle, A. R., 1996, Phys. Rev. D **54**, 7191, astro-ph/9607096.

Gruzinov, A., and Hu, W., 1998, Astrophys. J. **508**, 435, astro-ph/9803188.

Gunn, J. E., and Gott, J. R., 1972, Astrophys. J. **176**, 1.

Gunn, J. E., and Peterson, B. A., 1965, Astrophys. J. **142**, 1633.

Gunn, K. F., and Thomas, P. A., 1996, Mon. Not. R. Astron. Soc. **281**, 1133, astro-ph/9510082.

Guth, A. H., 1981, Phys. Rev. D **23**, 347.

Guth, A. H., and Pi, S.-Y., 1982, Phys. Rev. Lett. **49**, 1110.

Guth, A. H., and Pi, S.-Y., 1985, Phys. Rev. D **32**, 1899.

Gyuk, G., and Turner, M. S., 1994, Phys. Rev. D **50**, 6130, astro-ph/9403054.

Haehnelt, M. G., 1993, Mon. Not. R. Astron. Soc. **265**, 727.

Haiman, Z., and Loeb, A., 1997, Astrophys. J. **483**, 21, astro-ph/9611028.

Halliwell, J. J., 1989, Phys. Rev. D **39**, 2912.

Halyo, E., 1996, Phys. Lett. B **387**, 43, hep-ph/9606423.

Hamazaki, T., Sasaki, M., Tanaka, T., and Yamamoto, K., 1996, Phys. Rev. D **53**, 2045, gr-qc/9507006.

Hamilton, A. J. S., 1997, Mon. Not. R. Astron. Soc. **289**, 295, astro-ph/9701009.

Hamilton, A. J. S., Gott, J. R., and Weinberg, D. H., 1986, Astrophys. J. **309**, 1.

Harrison, R., 1970, Phys. Rev. D **1**, 2726.

Hata, N., Scherrer, R. J., Steigman, G., Thomas, D., Walker, T. P., Bludman, S., and Langacker, P., 1995, Phys. Rev. Lett. **75**, 3977, hep-ph/9505319.

Hawking, S. W., 1966, Astrophys. J. **145**, 544.

Hawking, S. W., 1982, Phys. Lett. B **115**, 339.

Hawking, S. W., and Turok, N., 1998, Phys. Lett. B **425**, 25, astro-ph/9802030.

Heavens, A. F., and Taylor, A. N., 1995, Mon. Not. R. Astron. Soc. **275**, 483, astro-ph/9409027.

Hendry, M. A., and Tayler, R. J., 1996, Contemp. Phys. **37**, 263.

Henry, J. P., 1997, Astrophys. J. **489**, L1.

Henry, J. P., and Arnaud, K. A., 1991, Astrophys. J. **372**, 410.

Hernquist, L., and Katz, N., 1989, Astrophys. J. Supp. **70**, 419.

Hernquist, L., Katz, N., Weinberg, D. H., and Miralda-Escudé, J., 1996, Astrophys. J. **457**, L51, astro-ph/9509105.

Hindmarsh, M. B., and Kibble, T. W. B., 1995, Rep. Prog. Phys. **58**, 477, hep-ph/9411342.

Hinshaw, G., et al., 1996, Astrophys. J. **464**, L25, astro-ph/9601058.

Hodges, H. M., and Blumenthal, G. R., 1990, Phys. Rev. D **42**, 3329.

Hodges, H. M., Blumenthal, G. R., Kofman, L., and Primack, J., 1990, Nucl. Phys. B **335**, 197.

Hogan, C. J., Kaiser, N., and Rees, M. J., 1982, Philos. Trans. R. Soc. London Ser. A **307**, 97.

Holman, R., Kolb, E. W., and Wang, Y., 1990, Phys. Rev. Lett. **65**, 17.

Holtzman, J., 1989, Astrophys. J. Supp. **71**, 1.

Hoyle, F., and Tayler, R. J., 1964, Nature **203**, 1108.

Hu, W., and Sugiyama, N., 1995, Phys. Rev. D **51**, 2599, astro-ph/9411008.

Hu, W., and White, M., 1996a, Astron. Astrophys. **315**, 33, astro-ph/9507060.

Hu, W., and White, M., 1996b, Astrophys. J. **471**, 30, astro-ph/9602019.

Hu, W., and White, M., 1997a, Astrophys. J. **479**, 568, astro-ph/9609079.

Hu, W., and White, M., 1997b, Astrophys. J. **486**, L1, astro-ph/9701210.

Hu, W., and White, M., 1997c, Phys. Rev. D **56**, 596, astro-ph/9702170.

Hu, W., and White, M., 1997d, New Astron. **2**, 323, astro-ph/9706147.

Hu, W., Scott, D., and Silk, J., 1994a, Phys. Rev. D **49**, 648, astro-ph/9305038.

Hu, Y., Turner, M. S., and Weinberg, E. J., 1994b, Phys. Rev. D **49**, 3830, astro-ph/9302002.

Hu, W., Scott, D., Sugiyama, N., and White, M., 1995, Phys. Rev. D **52**, 5498, astro-ph/9505043.

Hu, W., Sugiyama, N., and Silk, J., 1997, Nature **386**, 37, astro-ph/9504057.

Hudon, J. D., and Lilly, S. J., 1996, Astrophys. J. **469**, 519, astro-ph/9605056.

Ikeuchi, S., and Ostriker, J. P., 1986, Astrophys. J. **310**, 522.

Jacoby, G. H., et al., 1992, Pub. Astron. Soc. Pac. **104**, 599.

Jaffe, A. H., and Kamionkowski, M., 1998, Phys. Rev. D **58**, 043001, astro-ph/9801022.

Jedamzik, K., 1995, Astrophys. J., **448**, 1, astro-ph/9408080.

Jetzer, P., 1992, Phys. Rep. **220**, 165.

Jones, B. J. T., and Wyse, R. F. G., 1985, Astron. Astrophys. **149**, 144.

Jones, M., et al., 1993, Nature **365**, 320.

Jones, M., et al., 1997, Astrophys. J. **479**, L1, astro-ph/9611218.

Jungman, G., Kamionkowski, M., and Griest, K., 1996a, Phys. Rep. **267**, 195, hep-ph/9506380.

Jungman, G., Kamionkowski, M., Kosowsky, A., and Spergel, D. N., 1996b, Phys. Rev. Lett. **76**, 1007, astro-ph/9507080.

Jungman, G., Kamionkowski, M., Kosowsky, A., and Spergel, D. N., 1996c, Phys. Rev. D **54**, 1332, astro-ph/9512139.

Kaiser, N., 1984, Astrophys. J. **284**, L9.

Kaiser, N., 1987, Mon. Not. R. Astron. Soc. **227**, 1.

Kaiser, N., and Squires, G., 1993, Astrophys. J. **404**, 441.

Kaiser, N., Efstathiou, G., Ellis, R., Frenk, C., Lawrence, A., Rowan-Robinson, M., and Saunders, W., 1991, Mon. Not. R. Astron. Soc. **252**, 1.

Kaiser, N., Malaney, R. A., and Starkman, G. D., 1993, Phys. Rev. Lett. **71**, 1128, hep-ph/9302261.

Kamionkowski, M., and Loeb, A., 1997, Phys. Rev. D **56**, 4511, astro-ph/9703118.

Kamionkowski, M., Kosowsky, A., and Stebbins, A., 1997a, Phys. Rev. Lett. **78**, 2058, astro-ph/9609132.

Kamionkowski, M., Kosowsky, A., and Stebbins, A., 1997b, Phys. Rev. D **55**, 7368, astro-ph/9611125.

Karlin, S., and Taylor, H. M., 1975, *A First Course on Stochastic Processes*, Academic Press, New York.

Kashlinsky, A., 1988, Astrophys. J. **331**, L1.

Katgert, P., et al., 1996, Astron. Astrophys. **310**, 8, astro-ph/9511051.

Katz, N., Hernquist, L., and Weinberg, D. H., 1992, Astrophys. J. **399**, 109.

Katz, N., Weinberg, D. H., Hernquist, L., and Miralda-Escudé, J., 1996a, Astrophys. J. **457**, L57, astro-ph/9509106.

Katz, N., Weinberg, D. H., and Hernquist, L., 1996b, Astrophys. J. Supp. **105**, 19, astro-ph/9509107.

Kauffmann, G., and Charlot, S., 1994, Astrophys. J. **430**, L97, astro-ph/9402015.

Kauffmann, G., Nusser, A., and Steinmetz, M., 1997, Mon. Not. R. Astron. Soc. **286**, 795, astro-ph/9512009.

Kernan, P. J., and Sarkar, S., 1996, Phys. Rev. D **54**, 3861, astro-ph/9603045.

Kibble, T. W. B., 1976, J. Phys. **A9**, 1387.

Kim, A. G., et al., 1997, Astrophys. J. **476**, L63, astro-ph/9701188.

Kinney, W. H., and Riotto, A, 1998, Phys. Lett. B **435**, 272, hep-ph/9802443.

Klypin, A., and Kopylov, A. I., 1983, Sov. Astron. Lett. **9**, 41.

Klypin, A., Holtzman, J., Primack, J. R., and Regös, E., 1994, Astrophys. J. **416**, 1, astro-ph/9305011.

Klypin, A., Borgani, S., Holtzman, J., and Primack, J. R., 1995, Astrophys. J. **444**, 1, astro-ph/9405003.

Knox, L., 1995, Phys. Rev. D **52**, 4307, astro-ph/9504054.

Knox, L., Scoccimarro, R., and Dodelson, S., 1998, Phys. Rev. Lett. **81**, 2004, astro-ph/9805012.

Kochanek, C. S., 1996, Astrophys. J. **466**, 638, astro-ph/9510077.

Kodama, H., and Sasaki, M., 1984, Prog. Theor. Phys. Supp. **78**, 1.

Kodama, H., and Sasaki, M., 1987, Int. J. Mod. Phys. **A2**, 491.

Kofman, L., and Starobinsky, A. A., 1985, Sov. Astron. Lett. **11**, 271.

Kofman, L., Gnedin, N. Y., and Bahcall, N. A., 1993, Astrophys. J. **413**, 1.

Kofman, L., Linde, A. D., and Starobinsky, A. A., 1994, Phys. Rev. Lett. **73**, 3195, hep-th/9405187.

Kofman, L., Linde, A. D., and Starobinsky, A. A., 1997, Phys. Rev. D **56**, 3258, hep-ph/9704452.

Kolatt, T., and Dekel, A., 1997, Astrophys. J. **479**, 592, astro-ph/9512132.

Kolb, E. W., 1991, Phys. Scr. **T36**, 199.

Kolb, E. W., and Turner, M. S., 1990, *The Early Universe*, Addison-Wesley, Redwood City, California.

Kosowsky, A., and Turner, M. S., 1995, Phys. Rev. D **52**, 1739, astro-ph/9504071.

Krauss, L. M., and Turner, M. S., 1995, Gen. Rel. Grav. **27**, 1137, astro-ph/9504003.

Kung, J. H., and Brandenberger, R. H., 1990, Phys. Rev. D **42**, 1008.

La, D., and Steinhardt, P. J., 1989, Phys. Rev. Lett. **62**, 376.

La, D., Steinhardt, P. J., and Bertschinger, E., 1989, Phys. Lett. B **231**, 231.

Lacey, C., and Cole, S., 1994, Mon. Not. R. Astron. Soc. **271**, 676, astro-ph/9402069.

Lanzetta, K. M., Wolfe, A. M., and Turnshek, D. A., 1995, Astrophys. J. **440**, 435.

Lazarides, G., Panagiotakopoulos, C., and Vlachos, N. D., 1996, Phys. Rev. D **54**, 1369, hep-ph/9606297.

Lemaitre, G., 1927, Ann. Soc. Sci. Bruxelles **A47**, 49. [Translated in Mon. Not. R. Astron. Soc. **91**, 483 (1931).]

Liddle, A. R., 1994, Phys. Rev. D **49**, 739, astro-ph/9307020.

Liddle, A. R., 1995, Phys. Rev. D **51**, R5347, astro-ph/9410083.

Liddle, A. R., and Lyth, D. H., 1992, Phys. Lett. B **291**, 391, astro-ph/9208007.

Liddle, A. R., and Lyth, D. H., 1993a, Phys. Rep. **231**, 1, astro-ph/9303019.

Liddle, A. R., and Lyth, D. H., 1993b, Mon. Not. R. Astron. Soc. **265**, 379, astro-ph/9304017.

Liddle, A. R., and Lyth, D. H., 1995, Mon. Not. R. Astron. Soc. **273**, 1177, astro-ph/9409077.

Liddle, A. R., and Madsen, M. S., 1992, Int. J. Mod. Phys. **D1**, 101.

Liddle, A. R., and Wands, D., 1991, Mon. Not. R. Astron. Soc. **253**, 637.

Liddle, A. R., and Wands, D., 1992, Phys. Rev. D **45**, 2665.

Liddle, A. R., Lyth, D. H., and Sutherland, W., 1992, Phys. Lett. B **279**, 244.

Liddle, A. R., Parsons, P., and Barrow, J. D., 1994, Phys. Rev. D **50**, 7222, astro-ph/9408015.

Liddle, A. R., Lyth, D. H., Roberts, D., and Viana, P. T. P., 1996a, Mon. Not. R. Astron. Soc. **278**, 644, astro-ph/9506091.

Liddle, A. R., Lyth, D. H., Schaefer, R. K., Shafi, Q., and Viana, P. T. P., 1996b, Mon. Not. R. Astron. Soc. **281**, 531, astro-ph/9511057.

Liddle, A. R., Lyth, D. H., Viana, P. T. P., and White, M., 1996c, Mon. Not. R. Astron. Soc. **282**, 281, astro-ph/9512102.

Lidsey, J. E., 1991, Phys. Lett. B **273**, 42.

Lidsey, J. E., Liddle, A. R., Kolb, E. W., Copeland, E. J., Barreiro, T., and Abney, M., 1997, Rev. Mod. Phys. **69**, 373, astro-ph/9508078.

Lifshitz, E. M., 1946, J. Phys. Moscow **10**, 116.

Lin, C. C., Mestel, L., and Shu, F. H., 1965, Astrophys. J. **142**, 1431.

Lin, H., et al., 1996, Astrophys. J. **471**, 617, astro-ph/9606055.

Linde, A. D., 1982, Phys. Lett. B **108**, 389.

Linde, A. D., 1983, Phys. Lett. B **129**, 177.

Linde, A. D., 1986, Phys. Lett. B **175**, 395.

Linde, A. D., 1990a, *Particle Physics and Inflationary Cosmology*, Harwood, Chur, Switzerland.

Linde, A. D., 1990b, Phys. Lett. B **249**, 18.

Linde, A. D., 1991, Phys. Lett. B **259**, 38.

Linde, A. D., 1994, Phys. Rev. D **49**, 748, astro-ph/9307002.

Linde, A. D., 1995, Phys. Lett. B **351**, 99, hep-th/9503097.

Linde, A. D., 1997, in *Critical Dialogues in Cosmology*, ed. N. Turok, World Scientific, Singapore, astro-ph/9610077.

Linde, A. D., 1998, Phys. Rev. D **58**, 083514, gr-qc/9802038.

Linde, A. D., and Lyth, D. H., 1990, Phys. Lett. B **246**, 353.

Linde, A. D., and Mezhlumian, A., 1995, Phys. Rev. D **52**, 6789, astro-ph/9506017.

Linde, A. D., and Mukhanov, V., 1997, Phys. Rev. D **56**, R535, astro-ph/9610219.

Linde, A. D., Linde. D., and Mezhlumian, A., 1994, Phys. Rev. D **49**, 1783, gr-qc/9306035.

Lineweaver, C. H., 1997, in *Microwave Background Anisotropies*, eds. F. R. Bouchet, et al., Editions Frontiers, Git-sur-Yvette, France, astro-ph/9609034.

Lineweaver, C. H., Tenorio, L., Smoot, G. F., Keegstra, P., Banday, A. J., and Lubin, P., 1996, Astrophys. J. **470**, 38, astro-ph/9601151.

Loewenstein, M., and Mushotzky, R. F., 1996, Astrophys. J. **471**, L83, astro-ph/9608111.

Loveday, J., and Pier, J., 1998, "The Sloan Digital Sky Survey," astro-ph/9809179.

Lucchin, F., and Matarrese, S., 1985, Phys. Rev. D **32**, 1316.

Luppino, G. A., and Gioia, I. M., 1995, Astrophys. J. **445**, L77, astro-ph/9502095.

Lyth, D. H., 1984, Phys. Lett. B **147**, 403, (E) **150**, 465.

Lyth, D. H., 1985, Phys. Rev. D **31**, 1792.

Lyth, D. H., 1990, Phys. Lett. B **236**, 408.

Lyth, D. H., 1992, Phys. Rev. D **45**, 3394.

Lyth, D. H., 1997, Phys. Rev. Lett. **78**, 1861, hep-ph/9606387.

Lyth, D. H., and Mukherjee, M., 1988, Phys. Rev. D **38**, 485.

Lyth, D. H., and Riotto, A., 1997, Phys. Lett. B **412**, 28, hep-ph/9707273.

Lyth, D. H., and Riotto, A., 1998, Phys. Rep. **314**, 1, hep-ph/9807278.

Lyth, D. H., and Stewart, E. D., 1990a, Phys. Lett. B **252**, 336.

Lyth, D. H., and Stewart, E. D., 1990b, Astrophys. J. **361**, 343.

Lyth, D. H., and Stewart, E. D., 1992a, Phys. Lett. B **274**, 168.

Lyth, D. H., and Stewart, E. D., 1992b, Phys. Lett. B **283**, 189.

Lyth, D. H., and Stewart, E. D., 1992c, Phys. Rev. D **46**, 532.

Lyth, D. H., and Stewart, E. D., 1995, Phys. Rev. Lett. **75**, 201, hep-ph/9502417.

Lyth, D. H., and Stewart, E. D., 1996a, Phys. Rev. D **53**, 1784, hep-ph/9510204.

Lyth, D. H., and Stewart, E. D., 1996b, Phys. Rev. D **54**, 7186, hep-ph/9606412.

Lyth, D. H., and Woszczyna, A., 1995, Phys. Rev. D **52**, 3338, astro-ph/9501044.

Ma, C.-P., and Bertschinger, E., 1994, Astrophys. J. **434**, L5, astro-ph/9407085.

Ma, C.-P., and Bertschinger, E., 1995, Astrophys. J. **455**, 7, astro-ph/9506072.

Ma, C.-P., Bertschinger, E., Hernquist, L., Weinberg, D. H., and Katz, N., 1997, Astrophys. J. **484**, L1, astro-ph/9705113.

MacGibbon, J. H., 1987, Nature **329**, 308.

MacGibbon, J. H., and Carr, B. J., 1991, Astrophys. J. **371**, 447.

Maddox, S. J., Efstathiou, G., Sutherland, W. J., and Loveday, J., 1990, Mon. Not. R. Astron. Soc. **242**, 43p.

Maddox, S. J., Sutherland, W. J., Efstathiou, G., Loveday, J., and Peterson, B. A., 1991, Mon. Not. R. Astron. Soc. **247**, 1p.

Maeda, K., 1989, Phys. Rev. D **39**, 3159.

MAP Satellite, home page at http://map.gsfc.nasa.gov/

Margon, B., 1998, "The Sloan Digital Sky Survey," astro-ph/9805314.

Matarrese, S., Lucchin, L., Moscardini, L., and Saez, D., 1992, Mon. Not. R. Astron. Soc. **259**, 437.

Matarrese, S., Coles, P., Lucchin, F., and Moscardini, L., 1997, Mon. Not. R. Astron. Soc. **286**, 115, astro-ph/9608004.

Mather, J. C., et al., 1990, Astrophys. J. **354**, L37.

Mathews, G. J., Kajino, T., and Orito, M., 1996, Astrophys. J. **456**, 98.

McNally, S. J., and Peacock, J. A., 1995, Mon. Not. R. Astron. Soc. **277**, 143, astro-ph/9506075.

Meiksin, A., and Madau, P., 1993, Astrophys. J. **412**, L53.

Melott, A. L., 1990, Phys. Rep. **193**, 1.

Melott, A. L., Cohen, A. P., Hamilton, A. J. S., Gott, J. R., and Weinberg, D. H., 1989, Astrophys. J. **345**, 618.

Miralda-Escudé, J., Cen, R., Ostriker, J. P., and Rauch, M., 1996, Astrophys. J. **471**, 582, astro-ph/9511013.

Misner, C. W., Thorne, K. S., and Wheeler, J. A., 1973, *Gravitation*, Freeman, New York.

Mo, H. J., and Fukugita, M., 1996, Astrophys. J. **467**, L9, astro-ph/9604034.

Mo, H. J., and Miralda-Escudé, J., 1994, Astrophys. J. **430**, L25, astro-ph/9402014.

Mollerach, S., Matarrese, S., and Lucchin, F., 1994, Phys. Rev. D **50**, 4835, astro-ph/9309054.

Mukhanov, V. F., 1985, JETP Lett. **41**, 493.

Mukhanov, V. F., 1989, Phys. Lett. B **218**, 17.

Mukhanov, V. F., and Chibisov, G. V., 1981, JETP Lett. **33**, 532.

Mukhanov, V. F., and Chibisov, G. V., 1982, Sov. Phys. JETP **56**, 258.

Mukhanov, V. F., Feldman, H. A., and Brandenberger, R. H., 1992, Phys. Rep. **215**, 203.

Muslimov, A. G., 1990, Class. Quant. Grav. **7**, 231.

Myers, S. T., Baker, J. E., Readhead, A. C. S., Leitch, E. M., and Herbig, T., 1997, Astrophys. J. **484**, 1, astro-ph/9703123.

Nakamura, T. T., and Stewart, E. D., 1996, Phys. Lett. B **381**, 413, astro-ph/9604103.

Navarro, J. F., Frenk, C. S., and White, S. D. M., 1995, Mon. Not. R. Astron. Soc. **275**, 720, astro-ph/9408069.

Nichol, R. C., Collins, C. A., Guzzo, L., and Lumsden, S. L., 1992, Mon. Not. R. Astron. Soc. **255**, 21p.

Oh, S. P., Spergel, D. N., and Hinshaw, G., 1999, Astrophys. J. **510**, 551, astro-ph/9805339.

Olive, K., 1990, Phys. Rep. **190**, 307.

Olson, D. W., 1976, Phys. Rev. D **14**, 32.

Ostriker, J. P., and Steinhardt, P. J., 1995, Nature **377**, 600, astro-ph/9505066.

Ostriker, J. P., and Suto, Y., 1990, Astrophys. J. **348**, 378.

Ostriker, J. P., and Vishniac, E. T., 1986, Astrophys. J. Lett. **306**, L51.

Oukbir, J., and Blanchard, A., 1992, Astron. Astrophys. **262**, L21.

Padmanabhan, T., 1989, Phys. Rev. D **39**, 2924.

Padmanabhan, T., 1993, *Structure Formation in the Universe*, Cambridge University Press, Cambridge, UK.

Pagel, B. E. J., 1997, *Nucleosynthesis and Chemical Evolution of Galaxies*, Cambridge University Press, Cambridge, UK.

Partridge, R. B., and Peebles, P. J. E., 1967, Astrophys. J. **147**, 868.

Peacock, J. A., 1996, "Inflationary cosmology and structure formation," astro-ph/9601135.

Peacock, J. A., 1997, Mon. Not. R. Astron. Soc. **284**, 885, astro-ph/9608151.

Peacock, J. A., 1999, *Cosmological Physics*, Cambridge University Press, Cambridge, UK.

Peacock, J. A., and Dodds, S. J., 1994, Mon. Not. R. Astron. Soc. **267**, 1020, astro-ph/9311057.

Peacock, J. A., and Heavens, A. F., 1985, Mon. Not. R. Astron. Soc. **217**, 805.

Peacock, J. A., and Heavens, A. F., 1990, Mon. Not. R. Astron. Soc. **243**, 133.

Pearce, F. R., and Couchman, H. M. P., 1997, New Astron. **2**, 411, astro-ph/9703183.

Peebles, P. J. E., 1968, Astrophys. J. **153**, 1.

Peebles, P. J. E., 1980, *The Large Scale Structure of the Universe*, Princeton University Press, Princeton, NJ.

Peebles, P. J. E., 1982, Astrophys. J. **263**, L1.

Peebles, P. J. E., 1984, Astrophys. J. **284**, 439.

Peebles, P. J. E., 1993, *Principles of Physical Cosmology*, Princeton University Press, Princeton, NJ.

Peebles, P. J. E., and Juszkiewicz, R., 1998, Astrophys. J. **509**, 483, astro-ph/9804260.

Peebles, P. J. E., and Yu, J. T., 1970, Astrophys. J. **162**, 815.

Pen, U.-L., 1998, Astrophys. J. **498**, 60, astro-ph/9610147.

Pen, U.-L., Seljak, U., and Turok, N., 1997, Phys. Rev. Lett. **79**, 1611, astro-ph/9704165.

Pequignot, D., Petitjean, P., and Boisson, C., 1991, Astron. Astrophys. **251**, 680.

Perlmutter, S., et al., 1998, Nature **391**, 51, astro-ph/9712212.

Pierpaoli, E., and Bonometto, S. A., 1995, Astron. Astrophys. **300**, 12, astro-ph/9410059.

Pierpaoli, E., Coles, P., Bonometto, S. A., and Borgani, S., 1996, Mon. Not. R. Astron. Soc. **470**, 92, astro-ph/9603150.

Planck Surveyor satellite, home page at http://astro.estec.esa.nl/Planck

Pogosyan, D. Yu., and Starobinsky, A. A., 1995, Astrophys. J. **447**, 465, astro-ph/9409074.

Polnarev, A. G., 1985, Sov. Astron. **29**, 607.

Press, W. H., and Schechter, P., 1974, Astrophys. J. **187**, 452.

Press, W. H., Teukolsky, S. A., Vetterling, W. T., and Flannery, B. P., 1992, *Numerical Recipes*, 2nd ed., Cambridge University Press, Cambridge, UK.

Protogeros, Z. A. M., and Weinberg, D. H., 1997, Astrophys. J. **489**, 457, astro-ph/9701147.

Randall, L., Soljačić, M., and Guth, A. H., 1996, Nucl. Phys. B **472**, 377, hep-ph/9512439.

Ratra, B., and Peebles, P. J. E., 1994, Astrophys. J. **432**, L5.

Ratra, B., and Peebles, P. J. E., 1995, Phys. Rev. D **52**, 1837.

Raychaudhuri, A. K., 1955, Phys. Rev. **98**, 1123.

Raychaudhuri, A. K., 1979, *Theoretical Cosmology*, Clarendon, Oxford.

Reid, I. N., 1997, Astron. J. **114**, 161, astro-ph/9704078.

Rephaeli, Y., 1995, Ann. Rev. Astron. Astrophys. **33**, 541.

Richstone, D., Loeb, A., and Turner, E. L., 1992, Astrophys. J. **393**, 559.

Riess, A., Press, W., and Kirshner, R., 1996, Astrophys. J. **473**, 88, astro-ph/9604143.

Roberts, D., Liddle, A. R., and Lyth, D. H., 1995, Phys. Rev. D **51**, 4122, astro-ph/9411104.

Robertson, H. P., 1936, Astrophys. J. **83**, 187.

Rubakov, V.A., Sazhin, M.V., and Veryastin, A.V., 1982, Phys. lett. B **115**, 189.

Sachs, R. K., and Wolfe, A. M., 1967, Astrophys. J. **147**, 73.

Sahni, V., and Coles, P., 1995, Phys. Rep. **262**, 1, astro-ph/9505005.

Sakagami, M., 1988, Prog. Theor. Phys. **79**, 442.

Salopek, D. S., 1995, Phys. Rev. D **52**, 5563, astro-ph/9506146.

Salopek, D. S., and Bond, J. R., 1990, Phys. Rev. D **42**, 3936.

Salopek, D. S., Bond, J. R., and Bardeen, J. M., 1989, Phys. Rev. D **40**, 1753.

Salopek, D. S., Stewart, J. M., and Croudace, K. M., 1994, Mon. Not. R. Astron. Soc. **271**, 1005, astro-ph/9403053.

Sasaki, M., 1986, Prog. Theor. Phys. **76**, 1036.

Sasaki, M., and Stewart, E. D., 1996, Prog. Theor. Phys. **95**, 71, astro-ph/9507001.

Sasaki, M., and Tanaka, T., 1996, Phys. Rev. D **54**, R4705, astro-ph/9605104.

Sasaki, M., Tanaka, T., Yamamoto, K., and Yokoyama, J., 1993, Phys. Lett. B **317**, 510.

Sasaki, M., Tanaka, T., and Yamamoto, Y., 1995, Phys. Rev. D **51**, 2979, gr-qc/9412025.

Sasaki, M., Tanaka, T., and Yakushige, Y., 1997, Phys. Rev. D **56**, 616, astro-ph/9702174.

Sazhin, M. V., and Shulga, V. V., 1996, Vestn. MSU N3, 69; N4, 87.

Schaefer, R. K., and de Laix, A. A., 1996, Astrophys. J. Supp. **105**, 1, astro-ph/9507003.

Schaefer, R. K., and Shafi, Q., 1992, Nature **359**, 199.

Schaefer, R. K., and Shafi, Q., 1994, Phys. Rev. D **49**, 4990, astro-ph/9310025.

Schaefer, R. K., Shafi, Q., and Stecker, F. W., 1989, Astrophys. J. **347**, 575.

Schechter, P. L., et al., 1997, Astrophys. J. **475**, L85, astro-ph/9611051.

Schmalzing, J., and Buchert, T., 1997, Astrophys. J. **482**, L1, astro-ph/9702130.

Schmalzing, J., Kerscher, M., and Buchert, T., 1995, in *Proceedings of "International School of Physics Enrico Fermi,"* eds. S. Bonometto, J. Primack, and A. Provenzale, IOP Press, Amsterdam, astro-ph/9508154.

Schmidt, B. P., et al., 1998, Astrophys. J. **507**, 46, astro-ph/9805200.

Schramm, D. N., and Turner, M. S., 1998, Rev. Mod. Phys. **70**, 303, astro-ph/9706069.

Sciama, D. W., 1990, Astrophys. J. **364**, 549.

Sciama, D. W., 1993, *Modern Cosmology and the Dark Matter Problem*, Cambridge University Press, Cambridge, UK.

Scoccimarro, R., and Frieman, J. A., 1996, Astrophys. J. Supp. **105**, 37, astro-ph/9509047.

Scott, D., Silk, J., Kolb, E. W., and Turner, M. S., 1999, in *Astrophysical Quantities*, ed. A. N. Cox, Springer, Berlin.

Seljak, U., 1994, Astrophys. J. **435**, L87, astro-ph/9406050.

Seljak, U., 1996, Astrophys. J. **463**, 1, astro-ph/9505109.

Seljak, U., 1997, Astrophys. J. **482**, 6, astro-ph/9608131.

Seljak, U., and Zaldarriaga, M., 1996, Astrophys. J. **469**, 1, astro-ph/9603033 [see also the CMBFAST WWW site at http://www.sns.ias.edu/~matiasz/CMBFAST/cmbfast.html].

Seljak, U., and Zaldarriaga, M., 1997, Phys. Rev. Lett. **78**, 2054, astro-ph/9609169.

Shafi, Q., and Stecker, F. W., 1984, Phys. Rev. Lett. **53**, 1292.

Shandarin, S. F., and Zel'dovich, Ya. B., 1989, Rev. Mod. Phys. **61**, 185.

Shapiro, P. R., Giroux, M. L., and Babul, A., 1994, Astrophys. J. **427**, 25.

Shi, X., Widrow, L. M., and Dursi, L. J., 1996, Mon. Not. R. Astron. Soc. **281**, 565, astro-ph/9506120.

Shtanov, Y., Traschen, J., and Brandenberger, R., 1995, Phys. Rev. D **51**, 5438, hep-ph/9407247.

Smith, C. C., Klypin, A., Gross, M. A. K., Primack, J. R., and Holtzman, J., 1998, Mon. Not. R. Astron. Soc. **297**, 910, astro-ph/9702099.

Smoot G. F., et al., 1992, Astrophys. J. **396**, L1.

Somerville, R. S., Davis, M., and Primack, J. R., 1997a, Astrophys. J. **479**, 616, astro-ph/9604041.

Somerville, R. S., Primack, J. R., and Nolthenius, R., 1997b, Astrophys. J. **479**, 606, astro-ph/9604051.

Starobinsky, A. A., 1980, Phys. Lett. B **91**, 99.

Starobinsky, A. A., 1982a, Phys. Lett. B **117**, 175.

Starobinsky, A. A., 1982b, in *Quantum Gravity*, Inst. Nuc. Res. USSR Acad. Sci., Moscow, p. 58.

Starobinsky, A. A., 1983, Sov. Astron. Lett. **9**, 302.

Starobinsky, A. A., 1985a, Sov. Astron. Lett. **11**, 133.

Starobinsky, A. A., 1985b, JETP Lett. **42**, 152.

Starobinsky, A. A., 1986, in *Lecture Notes in Physics*, Vol. 242, eds. H. J. de Vega and N. Sanchez, Springer, Berlin.

Starobinsky, A. A., and Sahni, V., 1984, in *Modern Theoretical and Experimental Problems of General Relativity* (Moscow University Press, in Russian), quoted by Shandarin and Zel'dovich (1989).

Starobinsky, A. A., and Yokoyama, J., 1995, in *Fourth Workshop on General Relativity and Gravitation*, eds. K. Nakao, et al., astro-ph/9502002.

Steidel, C. C., Giavalisco, M., Pettini, M., Dickinson, M., and Adelberger, K. L., 1996a, Astrophys. J. **462**, L17, astro-ph/9602024.

Steidel, C. C., Giavalisco, M., Dickinson, M., and Adelberger, K. L., 1996b, Astron. J. **112**, 352, astro-ph/9604140.

Steidel, C. C., and Sargent, W. L. W., 1987, Astrophys. J. **318**, L11.

Steinhardt, P. J., and Accetta, F., 1990, Phys. Rev. Lett. **64**, 2740.

Stevens, D., Scott, D., and Silk, J., 1993, Phys. Rev. Lett. **71**, 20.

Stewart, E. D., 1995a, Phys. Rev. D **51**, 6847, hep-ph/9405389.

Stewart, E. D., 1995b, Phys. Lett. B **345**, 414, astro-ph/9407040.

Stewart, E. D., 1997a, Phys. Lett. B **391**, 34, hep-ph/9606241.

Stewart, E. D., 1997b, Phys. Rev. D **56**, 2019, hep-ph/9703232.

Stewart, E. D., and Lyth, D. H., 1993, Phys. Lett. B **302**, 171, gr-qc/9302019.

Stewart, J. M., 1990, Class. Quant. Grav. **7**, 1169.

Stompor, R., Górski, K. M., and Banday, A. J., 1996, Astrophys. J. **463**, 8, astro-ph/9511087.

Storrie-Lombardi, L. J., McMahon, R. G., and Irwin, M. J., 1996, Mon. Not. R. Astron. Soc. **283**, L79, astro-ph/9608147.

Strauss, M. A., and Willick, J. A., 1995, Phys. Rep. **261**, 271, astro-ph/9502079.

Suginohara, M., Suginohara, T., and Spergel, D. N., 1998, Astrophys. J. **495**, 511, astro-ph/9705134.

Sugiyama, N., 1995, Astrophys. J. Supp. **100**, 281, astro-ph/9412025.

Sugiyama, N., and Gouda, N., 1992, Prog. Theor. Phys. **88**, 803.

Sunyaev, R., and Zel'dovich, Ya. B., 1970, Astrophys. Space Sci. **9**, 368.

Sunyaev, R., and Zel'dovich, Ya. B., 1972, Commun. Astrophys. Space Phys. **4**, 173.

Sunyaev, R., and Zel'dovich, Ya. B., 1980, Ann. Rev. Astron. Astrophys. **18**, 537.

Sutherland, W., 1988, Mon. Not. R. Astron. Soc. **234**, 159.

Suto, Y., and Fujita, M., 1990, Astrophys. J. **360**, 7.

Suto, Y., Gouda, N., and Sugiyama, N., 1990, Astrophys. J. Supp. **74**, 665.

Suto, Y., Cen, R., and Ostriker, J. P., 1992, Astrophys. J. **395**, 1.

Tadros, H., and Efstathiou, G., 1995, Mon. Not. R. Astron. Soc. **276**, L45, astro-ph/9507050.

Tadros, H., and Efstathiou, G., 1996, Mon. Not. R. Astron. Soc. **282**, 1381, astro-ph/9603016.

Tanaka, T., and Sasaki, M., 1997, Prog. Theor. Phys. **97**, 243, astro-ph/9701053.

Taylor, A. N., and Rowan-Robinson, M., 1992, Nature **359**, 396.

Tegmark, M., 1997a, Astrophys. J. **480**, L87, astro-ph/9611130.

Tegmark, M., 1997b, Phys. Rev. D **55**, 5895, astro-ph/9611174.

Tegmark, M., and Bunn, E. F., 1995, Astrophys. J. **455**, 1, astro-ph/9412005.

Tegmark, M., and Efstathiou, G., 1996, Mon. Not. R. Astron. Soc. **281**, 1297, astro-ph/9507009.

Tegmark, M., and Hamilton, A. J. S., 1997, "Uncorrelated measurements of the cmb power spectrum," astro-ph/9702019.

Tegmark, M., and Silk, J., 1995, Astrophys. J. **441**, 458, astro-ph/9405042.

Tegmark, M., Silk, J., and Blanchard, A., 1994, Astrophys. J. **420**, 484, astro-ph/9307017.

Tegmark, M., Silk, J., Rees, M., Blanchard, A., Abel, T., and Palla, F., 1997a, Astrophys. J. **474**, 1, astro-ph/9603007.

Tegmark, M., Taylor, A., and Heavens, A., 1997b, Astrophys. J. **480**, 22, astro-ph/9603021.

Tegmark, M., Eisenstein, D. J., Hu, W., and Kron, R., 1998, "Cosmic complementarity," astro-ph/9805117.

Thomas, P. A., and Couchman, H. M. P., 1992, Mon. Not. R. Astron. Soc. **257**, 11.

Thomas, P. A., Couchman, H. M. P., and Pearce, F. R., 1996, "Release of data from cosmological N-body simulations," astro-ph/9610095.

Traschen, J. H., and Brandenberger, R. H., 1990, Phys. Rev. D **42**, 2491.

Turner, E. L., Cen, R., and Ostriker, J. P., 1992, Astron. J. **103**, 1427.

Turner, M. S., 1991, Phys. Rev. D **44**, 3737.

Turner, M. S., Steigman, G., and Krauss, L. M., 1984, Phys. Rev. Lett. **52**, 2090.

Turner, M. S., and White, M., 1997, Phys. Rev. D **56**, R4439, astro-ph/9701138.

Turok, N., and Hawking, S. W., 1998, Phys. Lett. B **432**, 271, hep-th/9803156.

Tytler, D., Burles, S., and Kirkham, D., 1996, "Annulling the case for high deuterium abundances," astro-ph/9612121.

van Haarlem, M. P., Frenk, C. S., and White, S. D. M., 1997, Mon. Not. R. Astron. Soc. **287**, 817, astro-ph/9701103.

Viana, P. T. P., and Liddle, A. R., 1996, Mon. Not. R. Astron. Soc. **278**, 644, astro-ph/9511007.

Viana, P. T. P., and Liddle, A. R., 1998, Phys. Rev. D **57**, 674, astro-ph/9708247.

Viana, P. T. P., and Liddle, A. R., 1999, Mon. Not. R. Astron. Soc. **303**, 535, astro-ph/9803244.

Vilenkin, A., 1998, Phys. Rev. D **57**, 7069, hep-th/9803084.

Vilenkin, A., and Shellard, E. P. S., 1994, *Cosmic Strings and Other Topological Defects*, Cambridge University Press.

Villumsen, J. V., Freundling, W., and da Costa, L. N., 1997, Astrophys. J. **481**, 578, astro-ph/9606084.

Vishniac, E. T., 1987, Astrophys. J. **322**, 597.

Vittorio, N., Matarrese, S., and Lucchin, F., 1988, Astrophys. J. **328**, 69.

von Hoerner, S., 1960, Zeit. Astrophys. **50**, 184.

Walker, A. G., 1936, Proc. London Math. Soc. **42**, 90.

Walker, T. P., Steigman, G., Schramm, D. N., Olive, K. A., and Kang, H.-S., 1991, Astrophys. J. **376**, 51.

Wands, D., 1994, Class. Quant. Grav. **11**, 269, gr-qc/9307034.

Webb, J. K., Barcons, X., Carswell, R. F., and Parnell, H. C., 1992, Mon. Not. R. Astron. Soc. **255**, 319.

Weinberg, E. J., 1989, Phys. Rev. D **40**, 3950.

Wess, J., and Bagger, J., 1983, *Supersymmetry and Supergravity*, Princeton University Press, Princeton, NJ.

White, D. A., and Fabian, A., 1995, Mon. Not. R. Astron. Soc. **273**, 72, astro-ph/9502092.

White, M., 1996, Phys. Rev. D **53**, 3011, astro-ph/9601158.

White, M., and Silk, J., 1996, Phys. Rev. Lett. **77**, 4704, astro-ph/9608177.

White, M., Scott, D., and Silk, J., 1994, Ann. Rev. Astron. Astrophys. **32**, 319.

White, M., Gelmini, G., and Silk, J., 1995a, Phys. Rev. D **51**, 2669, astro-ph/9411098.

White, M., Scott, D., Silk, J., and Davis, M., 1995b, Mon. Not. R. Astron. Soc. **276**, L69, astro-ph/9508009.

White, M., Viana, P. T. P., Liddle, A. R., and Scott D., 1996, Mon. Not. R. Astron. Soc. **283**, 107, astro-ph/9605057.

White, S. D. M., 1994, "Formation and evolution of galaxies," astro-ph/9410043.

White, S. D. M., Frenk, C. S., and Davis, M., 1983a, Astrophys. J. **274**, L1.

White, S. D. M., Frenk, C. S., and Davis, M., 1983b, Astrophys. J. **287**, 1.

White, S. D. M., Frenk, C. S., Davis, M., and Efstathiou, G., 1987, Astrophys. J. **313**, 505.

White, S. D. M., Efstathiou, G., and Frenk, C. S., 1993a, Mon. Not. R. Astron. Soc. **262**, 1023.

White, S. D. M., Navarro, J. F., Evrard, A. E., and Frenk, C. S., 1993b, Nature **366**, 429.

Whitt, B., 1984, Phys. Lett. B **145**, 176.

Will, C. M., 1993, *Theory and Experiment in Gravitational Physics*, Cambridge University Press, Cambridge, UK.

Willick, J. A., Courteau, S., Faber, S. M., Burstein, D., and Dekel, A., 1995, Astrophys. J. **446**, 12, astro-ph/9411046.

Willick, J. A., Courteau, S., Faber, S. M., Burstein, D., Dekel, A., and Kolatt, T., 1996, Astrophys. J. **457**, 460.

Willick, J. A., Courteau, S., Faber, S. M., Burstein, D., Dekel, A., and Strauss, M. A., 1997, Astrophys. J. Supp. **109**, 333.

Wright, E. L., 1996, "Scanning and mapping strategies for cmb experiments," astro-ph/9612006.

Wright, E. L., Bennett, C. L., Górski, K., Hinshaw, G., and Smoot, G. F., 1996, Astrophys. J. **464**, L25, astro-ph/9601059.

Wu, X.-P., and Fang, L.-Z., 1997, Astrophys. J. **483**, 62, astro-ph/9701196.

Yamamoto, K., and Bunn, E. F., 1996, Astrophys. J. **461**, 8, astro-ph/9508090.

Yamamoto, K., Sasaki, M., and Tanaka, T., 1996, Phys. Rev. D **54**, 5031, astro-ph/9605103.

Yano, T. H., Nagashima, M., and Gouda, N., 1996, Astrophys. J. **466**, 1, astro-ph/9504073.

Yi, I., and Vishniac, E. T., 1993, Phys. Rev. **47**, 5280.

Zaldarriaga, M., and Seljak, U., 1997, Phys. Rev. D **55**, 1830, astro-ph/9609170.

Zaldarriaga, M., Spergel, D., and Seljak, U., 1997, Astrophys. J. **488**, 1, astro-ph/9702157.

Zaroubi, S., Zehavi, I., Dekel, A., Hoffman, Y., and Kolatt, T., 1997, Astrophys. J. **486**, 21, astro-ph/9610226.

Zel'dovich, Ya. B., 1970, Astron. Astrophys. **5**, 84.

Zel'dovich, Ya. B., Kobzarev, I. Yu., and Okun, L. B., 1975, Sov. Phys. JETP **40**, 1.

Zhang, Y., Meiksin, A., Anninos, P., and Norman, M. L., 1998, Astrophys. J. **495**, 63, astro-ph/9706087.

Index